Textbook of Chemistry

Textbook of Chemistry

Edited by **Gerald Cole**

NY RESEARCH
P R E S S

New York

Published by NY Research Press,
23 West, 55th Street, Suite 816,
New York, NY 10019, USA
www.nyresearchpress.com

Textbook of Chemistry
Edited by Gerald Cole

International Standard Book Number: 978-1-63238-491-1 (Hardback)

The publisher's policy is to use permanent paper from mills that operate a sustainable forestry policy. Furthermore, the publisher ensures that the text paper and cover boards used have met acceptable environmental accreditation standards.

Trademark Notice: Registered trademark of products or corporate names are used only for explanation and identification without intent to infringe.

Printed in the United States of America.

Contents

Permissions

List of Contributors

Preface

Chemistry is the branch of science which studies the structure, composition and properties of all the matter around us. It is a broad discipline that branches out into various sub-fields like analytical chemistry, physical chemistry, inorganic chemistry, biochemistry, organic chemistry, etc. It is a discipline that has existed for a long time and has evolved to such a great extent that it is applicable in a broad spectrum of industries. This book attempts to understand the multiple branches that fall under the discipline of chemistry and how such concepts have practical applications. It studies, analyses and upholds the pillars of this subject and its utmost significance in modern times and serves as a valuable source of reference to researchers, academicians and students associated with any branch of chemistry.

All of the data presented henceforth, was collaborated in the wake of recent advancements in the field. The aim of this book is to present the diversified developments from across the globe in a comprehensible manner. The opinions expressed in each chapter belong solely to the contributing authors. Their interpretations of the topics are the integral part of this book, which I have carefully compiled for a better understanding of the readers.

At the end, I would like to thank all those who dedicated their time and efforts for the successful completion of this book. I also wish to convey my gratitude towards my friends and family who supported me at every step.

Editor

Hydroboration of Substituted Cyclopropane: A Density Functional Theory Study

Satya Prakash Singh[1] and Pompozhi Protasis Thankachan[2]

[1] Department of Chemical Sciences, Indian Institute of Science Education and Research, Knowledge City, Sector 81, Mohali, Panjab 140306, India
[2] Indian Institute of Technology Roorkee, Roorkee 247667, India

Correspondence should be addressed to Satya Prakash Singh; satyapiit@gmail.com

Academic Editor: Daniel Glossman-Mitnik

The hydroboration of substituted cyclopropanes has been investigated using the B3LYP density functional method employing 6-31G** basis set. Borane moiety approaching the cyclopropane ring has been reported. It is shown that the reaction proceeds via a three-centered, "loose" and "tight," transition states when boron added to the cyclopropane across a bond to a substituents. Single point calculations at higher levels of theory were also performed at the geometries optimized at the B3LYP level, but only slight changes in the barriers were observed. Structural parameters for the transition state are also reported.

1. Introduction

Hydroboration of substituted alkenes has been investigated theoretically and experimentally. Brown and Zweifel [1] have shown that the hydroboration of alkyl substituted olefins yields the anti-Markownikoff addition product predominantly and that addition takes place predominantly at β-carbon atom. For monosubstituted olefins, 93-94% of borane addition takes place at the terminal carbon atom. For di- and trisubstituted olefins the preference for the anti-Markownikoff product is 98-99%. They have also observed steric and electronic effects in the case of trans-2-pentene.

When electron withdrawing groups are attached to the alkene preferential formation of the Markownikoff addition products has been reported. Phillips and Stone [2] have shown that borane adds to 1,1,1-trifluoropropene giving the Markownikoff product with 87–92% selectivity in appropriate solvents. Graham et al. [3] have carried out studies on the substituent effect in hydroboration of propylene and cyanoethylene using the partial retention of diatomic differential overlap (PRDDO) method with application of linear synchronous transits (LSTs) and orthogonal optimizations to construct the reaction pathways for the Markownikoff

and anti-Markownikoff addition of borane to propylene and cyanoethylene. Villiers and Ephritikhine [4] have carried out the borane-catalysed hydroboration of substituted alkenes by lithium borohydride or sodium borohydride. They have shown the unusual order of decreasing reactivity: tetramethylethylene > 1-methylcyclohexene > cyclohexene. Xu et al. [5] theoretically studied the hydroboration of disilenes with borane. They investigated the reaction mechanism exhaustively and found the mechanism for hydroboration of disilenes to be interestingly different from that proposed for hydroboration of alkenes.

We have theoretically investigated the hydroboration of cyclopropane [6] in which the borane moiety was situated along the plane of cyclopropane ring. After that the possibility of borane moiety perpendicular to cyclopropane ring has been reported [7] at different method and basis set. The reaction is similar to hydroboration of cyclopropane, that is, hydroalumination of cyclopropane with alane (AlH_3) [8] reported. In this paper we present our computational studies on the reactions of BH_3 with cyclopropane in which the hydrogen atom is replaced with six different kinds of substituents (–F, –Cl, –CN, –NC, –CH$_3$, and –(CH$_3$)$_2$) at DFT level of theory using 6-31G** basis set in each case. The main

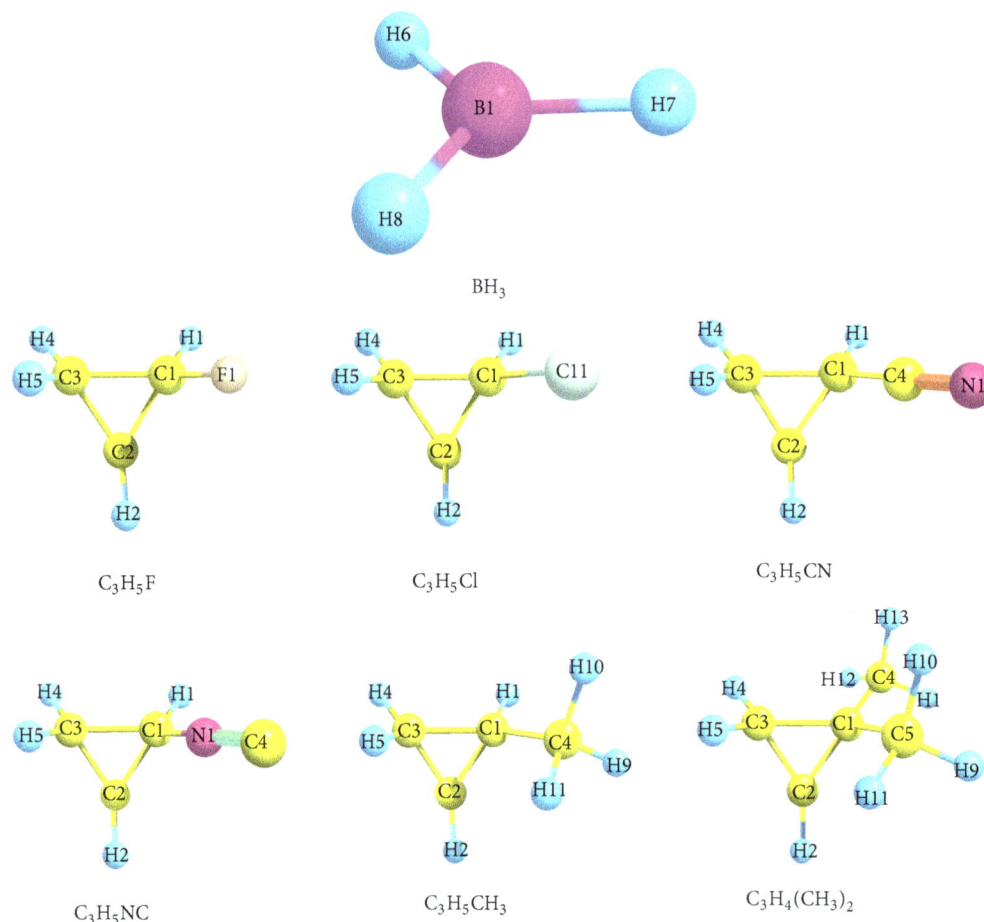

FIGURE 1: Optimized geometries of borane (BH_3) and six kinds of substituted cyclopropane at B3LYP/6-31G** level.

goal of work is to study the feasibility of reactions. Effect of substituents on the reaction mechanism and on energetics is investigated (see Supplementary Material available online at http://dx.doi.org/10.1155/2014/427396), and some calculations at higher levels of theory are also included.

2. Computational Methods

Optimization of all the geometries of stationary structures involved in the reaction was carried out using 6-31G** basis set at DFT/B3LYP [9] level using Gaussian 98 W software package [10]. The nature of each stationary point was probed by frequency calculations. Single point calculations at the DFT optimized geometry at higher ab initio levels of theory have also been performed. Single point calculations were done at CCSD, CCSD(T) [11–15], QCISD, QCISD(T) [15], MP2 [16–20], and MP4D [21] levels.

3. Results and Discussion

Six substituted cyclopropanes (Figure 1), namely, C_3H_5F, C_3H_5Cl, C_3H_5CN, C_3H_5NC, $C_3H_5(CH_3)$, and $C_3H_4(CH_3)_2$, were chosen for the study of their reaction with borane. Addition of BH_3 across bonds adjacent to the substituted

atom has been studied. The structures of all the reactants were optimized at the B3LYP/6-31G** level and are shown in Figure 1. Here we have discussed the results of fluorocyclopropane and the result of other substituted cyclopropanes furnished in Supplementary Material.

There are two possibilities to be considered in connection with each substituted cyclopropane: first with the carbon atom bearing substituents denoted by C1, addition takes place across the C1–C3 bond and second addition takes place across the C2–C3 bond.

Optimization led to two types of transition states. In one case the BH_3 group is closer to the ring ("tight" TS) and is on the side opposite the fluorine, whereas in the other the BH_3 group is somewhat farther ("loose" TS) and on the side of the fluorine. The BH_3 group has lost its planarity in both, but the distortion from planarity is more pronounced in the first case. These transition states are shown in Figure 2. IRC calculations from the transition states obtained show that the pathways through the transition states go towards the anti-Markownikoff and Markownikoff products, respectively. Figures 3 and 4 show the IRCs in these two cases.

On the other hand in the case of 1-chlorocyclopropane, we obtain both "tight" and "loose" transition states but in this case the path through the "tight" transition state leads to the

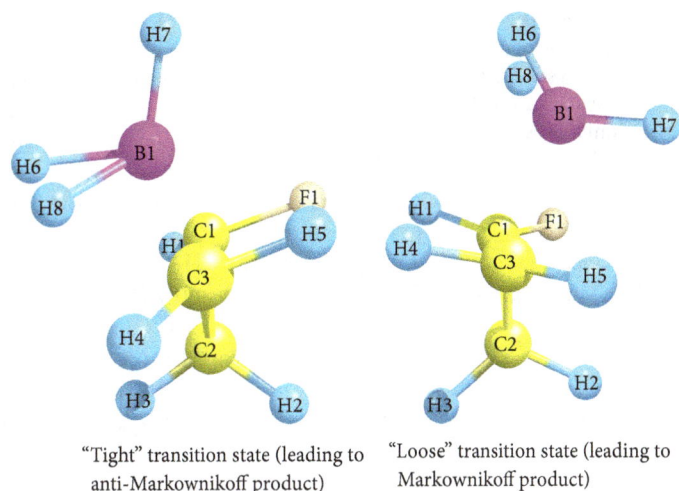

"Tight" transition state (leading to anti-Markownikoff product)

"Loose" transition state (leading to Markownikoff product)

FIGURE 2: Transition states optimized at B3LYP/6-31G** in the case of 1-fluorocyclopropane along the plane of cyclopropane ring.

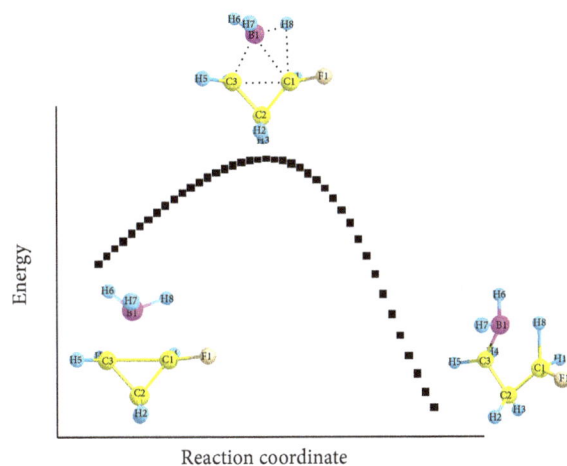

FIGURE 3: IRC plot for the loose transition state for the addition of borane to 1-fluorocyclopropane along the plane of cyclopropane ring.

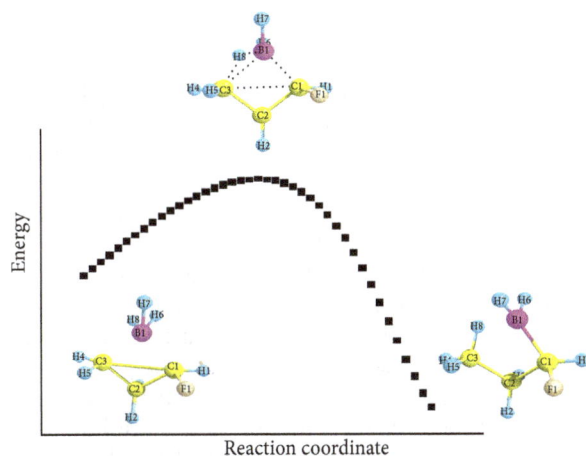

FIGURE 4: IRC plot for the tight transition state for the addition of borane to 1-fluorocyclopropane along the plane of cyclopropane ring.

Markownikoff product and the path through the "loose" transition state leads to the anti-Markownikoff product. Of the other substituted cyclopropanes studied it is found that cyano and isocyano cyclopropanes behave like chlorocyclopropane, while methyl and 1,1-dimethyl cyclopropanes behave like fluorocyclopropane in this respect (see the Supplementary Material).

For each of the substituted cyclopropanes, the structures of the transition structures at the B3LYP/6-31G** level were optimized. The "loose" transition state is preceded by an intermediate complex, while the "tight" transition state is apparently not. The structures of the complexes, "loose" transition structures, and products are shown in Figure S1 (Supplementary Material) and are denoted by LM-CX, TS, and LM, respectively. "LM" stands for local minimum on potential energy surface, "CX" stands for complex, and "TS" stands for transition state. The selected optimized structural parameters for these are shown in Table 1(a).

It is seen that the C1–B1 and C3–B1 distances found in the complexes are significantly longer in the case of all substituted cyclopropanes compared to the unsubstituted case; for example, in case of fluoro substitution the C1–B1 and C3–B1 distances are 3.094 and 2.922 Å against 2.905 Å in unsubstituted case pointing to weaker complexation. In the complex with –F, –Cl, –CN, and –NC substituted cyclopropanes, the boron is nearly symmetrically disposed with respect to C1 and C3 (Figure S1) whereas in the case of methyl-substituted compounds (Figure S1) there is pronounced asymmetry, probably due to increased steric effects.

The C1–C3 distance in the "loose" transition structure for the reaction between cyclopropane and BH_3 is 1.994 Å. In case of substitutions by –F, –Cl, –CN, and –NC this distance is less than this value but in the methyl and dimethyl case it is greater, pointing to a weaker C1–C3 bond in these cases.

In the case of unsubstituted cyclopropane, in the transition state the C1–B1 distance is greater than C3–B1 (see Table 1(a)) and both are equal in the complex, indicating

TABLE 1: (a) B3LYP/6-31G** optimized structural parameters (units in Å for bond length) for the –F substituted cyclopropanes, intermediate complexes (LM-CX), "loose" transition structures (TS), and products (LM) along the plane of cyclopropane ring. (b) B3LYP/6-31G** optimized structural parameters (units in Å for bond length and in degree for angle) for the "tight" transition structures (TS) and products (LM) along the plane of cyclopropane ring.

(a)

	C1–C3	C1–B1	C3–B1	B1–H8
C_3H_5F	1.493	—	—	—
LM-CX1	1.506	3.094	2.922	1.190
TS1	1.912	1.991	1.829	1.209
LM1	2.529	3.263	1.559	2.955
C_3H_5Cl	1.498	—	—	—
LM-CX2	1.505	3.350	3.141	1.190
TS2	1.893	1.883	1.929	1.206
LM2	2.498	1.565	3.417	3.166
C_3H_5CN	1.522	—	—	—
LM-CX3	1.528	3.325	3.209	1.190
TS3	1.958	1.863	1.989	1.206
LM3	2.546	1.590	3.231	2.914
C_3H_5NC	1.512	—	—	—
LM-CX4	1.519	3.309	3.127	1.190
TS4	1.947	1.926	1.910	1.207
LM4	2.532	1.592	3.273	2.970
$C_3H_5CH_3$	1.509	—	—	—
LM-CX5	1.522	3.166	2.894	1.191
TS5	2.012	1.767	2.042	1.215
LM5	2.580	3.303	1.558	2.960
$C_3H_4(CH_3)_2$	1.511	—	—	—
LM-CX6	1.519	3.493	2.942	1.192
TS6	2.040	2.216	1.764	1.216
LM6	2.591	3.304	1.558	2.915

(b)

	C1–C3	C1–B1	C3–B1	B1–H8
C_3H_5F				
TS7	2.251	1.620	1.759	1.264
LM7	2.515	1.572	3.430	3.186
C_3H_5Cl				
TS8	2.293	1.809	1.609	1.266
LM8	2.522	3.247	1.560	2.954
C_3H_5CN				
TS9	2.310	1.841	1.605	1.276
LM9	2.546	3.261	1.561	2.930
C_3H_5NC				
TS10	2.309	1.837	1.606	1.267
LM10	2.535	3.257	1.561	2.929
$C_3H_5CH_3$				
TS11	2.278	1.761	1.622	1.266
LM11	2.579	1.558	3.272	2.971
$C_3H_4(CH_3)_2$				
TS12	2.040	2.216	1.764	1.216
LM12	2.591	1.566	3.211	2.894

the tendency of "B1" to bond to "C3." In the substituted cases, the complexes themselves are unsymmetrical, with C1–B1 distances being longer than the C3–B1 distances in all cases. However the "loose" transition state that follows this situation continues only in cases of –F, –CH$_3$, and –(CH$_3$)$_2$ substituents, whereas with –Cl, –CN, and –NC in the TS, B1 is closer to "C1" suggesting that in these cases the "loose" transition state leads to the Markownikoff product whereas with –F, –CH$_3$, and –(CH$_3$)$_2$ these "loose" transition states lead to the anti-Markownikoff product.

The selected geometrical parameters for the "tight" transition structures are shown in Table 1(b). In these the C1–C3 distances are greater than in the corresponding "loose" transition structures whereas the C1–B1 and C3–B1 distances are shorter. The C1–B1 distance is less than the C3–B1 distance in the case of fluoro substitution alone. However in the case of methylcyclopropane and 1,1-dimethyl cyclopropane the C3–B1 distance is less in the TS, but in the product B1 gets attached to C1 (Markownikoff product).

The molecular orbital plots in Figure 5 of intermediate complexes and "loose" transition structures and Figure 6 of "tight" transition structures show the degradation of the C1–C3 partial π-system accompanied by the bond formation.

3.1. Reaction Energies. The energies of the optimized intermediate complex (LM-CX), the "loose" transition structures (TS), and the addition product along with the product type (Markownikoff or anti-Markownikoff) are shown in Table 2. The energies (in kcal/mol) relative to the reactants are given in Table 2. Free energy changes; ΔG and entropy change; ΔS has also been listed.

The reactants proceed without barrier to an intermediate complex and cross over a barrier between 23.52 and 30.74 kcal/mol to form the product, which is more stable than the reactants by 35.59 to 43.39 kcal/mol, depending on the substituent. The intermediate occurs at shallow minima, stabilized by 0.50 to 2.00 kcal/mol relative to the reactants.

The loose transition states in these cases all correspond to barrier comparable to the case of unsubstituted cyclopropane; the barriers are slightly lower (than for cyclopropane) in the case of fluoro, methyl, and dimethyl substitutions whereas they are slightly higher for the others and the nature of the products also differs in the two cases. It is thought that the high electronegativity of fluorine causes the carbon to which it is bonded to be more positive overall, hence facilitating the abstraction of a hydride (or hydrogen with net negative Mulliken charge), thus leading to the anti-Markownikoff product. In the case of methyl substitution steric influence of the methyl group(s) may be what causes the larger BH$_2$ moiety to move to the less substituted carbon.

In the case of "tight" transition structure no intermediate complex has been observed. The relative energies are listed in Table 3. In comparison to the "loose" transition structures "tight" transition structures are found to have low energy barriers varying between 6.60 and 10.64 kcal/mol.

Single point calculations at the DFT optimized geometries have also been carried out on all the key species studied at CCSD, CCSD(T), QCISD(T), MP2, and MP4D levels.

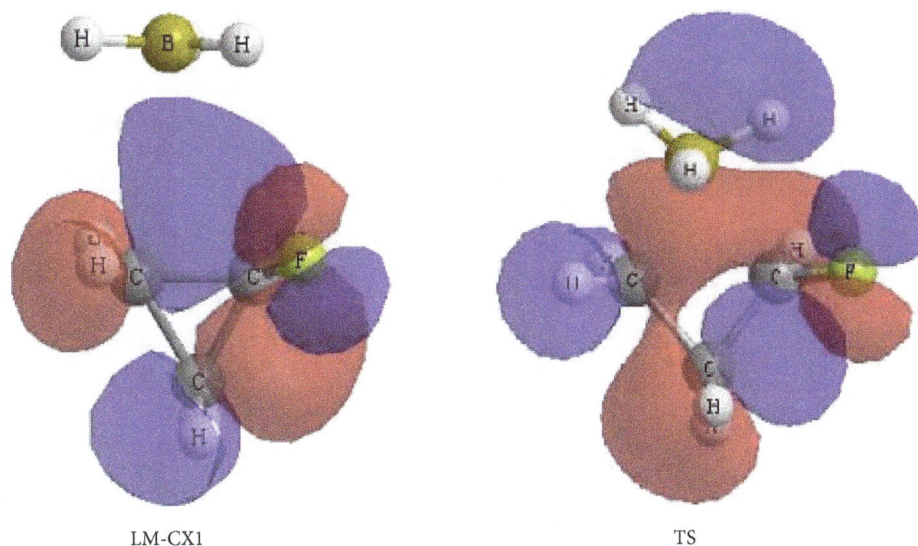

LM-CX1   TS

FIGURE 5: HOMOs of the complexes and "loose" transition structures for the hydroboration of –F substituted cyclopropanes at B3LYP/6-31G** level along the plane of cyclopropane ring.

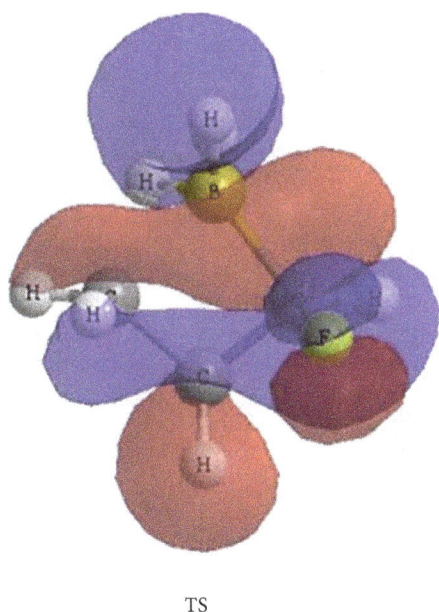

TS

FIGURE 6: HOMOs of the complexes and "tight" transition structures for the hydroboration of –F substituted cyclopropanes at B3LYP/6-31G** level along the plane of cyclopropane ring.

The single point energies obtained are shown in Tables S3 and S4 (Supplementary Material). It is observed that introduction of triples stabilizes the species, and the stabilization is most significant for the transition states. However one cannot conclude from this that the CCSD(T) barrier is lower than the CCSD barrier, since the points computed are not necessarily the true stationary points on the CCSD or CCSD(T) potential energy surfaces. Assuming that the true optima are not far from the DFT optima, one can say that the intermediate complexes are more stable relative to the reactants at these post-Hartree-Fock ab initio levels than at DFT levels. However, for the transition states the situation is reversed and the transition structures have higher relative energy. The products are again at comparable relative energies to the DFT case. Since optimization at these levels is not practicable, these observations do not provide any clear pointers to the relative efficacies of these methods, and for the moment the DFT results can be taken as a good indicator of the true energy barriers.

These "tight" transition structures are an interesting anomaly in that their energies are uncharacteristically low; that is, they correspond to very low barriers compared to the other cases, being comparable to reported values for hydroboration of ethylene. The tightly bound structure being stabler is to be expected and we find that the BH_3 moiety is distorted farther from planarity than in the "loose" structure. The hydrogen on the carbon atoms also assumes a nearly planar disposition.

4. Concluding Remarks

In summary, we have investigated the stationary structures involved in the hydroboration of substituted cyclopropanes with borane. Our study posits three-centered transition states for these reactions. It is also hoped that studies on reactions involving cyclopropane and its derivatives with other reagents will clarify the situation. Of the reactions studied three-centered transition states are encountered in the case of BH_3 adding to cyclopropane with an in-plane approach. We are led to suspect that the electronic structure and high reactivity of borane are the major causative factors involved.

Conflict of Interests

The authors declare that there is no conflict of interests regarding the publication of this paper.

Table 2: B3LYP/6-31G** optimized total energies (in kcal/mol) for the intermediate complex, "loose" transition structure, and product for substituted cyclopropanes for addition across Cl–C3 bond along the plane of cyclopropane ring.

	LM-CX	TS	Product	Product type	Gibbs energy (ΔG_{298}) (kcal/mol)	Entropy change (ΔS_{298}) (kcal/mol K)
$C_3H_5F + BH_3$	−1.24	23.62	−39.84	AM	−34.02	−0.008
$C_3H_5Cl + BH_3$	−0.58	28.62	−43.39	M	−36.05	−0.014
$C_3H_5CN + BH_3$	−0.41	30.74	−36.87	M	−30.46	−0.013
$C_3H_5NC + BH_3$	−0.51	29.65	−36.38	M	−30.13	−0.012
$C_3H_5CH_3 + BH_3$	−1.56	24.63	−37.45	AM	−31.78	−0.008
$C_3H_4(CH_3)_2 + BH_3$	−0.66	23.52	−35.59	AM	−30.28	−0.009

Relative energies for the parent cyclopropane are LM-CX = −1.97 kcal/mol, TS = 25.17 kcal/mol, LM = −40.15 kcal/mol, ΔG_{298} = −34.51 kcal/mol, and ΔS_{298} = −0.007 kcal/mol K.

Table 3: B3LYP/6-31G** optimized total energies (in kcal/mol) for the "tight" transition structure and product for substituted cyclopropanes for addition across Cl–C3 bond along the plane of cyclopropane ring.

	TS	Product	Product type	Gibbs energy (ΔG_{298}) (kcal/mol)	Entropy change (ΔS_{298}) (kcal/mol K)
$C_3H_5F + BH_3$	6.60	−36.87	M	−22.90	−0.034
$C_3H_5Cl + BH_3$	9.02	−40.65	AM	−26.97	−0.034
$C_3H_5CN + BH_3$	9.26	−37.73	AM	−24.39	−0.034
$C_3H_5NC + BH_3$	9.75	−38.70	AM	−25.15	−0.031
$C_3H_5CH_3 + BH_3$	8.97	−36.16	M	−22.57	−0.035
$C_3H_4(CH_3)_2 + BH_3$	10.64	−32.66	M	−19.18	−0.035

Acknowledgment

One of the authors (Satya Prakash Singh) is grateful to the Ministry of Human Resources and Development (MHRD), Government of India, for the award of a fellowship.

References

[1] H. C. Brown and G. Zweifel, "Hydroboration. VII. Directive effects in the hydroboration of olefins," *Journal of the American Chemical Society*, vol. 82, no. 17, pp. 4708–4712, 1960.

[2] J. R. Phillips and F. G. A. Stone, "Organoboron halides. Part VI. Hydroboration of 3,3,3-trifluoropropene," *Journal of the Chemical Society*, pp. 94–97, 1962.

[3] G. D. Graham, S. C. Freilich, and W. N. Lipscomb, "Substituent effects in hydroboration: reaction pathways for the Markownikoff and anti-Markownikoff addition of borane to propylene and cyanoethylene," *Journal of the American Chemical Society*, vol. 103, no. 10, pp. 2546–2552, 1981.

[4] C. Villiers and M. Ephritikhine, "Borane-catalyzed hydroboration of substituted alkenes by lithium borohydride or sodium borohydride," *Tetrahedron Letters*, vol. 44, no. 44, pp. 8077–8079, 2003.

[5] Y. J. Xu, Y. F. Zhang, and J. Q. Li, "Theoretical study of the hydroboration reaction of disilenes with borane," *Chemical Physics Letters*, vol. 421, no. 1–3, pp. 36–41, 2006.

[6] S. P. Singh and P. P. Thankachan, "Theoretical study of the hydroboration reaction of cyclopropane with borane," *Journal of Molecular Modeling*, vol. 18, no. 2, pp. 751–754, 2012.

[7] S. P. Singh and P. P. Thankachan, "Hydroboration of cyclopropane: a transition state study," *Chemical Science Transactions*, vol. 2, no. 2, pp. 479–484, 2013.

[8] S. P. Singh and P. P. Thankachan, "Hydroalumination of cyclopropane: a transition state study," *Chemical Science Transactions*, vol. 2, no. 3, pp. 1009–1015, 2013.

[9] A. D. Becke, "Density-functional thermochemistry. III. The role of exact exchange," *The Journal of Chemical Physics*, vol. 98, no. 7, pp. 5648–5652, 1993.

[10] M. J. Frisch, G. W. Trucks, H. B. Schlegel et al., *Gaussian 98, Revision A.7*, Gaussian, Pittsburgh, Pa, USA, 1998.

[11] J. Cizek, "On the use of the cluster expansion and the technique of diagrams in calculations of correlation effects in atoms and molecules," *Advances in Chemical Physics*, vol. 14, pp. 35–89, 1969.

[12] G. D. Purvis and R. J. Bartlett, "A full coupled-cluster singles and doubles model: the inclusion of disconnected triples," *The Journal of Chemical Physics*, vol. 76, no. 4, pp. 1910–1918, 1982.

[13] G. E. Scuseria, C. L. Janssen, and H. F. Schaefer III, "An efficient reformulation of the closed-shell coupled cluster single and double excitation (CCSD) equations," *Journal of Chemical Physics*, vol. 89, no. 12, pp. 7382–7387, 1988.

[14] G. E. Scuseria and H. F. Schaefer III, "Is coupled cluster singles and doubles (CCSD) more computationally intensive than quadratic configuration interaction (QCISD)?" *The Journal of Chemical Physics*, vol. 90, no. 7, pp. 3700–3703, 1989.

[15] J. A. Pople, M. Head-Gordon, and K. Raghavachari, "Quadratic configuration interaction. A general technique for determining electron correlation energies," *The Journal of Chemical Physics*, vol. 87, no. 10, pp. 5968–5975, 1987.

[16] M. Head-Gordon, J. A. Pople, and M. J. Frisch, "MP2 energy evaluation by direct methods," *Chemical Physics Letters*, vol. 153, no. 6, pp. 503–506, 1988.

[17] M. J. Frisch, M. Head-Gordon, and J. A. Pople, "A direct MP2 gradient method," *Chemical Physics Letters*, vol. 166, no. 3, pp. 275–280, 1990.

[18] M. J. Frisch, M. Head-Gordon, and J. A. Pople, "Semi-direct algorithms for the MP2 energy and gradient," *Chemical Physics Letters*, vol. 166, no. 3, pp. 281–289, 1990.

[19] M. Head-Gordon and T. Head-Gordon, "Analytic MP2 frequencies without fifth-order storage. Theory and application to bifurcated hydrogen bonds in the water hexamer," *Chemical Physics Letters*, vol. 220, no. 1-2, pp. 122–128, 1994.

[20] S. Saebo and J. Almlof, "Avoiding the integral storage bottleneck in LCAO calculation of electron correlation," *Chemical Physics Letters*, vol. 154, pp. 83–89, 1989.

[21] D. E. Woon and T. H. Dunning Jr., "Gaussian basis sets for use in correlated molecular calculations. III. The atoms aluminum through argon," *The Journal of Chemical Physics*, vol. 98, no. 2, pp. 1358–1371, 1993.

Philicity and Fugality Scales for Organic Reactions

Rodrigo Ormazábal-Toledo[1] and Renato Contreras[2]

[1] *Departamento de Física, Facultad de Ciencias, Universidad de Chile, Casilla, 653 Santiago, Chile*
[2] *Departamento de Química, Facultad de Ciencias, Universidad de Chile, Casilla, 653 Santiago, Chile*

Correspondence should be addressed to Renato Contreras; rcontrer@uchile.cl

Academic Editor: Yusuf Atalay

Theoretical scales of reactivity and selectivity are important tools to explain and to predict reactivity patterns, including reaction mechanisms. The main achievement of these efforts has been the incorporation of such concepts in advanced texts of organic chemistry. In this way, the modern organic chemistry language has become more quantitative, making the classification of organic reactions an easier task. The reactivity scales are also useful to set up a number of empirical rules that help in rationalizing and in some cases anticipating the possible reaction mechanisms that can be operative in a given organic reaction. In this review, we intend to give a brief but complete account on this matter, introducing the conceptual basis that leads to the definition of reactivity indices amenable to build up quantitative models of reactivity in organic reactions. The emphasis is put on two basic concepts describing electron-rich and electron-deficient systems, namely, nucleophile and electrophiles. We then show that the regional nucleophilicity and electrophilicity become the natural descriptors of electrofugality and nucleofugality, respectively. In this way, we obtain a closed body of concepts that suffices to describe electron releasing and electron accepting molecules together with the description of permanent and leaving groups in addition, nucleophilic substitution and elimination reactions.

1. Introduction

The development of reactivity indices to describe organic reactivity has been an active area of research from the dawn of theoretical physical organic chemistry [1–4]. From the earlier semiempirical models proposed in Hückel molecular orbital (HMO) theory, reactivity was described with the aid of static first order reactivity descriptors like atomic charges, free valence index, and bond orders [5, 6]. These reactivity indices were formerly developed around the ground state of reactants. The next generation of reactivity indices began with the elegant theory proposed by Coulson and Longuet-Higgins, in a series of papers describing the response functions, including second order quantities like atomic and bond polarizabilities [7–9]. It is worth emphasizing that, at that time, HMO and Coulson-Longuet-Higgins theories were conceived not as methods to approximately solve the Schröedinger equation but as models of chemical bond. The third generation of reactivity indices started after the pioneering work of Gilles Klopman, who introduced the concept of charge and frontier controlled reactions, including

solvation effects [4]. Nowadays, the treatment of chemical reactivity is mostly performed using the frame of the conceptual density functional theory developed by Parr et al. [10–14], Pearson [15–18] and Pearson and Songstad [19], and other authors [20–22]. This reactivity model converted classical chemical concepts like electronegativity, hardness, and softness into numbers. In this way, atoms, molecules, and charged system were classified into quantitative scales of reactivity. This historical description does not illustrate different and independent approaches to describing chemical reactivity on quantitative basis. As a matter of fact, those response functions defined within the Coulson-Longuet-Higgins theory can easily be cast into the form of response function of DFT [23, 24].

This brief review has been organized as follows: in the first section we introduce the basic definitions leading to the quantitative description of electronegativity, hardness, and softness. The key piece to achieve these definitions lies in the concept of electronic chemical potential and its derivatives introduced by Parr and Yang [13, 27] and Pearson and Songstad [19]. The electronic chemical potential μ is defined

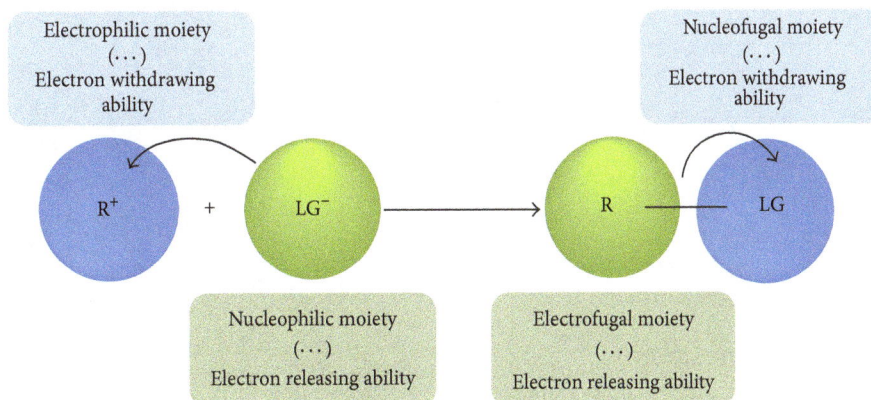

SCHEME 1: General heterolytic bond breaking process defining the nucleofuge (LG) and the electrofuge (PG). Adapted with permission from [25]. Copyright (2011) American Chemical Society.

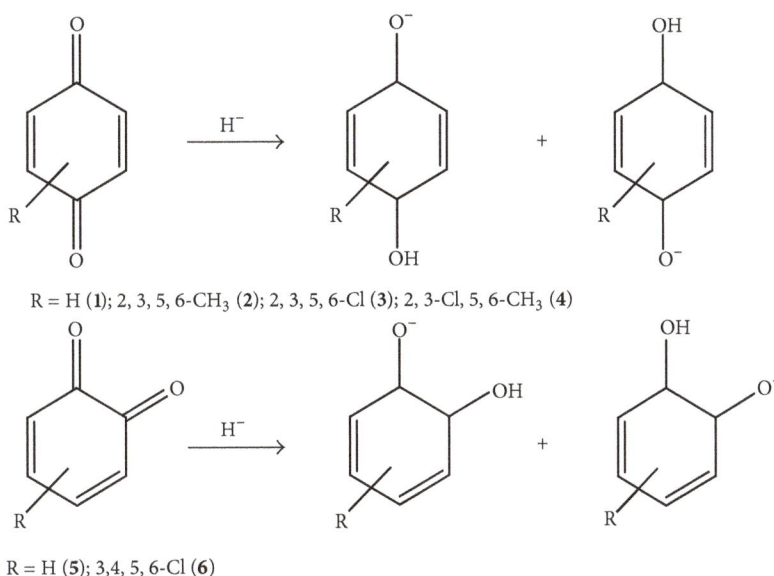

R = H (**1**); 2, 3, 5, 6-CH$_3$ (**2**); 2, 3, 5, 6-Cl (**3**); 2, 3-Cl, 5, 6-CH$_3$ (**4**)

R = H (**5**); 3,4, 5, 6-Cl (**6**)

SCHEME 2: Addition of H$^-$ to quinones. Adapted with permission from [26]. Copyright (2009) Elsevier.

therein as the first derivative of the energy E with respect to the number of electrons N [27]:

$$\mu = \left[\frac{\partial E}{\partial N}\right]_{v(\mathbf{r})} \approx -\frac{(I - A)}{2}. \qquad (1)$$

In (1), the derivative is taken at constant external potential $v(\mathbf{r})$ (i.e., the potential due to the compensating nuclear charges in the system). I and A are the vertical ionization potential and electron affinity, respectively. The electronic chemical potential as given by (1) becomes the negative of Mulliken electronegativity, and therefore it becomes a natural description of the direction of the electronic flux during a chemical interaction [27]. For this reason, Parr et al. proposed the electronic chemical potential as a quantity measuring the tendency of electrons to escape from the system. This result is relevant for it gives a first appraisal of the global electron-donating/electron accepting pattern of an interacting pair of atoms or molecules. If the chemical potential of a species A, say, is greater than its partner B, then the electronic flux will

take place from A to B, thereby suggesting that, during the interaction, A will act as nucleophile and B as electrophile. We will return to this point afterwards to introduce more refined models of electrophilicity and nucleophilicity.

A useful computational definition of the electronic chemical potential may be obtained using Koppmans's theorem [28] that leaves (1) in terms of the one-electron energy levels of the frontier molecular orbitals HOMO and LUMO:

$$\mu \cong \frac{\left(\epsilon_{\text{HOMO}} + \epsilon_{\text{LUMO}}\right)}{2}. \qquad (2)$$

The electronic chemical potential of stable species is a negative semidefinite quantity. This remark will be of importance later.

Another pertinent quantity is the chemical hardness, defined as the first derivative of the electronic chemical potential with respect to N (or the second derivative of the

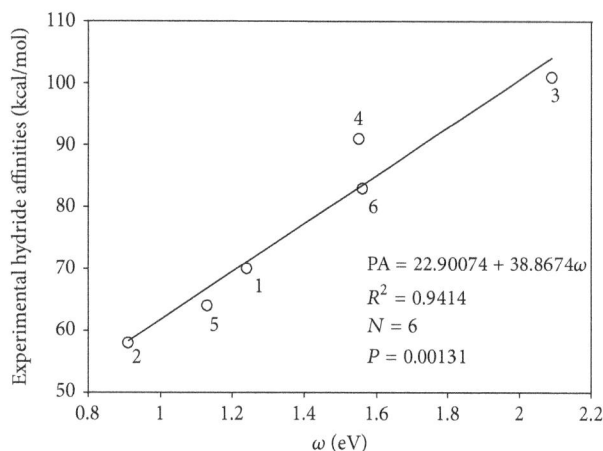

FIGURE 1: Comparison between the experimental hydride affinities of the quinone derivatives and their electrophilicity index (ω). Reprinted with permission from [26]. Copyright (2009) Elsevier.

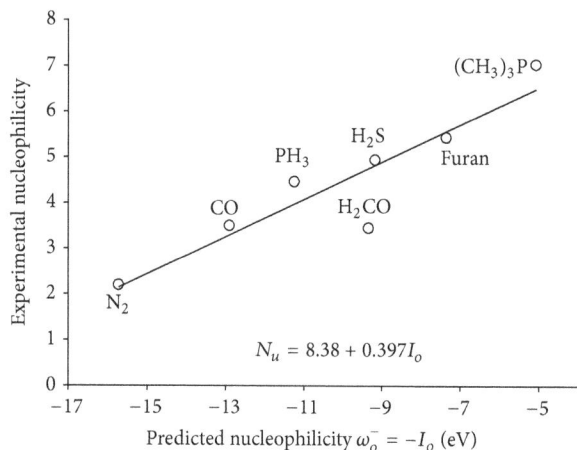

FIGURE 3: Comparison between experimental gas phase nucleophilicity and the negative of the gas phase ionization potential Io. Reprinted with permission from [30]. Copyright (2003) American Chemical Society.

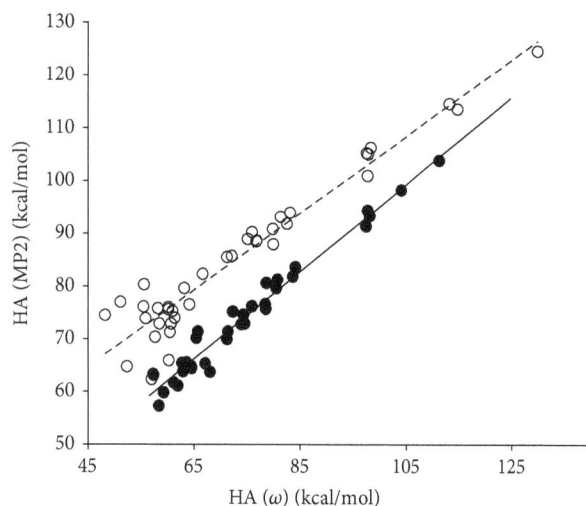

FIGURE 2: Comparison between hydride affinity of quinone derivatives from [29] and the hydride affinity obtained by electrophilicity index. The empty circles correspond to *ortho*-quinones derivatives (dashed line) and the full circles correspond to *para*-quinone derivatives (solid line). Reprinted with permission from [26]. Copyright (2009) Elsevier.

TABLE 1: Nucleophilic sites for electron donors and contributions to the regional nucleophilicity index. Adapted with permission from [30].

Species	Nucleophilic sites (k)	ω^-	f_k^-	ω_k^-
HO^-	O	−3.93	0.99	−3.89
HOO^-	O	−3.48	0.74	−2.58
N_3^{-a}	N_1		0.49	−2.84
	N_2	−5.79	0.02	−0.12
	N_3		0.49	−2.84
CH_3O^-	O	−3.99	0.73	−2.91
$CF_3CH_2O^-$	O	−4.91	0.75	−3.68
$C_6H_5S^-$	S	−5.45	0.62	−3.38
CN^-	C	−6.97	0.39	−2.72
	N		0.61	−4.25
NH_2OH	N	−3.72	0.71	−2.64
	O		0.27	−1.00
NH_3	N	−4.58	0.97	−4.44
$NH_2CONHNH_2$	O	−5.41	0.48	−2.60
H_2O	O	−6.65	0.98	−6.52
$CF_3CH_2NH_2$	N	−4.83	0.78	−3.77
CH_3ONH_2	O	−3.84	0.68	−2.61
	N		0.23	0.88
$CH_3CH_2CH_2S^-$	S	−3.46	0.94	−3.25
$CH_3CH_2S^-$	S	−3.43	0.95	−3.26
$OHCH_2CH_2S^-$	S	−3.15	0.94	−2.96
Piperidine	N	−4.32	0.66	−2.85
Morpholine	N	−4.46	0.49	−2.19
	O		0.13	−0.58

$^a N_2$ is the central atom.

energy with respect to the number of electrons) [27]. Its finite difference version is as follows:

$$\eta \cong I - A \cong \epsilon_{LUMO} - \epsilon_{HOMO}. \qquad (3)$$

Note that chemical hardness is a positive definite quantity. Its associated inverse is the chemical softness $S = 1/\eta$.

The functional derivative of the electronic chemical potential with respect to the external potential $v(\mathbf{r})$ at constant N defines a local response function called the Fukui function $f(\mathbf{r})$ [27]:

$$f(\mathbf{r}) = \left[\frac{\delta \mu}{\delta v(\mathbf{r})} \right]_N. \qquad (4)$$

The Fukui function has a more workable operative form obtained after using a Maxwell relationship [27]:

$$f(\mathbf{r}) = \left[\frac{\delta \rho(\mathbf{r})}{\delta N} \right]_{v(\mathbf{r})}. \qquad (5)$$

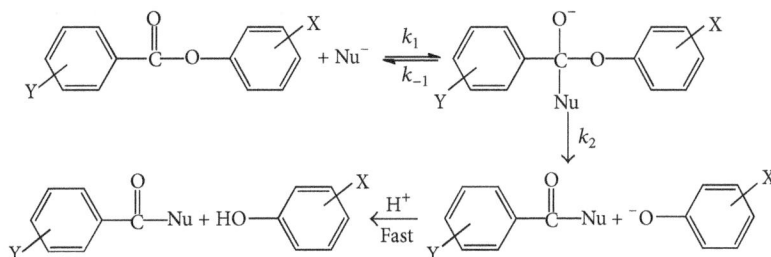

SCHEME 3: General picture of the accepted reaction mechanism of aryl benzoates. Adapted with permission from [31]. Copyright (2010) Elsevier.

The Fukui function written as in (5) defines it as the change in the electron density at point $\mathbf{r}$ in space $\rho(\mathbf{r})$ after the system accepts or releases one electron unit. The Fukui function is a reactivity index itself (a normalized softness) [27]. However, the most relevant role of the Fukui function is to act as a distribution function that may be used to project any global quantity G [35]. The most useful form is however its condensed to atom version, f_k, obtained after a regional integration around the atomic center k in a molecule [36–38].

The condensed to atom Fukui function may be approached from a three-point interpolation finite difference formula, or using a single-point calculation using a Mulliken-like population analysis. Other approaches have also been proposed [39]. The finite difference approximation leads to the definition of electrophilic, nucleophilic, and radical Fukui functions that will play a key role in what follows [36, 37].

2. Philicity and Fugality Indices

Based on a proposal by Maynard et al. [40], Parr et al. derived global electrophilicity, defined as the stabilization in energy that an electron acceptor atom or molecule undergoes, when it is embedded in an electron bath at constant electronic chemical potential [12]. The global electrophilicity index was given the following working expression:

$$\omega = \frac{\mu^2}{2\eta} = \frac{\mu^2}{2} S. \tag{6}$$

Equation (6) shows that the best electrophile will be the species displaying a high value of the electronic chemical potential and a low value of chemical hardness or high electronegativity and high softness (or high polarizability). We have implemented Parr's global electrophilicity index in the form of reactivity scales for a series of classical reactions in organic chemistry [41] that include cycloaddition reactions [42], elimination reactions [43], nucleophilic reactions (both aromatic and aliphatic) [44, 45], addition reactions [46], epoxidation reactions [47], redox, and biochemical processes [46, 48], including the chemistry of carbenes [49].

Nucleophilicity on the other hand cannot be derived within the same model leading to the definition of the electrophilicity index [30, 50, 51]. This problem arises because,

for the right-hand side of the parabola model used by Parr et al., the electronic chemical potential becomes positive semidefinite. This drawback of the philicity model is important for it is related to an empirical rule stating that big electrophilicity/nucleophilicity differences can be related to stepwise reaction mechanism with a high polar transition state, whereas small differences are related to nonpolar concerted mechanisms [52]. However, it is still possible to propose a nucleophilicity index based on the same energy expansion leading to the electrophilicity index. This index has been derived for the limit case where the charge released by the nucleophile is exactly equal to one electron unit. In this case the nucleophilicity index can be simply represented as the negative of the ionization potential; namely [30],

$$\omega^- = -I \approx \epsilon_{\text{HOMO}}. \tag{7}$$

This definition can intuitively be justified for it suggests that the best nucleophile will be the species that destabilizes to a lowest extent in the process of releasing one electron unit of charge. In the following sections, we shall illustrate the reliability and usefulness of the electrophilicity and nucleophilicity indices for a series of classic organic reactions.

The definition of fugality indices requires the introduction of local reactivity indices, where the electrophilic (+) and nucleophilic (−) Fukui functions play a key role [25]. For instance, using (6) together with the additive property of global softness, one of us introduced the concept of semilocal or regional electrophilicity condensed to atom k as follows [43, 53]:

$$\omega_k = \omega f_k^+, \tag{8}$$

where f_k^+ is the condensed to atom k electrophilic Fukui function. In a similar way, the condensed to atom k nucleophilicity index can be expressed as [25]

$$\omega_k^- = \omega^- f_k^-, \tag{9}$$

where f_k^- is the condensed to atom k nucleophilic Fukui function.

The relationship with fugality concepts is framed on the regional philicity concepts sketched in Scheme 1 [25].

First of all, fugality quantities are group properties of a molecule [25]. This means that the propensity of a fragment

TABLE 2: Electrophilicity index of X-substituted phenyl benzoated. Adapted with permission from [31].

Compound	ω (eV)	X	ω_{LG} (eV)	$\Delta\omega_{LG}$ (eV)	σ	$k_N{}^a$ (M^{-1}s^{-1})
1	0.94	3,4-diNO$_2$	0.90	0.89	0.83	1.92
2	0.75	4-NO$_2$	0.72	0.71	0.78	0.231
3	0.61	4-CN	0.61	0.60	0.66	—
4	0.61	4-CHO	0.35	0.34	0.42	0.103
5	0.57	4-COCH$_3$	0.20	0.19	0.50	0.0768
6	0.56	4-CO$_2$Et	0.09	0.08	0.45	—
7	0.77	3-NO$_2$	0.77	0.76	0.71	0.141
8	0.56	3-COCH$_3$	0.21	0.20	0.38	0.0359
9	0.53	4-Cl	0.01	0.00	0.23	—
10	0.66	H	0.01	0.00	0.00	0.0102
11	0.47	4-CH$_3$	0.01	0.00	−0.17	0.00783
12	0.83	4-OCH$_3$	0.01	0.00	−0.27	0.00843

[a]Data from references [32–34].

TABLE 3: Experimental electrophilicity E of the benzhydryl cations; regional nucleophilicity at fragment R, N(R), and at fragment LG, N(LG), in the complex R-LG; experimental and predicted electrofugality of benzhydryl cations. Adapted with permission from [25].

Entry	E	N(R) (eV)	N(LG) (eV)	%N(R)	ω^{-b}
1	5.90	−8.67	−0.08	99.1	−6.05
2	3.63	−8.37	−0.06	99.3	−3.47
3	2.90	−8.29	−0.03	99.6	−3.55
4	2.11	−8.16	−0.03	99.7	−2.06
5	1.48	−8.10	−0.03	99.6	−1.29
6	0.61	−8.13	−0.05	99.3	−0.81
7	0.00	−8.03	−0.04	99.5	0.00
8	−0.56	−7.94	−0.03	99.6	0.33^c
9	−1.36	−7.91	−0.04	99.5	0.60^c
10	−3.14	−7.75	−0.03	99.6	2.05^c
11	−3.89	−7.91	−0.03	99.6	0.63^c
12^a	−4.72	—	—	—	—
13	−5.53	−7.94	−0.03	99.6	0.36^c
14	−5.89	−7.38	−0.02	99.7	5.46^c
15	−7.02	−7.30	−0.02	99.7	6.22^c
16	−7.69	−7.02	−0.02	99.7	8.78^c
17	−8.22	−7.21	−0.02	99.7	6.99^c
18	−8.76	−7.30	−0.02	99.7	6.22^c
19	−9.45	−7.11	−0.02	99.7	7.91^c
20	−10.04	−7.17	−0.02	99.7	7.34^c

[a]For this compound the algorithm used to evaluate the nucleophilic Fukui function produces negative values. [b]Experimental electrofugality from [33]. [c]Predicted values using the empirical equation included in Figure 2(b).

to detach during a heterolytic bond breaking process may be safely described using a regional property of that fragment embedded in the chemical ambient of the remaining moiety of the molecule. It is important to stress this point that fugality quantities are not intrinsic properties of the isolated fragment [25, 43, 53]. With this model in mind, we can readily define nucleofugality, as the group electrophilicity evaluated on the whole molecule, where the highest values of group electrophilicity are expected to be mostly concentrated at the leaving group (LG) moiety. This is a reasonable representation of nucleofugality number, as the nucleofuge is the group that departs from the molecule bearing the bond electron pair

during the heterolytic bond cleavage [53]. At the same time, the electrofuge (or permanent group R) is expected to act as an electron releasing fragment, and, therefore, electrofugality may reasonably be described by the group nucleophilicity [25]. The working formulae to quantify nucleofugality ν^+ and electrofugality ν^- numbers are given by

$$\nu^+ = \sum_{k\varepsilon G} \omega_k^+, \tag{10}$$

$$\nu^- = \sum_{k\varepsilon G} \omega_k^-, \tag{11}$$

respectively.

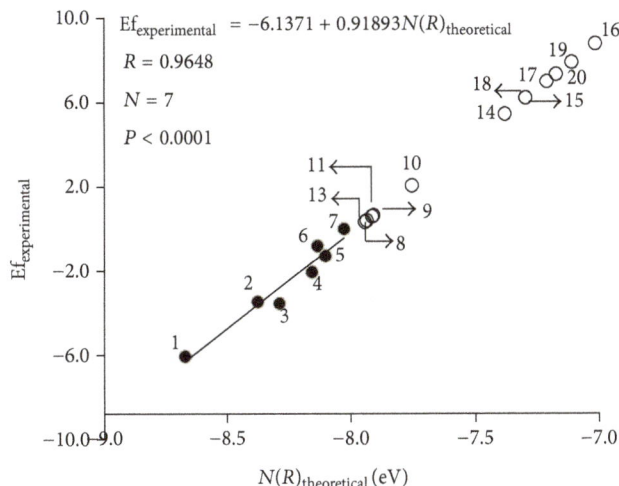

FIGURE 4: Comparison between experimental electrofugality (●) and the regional nucleophilicity of the permanent group R for a series of benzhydryl phenyl sulfinates derivatives. Predicted electrofugalities values are included (○). Reprinted with permission from [25]. Copyright (2011) American Chemical Society.

TABLE 4: Global and regional indices for the IMDA reaction leading to diterpenoid elisabethin A.

μ	η	ω	ω^-	ω_{Dp}	ω_D	ω^-_{Dp}	ω^-_D
−0.1607	0.0940	3.65	3.46	3.64	0.00	0.00	3.44

3. Applications

The phenomenological reactivity theory described in the previous sections may be applied at the ground state of atoms and molecules as well as at the transition state stage of reactions. Note that, in doing so, both global and semilocal (regional) quantities may be used to build up "activation" properties. In what follows we present some of the applications that we and other authors have used to illustrate the reliability and usefulness of this theoretical model of reactivity that embodies reactivity, (regio)selectivity, and site activation [45, 54, 55].

3.1. The Electrophilicity Index. The electrophilicity index developed by Parr et al. (6) has been widely used to explain many organic reactions. The main purpose of these studies was traced to relate them to experimental data that include hydride affinity [26], rate coefficients [44, 56, 57], toxicity indexes [58–60], and many other applications [61–64]. For instance, Campodónico et al. demonstrated that the electrophilicity of quinones may be concisely used to deduce a hydride affinity (HA) scale in the gas phase [26]. The opportunity of having a global electrophilicity hierarchy related to HA is useful because experimental HA data are scarce. Because the measurement of HA is not direct, the electrophilicity-HA relationships provide a simple way to establish a sound HA scale. The gas phase hydride affinity may be obtained as the negative enthalpy change for the reaction $Q + H^- \rightarrow QH^-$, where Q is an oxidized molecule and QH^- is its reduced form. The hydride affinity has been considered as a descriptor for Lewis's acidity [65] since it may be a characteristic of the electron accepting capability of the electrophile [29]. Within this model, the ability of quinones to bind an H^- ion will be related to their electrophilic response and may be modulated by the presence of a series

of substituent R that can stabilize an extra negative charge, as illustrated in Scheme 2 [26].

The comparison between the experimentally observed hydride affinity and the electrophilicity index is shown in Figure 1. The regression is the result of the comparison of six experimentally obtained hydride affinities.

The empirical regression equation is

$$HA = 22.900 + 38.867\omega; \qquad R^2 = 0.94 \qquad (12)$$

and is a suitable way to obtain the hydride affinity of quinones not established up to date. First of all, from the regression equation (12), it is possible to note that the hydride affinities are well correlated with the ω index. The relationship with the substituent effect is also described by ω in the sense that electron-donating groups enhance the hydride affinity of quinones, while electroattracting groups diminish the hydride affinity values. When the substituent is the CN or Cl groups (moderate electron-withdrawing groups), results in an electrophilic activation ($\omega = 2.09$ eV and $\omega = 1.56$ eV, resp.) with respect to the reference non-substituted quinone ($\omega = 1.24$ eV). On the other hand, the presence of the marginal donating methyl group results in an electrophilic deactivation ($\omega = 0.91$ eV) [26]. The regression between the electrophilicity index and the predicted hydride affinities [66] is presented in Figure 2.

The comparison in Figure 2 shows two families that cannot be accommodated in a single correlation line. The regression corresponds to the *ortho*-quinone derivatives (dashed line) and the *para*-quinone derivatives (solid line). These results represent another useful application of the ω index. The hydride attachment strongly depends on the relative position of the carbonyl groups on the quinine: in the *ortho*-like derivatives, the electrophilicity index well describes the alpha-like effect promoted in this system by the presence of an adjacent electron-rich atom [67, 68]. These systems are predicted as nucleophiles or marginal electrophiles [69, 70].

3.2. Nucleophilicity. The usefulness of the nucleophilicity index (7) is discussed here for a series of neutral and charged electron donors [30]. The series of neutral nucleophiles

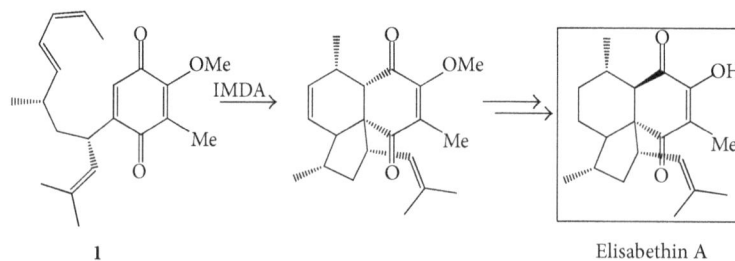

SCHEME 4: Synthesis of diterpenoid elisabethin A via an IMDA reaction.

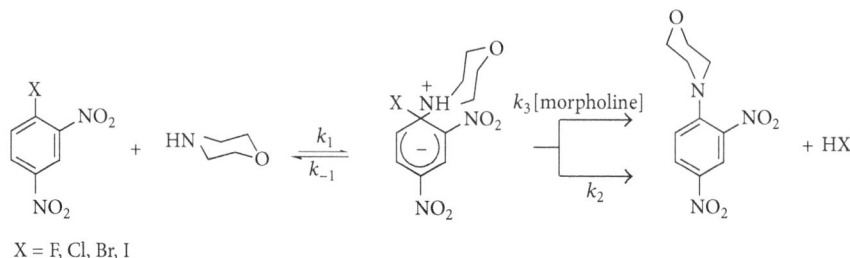

X = F, Cl, Br, I

SCHEME 5: General S_NAr reaction mechanism. Reprinted with permission from [45]. Copyright (2013) American Chemical Society.

relative to water include N_2, CO, PH_3, H_2CO, H_2S, furan, $(CH_3)_3P$, H_2O, NH_3, and CH_3OCH_3 [71, 72].

This series was experimentally studied by Legon and Millen who proposed a spectroscopic scale of electrophilicity and nucleophilicity based on the intermolecular stretching force constant for the interaction of these neutral nucleophiles [71, 72] towards the hydrogen fluoride HF as probe. In the gas phase, the facile formation of a hydrogen bonded nucleophile-HF complex permits the intermolecular stretching of the nucleophile/HF moieties to be evaluated. The nucleophilicity scale is simply given by the ordered hierarchy of the respective force constants. The comparison between the experimental nucleophilicity numbers reported by Legon and Millen with those obtained from our model equation (7) for the whole series of neutral nucleophiles was fair, because ammonia and dimethyl ether strongly deviated from linearity. A deep sight to the available IR data for the complexes H_3N-HF and $(CH_3)_2O$-HF revealed that the normal mode assignments for these species were not as clean as desired. The calculated intermolecular stretching around the hydrogen bond was hardly contaminated by torsion and bond deformation (wagging) of the sp^3 groups attached to the heteroatoms N and O. However, when the regression was made with the 7 remaining nucleophiles, a reasonably good correlation was obtained. The comparison is shown in Figure 3.

It is worth mentioning that the model equation (7) also works reasonably well for ionization potential in solution phase [30].

As we will show later, the semilocal or regional nucleophilicity patterns of molecules become of great relevance to describe another property, namely, electrofugality, a group property describing permanent group abilities in heterolytic bond cleavage processes. The model equation describing

regional or group nucleophilicity is that quoted as (9). Site reactivity is also useful for predicting regioselectivity, hydrogen bond basicity, and Lewis molecular basicity in polyfunctional species. In order to evaluate the regional nucleophilicity index given by (9), a series of charged and neutral nucleophiles in the gas phase were evaluated. The results are compiled in Table 1.

From the data collected in Table 1, it may be seen that the ω_k^- index consistently distributes the global nucleophilicity values on those atoms that are expected to be more nucleophilic. As expected, the most nucleophilic center is at the heteroatom (N, O, or S) site.

3.3. Nucleofugality. As defined at the beginning of this review, the nucleofugality index as stated in (10) may be used to quantitatively characterize the leaving group abilities of several fragments commonly present as nucleofuges in substitution and elimination reactions in organic chemistry. The basic ideas and concepts are those summarized in Scheme 1. In order to illustrate the reliability and usefulness of the nucleofugality index we have evaluated it for a series of aryl benzoates and discuss the usefulness of the resulting scales to assist in the rationalization of their reaction mechanism in nucleophilic substitution reactions [31]. The acyl group in aryl benzoates may react along stepwise or concerted nucleophilic substitution channels to yield the corresponding amide [32–34]. A general route is sketched in Scheme 3 [31].

In order to assess the effect of the leaving group on the reaction mechanism, we selected a series of reactions where the permanent group is kept fixed. The series is depicted in Table 2 [31].

From the data summarized in Table 2, it is possible to note that the electrophilicity index at the leaving group is mainly

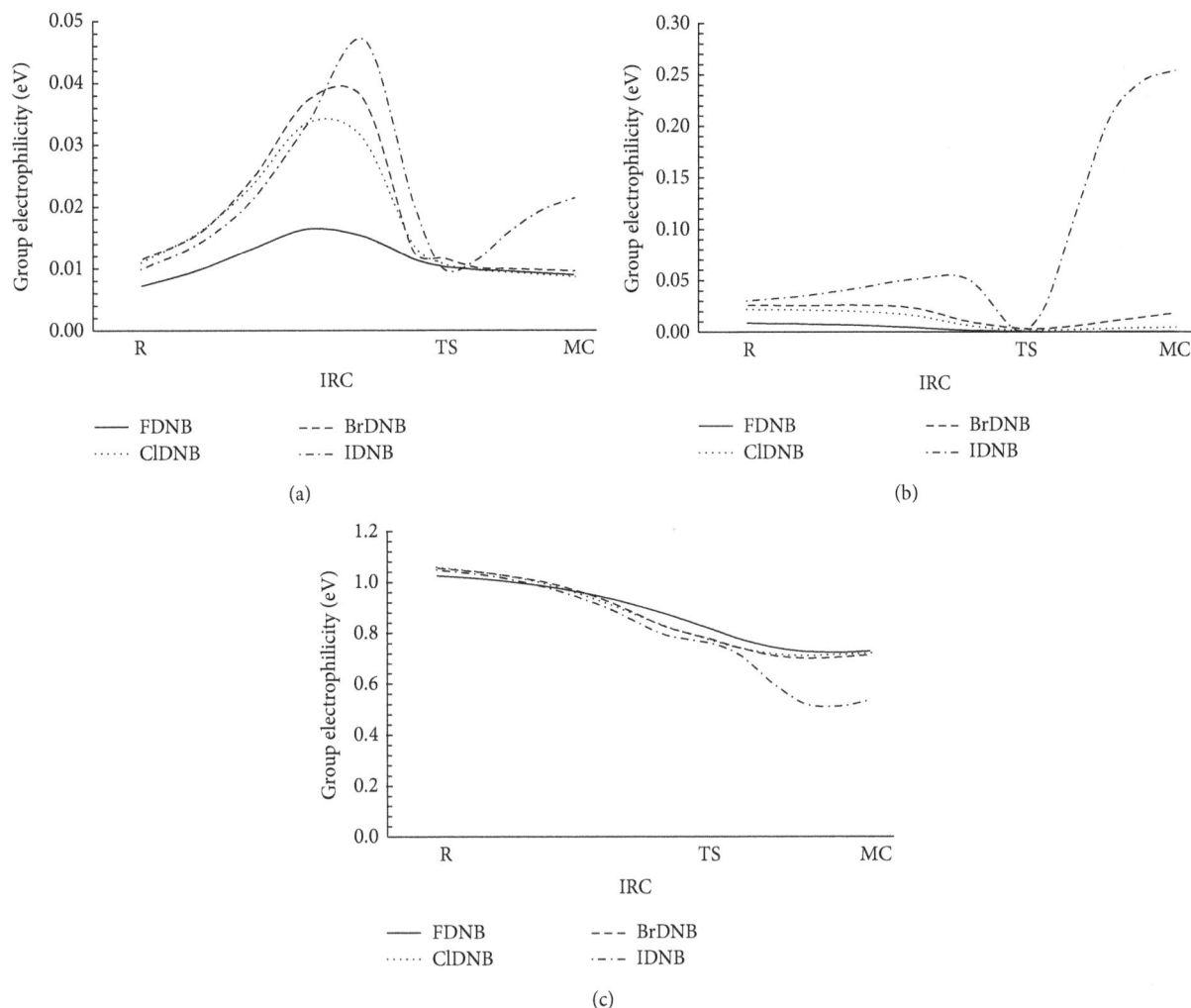

FIGURE 5: Profiles of group electrophilicity of the fragments centered in the moieties corresponding to (a) nucleophile, (b) leaving group, and (c) permanent group in the reaction between morpholine toward the XDNB series (X = F, Cl, Br, and I). Adapted with permission from [45].Copyright (2013) American Chemical Society.

driven by the presence of electron-withdrawing groups. The presence of these groups is related to an enhancement of the nucleofugality (leaving group ability) of this group. It is important to emphasize that the trends in nucleofugality coherently compare with the experimental rate coefficients measured by Um and coworkers [32–34]. The addition of electron-attracting groups on the leaving group may be responsible for the enhancement of the rate constant, since these groups contribute to the net destabilization of the intermediate from which the leaving group detaches.

3.4. Electrofugality. The electrofugality index given by (11) has been applied to a set of 20 benzhydryl sulfinates experimentally studied by Baidya et al. to establish a quantitative hierarchy of electrofugality [73]. According to Scheme 1, the best electrofuge is the fragment that displays the highest regional nucleophilicity [25]. During the bond cleavage the electrofuge is the fragment that releases the electron density,

thereby acting as an electron donor. The results are summarized in Table 3.

Table 3 reveals that the regional nucleophilicity is mainly concentrated at fragment R which corresponds to the permanent group. It is important to remark that the electrofugality is fairly ordered in terms of inductive effects promoted by the substituents. The usefulness of this model is illustrated in Figure 4, in the perspective of the predictive potential of this tool: with known electrofugality data for a reduced set of molecules at hand, it becomes possible to anticipate the electrofugality number by simply evaluating the group nucleophilicity of the permanent group moiety at the intermediate complexes [25].

3.5. Site Activation: Fragment Reactivity Analysis. In equilibrium thermodynamics, the electronic chemical potential of density functional theory is a global response function, and therefore it is expected to have a uniform distribution

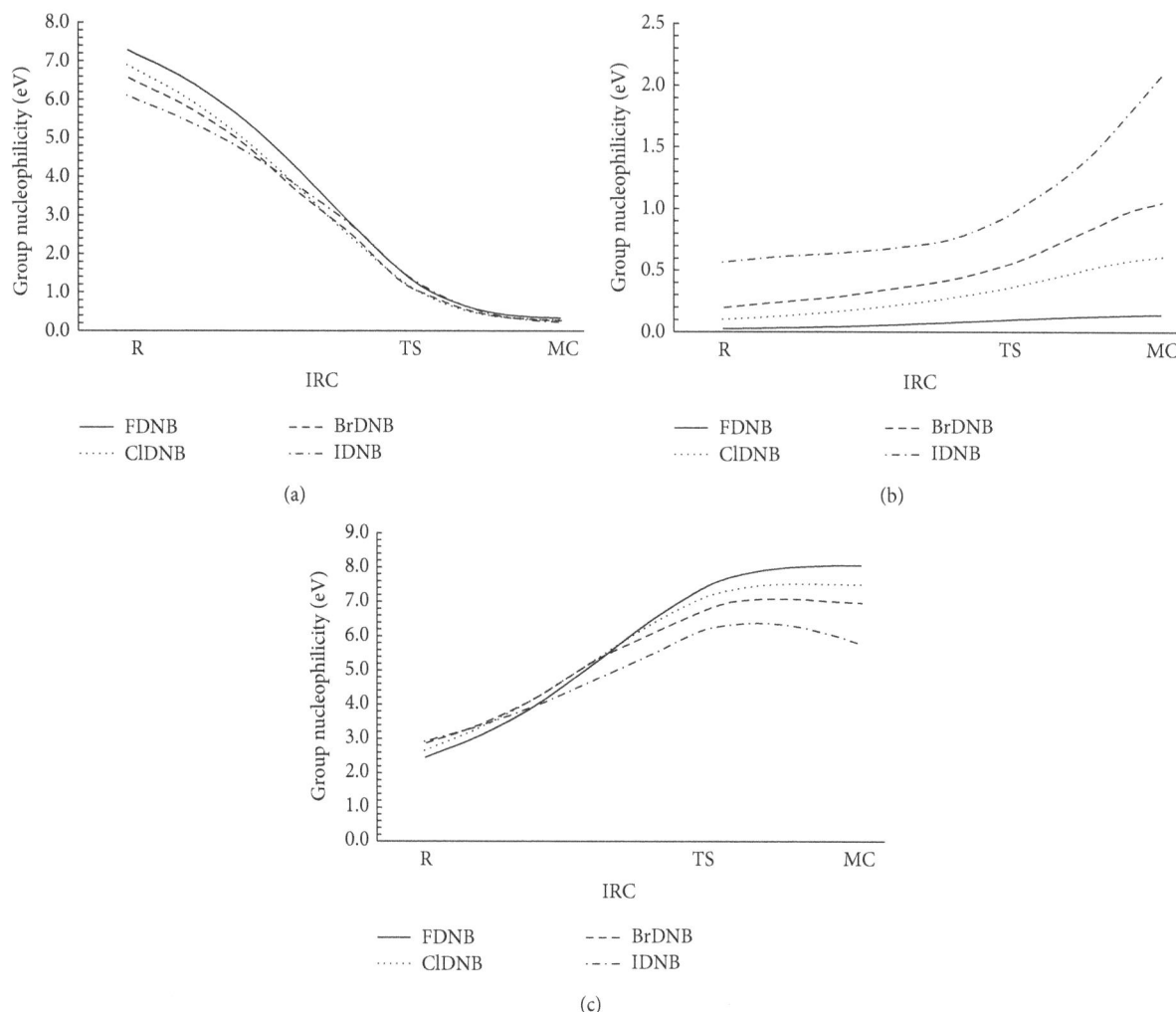

FIGURE 6: Group nucleophilicity profiles centered in the (a) nucleophile, (b) leaving group, and (c) permanent group for the reaction between morpholine towards 1-X-2,4-dinitrobenzene series (X = F, Cl, Br, and I). Adapted with permission from [45]. Copyright (2013) American Chemical Society.

within the whole molecular structure. However, intramolecular reactivity put a serious challenge for the application of electronic descriptors of reactivity, and some adaptations are to be introduced in order to account for how two or more fragments, within the same molecule, may interact to produce bond making/breaking processes. We proposed a model framed on nonequilibrium thermodynamics allowing two molecular fragments to be described by different electronic chemical potential, in such a way that they can exchange work, heat, or energy [74, 75]. We have used the intramolecular Diels-Alder (IMDA) reaction of Quinone systems to illustrate this model. The benchmark reaction used was the synthesis of the diterpenoid elisabethin A. The reaction is sketched in Scheme 4 [76].

The fragmentation scheme together with the global and regional indices associated with the diene (D) and Dienophile (Dp) fragments is summarized in Table 4.

The global electrophilicity of quinone **1** is within the range of strong electrophiles [69, 70]. The most electrophilic center of **1** that may react is the C1 carbon. Consequently, the favored IMDA reaction is that involving the diene fragment and the C1-C2 double bond. Note the remarkable resolution of the philicity patterns, a job that is performed by the Fukui function (see (8) and (9)). The fragmentation scheme used put over 98% of electrophilicity on the Dp fragment and over 99% of nucleophilicity on the D moiety.

3.6. Quasi-Static Approach: Reactivity Indices along a Reaction Pathway. A final word concerning the reactivity indices in organic reactions is a brief discussion on a nucleophilic aromatic substitution reaction, within a quasi-static scheme, obtained by following the changes in electrophilicity and nucleophilicity along the intrinsic reaction coordinates (IRC). The model reaction is that sketched in Scheme 5 [45, 55].

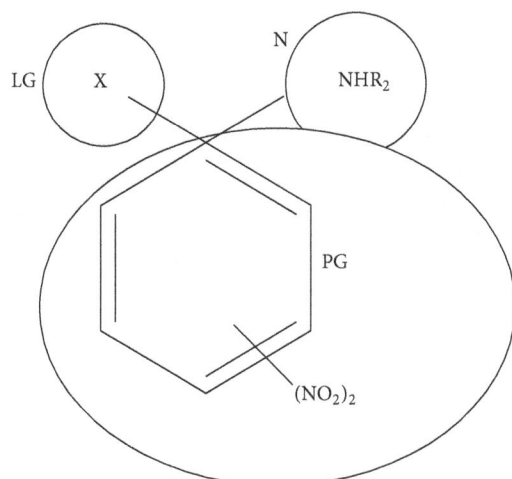

SCHEME 6: General fragmentation model of the electrophile-nucleophile pair. LG, PG, and N stand for leaving group, permanent group, and nucleophile, respectively. Adapted with permission from [45]. Copyright (2013) American Chemical Society.

The model reaction is the $S_N Ar$ involving morpholine and 1-X-2,4-dinitrobenzenes (XDNB, X = F, Cl, Br, and I) [77, 78]. Figures 5 and 6 display the group electrophilicity and nucleophilicity profiles using the arbitrary fragmentation scheme below [25, 45].

As expected, the electrophilicity of the amine moiety (N) is marginal (Figure 5(a)). The electrophilicity condensed at the LG fragment shows a sudden enhancement for the iodine derivative. This result suggests that iodine may detach in the first stage of the reaction depicted in Scheme 5, in agreement with the experimental reports [77, 78]. Figure 5(c) shows the role of the permanent group on the reaction mechanism. In the region of the Meisenheimer complex (MC) all the substrates become electronically saturated. This result may be associated with the end of the charge transfer process, except for iodine which began to detach from the structure as iodide.

Figure 6 shows the nucleophilicity profile within the partitioning Scheme 6. When the index is condensed over the nucleophilic moiety, it seems that it dramatically diminishes downward the MC formation. This result suggests that the charge transfer reaches its minimum after the nucleophilic attack. It is important to note that the property condensed at the LG and PG reaches a maximum value near the MC: the charge is transferred from the nucleophile and accepted for the PG and then redirected to the LG. However, the maximum values for morpholine are different: 7.29, 6.91, 6.56, and 6.11 eV for X = F, Cl, Br, and I, respectively. These values suggest that the substrates interact in different way depending on the LG present. The results reported are in good agreement with the experimental observations, since the leaving group abilities are in the order F > Cl > Br > I when the nucleophile is morpholine [45, 55, 78].

4. Concluding Remarks and Perspectives

In this review, we have shown how conceptual aspects of density functional theory lead to the definition of reactivity indices amenable to build up quantitative models of reactivity in organic reactions. The emphasis has been put on two basic concepts describing electron-rich and electron-deficient systems, namely, nucleophile and electrophiles. We then show that the regional nucleophilicity and electrophilicity become the natural descriptors of electrofugality and nucleofugality, respectively. In this way, we have obtained a closed body of concepts that suffices to describe electron releasing and electron accepting molecules together with the ordering of leaving group ability of nucleofuges present in addition, nucleophilic substitution and elimination reactions. A natural perspective of these models is their use along a reaction coordinate following that result in an additional tool to rationalize reaction mechanisms in organic chemistry.

Conflict of Interests

The authors declare that there is no conflict of interests regarding the publication of this paper.

References

[1] J. D. Roberts and A. Streitwieser Jr., "Quantum mechanical calculations of orientation in aromatic substitution," *Journal of the American Chemical Society*, vol. 74, no. 18, pp. 4723–4725, 1952.

[2] A. A. Frost and B. Musulin, "A mnemonic device for molecular orbital energies," *The Journal of Chemical Physics*, vol. 21, no. 3, pp. 572–573, 1953.

[3] L. Salem, "Intermolecular orbital theory of the interaction between conjugated systems. I. General theory," *Journal of the American Chemical Society*, vol. 90, no. 3, pp. 543–552, 1968.

[4] G. Klopman, "Chemical reactivity and the concept of charge- and frontier-controlled reactions," *Journal of the American Chemical Society*, vol. 90, no. 2, pp. 223–234, 1968.

[5] E. Z. Hückel, "Quantentheoretische Beiträge zum Benzolproblem," *Zeitschrift für Physik*, vol. 70, pp. 204–286, 1931.

[6] W. Heitler and F. London, "Wechselwirkung neutraler Atome und homöopolare Bindung nach der Quantenmechanik," *Zeitschrift für Physik*, vol. 44, no. 6-7, pp. 455–472, 1927.

[7] C. A. Coulson and H. C. Longuet-Higgins, "The electronic structure of conjugated systems. I. General theory," *Proceedings of the Royal Society of London Series A: Mathematical and Physical Sciences*, vol. 191, no. 1024, pp. 39–60, 1947.

[8] C. A. Coulson and H. C. Longuet-Higgins, "The electronic structure of conjugated systems. II. Unsaturated hydrocarbons and their hetero-derivatives," *Proceedings of the Royal Society of London. Series A. Mathematical and Physical Sciences*, vol. 192, no. 1028, pp. 16–32, 1947.

[9] C. A. Coulson and H. C. Longuet-Higgins, "The electronic structure of conjugated systems. III. Bond orders in unsaturated molecules; IV. Force constants and interaction constants in unsaturated hydrocarbons," *Proceedings of the Royal Society of London A*, vol. 193, no. 1035, pp. 447–464, 1948.

[10] R. G. Parr and P. K. Chattaraj, "Principle of maximum hardness," *Journal of the American Chemical Society*, vol. 113, no. 5, pp. 1854–1855, 1991.

[11] R. G. Parr and R. G. Pearson, "Absolute hardness: companion parameter to absolute electronegativity," *Journal of the American Chemical Society*, vol. 105, no. 26, pp. 7512–7516, 1983.

[12] R. G. Parr, L. V. Szentpály, and S. Liu, "Electrophilicity index," *Journal of the American Chemical Society*, vol. 121, no. 9, pp. 1922–1924, 1999.

[13] R. G. Parr and W. Yang, "Density functional approach to the frontier-electron theory of chemical reactivity," *Journal of the American Chemical Society*, vol. 106, no. 14, pp. 4049–4050, 1984.

[14] J. P. Perdew, R. G. Parr, M. Levy, and J. L. Balduz Jr., "Density-functional theory for fractional particle number: derivative discontinuities of the energy," *Physical Review Letters*, vol. 49, no. 23, pp. 1691–1694, 1982.

[15] R. G. Pearson, "Hard and soft acids and bases," *Journal of the American Chemical Society*, vol. 85, no. 22, pp. 3533–3539, 1963.

[16] R. G. Pearson, "Absolute electronegativity and hardness: application to inorganic chemistry," *Inorganic Chemistry*, vol. 27, no. 4, pp. 734–740, 1988.

[17] R. G. Pearson, "The principle of maximum hardness," *Accounts of Chemical Research*, vol. 26, no. 5, pp. 250–255, 1993.

[18] R. G. Pearson, "Absolute electronegativity and hardness: applications to organic chemistry," *Journal of Organic Chemistry*, vol. 54, no. 6, pp. 1423–1430, 1989.

[19] R. G. Pearson and J. Songstad, "Application of the principle of hard and soft acids and bases to organic chemistry," *Journal of the American Chemical Society*, vol. 89, no. 8, pp. 1827–1836, 1967.

[20] P. K. Chattaraj, H. Lee, and R. G. Parr, "HSAB principle," *Journal of the American Chemical Society*, vol. 113, no. 5, pp. 1855–1856, 1991.

[21] P. W. Ayers and M. Levy, "Perspective on "Density functional approach to the frontier-electron theory of chemical reactivity"," *Theoretical Chemistry Accounts*, vol. 103, no. 3-4, pp. 353–360, 2000.

[22] P. W. Ayers and R. G. Parr, "Variational principles for describing chemical reactions: the Fukui function and chemical hardness revisited," *Journal of the American Chemical Society*, vol. 122, no. 9, pp. 2010–2018, 2000.

[23] M. Gonzalez-Suarez, A. Aizman, and R. Contreras, "Phenomenological chemical reactivity theory for mobile electrons," *Theoretical Chemistry Accounts*, vol. 126, no. 1, pp. 45–54, 2010.

[24] M. Gonzalez-Suarez, A. Aizman, J. Soto-Delgado, and R. Contreras, "Bond Fukui functions as descriptor of the electron density reorganization in π conjugated systems," *The Journal of Organic Chemistry*, vol. 77, no. 1, pp. 90–95, 2011.

[25] R. Ormazábal-Toledo, P. R. Campodónico, and R. Contreras, "Are electrophilicity and electrofugality related concepts? A density functional theory study," *Organic Letters*, vol. 13, pp. 822–824, 2011.

[26] P. R. Campodónico, A. Aizman, and R. Contreras, "Electrophilicity of quinones and its relationship with hydride affinity," *Chemical Physics Letters*, vol. 471, no. 1–3, pp. 168–173, 2009.

[27] R. G. Parr and W. Yang, *Density-Functional Theory of Atoms and Molecules*, Oxford University Press, New York, NY, USA, 1989.

[28] T. Koopmans, "Über die Zuordnung von Wellenfunktionen und Eigenwerten zu den Einzelnen Elektronen Eines Atoms," *Physica*, vol. 1, no. 1–6, pp. 104–113, 1934.

[29] R. Vianello, N. Peran, and Z. B. Maksić, "Hydride affinities of some substituted alkynes: prediction by DFT calculations and rationalization by triadic formula," *Journal of Physical Chemistry A*, vol. 110, no. 47, pp. 12870–12881, 2006.

[30] R. Contreras, J. Andres, V. S. Safont, P. Campodonico, and J. G. Santos, "A theoretical study on the relationship between nucleophilicity and ionization potentials in solution phase," *Journal of Physical Chemistry A*, vol. 107, no. 29, pp. 5588–5593, 2003.

[31] P. R. Campodónico, R. Ormazábal-Toledo, A. Aizman, and R. Contreras, "Permanent group effect on nucleofugality in aryl benzoates," *Chemical Physics Letters*, vol. 498, pp. 221–225, 2010.

[32] I. Um, H. Han, J. Ahn, S. Kang, and E. Buncel, "Reinterpretation of curved hammett plots in reaction of nucleophiles with aryl benzoates: change in rate-determining step or mechanism versus ground-state stabilization," *Journal of Organic Chemistry*, vol. 67, no. 24, pp. 8475–8480, 2002.

[33] I. Um, S. Jeon, and J. Seok, "Aminolysis of 2,4-dinitrophenyl X-substituted benzoates and Y-substituted phenyl benzoates in MeCN: Effect of the reaction medium on rate and mechanism," *Chemistry*, vol. 12, no. 4, pp. 1237–1243, 2006.

[34] I. H. Um, J. Y. Lee, M. Fujio, and Y. Tsuno, "Structure-reactivity correlations in nucleophilic substitution reactions of Y-substituted phenyl X-substituted benzoates with anionic and neutral nucleophiles," *Organic & Biomolecular Chemistry*, vol. 4, no. 15, pp. 2979–2985, 2006.

[35] R. Contreras, J. Andrés, P. Pérez, A. Aizman, and O. Tapia, "Theory of non-local (pair site) reactivity from model static-density response functions," *Theoretical Chemistry Accounts*, vol. 99, pp. 183–191, 1998.

[36] W. Yang and W. J. Mortier, "The use of global and local molecular parameters for the analysis of the gas-phase basicity of amines," *Journal of the American Chemical Society*, vol. 108, no. 19, pp. 5708–5711, 1986.

[37] R. R. Contreras, P. Fuentealba, M. Galván, and P. Pérez, "A direct evaluation of regional Fukui functions in molecules," *Chemical Physics Letters*, vol. 304, no. 5-6, pp. 405–413, 1999.

[38] P. Fuentealba, P. Pérez, and R. Contreras, "On the condensed Fukui function," *Journal of Chemical Physics*, vol. 113, no. 7, pp. 2544–2551, 2000.

[39] P. Bultinck, C. Cardenas, P. Fuentealba, P. A. Johnson, and P. W. Ayers, "How to compute the Fukui matrix and function for systems with (Quasi-)degenerate states," *Journal of Chemical Theory and Computation*, vol. 10, no. 1, pp. 202–210, 2014.

[40] A. T. Maynard, M. Huang, W. G. Rice, and D. G. Covell, "Reactivity of the HIV-1 nucleocapsid protein p7 zinc finger domains from the perspective of density-functional theory," *Proceedings of the National Academy of Sciences of the United States of America*, vol. 95, no. 20, pp. 11578–11583, 1998.

[41] P. Pérez, A. Aizman, and R. Contreras, "Comparison between experimental and theoretical scales of electrophilicity based on reactivity indexes," *Journal of Physical Chemistry A*, vol. 106, no. 15, pp. 3964–3966, 2002.

[42] P. Pérez, L. R. Domingo, M. J. Aurell, and R. Contreras, "Quantitative characterization of the global electrophilicity pattern of some reagents involved in 1,3-dipolar cycloaddition reactions," *Tetrahedron*, vol. 59, no. 17, pp. 3117–3125, 2003.

[43] P. R. Campodónico, J. Andrés, A. Aizman, and R. Contreras, "Nucleofugality index in α-elimination reactions," *Chemical Physics Letters*, vol. 439, pp. 177–182, 2007.

[44] P. R. Campodónico, P. Fuentealba, E. A. Castro, J. G. Santos, and R. Contreras, "Relationships between the electrophilicity index and experimental rate coefficients for the aminolysis of thiolcarbonates and dithiocarbonates," *Journal of Organic Chemistry*, vol. 70, no. 5, pp. 1754–1760, 2005.

[45] R. Ormazábal-Toledo, R. Contreras, and P. R. Campodónico, "Reactivity indices profile: a companion tool of the potential

energy surface for the analysis of reaction mechanisms. Nucleophilic aromatic substitution reactions as test sase," *Journal of Organic Chemistry*, vol. 78, pp. 1091–1097, 2013.

[46] P. Pérez and R. Contreras, "A theoretical analysis of the gas-phase protonation of hydroxylamine, methyl-derivatives and aliphatic amino acids," *Chemical Physics Letters*, vol. 293, no. 3-4, pp. 239–244, 1998.

[47] K. Neimann and R. Neumann, "Electrophilic activation of hydrogen peroxide: Selective oxidation reactions in perfluorinated alcohol solvents," *Organic Letters*, vol. 2, no. 18, pp. 2861–2863, 2000.

[48] R. Das, J. L. Vigneresse, and P. K. Chattaraj, "Redox and Lewis acid–base activities through an electronegativity-hardness landscape diagram," *Journal of Molecular Modeling*, vol. 19, no. 11, pp. 4857–4864, 2013.

[49] W. Zhang, Y. Zhu, D. Wei, Y. Li, and M. Tang, "Theoretical investigations toward the [4 + 2] cycloaddition of ketenes with N-benzoyldiazenes catalyzed by N-heterocyclic carbenes: mechanism and enantioselectivity," *Journal of Organic Chemistry*, vol. 77, no. 23, pp. 10729–10737, 2012.

[50] A. Cedillo, R. Contreras, M. Galván, A. Aizman, J. Andrés, and V. S. Safont, "Nucleophilicity index from perturbed electrostatic potentials," *Journal of Physical Chemistry A*, vol. 111, no. 12, pp. 2442–2447, 2007.

[51] P. Jaramillo, P. Pérez, R. Contreras, W. Tiznado, and P. Fuentealba, "Definition of a nucleophilicity scale," *Journal of Physical Chemistry A*, vol. 110, no. 26, pp. 8181–8187, 2006.

[52] P. Campodonico, J. G. Santos, J. Andres, and R. Contreras, "Relationship between nucleophilicity/electrophilcity indices and reaction mechanisms for the nucleophilic substitution reactions of carbonyl compounds," *Journal of Physical Organic Chemistry*, vol. 17, no. 4, pp. 273–281, 2004.

[53] P. R. Campodonico, A. Aizman, and R. Contreras, "Group electrophilicity as a model of nucleofugality in nucleophilic substitution reactions," *Chemical Physics Letters*, vol. 422, no. 4-6, pp. 340–344, 2006.

[54] S. Jorge, A. Aizman, R. Contreras, and L. R. Domingo, "On the catalytic effect of water in the intramolecular diels-alder reaction of quinone systems: a theoretical study," *Molecules*, vol. 17, no. 11, pp. 13687–13703, 2012.

[55] R. Ormazábal-Toledo, R. Contreras, R. A. Tapia, and P. R. Campodónico, "Specific nucleophile-electrophile interactions in nucleophilic aromatic substitutions," *Organic and Biomolecular Chemistry*, vol. 11, pp. 2302–2309, 2013.

[56] A. Aizman, R. Contreras, and P. Pérez, "Relationship between local electrophilicity and rate coefficients for the hydrolysis of carbenium ions," *Tetrahedron*, vol. 61, no. 4, pp. 889–895, 2005.

[57] R. Contreras, J. Andrés, L. R. Domingo, R. Castillo, and P. Pérez, "Effect of electron-withdrawing substituents on the electrophilicity of carbonyl carbons," *Tetrahedron*, vol. 61, pp. 417–422, 2005.

[58] S. Karabunarliev, O. G. Mekenyan, W. Karcher, C. L. Russom, and S. P. Bradbury, "Quantum-chemical descriptors for estimating the acute toxicity of electrophiles to the bathed minnow (*Pimephales promelas*): an analysis based on molecular mechanisms," *Quantitative Structure-Activity Relationships*, vol. 15, no. 4, pp. 302–310, 1996.

[59] P. J. O'Brien, "Molecular mechanisms of quinone cytotoxicity," *Chemico-Biological Interactions*, vol. 80, no. 1, pp. 1–41, 1991.

[60] D. R. Roy, R. Parthasarathi, B. Maiti, V. Subramanian, and P. K. Chattaraj, "Electrophilicity as a possible descriptor for toxicity prediction," *Bioorganic and Medicinal Chemistry*, vol. 13, no. 10, pp. 3405–3412, 2005.

[61] P. K. Chattaraj, U. Sarkar, and D. R. Roy, "Electrophilicity index," *Chemical Reviews*, vol. 106, no. 6, pp. 2065–2091, 2006.

[62] P. Thanikaivelan, V. Subramanian, J. Raghava Rao, and B. Unni Nair, "Application of quantum chemical descriptor in quantitative structure activity and structure property relationship," *Chemical Physics Letters*, vol. 323, no. 1-2, pp. 59–70, 2000.

[63] R. Parthasarathi, V. Subramanian, D. R. Roy, and P. K. Chattaraj, "Electrophilicity index as a possible descriptor of biological activity," *Bioorganic & Medicinal Chemistry*, vol. 12, no. 21, pp. 5533–5543, 2004.

[64] A. Cerda-Monje, A. Aizman, R. A. Tapia, C. Chiappe, and R. Contreras, "Solvent effects in ionic liquids: empirical linear energy-density relationships," *Physical Chemistry Chemical Physics*, vol. 14, no. 28, pp. 10041–10049, 2012.

[65] K. B. Yatsimirskii, "Hydride affinity as a measure of acidity (electrophilicity)," *Theoretical and Experimental Chemistry*, vol. 17, no. 1, pp. 75–79, 1981.

[66] X. Zhu, C. Wang, H. Liang, and J. Cheng, "Theoretical prediction of the hydride affinities of various p- and o-quinones in DMSO," *Journal of Organic Chemistry*, vol. 72, no. 3, pp. 945–956, 2007.

[67] E. Buncel and I. Um, "The α-effect and its modulation by solvent," *Tetrahedron*, vol. 60, no. 36, pp. 7801–7825, 2004.

[68] N. J. Fina and J. O. Edwards, "The alpha effect. A review," *International Journal of Chemical Kinetics*, vol. 5, no. 1, pp. 1–26, 1973.

[69] L. R. Domingo, M. J. Aurell, P. Pérez, and R. Contreras, "Quantitative characterization of the global electrophilicity power of common diene/dienophile pairs in Diels–Alder reactions," *Tetrahedron*, vol. 58, no. 22, pp. 4417–4423, 2002.

[70] L. R. Domingo, M. J. Aurell, P. Pérez, and R. Contreras, "Quantitative characterization of the local electrophilicity of organic molecules. Understanding the regioselectivity on Diels-Alder reactions," *The Journal of Physical Chemistry A*, vol. 106, no. 29, pp. 6871–6875, 2002.

[71] A. C. Legon, "Quantitative gas-phase electrophilicities of the dihalogen molecules XY = F2, Cl2, Br2, BrCl and ClF," *Chemical Communications*, no. 23, pp. 2585–2586, 1998.

[72] A. C. Legon and D. J. Millen, "Hydrogen bonding as a probe of electron densities: limiting gas-phase nucleophilicities and electrophilicities of B and HX," *Journal of the American Chemical Society*, vol. 109, no. 2, pp. 356–358, 1987.

[73] M. Baidya, S. Kobayashi, and H. Mayr, "Nucleophilicity and nucleofugality of phenylsulfinate (PhSO$_2^-$): A key to understanding its ambident reactivity," *Journal of the American Chemical Society*, vol. 132, no. 13, pp. 4796–4805, 2010.

[74] J. Soto-Delgado, A. Aizman, L. R. Domingo, and R. Contreras, "Invariance of electrophilicity of independent fragments. Application to intramolecular Diels-Alder reactions," *Chemical Physics Letters*, vol. 499, no. 4-6, pp. 272–277, 2010.

[75] J. Soto-Delgado, L. R. Domingo, and R. Contreras, "Quantitative characterization of group electrophilicity and nucleophilicity for intramolecular Diels-Alder reactions," *Organic and Biomolecular Chemistry*, vol. 8, no. 16, pp. 3678–3683, 2010.

[76] T. J. Heckrodt and J. Mulzer, "Total synthesis of elisabethin A: Intramolecular Diels-Alder reaction under biomimetic conditions," *Journal of the American Chemical Society*, vol. 125, no. 16, pp. 4680–4681, 2003.

[77] B. G. Cox and A. J. Parker, "Solvation of ions. XVIII. Protic-dipolar aprotic solvent effects on the free energies, enthalpies, and entropies of activation of an SNAr reaction," *Journal of the American Chemical Society*, vol. 95, no. 2, pp. 408–410, 1973.

[78] I. Um, L. Im, J. Kang, S. S. Bursey, and J. M. Dust, "Mechanistic assessment of S NAr displacement of halides from 1-Halo-2,4-dinitrobenzenes by selected primary and secondary amines: Brønsted and Mayr analyses," *Journal of Organic Chemistry*, vol. 77, no. 21, pp. 9738–9746, 2012.

3

Complex Molecules at Liquid Interfaces: Insights from Molecular Simulation

David L. Cheung

Department of Pure and Applied Chemistry, University of Strathclyde, Glasgow G1 1XL, UK

Correspondence should be addressed to David L. Cheung; david.cheung@strath.ac.uk

Academic Editor: Fuke Wang

The behaviour of complex molecules, such as nanoparticles, polymers, and proteins, at liquid interfaces is of increasing importance in a number of areas of science and technology. It has long been recognised that solid particles adhere to liquid interfaces, which provides a convenient method for the preparation of nanoparticle structures or to modify interfacial properties. The adhesion of proteins at liquid interfaces is important in many biological processes and in a number of materials applications of biomolecules. While the reduced dimensions of these particles make experimental investigation challenging, molecular simulations provide a natural means for the study of these systems. In this paper I will give an overview of some recent work using molecular simulation to investigate the behaviour of complex molecules at liquid interfaces, focusing on the relationship between interfacial adsorption and molecular structure, and outline some avenues for future research.

1. Introduction

Dating back to the ancient Greeks the interface between immiscible fluids, in particular between oil and water, has been of both scientific and practical interest [1]. As the two liquid components can have radically different properties, the behaviour of molecules in an interfacial environment can differ from bulk solution. For example, molecular recognition driven by hydrogen bonding can be orders of magnitude stronger at the air-water interface than in solution [2, 3], which may be exploited in the formation of ordered materials [4] or the operation of molecular machines at the air-water interface [5]. When additional species are added to the system, these may accumulate at the interface; this is exemplified by amphiphilic surfactants, the adsorption of which to interfaces can be used to modify interfacial properties, such as surface tension. This adsorption of molecules also allows the interfaces to be used as platforms for chemical synthesis [6] and formation of nanoparticles [7]. It has long been recognised that liquid interfaces potentially provide elegant and convenient templates for the construction of dense and ordered structures. While the adsorption and self-assembly of small surfactant molecules have attracted attention over many years, recently the adsorption of larger and more

complex molecules, such as nanoparticles, polymers, and proteins, and other biomolecules has attracted attention.

For over 100 years it has been known that solid particles can adhere to interfaces between immiscible liquids [8]. This has been exploited in the formation of two-dimensional colloidal crystals [9] or in the use of colloidal particles to stabilise structures formed through phase separation of immiscible liquids, such as the so-called Pickering emulsions [8] and bijels [10] (particle-stabilised gels). More recently attention has turned to nanometre-sized particles [11, 12]. The adsorption of nanoparticles onto liquid interfaces has been used by a number of groups to create dense nanoparticle monolayers [13] and to create nanoparticle stabilised emulsions (colloidosomes) [14]. Adsorption of metallic or semiconducting nanoparticles at electrified interfaces has the potential to create novel electrooptical devices [15], such as nanoplasmonic sensors or nanoparticle mirrors.

The adsorption of polymers to liquid interfaces has also been used to create highly ordered structures, for both simple and more complex polymer architectures. In particular dendrimers, polymers with a branched, tree-like structure have been shown to self-assemble at liquid interfaces [16], creating highly ordered thin films with applications in organic electronics. The self-assembled structures can also be controlled through changes to the solvent phases. In recent work Richmond and coworkers have shown that polyelectrolytes may self-assemble at oil-water interfaces and that the structures formed may be controlled through changes to the pH [17, 18] or the addition of ions [19] to the water subphase. In particular, for polyacrylic acid, desorption from the interface can be triggered by increasing the pH of the water phase above 4.5 [17], demonstrating the potential for the control of polymer structures at oil-water interfaces.

As well as synthetic molecules, proteins and other biomolecules may adsorb onto liquid interfaces. While in most biological contexts protein adsorption at interfaces is avoided, due to the denaturing effect of the interface, for a number of proteins the interface is their native environment [20] and they fulfil a number of biological functions, including acting as surfactants, catalysis, and immune response. The behaviour of proteins at interfaces is also of interest for many industrial and technological applications. Many foods [21], such as ice cream, mayonnaise, and confectionary products, are emulsions or foams and a large number of proteins are being used as stabilisers for these disperse systems. It has also been shown that oil-water interfaces may be used to prepare highly ordered protein crystals [22].

Despite this intense interest there are still a number of open questions regarding the behaviour of complex molecules, both synthetic and biological, at liquid interfaces. These include the following.

(1) How do complex molecules interact with liquid interfaces?

As the adsorption of molecules, either individually or in larger aggregates, is the initial step in interfacial self-assembly, understanding how these molecules interact with the interface is vital. In addition to giving information regarding the adsorption strength (hence thermodynamic stability), the presence of a barrier to adsorption can lead to reduction in the rate of adsorption of molecules at interfaces.

(2) How do complex molecules assemble at interfaces?

While the formation of dense nanoparticle structures on liquid interfaces has been demonstrated, the controlled formation of more complex structures is still a matter of much investigation. The confinement of molecules to two dimensions can change the structures formed (compared to a fully three-dimensional system) and the different properties of the two liquid components can have a strong influence on the interaction between molecules confined to the interface between them. In particular capillary interactions (interactions mediated by the interface) can themselves be affected by pinning of the contact line to the particles [23].

(3) How does adsorption at interfaces affect the dynamics of complex molecules?

Due to the high adsorption strengths at liquid interfaces the motion of adsorbed molecules becomes essentially two-dimensional, leading to transport of

properties that can be quite different to bulk solution. This can be particularly complex when the different fluid phases have significantly different viscosities, for instance, when one of the components is a polymer solution or melt.

(4) What conformations do complex molecules adopt at liquid interfaces?

For molecules that possess internal degrees of freedom, such as polymers and proteins, adsorption onto liquid interfaces can alter the molecular conformation. Due to the close relation between their structure and function this is particularly important for proteins, with adsorption on interfaces typically (though not always) associated with a decrease or loss of functionality.

Many of these questions remain unresolved due to the difficulty of studying interfacial systems experimentally; by definition these systems have multiple components, which can be complex to prepare and make separating generic effects from those specific to certain systems difficult, and the properties of interfacial systems can change over length scales as small as Angstroms, which requires a level of spatial resolution that is hard to achieve experimentally. By contrast molecular simulation operates directly on the molecular level, giving a high level of spatial resolution, and simulations can be performed on simple model systems, allowing for the systematic study of effects of, for example, size of colloidal particle or length of polymer chain in a controlled manner.

In this paper I will outline some of my recent work, which has focused using molecular simulation on investigating complex molecules at liquid interfaces, typically focusing on the underlying statistical mechanics of adsorption and on how this can be related to molecular structure and properties. This paper is not designed to be a comprehensive overview of the behaviour of complex molecules at interfaces, rather a short personal account of some recent work and future challenges. The interested reader may find more detailed accounts in a number of excellent review articles (e.g., [28–31]).

2. Nanoparticles

As outlined in the introduction, it has long been known that solid particles can adhere to liquid interfaces. For colloidal particles ($R \sim \mu m$) the adsorption strength has been rationalised through changes to the liquid-liquid and particle-liquid surface areas and the adsorption free energy may be written in terms of macroscopic surface tensions as [9]

$$E = -\gamma_{12}\delta A_{12} + \gamma_{1P}A_{1P} + \gamma_{2P}A_{2P}, \tag{1}$$

where γ_{ij} is the surface tension between components i and j (where i, j are liquid 1, liquid 2, or colloid/nanoparticle P), δA_{12} is the change in surface area between the liquid components, and A_{iP} is the surface area of the particle exposed to liquid component i (illustrated in Figure 1). Using typical values of the surface tension this predicts that the

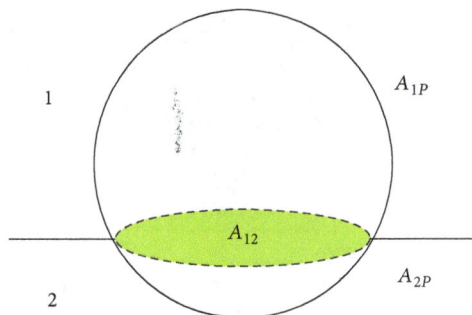

FIGURE 1: Schematic diagram illustrating nanoparticle-solvent contact areas.

adsorption energy of a micron sized colloidal particle is of the order $10^6 k_B T$, explaining the essentially irreversible adsorption of colloidal particles, the formation of colloidal arrays at liquid interfaces, and the using of colloids as emulsion stabilisers (Pickering emulsions). Later, this was generalised by Aveyard and Clint [32] to include the effect of line tension. Negative line tensions were found to increase the particle stability, whereas positive line tension decreased it. For large enough positive line tensions it is possible for this to lead to complete destabilisation of the particle from the interface.

The first attempts to test the applicability of these macroscopic theories to nanoparticles were performed by Bresme and Quirke [33–35]. They found that, under the assumption that particle-liquid surface tensions and line tension depend on particle size, Young's equation

$$\gamma_{13} - \gamma_{23} - \gamma_{12} \cos\theta + \frac{\tau \cos\theta}{R \sin\theta} = 0 \qquad (2)$$

could be used to predict the contact angle for nanoparticles at liquid interfaces.

While this tells us about the stability of particles at interfaces, they gave no information on the form of the particle-interface interaction. In some of my recent work, I used molecular simulations to calculate the nanoparticle-interface interaction for a number of simple, model systems. Initially this studied the Widom-Rowlinson mixture [24] (a limiting case of the nonadditive hard sphere mixture):

$$V_{ij}(r) = \left(1 - \delta_{ij}\right) V_{HS}(r, \sigma), \qquad (3)$$

where δ_{ij} is the Kronecker delta function and $V_{HS}(r, \sigma)$ is the hard sphere potential. Monte Carlo simulations, using Wang-Landau sampling [36, 37], were used to calculate the nanoparticle-interface interaction. Comparison with the prediction of Pieranski theory [9] shows that this underestimates both the adsorption strength and the range of the interaction (Figure 2(a)). While it is possible that the underestimation of the barrier is due to the neglect of line tension in Pieranski theory, simply adding this on would not change the interaction range. Rather this underestimation of interaction range arises due to the assumption of a flat, infinitely thin interface in the macroscopic theory (giving an interaction range strictly equal to the particle radius), whereas it is

well known that fluctuations in the interface position, from thermal motion (capillary waves) and fluctuations in the bulk density, can lead to an interface that is rough and broad on a microscopic scale. From simulation snapshots it can be seen that these interface fluctuations give an interface width that is comparable in size to the studied nanoparticles and can lead to bridging between the nanoparticle and the interface (Figure 2(b)). Similar results were also by Fan et al. in atomistic simulations of silica nanoparticles at water-decane interface [38]. In comparison with the simulation results the continuum model also predicts a more rapid variation in the nanoparticle-interface interaction near the interface. It has been shown that inclusion of capillary waves into the continuum models may also lead to a softening in the interaction [23].

In order to test how transferable these results were to other systems more recently I used molecular dynamics simulations (with umbrella sampling [39]) to determine the nanoparticle-interface interaction for a nanoparticle at an interface in a binary Lennard-Jones mixture [25]. As for the Widom-Rowlinson mixture the nanoparticle-interface interaction is both longer ranged and stronger than predicted from macroscopic theories (Figure 3(a)), indicating that this increased interaction range is a general feature of nanoparticles at interfaces. In this case as the particle size and interfacial tension increase, a maximum appears in the nanoparticle-interface interaction, indicating a slight ($\sim 2 k_B T$) barrier to particle adsorption at the interface. Recent experimental work studying gold nanoparticles that the water-toluene interface has, for certain ligands, also found a similar barrier to interfacial adsorption [40]. It was suggested that the barrier in the experimental system arose due to rearrangement of ligands or to electrostatic interactions. In the simulations, however, these effects are absent and the free energy barrier to adsorption potentially arises due to deformation of the interface, which may be seen from simulation snapshots (Figure 3(b)); for particles at the interface and in bulk solvent, the interface is approximately flat, whereas at intermediate separations the interface noticeably deforms due to the presence of the particles. It should also be noted that, in molecular dynamics simulations of surfactant functionalised nanoparticles by Udayana Ranatunga et al. [41], no barrier to adsorption was seen, indicating that the presence of stabilising ligands is not a sufficient condition for an adsorption barrier.

As well as uniform nanoparticles, particles with anisotropy, in their surface structure and/or shape, have attracted attention. The behaviour of Janus particles, particles with hemispheres of differing functionality, at interface has attracted interest [42]; as they can possess hydrophobic and hydrophilic faces, these can have enhanced stability over uniform particles. Based on macroscopic models, Binks and Fletcher [43] showed that the desorption energy of a Janus particle can be up to three times that of a uniform particle.

Using a modified version of the hard sphere system I investigated the stability of Janus nanoparticles at liquid interfaces [26]. In agreement with expectation, the Janus particles were more stable than uniform particles, with the stability increasing with the surface tension difference ($\Delta\gamma$)

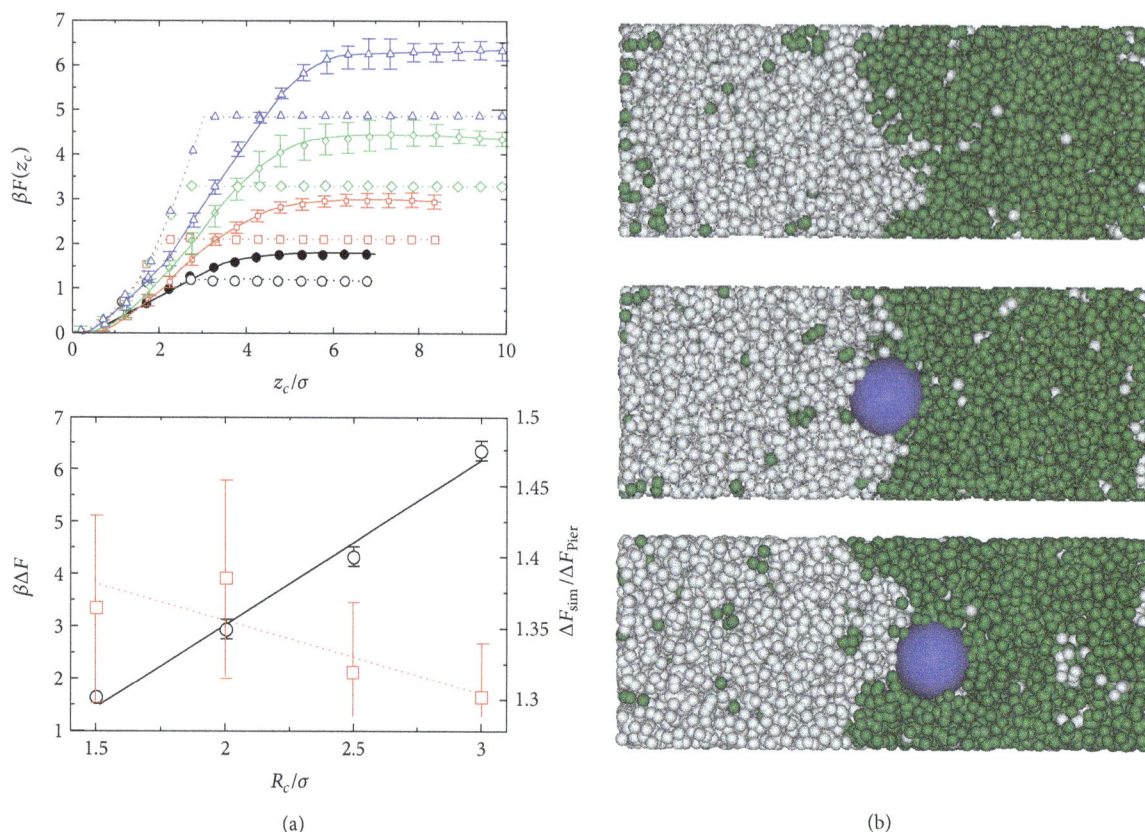

FIGURE 2: (a) (Top) Free energy profile for hard sphere nanoparticle at Widom-Rowlinson interface. Data from nanoparticle radius $R_c = 1.5\sigma, 2.0\sigma, 2.5\sigma,$ and 3.0σ denoted by black, red, green, and blue lines, respectively (solid denotes simulation data and dotted line Pieranski approximation). (Bottom) Simulation desorption free energy (black line) and the ratio of the desorption free energies from simulation and Pieranski theory (red). (b) Simulation snapshots showing interface in absence of nanoparticle (top), nanoparticle with $z_c = -0.13\sigma$ (middle), and nanoparticle with $z_c = 3.7\sigma$. Figure adapted from [24] © American Physical Society.

between the two faces (Figure 4(a)). A similar increase in stability of Janus particles over uniform ones was seen by Fan et al. in the study of silica nanoparticles with varying numbers and locations of hydroxyl groups on the nanoparticle surface [38] and in recent work by Razavi et al. [44]. The stability of the most amphiphilic particle (with faces completely wet by the appropriate solvent), however, was substantially less than three times that of a uniform particle as predicted by Binks and Fletcher. Study of the particle motion showed that the particles possessed a considerable degree of orientational freedom (Figure 4(b)), which was absent in the macroscopic model (in which the particle orientation was assumed to be fixed in the preferred orientation). By performing simulations with the particle orientation fixed in the ideal state, it was found that the particle stability was increased (Figure 4(c)).

As well as the stability of nanoparticles at interfaces, their dynamics and transport properties have attracted interest. The motion of nanoparticles at interfaces is important in understanding their use as tracers in microrheology [45]. Using molecular dynamics simulations, I investigated the diffusion of small nanoparticles at fluid interfaces [46]. Due to the high adsorption energy, the motion of these particles was indeed two-dimensional. Once the two-dimensional confinement of the particles was accounted for, it was found

that the particle diffusion was increased relative to bulk solution, which may be understood due to the lower effective viscosity of the interfacial region. The results of this work differed from those of Song et al. [47], studying carbon nanoparticles at a water-PDMS interface. This difference may be due to the larger particles used in that work or the difference in viscosity between the water and polymer components in that work; the diffusion of particles on a fluid interface increases when the viscosity difference between the two fluid components is small [48]. The rotational motion of the particles was also probed and it was found that this differed little from bulk solution; understanding the rotational motion of the nanoparticles at interfaces is important both for their potential application as catalysts and in the synthesis of Janus nanoparticles.

3. Dendrimers

Since they were first synthesized, dendrimers [49] have attracted much interest with potential applications in areas including drug delivery [50] and organic electronics [51]. They possess a high degree of symmetry and through a variety of synthetic routes, it is possible to selectively functionalise the dendrimer end groups [52]. This allows the

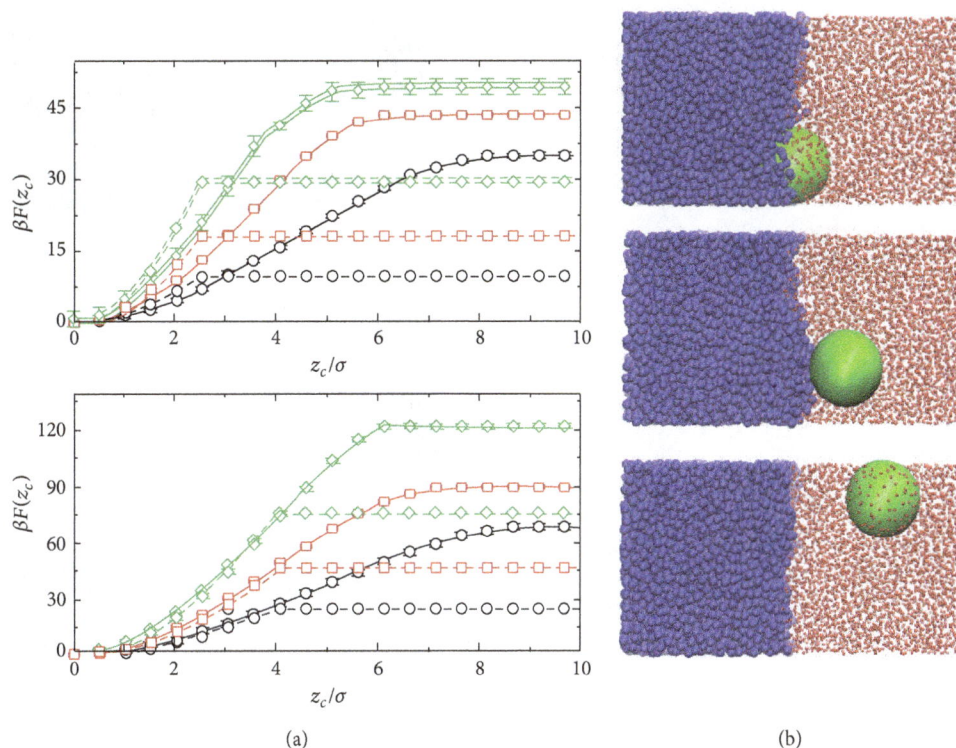

FIGURE 3: (a) Free energy profiles for LJ-nanoparticle with $R_c = 2.5\sigma$ (top) and 4.0σ (bottom) at liquid-liquid interface. Black, red, and green lines denote solvent density $\rho\sigma^3$ = 0.50, 0.60, and 0.69, respectively. (b) Simulation snapshots showing interface structure for nanoparticle with $z_c = 0$ (top), $z_c = 6\sigma$ (middle), and $z_c = 9\sigma$ (bottom). Figure adapted from [25] © American Institute of Physics.

preparation of amphiphilic or Janus dendrimers, which have been shown to self-assemble into a range of structures in solution [53]. The large number of end groups (even for relatively low generation dendrimers) means that this selective functionalization can have a large effect of the interfacial properties of dendrimers. They have recently been shown to be effective components for interfacial materials, having a range of interesting self-assembly processes [16]. Poly(amino amide) (PAMAM) or poly(propylene imine) dendrimers have attracted particular attention having been shown to form a Langmuir-Blodgett monolayers at the surfaces or interfaces [54].

Using atomistic simulations Nawaz and Carbone studied the behaviour of alkyl-modified PAMAM dendrimers at the air-water interface [55]. In comparison with the dendrimer in bulk water at the interface, it adopts an oblate structure (maximising the decrease in the air-water interfacial area), similar to that predicted for ligand-decorated nanoparticles. The stability of the dendrimer, measured through the number of dendrimer solvent hydrogen bonds, was found to be larger for dendrimers that were fully functionalized with alkyl chains compared to unfunctionalized and semifunctionalized dendrimers. This increase in stability arises due to the high degree of flexibility in these molecules, allowing the hydrophobic end groups to leave the water phase while the hydrophilic core remains in solution.

Following on from this I investigated the adsorption strength of a dendrimer at a model liquid-liquid interface

[27]. On approaching the interface, the dendrimer was found to undergo a similar shape change as the atomistic dendrimers, roughly spherical in bulk, disk-shaped at the interface, and rod-like between these (Figure 5(a)). Even a uniform dendrimer was found to adsorb strongly to the interface, with an adsorption strengths ~100–150k_BT (Figure 5(b)). Comparison with nanoparticles showed that this conformational change leads to an increase in stability. Selective functionalization of the dendrimer was found to increase the stability of the dendrimer. As in the atomistic simulations, core-shell dendrimer was more stable at the interface (Figure 5(c)). However, the barrier to enter the favoured solvent for the interior beads was lower than that for the favoured solvent for the terminal beads, demonstrating the flexibility of dendrimer and considerable interaction between the dendrimer interior and the solvent. Simulations of Janus dendrimers (end groups with different solvent affinities) demonstrated that this increased the interfacial adsorption strength, as for Janus nanoparticles [26]; this increase in adsorption strength depended on the placement of the end groups, with dendrimers with these end groups segregated adsorbing most strongly to the interface and dendrimers where these end groups alternate having lower adsorption strengths (dendrimers with randomly distributed end groups have an adsorption strength intermediate between these), demonstrating the importance of the placement of these groups for optimizing the interfacial behaviour of dendrimers (Figure 5(d)).

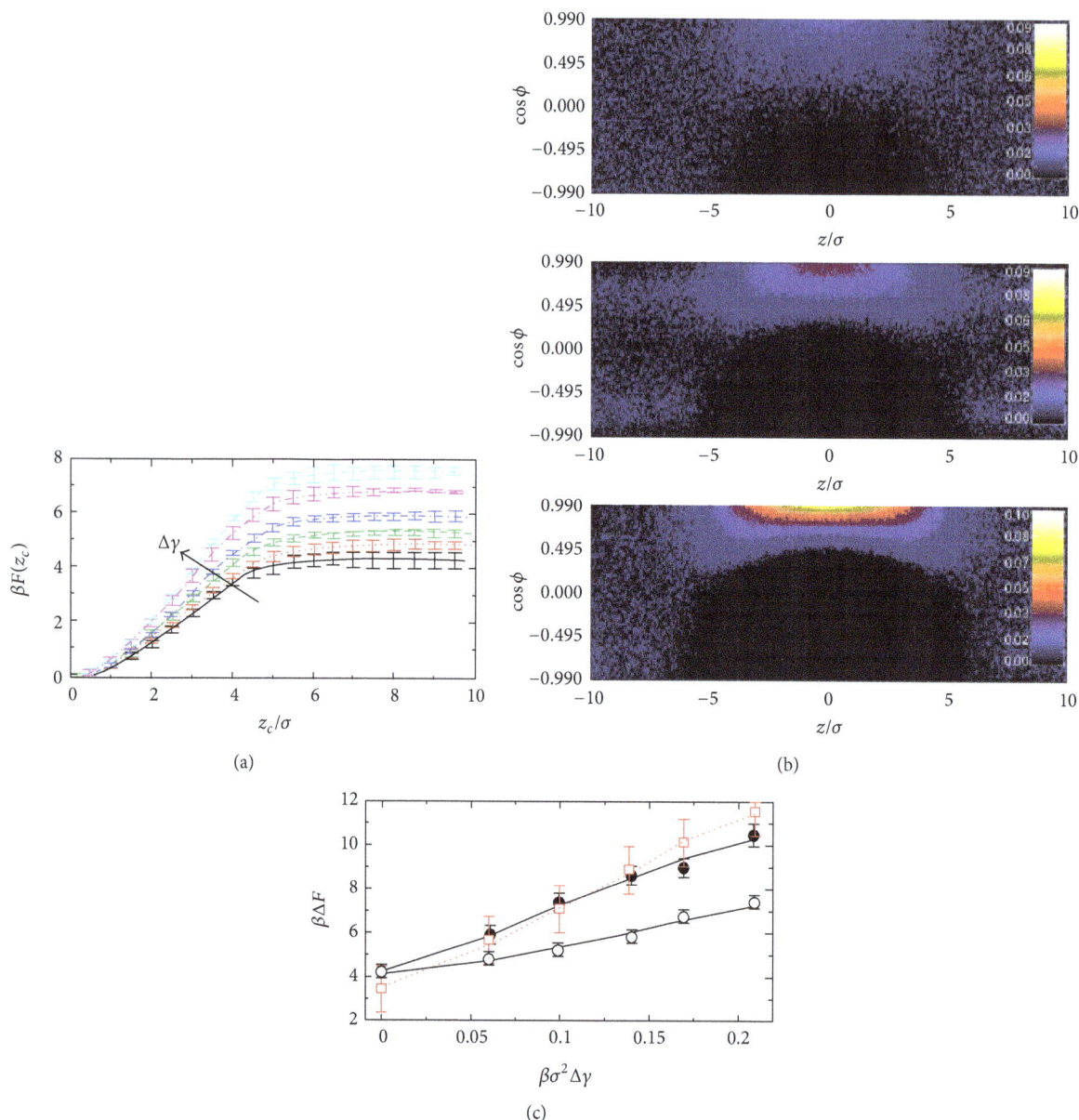

FIGURE 4: (a) Free energy profiles for Janus nanoparticles at liquid-liquid interface. Data for $\beta\sigma^2\Delta\gamma = 0$, 0.06, 0.10, 0.14, 0.17, and 0.21 denoted by black, red, green, blue, magenta, and turquoise lines, respectively. (b) Joint position-orientation probability distributions for Janus nanoparticles with $\beta\sigma^2\Delta\gamma = 0.06$ (top), $\beta\sigma^2\Delta\gamma = 0.10$ (middle), and $\beta\sigma^2\Delta\gamma = 0.21$ (bottom). (c) Desorption free energy for Janus nanoparticles from simulation (black, circles) and Binks-Fletcher (red, squares). For the simulation data, open symbols denote particles that are free to rotate and filled symbols particles of fixed orientation. Adapted from [26] with permission from The Royal Society of Chemistry.

4. Proteins

While proteins may be considered naturally amphiphilic, being composed of a mixture of hydrophobic and hydrophilic amino acids, due to its denaturing effect, adsorption of proteins at interfaces is typically detrimental to their function and in most cases is avoided. For a variety of proteins, however, adsorption at liquid interfaces is an intrinsic part of their function. While these surface-active proteins fulfil a number of functions, including catalysis [56] (lipases) and immune response [57] (lung surfactant proteins), the most common use is as protein surfactants (biosurfactants) [20]. As well as needing to lower surface tension, protein biosurfactants have additional design constraints, such as the desire to minimise aggregation in solution and avoid disruption to the cell membrane. The need to satisfy these somewhat contradictory properties has led to the evolution protein surfactants with a wide range of different structures.

One of the most notable examples of protein surfactants is hydrophobins [58]. These are a class of proteins expressed by certain species of filamentous fungi, which possess a distinctive surface structure. While the majority of the surface

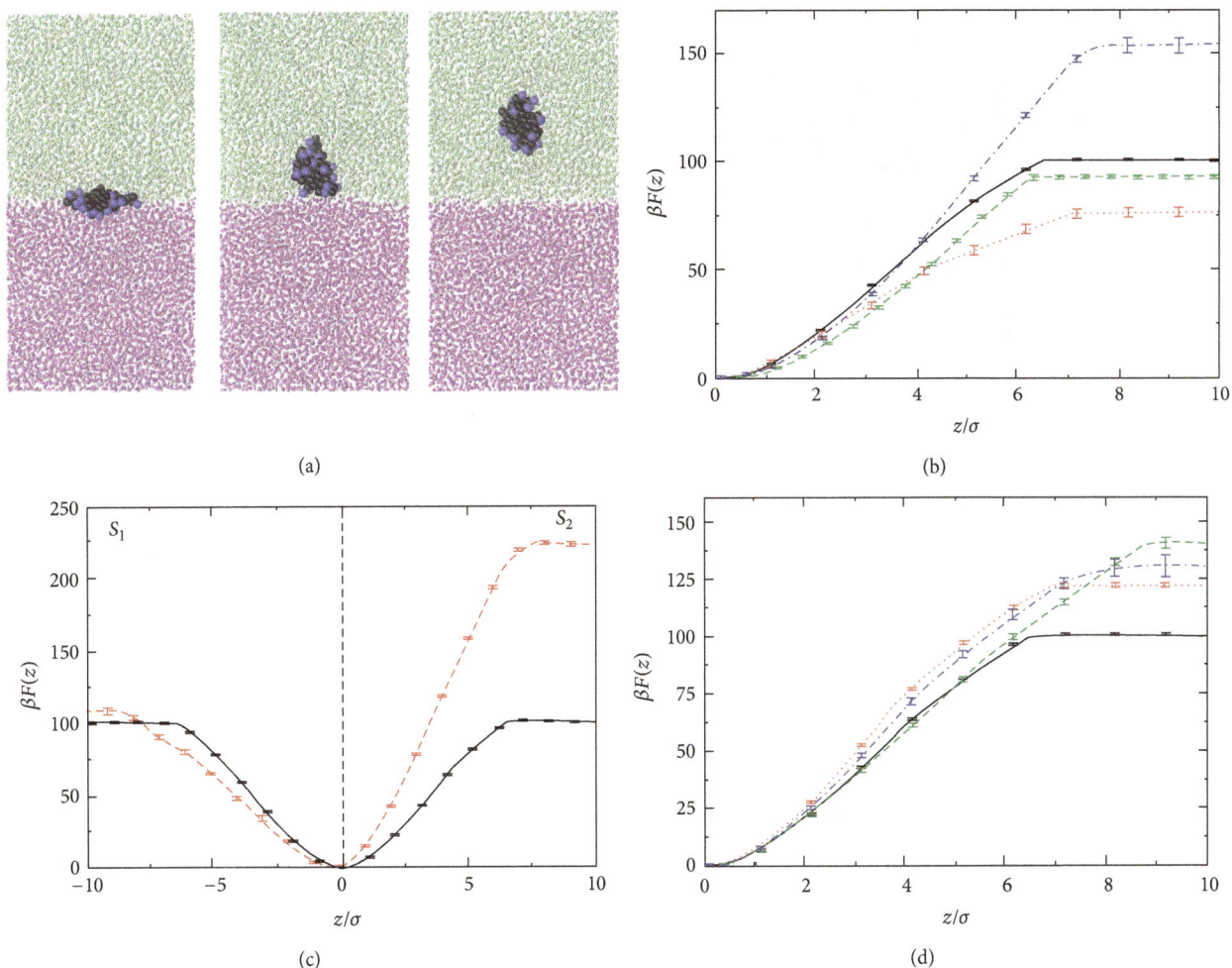

(a)

(b)

(c)

(d)

FIGURE 5: (a) Simulation snapshots showing uniform dendrimer at liquid-liquid interface with nanoparticle-interface separations 0 (left), 6σ (middle), and 10σ (right). (b) Free energy profile for uniform dendrimers at liquid-liquid interface. Solid line (black) denotes dendrimer in poor solvent, dotted line (red) denotes dendrimer in good solvent, and dashed line (green) denotes nanoparticle in poor solvent, and dot-dashed line (blue) denotes nanoparticle in good solvent. (c) Free energy profiles for uniform dendrimer (solid line, black) and core-shell dendrimer (dashed line, red). (d) Free energy profiles for uniform dendrimer (solid line, black), Janus dendrimer with alternating end groups (dotted line, red), Janus dendrimer with segregated end groups (dashed line, green), and Janus dendrimer with randomly distributed end groups (dot-dashed line, blue). Adapted from [27] with permission from The Royal Society of Chemistry.

is hydrophilic, they typically have a patch of hydrophobic residues covering ~20% of their surface. This amphiphilic structure, essentially making them naturally occurring Janus particles, makes them highly surface active, with them being amongst the most surface active proteins known. Beyond this common surface structure and a well-defined disulphide bond network, the hydrophobins form a diverse group of proteins and are commonly grouped into two classes, differentiated by their solubility and self-assembly at interfaces. As well as fulfilling a number of biological functions, such as formation of aerial structures or mediating fungal adhesion to hydrophobic surfaces, they are being investigated for a number of materials applications, such as foam and emulsion stabilisers [59, 60] or biocompatible coatings on drug particles [61]. While such applications, along with their unusual behaviour, have prompted a large number of experimental studies until recent years, simulation studies of hydrophobins

have been rarer. Early simulations by Mark et al. examined the structure of SC3, a fibril forming (class-I) hydrophobin in bulk water and at water-hydrophobic interfaces [62, 63].

While the adsorption of hydrophobins at oil-water interfaces may be qualitatively understood due to the anisotropic surface structure, it is useful to understand how the adsorption strength is controlled by their sequence and surface structure. Using coarse-grain molecular dynamics simulations, I investigated the adsorption strength of two common type-II hydrophobins (HFBII and HFBI) at the water-octane interface [64]. Unlike class-I hydrophobins, which often undergo significant conformational change at interfaces, class-II hydrophobins have rigid globular structures that remain largely unchanged (Figure 6(a)). Both of these were found to adsorb strongly at the interface with adsorption strengths ~$100k_BT$ (comparable to those for nanoparticles), indicating that these essentially adsorb irreversibly onto

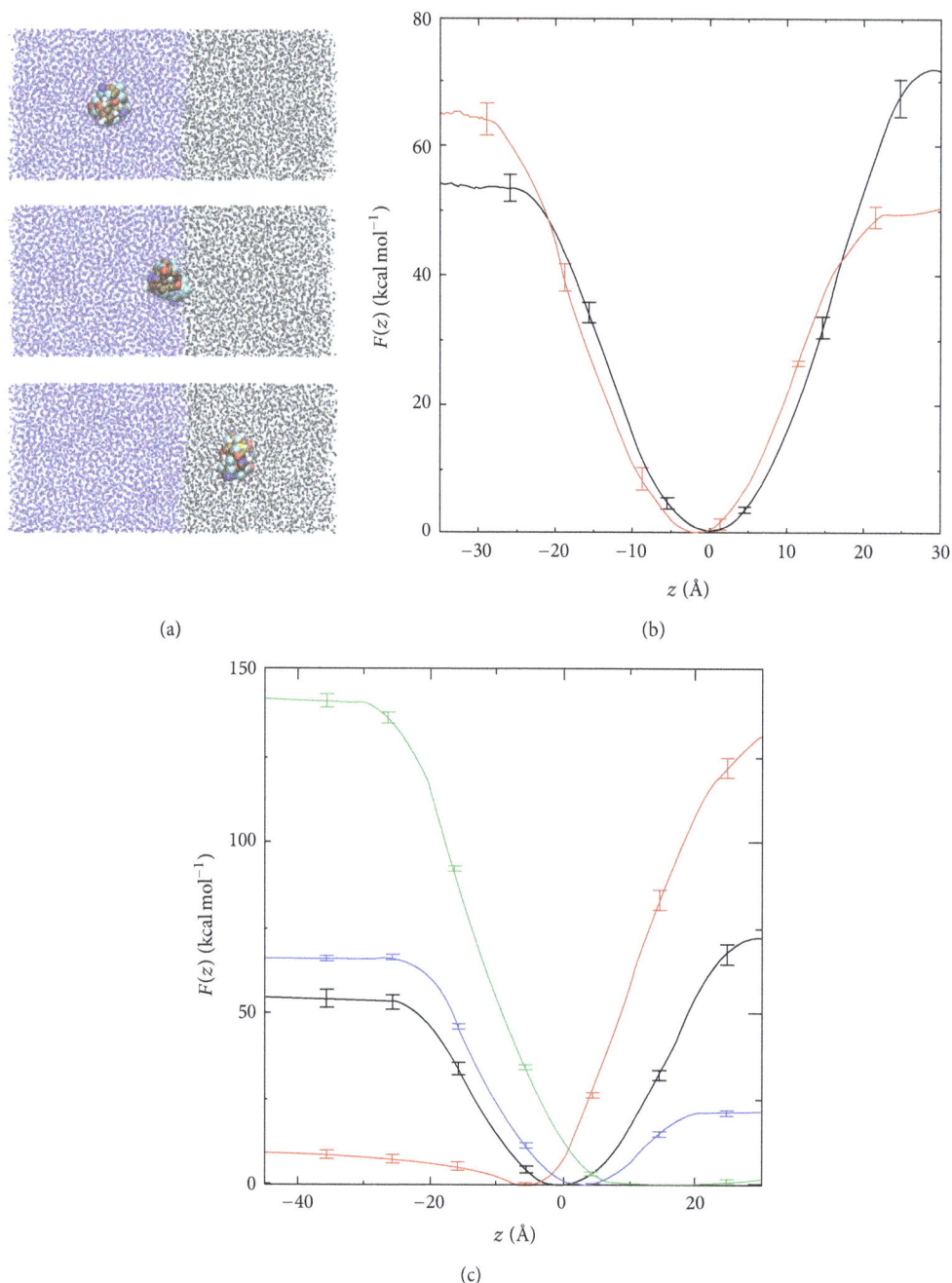

FIGURE 6: (a) Simulation snapshots showing CG HFBII in bulk water (top), at water-octane interface (middle), and in bulk octane (bottom). (b) Free energy profile for HFBII (black) and HFBI (red). (c) Free energy profiles for native HFBII (black) and hydrophilic (red), hydrophobic (green), and average (blue) HFBII pseudoproteins. Reprinted with permission from D. L. Cheung, Langmuir, 28, 8730 (2012). Copyright 2012 American Chemical Society.

the interface (Figure 6(b)). Comparison between these two showed that HFBII was slightly hydrophilic (the free energy barrier for desorption into the water component was slightly lower than going into octane), whereas HFBI was slightly hydrophobic. While this difference in behaviour is consistent with both the differing hydrophobic surface areas of these proteins and their different biological functions, it is notable that these proteins have both very similar structures and a

large degree of sequence similarity. In order to go beyond this simple observational approach, I performed simulations of pseudo-HFBII proteins, where the interactions with the solvent beads were taken to be identical for all the protein beads. Three different cases were considered: the protein-solvent interactions being uniformly hydrophilic, uniformly hydrophobic, and an average interaction, where the interaction strength between the protein and solvent beads was taken to be the average of the interaction strengths for the beads in the native protein (Figure 6(c)). For the first two cases, the pseudo-HFBII was found to be completely destabilised from the interface, with the protein residing in either the water (uniformly hydrophilic) or the octane (uniformly hydrophobic) regions. In contrast, the average pseudo-HFBII was found to be surface active (with a free energy minimum at the interface) but unlike the native protein it is hydrophobic (the barrier to move into the octane region is lower than that for the water region). This can be understood as the hydrophobic patch contains many highly hydrophobic residues that, together with the hydrophobic residues in the core of protein, made the averaged interaction overall hydrophobic (simulations of a single bead from the average protein give an octane-water transfer free energy of ~ 0.8 kcal mol^{-1}). Comparison between the flexible proteins and proteins that are held rigid showed that even in the case of these rigid proteins structural flexibility can still play a significant role in their interfacial adsorption.

The use of a coarse-grain model in this previous work was justified due to the limited conformational change seen in class-II hydrophobins; it is less applicable to class-I hydrophobins that often undergo significant structural change at interfaces. As shown by Schulz et al. [65] this change in conformational behaviour can strongly influence the potential application of these proteins, specifically finding that EAS (a class-I hydrophobin) induced the formation of hydroxyapatite at water-hexane interface, whereas HFBII did not, which can be explained by the larger conformational change exhibited by EAS at the interface and in the presence of Ca^{2+} and HPO_4^- ions. Simulation of EAS [66] using a multiscale model also showed that its conformational change is responsible for the aggregation of these proteins into fibrils, underlining the importance of accurately representing conformational change in these proteins. Going beyond hydrophobins, Euston has also compared atomistic and coarse-grain models for the simulation of proteins at fluid interfaces [67], finding that CG models gave results similar to atomistic models, although the CG model used underestimated the penetration of the protein (Barley LTP) into the oil phase.

5. Outlook

As outlined above over the past few years, much of my work has been aimed at using molecular simulation to examine the behaviour of complex molecules at liquid interfaces, with a particular focus on understanding how the structure and properties of these molecules affect their adsorption strength. Together with work performed by other groups, this has dramatically improved our understanding of the adsorption of nanoparticles at interfaces, in particular on the breakdown of continuum approximations for nanoscale particles, the role of capillary waves, and other microscopic interface deformations on the behaviour of particles, and the effect of nanoparticle surface structure. This has already improved our understanding of the adsorption processes, with current

work focusing on understanding the interactions between nanoparticles adsorbed on liquid interfaces and their self-assembly.

While my research to date has largely focused on simple systems, with even the hydrophobins being amongst the simplest possible proteins, there is increasing interest in the behaviour of more complex systems. With improvements to synthetic methodologies it is becoming possible to create nanoparticles [68] and polymers with increasingly complex structures. Notwithstanding recent controversy over stripy nanoparticles [69, 70], it is becoming apparent that it is possible to create patterned surfaces with features on the nanometre length scale [71, 72]. These increasingly complex particles will exhibit interesting behaviour at interfaces, with potential applications that extend beyond simple emulsion stabilisers or as templates for the formation of nanostructured materials. For instance Crossley et al. have used structured nanoparticles as phase transfer catalysts for upgrading biofuels [73], with this recently being extended to zeolite nanoparticles [74].

Due to their lack of conformational change at interface, hydrophobins may be regarded as low-hanging fruit compared to other proteins. Understanding conformational change of proteins at oil-water interfaces is a challenging problem for both simulation and experiment. Tensiometry and rheological measurement give indirect insight into changes in the protein conformation [75], neutron/X-ray reflectivity can give average protein sizes and thicknesses of adsorbed layers [76], and circular dichroism and IR spectroscopy can give information about protein secondary structure [77]. Recently the development of synchrotron radiation circular dichroism (SRCD) spectroscopy [78] has allowed the quantitative determine of protein secondary structure at oil-water interfaces. This has been applied to a number of proteins at oil-water interfaces, showing that, in general, there is a tendency for the α-helical content of proteins to decrease upon interfacial adsorption [30, 79], although this is, by no means, a universal phenomenon. In contrast to experimental studies simulation may allow the direct visualisation of protein structure at oil-water interfaces. For many years, however, simulations were unable to access timescales necessary to study protein conformations at interfaces. With recent advances in GPU computing [80], CG [81, 82] and hybrid [83] protein models and simulation methods, such as replica exchange [84] or metadynamics [85], will in the coming years make this more feasible. While to date more simulations have focused on short peptides [86] or food proteins [87, 88], these advances in computational power and methodologies will allow for the study of more complex proteins at interfaces, including proteins with complex functionality such as enzymes.

Conflict of Interests

The author declares that there is no conflict of interests regarding the publication of this paper.

Acknowledgments

The research outlined in this paper has benefited from collaboration and helpful conversations with Michael Allen, Stefan Bon, Fernando Bresme, Paola Carbone, Phillip Cox, Russell DeVane, and Martin Oettel. It was supported by UK EPSRC, ERC, and the Leverhulme trust.

References

[1] C. L. McFearin, D. K. Beaman, F. G. Moore, and G. L. Richmond, "From franklin to today: toward a molecular level understanding of bonding and adsorption at the Oil-Water interface," *Journal of Physical Chemistry C*, vol. 113, no. 4, pp. 1171–1188, 2009.

[2] X. Cha, K. Ariga, M. Onda, and T. Kunitake, "Molecular recognition of aqueous dipeptides by noncovalently aligned oligoglycine units at the air/water interface," *Journal of the American Chemical Society*, vol. 117, no. 48, pp. 11833–11838, 1995.

[3] K. Ariga, T. Nakanishi, J. P. Hill et al., "Tunable pK of amino acid residues at the air-water interface gives an L-zyme (Langmuir Enzyme)," *Journal of the American Chemical Society*, vol. 127, no. 34, pp. 12074–12080, 2005.

[4] K. Ariga, M. V. Lee, T. Mori, X.-Y. Yu, and J. P. Hill, "Two-dimensional nanoarchitectonics based on self-assembly," *Advances in Colloid and Interface Science*, vol. 154, no. 1-2, pp. 20–29, 2010.

[5] K. Ariga, T. Mori, and J. P. Hill, "Evolution of molecular machines: from solution to soft matter interface," *Soft Matter*, vol. 8, no. 1, pp. 15–20, 2012.

[6] A. Fallah-Araghi, K. Meguellati, J. Baret et al., "Enhanced chemical synthesis at soft interfaces: a universal reaction-adsorption mechanism in microcompartments," *Physical Review Letters*, vol. 112, no. 2, Article ID 028301, 2014.

[7] A. I. Campbell, R. A. Dryfe, and M. D. Haw, "Deposition and aggregation of au at the liquid/liquid interface," *Analytical Sciences*, vol. 25, no. 2, pp. 307–310, 2009.

[8] S. U. Pickering, "CXCVI.—emulsions," *Journal of the Chemical Society, Transactions*, vol. 91, pp. 2001–2021, 1907.

[9] P. Pieranski, "Two-dimensional interfacial colloidal crystals," *Physical Review Letters*, vol. 45, p. 569, 1980.

[10] P. S. Clegg, "Fluid-bicontinuous gels stabilized by interfacial colloids: low and high molecular weight fluids," *Journal of Physics: Condensed Matter*, vol. 20, no. 11, Article ID 113101, 2008.

[11] Y. Lin, H. Skaff, T. Emrick, A. D. Dinsmore, and T. P. Russell, "Nanoparticle assembly and transport at liquid-liquid interfaces," *Science*, vol. 299, no. 5604, pp. 226–229, 2003.

[12] Y. Lin, A. Böker, H. Skaff et al., "Nanoparticle assembly at fluid interfaces: structure and dynamics," *Langmuir*, vol. 21, no. 1, pp. 191–194, 2005.

[13] A. Böker, J. He, T. Emrick, and T. P. Russell, "Self-assembly of nanoparticles at interfaces," *Soft Matter*, vol. 3, no. 10, pp. 1231–1248, 2007.

[14] A. D. Dinsmore, M. F. Hsu, M. G. Nikolaides, M. Marquez, A. R. Bausch, and D. A. Weitz, "Colloidosomes: selectively permeable capsules composed of colloidal particles," *Science*, vol. 298, no. 5595, pp. 1006–1009, 2002.

[15] M. E. Flatté, A. A. Kornyshev, and M. Urbakh, "Nanoparticles at electrified liquid-liquid interfaces: new options for electro-optics," *Faraday Discussions*, vol. 143, pp. 109–115, 2009.

[16] M. J. Felipe, N. Estillore, R. B. Pernites, T. Nguyen, R. Ponnapati, and R. C. Advincula, "Interfacial behavior of OEG-linear dendron monolayers: aggregation, nanostructuring, and electropolymerizability," *Langmuir*, vol. 27, no. 15, pp. 9327–9336, 2011.

[17] D. K. Beaman, E. J. Robertson, and G. L. Richmond, "Unique assembly of charged polymers at the oil-water interface," *Langmuir*, vol. 27, no. 6, pp. 2104–2106, 2011.

[18] E. J. Robertson and G. L. Richmond, "Chunks of charge: effects at play in the assembly of macromolecules at fluid surfaces," *Langmuir*, vol. 29, no. 35, pp. 10980–10989, 2013.

[19] D. K. Beaman, E. J. Robertson, and G. L. Richmond, "Metal ions: driving the orderly assembly of polyelectrolytes at a hydrophobic surface," *Langmuir*, vol. 28, no. 40, pp. 14245–14253, 2012.

[20] A. Cooper and M. W. Kennedy, "Biofoams and natural protein surfactants," *Biophysical Chemistry*, vol. 151, no. 3, pp. 96–104, 2010.

[21] J. Ubbink, "Soft matter approaches to structured foods: from "cook-and-look" to rational food design?" *Faraday Discussions*, vol. 158, pp. 9–35, 2012.

[22] B. R. Silver, V. Fülöp, and P. R. Unwin, "Protein crystallization at oil/water interfaces," *New Journal of Chemistry*, vol. 35, no. 3, pp. 602–606, 2011.

[23] H. Lehle and M. Oettel, "Stability and interactions of nanocolloids at fluid interfaces: effects of capillary waves and line tensions," *Journal of Physics: Condensed Matter*, vol. 20, no. 40, Article ID 404224, 2008.

[24] D. Cheung and S. Bon, "Interaction of nanoparticles with ideal liquid-liquid interfaces," *Physical Review Letters*, vol. 102, Article ID 066103, 2009.

[25] D. L. Cheung, "Molecular dynamics study of nanoparticle stability at liquid interfaces: effect of nanoparticle-solvent interaction and capillary waves," *The Journal of Chemical Physics*, vol. 135, Article ID 054704, 2011.

[26] D. L. Cheung and S. A. F. Bon, "Stability of Janus nanoparticles at fluid interfaces," *Soft Matter*, vol. 5, no. 20, pp. 3969–3976, 2009.

[27] D. L. Cheung and P. Carbone, "How stable are amphiphilic dendrimers at the liquid-liquid interface?" *Soft Matter*, vol. 9, no. 29, pp. 6841–6850, 2013.

[28] F. Bresme and M. Oettel, "Nanoparticles at fluid interfaces," *Journal of Physics Condensed Matter*, vol. 19, no. 41, Article ID 413101, 2007.

[29] M. E. Flatté, A. A. Kornyshev, and M. Urbakh, "Understanding voltage-induced localization of nanoparticles at a liquid-liquid interface," *Journal of Physics Condensed Matter*, vol. 20, no. 7, Article ID 073102, 2008.

[30] J. L. Zhai, L. Day, M.-I. Aguilar, and T. J. Wooster, "Protein folding at emulsion oil/water interfaces," *Current Opinion in Colloid and Interface Science*, vol. 18, no. 4, pp. 257–271, 2013.

[31] S. Razavi, J. Koplik, and I. Kretzschmar, "Molecular dynamics simulations: insight into molecular phenomena at interfaces," *Langmuir*, vol. 30, no. 38, pp. 11272–11283, 2014.

[32] R. Aveyard and J. H. Clint, "Particle wettability and line tension," *Journal of the Chemical Society—Faraday Transactions*, vol. 92, no. 1, pp. 85–89, 1996.

[33] F. Bresme and N. Quirke, "Computer simulation study of the wetting behavior and line tensions of nanometer size particulates at a liquid-vapor interface," *Physical Review Letters*, vol. 80, no. 17, pp. 3791–3794, 1998.

[34] F. Bresme and N. Quirke, "Computer simulation of wetting and drying of spherical particulates at a liquid-vapor interface," *Journal of Chemical Physics*, vol. 110, no. 7, pp. 3536–3547, 1999.

[35] F. Bresme and N. Quirke, "Nanoparticulates at liquid/liquid interfaces," *Physical Chemistry Chemical Physics*, vol. 1, pp. 2149–2155, 1999.

[36] F. Wang and D. P. Landau, "Efficient, multiple-range random walk algorithm to calculate the density of states," *Physical Review Letters*, vol. 86, no. 10, pp. 2050–2053, 2001.

[37] F. Calvo, "Sampling along reaction coordinates with the Wang-Landau method," *Molecular Physics*, vol. 100, no. 21, pp. 3421–3427, 2002.

[38] H. Fan, D. E. Resasco, and A. Striolo, "Amphiphilic silica nanoparticles at the decane-water interface: insights from atomistic simulations," *Langmuir*, vol. 27, no. 9, pp. 5264–5274, 2011.

[39] G. M. Torrie and J. P. Valleau, "Nonphysical sampling distributions in Monte Carlo free-energy estimation: umbrella sampling," *Journal of Computational Physics*, vol. 23, no. 2, pp. 187–199, 1977.

[40] K. Du, E. Glogowski, T. Emrick, T. P. Russell, and A. D. Dinsmore, "Adsorption energy of nano- and microparticles at liquid-liquid interfaces," *Langmuir*, vol. 26, no. 15, pp. 12518–12522, 2010.

[41] R. J. K. Udayana Ranatunga, R. J. B. Kalescky, C.-C. Chiu, and S. O. Nielsen, "Molecular dynamics simulations of surfactant functionalized nanoparticles in the vicinity of an oil/water interface," *Journal of Physical Chemistry C*, vol. 114, no. 28, pp. 12151–12157, 2010.

[42] N. Glaser, D. J. Adams, A. Böker, and G. Krausch, "Janus particles at liquid-liquid interfaces," *Langmuir*, vol. 22, no. 12, pp. 5227–5229, 2006.

[43] B. P. Binks and P. D. I. Fletcher, "Particles adsorbed at the oil-water interface: a theoretical comparison between spheres of uniform wettability and "Janus" particles," *Langmuir*, vol. 17, no. 16, pp. 4708–4710, 2001.

[44] S. Razavi, J. Koplik, and I. Kretzschmar, "The effect of capillary bridging on the Janus particle stability at the interface of two immiscible liquids," *Soft Matter*, vol. 9, no. 18, pp. 4585–4589, 2013.

[45] J. Wu and L. L. Dai, "One-particle microrheology at liquid-liquid interfaces," *Applied Physics Letters*, vol. 89, Article ID 094107, 2006.

[46] D. L. Cheung, "Molecular simulation of nanoparticle diffusion at fluid interfaces," *Chemical Physics Letters*, vol. 495, no. 1–3, pp. 55–59, 2010.

[47] Y. Song, M. Luo, and L. L. Dai, "Understanding nanoparticle diffusion and exploring interfacial nanorheology using molecular dynamics simulations," *Langmuir*, vol. 26, no. 1, pp. 5–9, 2010.

[48] M. Negishi, T. Sakaue, and K. Yoshikawa, "Mismatch of bulk viscosity reduces interfacial diffusivity at an aqueous-oil system," *Physical Review E*, vol. 81, no. 2, Article ID 020901, 2010.

[49] D. A. Tomalia and J. M. Fréchet, "Introduction to 'dendrimers and dendritic polymers'," *Progress in Polymer Science*, vol. 30, no. 3-4, pp. 217–219, 2005.

[50] E. R. Gillies and J. M. J. Fréchet, "Dendrimers and dendritic polymers in drug delivery," *Drug Discovery Today*, vol. 10, no. 1, pp. 35–43, 2005.

[51] J. Li and D. Liu, "Dendrimers for organic light-emitting diodes," *Journal of Materials Chemistry*, vol. 19, pp. 7584–7591, 2009.

[52] B. M. Rosen, C. J. Wilson, D. A. Wilson, M. Peterca, M. R. Imam, and V. Percec, "Dendron-mediated self-assembly, disassembly, and self-organization of complex systems," *Chemical Reviews*, vol. 109, no. 11, pp. 6275–6540, 2009.

[53] V. Percec, D. A. Wilson, P. Leowanawat et al., "Self-assembly of janus dendrimers into uniform dendrimersomes and other complex architectures," *Science*, vol. 328, no. 5981, pp. 1009–1014, 2010.

[54] K. W. Chooi, A. I. Gray, L. Tetley, Y. Fan, and I. F. Uchegbu, "The molecular shape of poly(propylenimine) dendrimer amphiphiles has a profound effect on their self assembly," *Langmuir*, vol. 26, no. 4, pp. 2301–2316, 2010.

[55] S. Nawaz and P. Carbone, "Stability of amphiphilic dendrimers at the water/air interface," *Journal of Physical Chemistry B*, vol. 115, no. 42, pp. 12019–12027, 2011.

[56] P. Reis, K. Holmberg, H. Watzke, M. E. Leser, and R. Miller, "Lipases at interfaces: a review," *Advances in Colloid and Interface Science C*, vol. 147-148, pp. 237–250, 2009.

[57] J. R. Wright, "Immunomodulatory functions of surfactant," *Physiological Reviews*, vol. 77, no. 4, pp. 931–962, 1997.

[58] M. B. Linder, "Hydrophobins: proteins that self assemble at interfaces," *Current Opinion in Colloid & Interface Science*, vol. 14, no. 5, pp. 356–363, 2009.

[59] F. L. Tchuenbou-Magaia, I. T. Norton, and P. W. Cox, "Hydrophobins stabilised air-filled emulsions for the food industry," *Food Hydrocolloids*, vol. 23, no. 7, pp. 1877–1885, 2009.

[60] A. R. Cox, D. L. Aldred, and A. B. Russell, "Exceptional stability of food foams using class II hydrophobin HFBII," *Food Hydrocolloids*, vol. 23, no. 2, pp. 366–376, 2009.

[61] H. K. Valo, P. H. Laaksonen, L. J. Peltonen, M. B. Linder, J. T. Hirvonen, and T. J. Laaksonen, "Multifunctional hydrophobin: toward functional coatings for drug nanoparticles," *ACS Nano*, vol. 4, no. 3, pp. 1750–1758, 2010.

[62] R. Zangi, M. L. De Vocht, G. T. Robillard, and A. E. Mark, "Molecular dynamics study of the folding of hydrophobin SC3 at a hydrophilic/hydrophobic interface," *Biophysical Journal*, vol. 83, no. 1, pp. 112–124, 2002.

[63] H. Fan, X. Wang, J. Zhu, G. T. Robillard, and A. E. Mark, "Molecular dynamics simulations of the hydrophobin SC3 at a hydrophobic/hydrophilic interface," *Proteins: Structure, Function and Genetics*, vol. 64, no. 4, pp. 863–873, 2006.

[64] D. L. Cheung, "Molecular simulation of hydrophobin adsorption at an oil-water interface," *Langmuir*, vol. 28, no. 23, pp. 8730–8736, 2012.

[65] A. Schulz, M. Fioroni, M. B. Linder et al., "Exploring the mineralization of hydrophobins at a liquid interface," *Soft Matter*, vol. 8, no. 44, pp. 11343–11352, 2012.

[66] A. de Simone, C. Kitchen, A. H. Kwan, M. Sunde, C. M. Dobson, and D. Frenkel, "Intrinsic disorder modulates protein self-assembly and aggregation," *Proceedings of the National Academy of Sciences of the United States of America*, vol. 109, no. 18, pp. 6951–6956, 2012.

[67] S. R. Euston, "Molecular dynamics simulation of protein adsorption at fluid interfaces: a comparison of all-atom and coarse-grained models," *Biomacromolecules*, vol. 11, no. 10, pp. 2781–2787, 2010.

[68] S. C. Glotzer and M. J. Solomon, "Anisotropy of building blocks and their assembly into complex structures," *Nature Materials*, vol. 6, no. 8, pp. 557–562, 2007.

[69] Y. Cesbron, C. P. Shaw, J. P. Birchall, P. Free, and R. Lévy, "Stripy nanoparticles revisited," *Small*, vol. 8, no. 24, pp. 3714–3726, 2012.

[70] M. Yu and F. Stellacci, "Response to "stripy nanoparticles revisited"," *Small*, vol. 8, no. 24, pp. 3720–3726, 2012.

[71] C. P. Shaw, D. G. Fernig, and R. Lévy, "Gold nanoparticles as advanced building blocks for nanoscale self-assembled systems," *Journal of Materials Chemistry*, vol. 21, no. 33, pp. 12181–12187, 2011.

[72] C. Singh, P. K. Ghorai, M. A. Horsch et al., "Entropy-mediated patterning of surfactant-coated nanoparticles and surfaces," *Physical Review Letters*, vol. 99, no. 22, Article ID 226106, 2007.

[73] S. Crossley, J. Faria, M. Shen, and D. E. Resasco, "Solid nanoparticles that catalyze biofuel upgrade reactions at the water/oil interface," *Science*, vol. 327, no. 5961, pp. 68–72, 2010.

[74] P. A. Zapata, J. Faria, M. P. Ruiz, R. E. Jentoft, and D. E. Resasco, "Hydrophobic zeolites for biofuel upgrading reactions at the liquid-liquid interface in water/oil emulsions," *Journal of the American Chemical Society*, vol. 134, no. 20, pp. 8570–8578, 2012.

[75] B. S. Murray, "Interfacial rheology of food emulsifiers and proteins," *Current Opinion in Colloid & Interface Science*, vol. 7, pp. 426–431, 2002.

[76] X. Zhao, F. Pan, and J. R. Lu, "Interfacial assembly of proteins and peptides: recent examples studied by neutron reflection," *Journal of the Royal Society Interface*, vol. 6, no. 5, pp. S659–S670, 2009.

[77] J. T. Pelton and L. R. McLean, "Spectroscopic methods for analysis of protein secondary structure," *Analytical Biochemistry*, vol. 277, no. 2, pp. 167–176, 2000.

[78] L. Day, J. Zhai, M. Xu, N. C. Jones, S. V. Hoffmann, and T. J. Wooster, "Conformational changes of globular proteins adsorbed at oil-in-water emulsion interfaces examined by synchrotron radiation circular dichroism," *Food Hydrocolloids*, vol. 34, pp. 78–87, 2014.

[79] J. Zhai, A. J. Miles, L. K. Pattenden et al., "Changes in β-lactoglobulin conformation at the oil/water interface of emulsions studied by synchrotron radiation circular dichroism spectroscopy," *Biomacromolecules*, vol. 11, no. 8, pp. 2136–2142, 2010.

[80] J. A. Anderson, C. D. Lorenz, and A. Travesset, "General purpose molecular dynamics simulations fully implemented on graphics processing units," *Journal of Computational Physics*, vol. 227, no. 10, pp. 5342–5359, 2008.

[81] L. Monticelli, S. K. Kandasamy, X. Periole, R. G. Larson, D. P. Tieleman, and S.-J. Marrink, "The MARTINI coarse-grained force field: extension to proteins," *Journal of Chemical Theory and Computation*, vol. 4, no. 5, pp. 819–834, 2008.

[82] R. Devane, W. Shinoda, P. B. Moore, and M. L. Klein, "Transferable coarse grain nonbonded interaction model for amino acids," *Journal of Chemical Theory and Computation*, vol. 5, no. 8, pp. 2115–2124, 2009.

[83] W. Han, C.-K. Wan, F. Jiang, and Y.-D. Wu, "PACE force field for protein simulations. 1. full parameterization of version 1 and verification," *Journal of Chemical Theory and Computation*, vol. 6, no. 11, pp. 3373–3389, 2010.

[84] D. J. Earl and M. W. Deem, "Parallel tempering: theory, applications, and new perspectives," *Physical Chemistry Chemical Physics*, vol. 7, pp. 3910–3916, 2005.

[85] A. Laio and F. L. Gervasio, "Metadynamics: a method to simulate rare events and reconstruct the free energy in biophysics, chemistry and material science," *Reports on Progress in Physics*, vol. 71, no. 12, Article ID 126601, 2008.

[86] V. Knecht, "β-Hairpin folding by a model amyloid peptide in solution and at an interface," *The Journal of Physical Chemistry B*, vol. 112, no. 31, pp. 9476–9483, 2008.

[87] S. R. Euston, P. Hughes, M. A. Naser, and R. E. Westacott, "Comparison of the adsorbed conformation of barley lipid transfer protein at the decane-Water and vacuum-Water interface: a molecular dynamics simulation," *Biomacromolecules*, vol. 9, no. 5, pp. 1443–1453, 2008.

[88] S. R. Euston, U. Bellstedt, K. Schillbach, and P. S. Hughes, "The adsorption and competitive adsorption of bile salts and whey protein at the oil-water interface," *Soft Matter*, vol. 7, no. 19, pp. 8942–8951, 2011.

4

Cobalt(II) and Manganese(II) Complexes of Novel Schiff Bases, Synthesis, Charcterization, and Thermal, Antimicrobial, Electronic, and Catalytic Features

Selma Bal and Sedat Salih Bal

Chemistry Department, Faculty of Arts and Science, Kahramanmaras Sutcu Imam University, Avsar Kampusu, 46100 Kahramanmaras, Turkey

Correspondence should be addressed to Selma Bal; selmadagli9@hotmail.com

Academic Editor: Alessandro D'Annibale

Carbazoles containing two new Schiff bases (Z,Z)-N,N′-bis[(9-ethyl-9H-carbazole-3-yl)methylene]propane-1,3 diamine (L¹) and (Z,Z)-N,N′-bis[(9-ethyl-9H-carbazole-3-yl)methylene]-2,2-dimethylpropane-1,3-diamine (L²) and their Co(II) and Mn(II) complexes were synthesized and characterized using various spectroscopic methods and thermal analysis, which gave high thermal stability results for the ligands and their cobalt complexes. The title compounds were examined for their antimicrobial and antifungal activities, which resulted in high activity values for the ligands and their manganese complexes. Oxidation reactions carried out on styrene and cyclohexene revealed that the complex compounds were the most effective catalysts for styrene oxidation, giving good selectivities than those of cyclohexene oxidation. Electronic features of the synthesized compounds were also reported within this work.

1. Introduction

Schiff bases and their complexes have been and are being employed to many reactions in synthetic chemistry. In particular, the oxidation of alkenes is important intermediate to get new, industrially important chemicals for both organic synthesis and pharmaceutical industry. Catalytic transformations of hydrocarbons into valuable oxygenated derivatives such as alcohols, aldehydes, and epoxides using peroxides as oxidants have been extensively studied over the last few decades [1–5]. In particular, the catalysis of alkene oxidation by soluble transition metal complexes is of great interest in both biomimetic chemistry and synthetic chemistry [6]. So far various Schiff base complexes have been employed to catalytic oxidation of olefins to epoxides and aldehydes, and it has been proved that many Schiff base complexes gave improved results as catalysts for these kinds of oxidation reactions [7–20]. In our research, the synthesized Schiff base complexes have been searched for their potential use as catalysts in oxidation reactions for both cyclohexene and

styrene. Not only for these oxidation reactions but also for many kinds, it is important to use eco-friendly with easy recoverability oxidants such as H₂O₂ and air. It is also important that the catalysts are thermally stable enough to carry out these kinds of reactions which generally require elevated temperatures [21, 22].

In addition their catalytic activities, various Schiff bases have also been examined for their biological activities in many previous studies [23–25]. This interest comes from the fact that their metal complexes can be used as antimicrobial, antifungal, and antitumor agents [26–28].

In previous studies various carbazoles containing Schiff bases and their coordination compounds have been synthesized, characterized, and examined for their different features such as luminescence, thermal property [29–33], biological activity [34], and electrochemical and optical behaviour [35–37] and for their use as Langmuir-Blodgett film [38].

We report here total synthesis, spectral and thermal characterization of two carbazole derived novel Schiff bases and their copper(II) and manganese(II) complexes, their thermal, electrochemical, antimicrobial features, and their effect as catalysts in the oxidation reactions of cyclohexene and styrene.

2. Experiment

2.1. Materials and Instrumentation. 9-Ethylcarbazole, phosphorus(V) oxychloride, 1,3-diaminopropane, and 2,2-dimethyl-1,3-diaminopropane, all the solvents used, and acetate salts of copper(II) and manganese(II) were purchased from Sigma Aldrich. Nuclear magnetic resonance spectra of the synthesized ligands were recorded on a Bruker AV 400 MHz spectrometer in the solvent $CDCl_3$. Infrared spectra were obtained using KBr discs on a Shimadzu 8300 FTIR spectrophotometer in the region of 400–4000 cm^{-1}. Ultraviolet spectra were run in ethanol on a Schimadzu UV-160A spectrophotometer. Magnetic measurements were carried out by the Gouy method using $Hg[Co(SCN)_4]$ as a calibrant. Molar conductances of the ligands and their transition metal complexes were determined in MeOH ($\sim10^{-3}$) at room temperature using a Jenway Model 4070 conductivity meter. Mass spectra of the ligand were recorded on a LC/MS APCI Agilent 1100 MSD spectrophotometer. The oxidation products were analyzed with a gas chromatograph (Shimadzu, GC-14B) equipped with a SAB-5 capillary column and a flame ionization detector. Elemental analyses were performed on a LECO CHNS 932 elemental analyzer and the metal analyses were carried out on an Ati Unicam 929 Model AA spectrometer in solutions prepared by decomposing the compounds in aqua regia and subsequently digesting them in concentrated HCl. Thermal analyses of synthesized ligands and their metal complexes were carried out on a Perkin-Elmer Thermogravimetric Analyzer TG/DTA 6300 instrument under nitrogen atmosphere between the temperature ranges 30°C and 988°C at a heating rate of 10°C/min. Cyclic voltammetry was performed using IviumStat electrochemical workstation equipped with a low current module (BAS PA-1) recorder.

2.2. Synthesis of 9-Ethyl-9H-carbazole-3-carbaldehyde. Formulation of 9-ethylcarbazole was done by using Vilsmeier formulating agents DMF and $POCl_3$. Inside a fume cupboard, DMF (32 mL, 0.4 mol) was put into a 250 mL round-bottom flask placed in an ice bath. Over DMF at 0°C 32 mL (0.32 mol) $POCl_3$ was added dropwise through a dropping funnel. Resulting solution was stirred at room temperature for 2 hours. To the stirring mixture, 9-ethyl-9H-carbazole (8 g, 0.04 mol) dissolved in 32 mL DMF was slowly added. The reaction mixture was heated at 80°C and left stirring for 24 hours. The resulting dark coloured mixture was poured into slurry of crushed ice and water (250 mL). The precipitate was washed with water and extracted by $CHCl_3$ and then washed with *n*-hexane. The dirty yellow precipitate was subjected to flash column chromatography with the eluent of ethyl acetate/hexane (1:10). Vilsmeier reaction always gives both mono- and dialdehyde of the formulated carbazole [39, 40]. The monoaldehyde was the first product eluted as yellowish-white crystals. The dialdehyde obtained was

white solid. TLC chromatography, elemental analysis, and spectral data confirm the purity and structure of synthesized mono- and dialdehyde products. Monoaldehyde 9-ethyl-9H-carbazole-3-carbaldehyde: yield: 40%. m.p.: 85–87°C. UV-Vis (ethanol) (λ_{max}, nm) (ε, M^{-1} cm^{-1}): 202(50000), 210(80000), 226(30000), 264(36000), 284(95000). FT-IR (KBr, cm^{-1}): 1591(w), 1622(w), 2850(w), 2922(w)(Ar-H and C-H); 1681(s)(CHO). ^{1}H NMR(400 MHz, $CDCl_3$): 10.1(CHO), 8.6 (H-4, d, J = 1.3), 8.16 (H-5, brd, J = 7.8), 8.02 (H-2, dd, J = 8.5&1.58), 7.46 (H-1&H-8, d, J = 8.4), 7.56 (H-7, dt, J = 8.2&1.2), 7.35 (H-6, dt, J = 8.0&1.03), 4.39 (2H-14, q, J = 7.2), 1.47 (3H-15, t, J = 7.2). ^{13}C NMR(400 MHz, $CDCl_3$): 191.8(CHO), 127.2(C-1), 126.8(C-2), 128.5(C-3), 124.01(C-4), 120.8(C-5), 120.3(C-6), 108.7(C-7), 109.2(C-8), 143.5(C-10), 123.1(C-11), 123.0(C-12), 140.7(C-13), 37.9(C-14), 13.9(C-15). Mass spectrum (LC/MS APCI): m/z 223.7 [M]$^+$.

2.3. Synthesis of the Ligands (Z,Z)-N,N'-Bis[(9-ethyl-9H-carbazol-3-yl)methylene]propane-1,3-diamine (L^1) and (Z,Z)-N,N'-Bis[(9-ethyl-9H-carbazol-3-yl)methylene]-2,2-dimethyl-propane-1,3-diamine (L^2). Inside a 100 mL round-bottom flask, 0.45 g (2 mmol) 9-ethyl-9H-carbazole-3-carbaldehyde was put and dissolved in enough amount of ethanol. Over this, 0.1 g (1 mmol) diamine was added dropwise. The resulting solution was heated under reflux for four hours and left overnight. The white precipitate was recrystallized from ethanol (Figure 1).

2.3.1. (Z,Z)-N,N'-Bis[(9-ethyl-9H-carbazole-3-yl)methylene]-propane-1,3-diamine (L^1). Yield: 80%, m.p.: 167°C, elemental analysis found % (calculated %): C 82.34(81.78) H 6.80(6.66) N 12.01(11.56). UV-Vis (ethanol) (λ_{max}, nm) (ε, M^{-1} cm^{-1}): 216(175000), 250(105000), 266(120000), 296(210000), 310(161000). FT-IR (KBr, cm^{-1}): 2843(w), 2920(w), 2971(w), 1598(w)(Ar-H and C-H); 806(s), 740(s)(Ar-H) 1637(s)(imine). Mass spectrum (LC/MS APCI): m/z 485.2 [M + 1]$^+$, 280.1 [M − 204.53], which arises from the loss of ten hydrogen and one carbazole units leaving $C_{19}H_{10}N_3$. ^{1}H NMR(400 MHz, $CDCl_3$): 7.44 (H-1&H-8, d, J = 8.5), 7.92 (H-2, dd, J = 8.5&1.5), 8.49 (H-4, d, J = 1.4), 8.15 (H-5, d, J = 7.6), 7.28 (H-6, dt, J = 1.2&7.6), 7.51 (H-7, dt, J = 1.2&7.1), 4.41 (H-14, q, J = 7.2), 1.47 (H-15, t, J = 7.2), 3.84 (H-16&H-18, q, J = 7), 2.26 (H-17, m, J = 7), 8.5 (imine, s). ^{13}C NMR(400 MHz, $CDCl_3$): 126.03(C-1), 125.9(C-2), 127.7(C-3), 123.1(C-4), 120.7(C-5), 119.4(C-6), 108.5(C-7), 108.7(C-8), 141.5(C-10), 127.2(C-11), 124.0(C-12), 140.5(C-13), 37.7(C-14), 13.8(C-15), 59.5(C-16&C-18), 32.3(C-17), 162.06(imine).

2.3.2. (Z,Z)-N,N'-Bis[(9-ethyl-9H-carbazol-3-yl)methylene]-2,2-dimethylpropane-1,3-diamine (L^2). Yield: 75%, m.p.: 115°C, elemental analysis found % (calculated %): C 82.11 (81.99) H 7.530(7.08) N 11.01(10.93). UV-Vis (ethanol) (λ_{max}, nm) (ε, M^{-1} cm^{-1}): 268(115000), 300(78000), 306(61000), 334(99000). FT-IR (KBr, cm^{-1}): 2973(w), 2949(w), 2867(w), 2821(w), 1595(w) (Ar-H and C-H); 809(s), 747(s)(Ar-H) 1646(s)(imine). Mass spectrum (LC/MS APCI): m/z 514.3

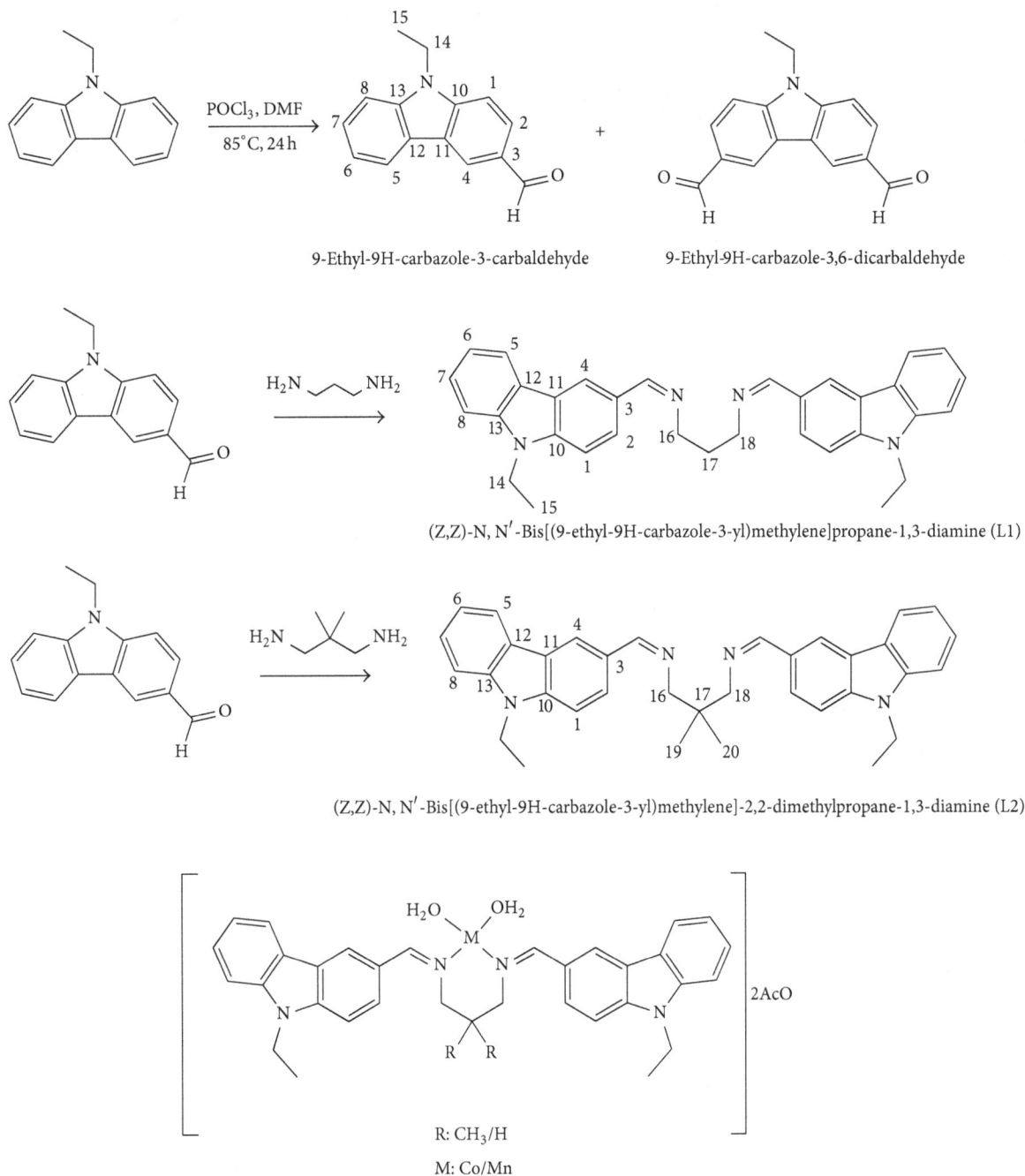

FIGURE 1: Synthesis scheme of the synthesized compounds and the proposed structure for the metal complexes.

$[M + 1]^+$, 308.1 $[M - 204]$, which arises from the loss of ten hydrogen and one carbazole units leaving $C_{21}H_{14}N_3$. ^1H NMR(400 MHz, CDCl$_3$): 7.44(H-1&H-8, d, $J = 8.4$), 7.99 (H-2, dd, $J = 8.5\&1.5$), 8.49 (H-4, d, $J = 1.3$), 8.16 (H-5, d, $J = 7.6$), 7.29 (H-6, dt, $J = 1.1\&7.1$), 7.53 (H-7, dt, $J = 1.1\&7.0$), 4.42 (H-14, q, $J = 7.2$), 1.48 8H-15, t, $J = 7.2$), 3.67 (H-16&H-18, s), 1.2 (H-19&H-20, s), 8.48 (imine, s). ^{13}C NMR(400 MHz, CDCl$_3$): 125.9(C-1), 125.9(C-2), 128.0(C-3), 123.05(C-4), 120.8(C-5), 119.4(C-6), 108.5(C-7), 108.7(C-8), 141.4(C-10), 127.2(C-11), 123.1(C-12), 140.7(C-13), 37.7(C-14), 13.9(C-15), 58.2(C-16&C-18), 37.3(C-17), 24.8 (C-19&C-20).

2.4. Synthesis of Complex Compounds. The ratio of the metal salts and the ligands was taken as 1:1 (Figure 1). A solution of the metal salt (1 mmol) in 15 mL absolute ethanol was added into the solution of the ligand L^1/L^2 (1 mmol) in 15 mL ethanol. The mixtures were stirred under reflux overnight. The precipitates were filtered, washed with distilled water to get rid of the excess salt, and dried in vacuum.

2.4.1. Cobalt(II) Complex of L^1. [CoL^1(H$_2$O)$_2$]·2AcO complex: (C$_{33}$H$_{36}$CoN$_4$O$_2$), brown coloured, yield: 72%, m.p.: 205.5°C, Elemental Analysis found % (calculated %):

C 63.65(63.38), H 6.71(6.26), N 9.66(9.67), Co 10.68(10.17). UV-Vis (ethanol) (λ_{max}, nm) (ε, $M^{-1}cm^{-1}$): 256(51000), 282(11600), 330(87000), 367(75000), 482(30000), 560(33000). FT-IR (KBr, cm^{-1}): 3350, 3220(O-H), 2972, 2822, 2745(Ar-H, C-H), 1575(imine), 439(M-N), 537(M-O). Λ_M ($\Omega^{-1}cm^2mol^{-1}$): 80.1, μ_{eff} B.M.: 4.35.

2.4.2. Manganese(II) Complex of L^1.
[$MnL^1(H_2O)_2$]·2AcO complex: ($C_{33}H_{36}MnN_4O_2$), brown coloured, yield: 64%, m.p.: 198°C, Elemental Analysis found % (calculated %): C 69.23(68.86), H 6.21(6.30), N 10.15(9.73), Mn 10.02(9.54). UV-Vis (ethanol) (λ_{max}, nm) (ε, $M^{-1}cm^{-1}$): 246(86000), 291(51000), 375(24000), 640(12000). FT-IR (KBr, cm^{-1}): 3342(O-H), 2847, 2730(Ar-H, C-H), 1589(imine), 481(M-N), 604(M-O). Λ_M ($\Omega^{-1}cm^2mol^{-1}$): 28.7, μ_{eff} B.M.: 5.85.

2.4.3. Cobalt(II) Complex of L^2.
[$CoL^2(H_2O)_2$]·2AcO complex: ($C_{35}H_{40}CoN_4O_2$), black coloured, yield: 60%, m.p.: 146.5°C, Elemental Analysis found % (calculated %): C 70.09(69.18), H 6.92(6.63), N 9.57(9.22), Co 10.01(9.70). UV-Vis (ethanol) (λ_{max}, nm) (ε, $M^{-1}cm^{-1}$): 230(120000), 284(61000), 384(43000), 620(10200). FT-IR (KBr, cm^{-1}): 3306(O-H), 2919, 2812(Ar-H, C-H), 1550(imine), 553(M-N), 654(M-O). Λ_M ($\Omega^{-1}cm^2mol^{-1}$): 67.3, μ_{eff} B.M.: 4.89.

2.4.4. Manganese(II) Complex of L^2.
[$MnL^2(H_2O)_2$]·2AcO complex: ($C_{33}H_{40}MnN_4O_2$), dark green coloured, yield: 54%, m.p.: 119°C, elemental analysis found % (calculated %): C 70.03(69.64), H 7.14(6.68), N 9.67(9.28), Mn 9.25(9.10). UV-V is (ethanol) (λ_{max}, nm) (ε, $M^{-1}cm^{-1}$): 265(90000), 395(71000), 567(14000). FT-IR (KBr, cm^{-1}): 3252(O-H), 2920, 2835(Ar-H, C-H), 1540(imine), 490(M-N), 601(M-O). Λ_M ($\Omega^{-1}cm^2mol^{-1}$): 35.2, μ_{eff} B.M.: 6.23.

2.5. Preparation of Microorganism Culture.
The growth inhibitory activity of the synthesized compounds was tested against 4 gram negative, 4 gram positive bacteria (*Klebsiella pneumoniae* FMC 5, *Escherichia coli* DM, and *Enterobacter faecium* (clinic isolate) and *Enterobacter aerogenes* ATCC 13048, *Bacillus subtilis* IMG 22, *Bacillus megaterium* DSM 32, *Staphylococcus aureus* ATCC 25923, and *Streptococcus faecalis*) and 3 yeasts (*Candida albicans* ATCC 1023, *Candida utilis*, and *Saccharomyces cerevisiae* WET 136). These microorganisms were provided from Microbiology Laboratory Culture Collection, Department of Biology, Kahramanmaraş Sütçü İmam University, Turkey.

Antimicrobial activities of the compoundswere determined using the hollow agar. The bacteria were first incubated at 37 ± 0.1°C for 24 h in nutrient broth (Difco), and the yeasts were incubated in Sabouraud dextrose broth (Difco) at 25 ± 0.1°C for 24 h. The cultures of the bacteria and yeast were injected into the Petri dishes (9 cm) in the amount of 0.1 mL (McFarland OD: 0.5, 1.5×10^8 bacteria/mL and 1.5×10^6 yeast/mL) [38, 39, 41, 42]. Then, Mueller Hinton agar and Sabouraud dextrose agar (sterilized in a flask and cooled to 45–50°C) were homogeneously distributed onto the sterilized Petri dishes in the amount of 25 mL. Finally, 2 mg of each chemical compound dissolved in ethanol was placed inside the sterilised antibiotic discs. The prepared antibiotic discs were then placed in the bacterial medium.

Afterwards, the plates combined with the discs were left at 4°C for 2 h, the plates injected with yeast were incubated at 25 ± 0.1°C for 24 h, and ones injected with bacteria were incubated at 37 ± 0.1°C for 24 h. After 24 h, inhibition zones appearing around the disks were measured and recorded in mm [41–44].

2.6. Determination of Minimal Inhibitory Concentration (MIC).
A broth microdilution broth susceptibility assay was used, as recommended by NCCLS, for the determination of the MIC of the ligand and the complexes and some reference components [38, 41]. All tests were performed in Mueller Hinton broth (MHB) supplemented with Tween 80 detergent (final concentration of 0.5% (v/v)), with the exception of the yeasts (Sabouraud dextrose broth (SDB) + Tween 80). Bacterial strains were cultured overnight at 37°C in MHB, and the yeasts were cultured overnight at 30°C in SDB. Geometric dilutions ranging from 200 to 2500 μg/mL of the compounds were prepared including one growth control (MHB + Tween 80) and one sterility control (MHB + Tween 80 + test oil). Test tubes were incubated under normal atmospheric conditions at 37°C for 24 h for bacteria and at 30°C for 48 h for the yeasts. The microbial growth was determined by turbidimetric methods.

2.7. Oxidation Procedure.
A mixture of 1.10^{-3} mol catalyst, 20 mL solvent (CH_3CN), and 10 mmol cyclohexene/styrene was stirred under nitrogen atmosphere in a 50 mL round-bottom flask equipped with a condenser and a dropping funnel at room temperature for 30 min. Then 20 mmol hydrogen peroxide (30% in water) was added (catalyst : substrate : oxidant ratio is 1 : 10 : 20). The resulting mixture was then refluxed for 8 h under nitrogen atmosphere at 90°C. After filtration and washing with solvent, the filtrate was concentrated and then subjected to GC analysis. The yields were recorded as the GC yield based on the starting styrene or cyclohexene. The identity of the oxidation products was confirmed by GC-MS. A blank reaction was also carried out without any catalyst for both oxidation reactions.

3. Results and Discussion

3.1. IR and UV Spectra of the Ligands and the Complexes.
IR spectrum of 9-ethyl-3-formylcarbazole revealed the characteristic weak C-H stretching bands belonging to aromatic rings at 1591 cm^{-1}, 1622 cm^{-1}, 2850 cm^{-1}, and 2922 cm^{-1}. The latter two bands might also be the C-H stretching frequencies generally observed for saturated methyl or methylene groups. The aromatic aldehyde was observed at 1681 cm^{-1}. Following the imine formation, this band was replaced for the imine stretching frequencies at 1638 cm^{-1} for L^1 and at 1646 cm^{-1} for L^2. In the IR spectra of L^1 and L^2, the weak bands between 2821 and 2973 cm^{-1} belong to weak aromatic and saturated C-H stretchings. The medium strength bands at 1598 cm^{-1} for L^1 and at 1595 cm^{-1} for L^2 belong to aryl-H vibrations. If we examine the substitution patterns of the

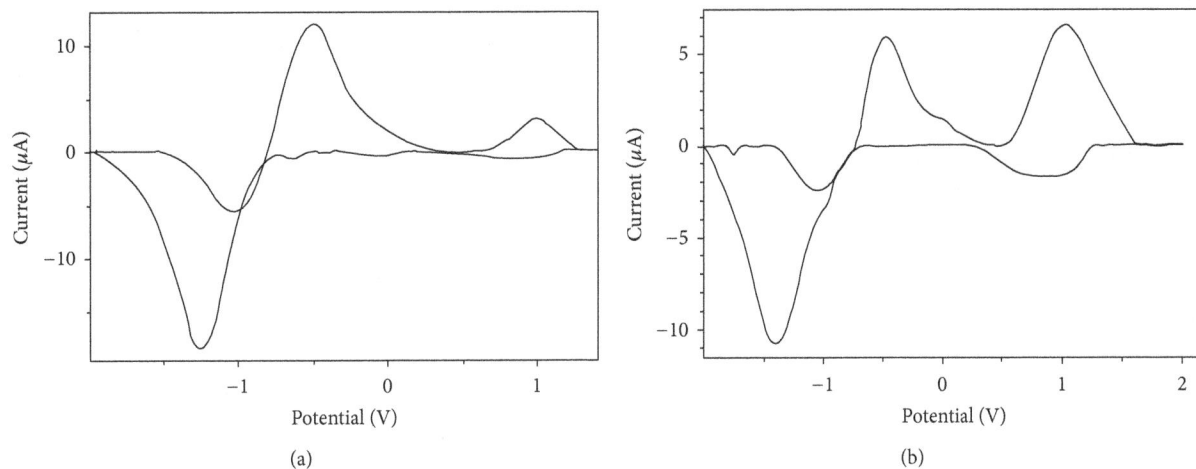

FIGURE 2: (a) Cyclic voltammogram of L^1 in the presence of $0.1\,M\ NBu_4BF_4$ in DMF solution at $1.10^{-3}\,M$ at $100\,mVs^{-1}$. (b) Cyclic voltammogram of L^1-Co in the presence of $0.1\,M\ NBu_4BF_4$ in DMF solution at $1.10^{-3}\,M$ at $500\,mVs^{-1}$.

benzene rings, we can say that one metadisubstituted (strong aryl-H vibrations at $806\,cm^{-1}$ for L^1 and at $809\,cm^{-1}$ for L^2) and one orthodisubstituted benzene ring (strong aryl-H vibrations at $740\,cm^{-1}$ for L^1 and at $747\,cm^{-1}$ for L^2) are present in our ligands.

Following the coordination, the imine frequencies recorded for the ligands were red shifted and appeared at around $1575\,cm^{-1}$ for L^1-Co, $1589\,cm^{-1}$ for L^1-Mn, $1550\,cm^{-1}$ for L^2-Co, and $1540\,cm^{-1}$ for L^2-Mn complexes as weak signals. All the coordination compounds showed coordinated water frequencies between 3252 and $3306\,cm^{-1}$ [45]. The M-N stretching vibrations observed between 439 and $553\,cm^{-1}$ proved that the nitrogens of the imine groups coordinate with the central metal atoms.

The UV spectra of all ligands and the complexes were taken in ethanol in the range between 200 nm and 700 nm. For the complex compounds the d-d transitions were observed between 560 nm and 640 nm, and the bands observed in the 395–367 nm range for these complexes can be attributed to the charge transfer bands from ligand to metal or from metal to ligand centre [46, 47]. The n-π^* and π-π^* transitions of the synthesized ligands and complexes were observed in 202–334 nm region. The O-H stretching frequencies in water molecules assisting in coordination were seen as broad singlets between $3306\,cm^{-1}$ and $3350\,cm^{-1}$ [45]. All the infrared and u.v. results are given in the experimental Sections 2.2–2.4. The magnetic moment values for cobalt complexes were recorded as 4.35 B.M. and 4.89 B.M., which are characteristic values for tetrahedral cobalt complexes [46, 47]. The manganese(II) complexes gave 5.85 B.M. and 6.23 B.M. values, indicating a high spin complex and suggesting tetrahedral geometry [48, 49].

3.2. Electrochemical Properties of the Synthesized Compounds.
Cyclic voltammogram studies were run in DMF (1×10^{-3} M)–$0.1\,M\ NBu_4BF_4$ as supporting electrolyte at 293 K. Electronic spectra of all the compounds were taken for scan rates 100,

250, 500, 750, and $1000\,mVs^{-1}$ against an internal ferrocene-ferrocenium standard and we have obtained different oxidation and reduction processes for different scan rates. Examination of cv graphics of L^1 shows reversible oxidation-reduction processes for the scan rate $100\,mVs^{-1}$ (Figure 2(a)) at $0.87\,V(E_{pc})$ and $0.88\,V(E_{pa})$ ($I_{pc} : I_{pa} = 1$) and for the scan rate $500\,mVs^{-1}$ at $0.94\,V(E_{pc})$ and $1.04\,V(E_{pa})$ ($I_{pc} : I_{pa} \cong 1$). The quasireversible processes were observed for $250\,mVs^{-1}$ scan rate at $0.87\,V(E_{pc})$ and $1.0\,V(E_{pa})$ and for $750\,mVs^{-1}$ at $0.9\,V(E_{pc})$ and $1.12\,V(E_{pa})$. The rest of the peaks observed for different scan rates can be said to be irreversible. The cv graphics of cobalt(II) complex of this ligand shows reversible redox processes for the scan rates 100, 500, and $750\,mVs^{-1}$; for example, for $500\,mVs^{-1}$ (Figure 2(b)), the $I_{pc} : I_{pa}$ ratio equals $0.9 \cong 1$ ($E_{pc} = 0.93\,V$ and $E_{pa} = 1.03\,V$). The other peak potentials recorded for this complex gave either quasireversible or irreversible redoxes. The manganese(II) complex of the same ligand gave reversible processes for the scan rates 100, 250, and $500\,mVs^{-1}$; for example, for $100\,mVs^{-1}$ at $0.98\,V(E_{pc})$ and at $1.07\,V(E_{pa})$, $I_{pc} : I_{pa}$ ratio equals around 1. The other ligand, L^2, revealed reversible redoxes for scan rates 250, 500, 750, and $1000\,mVs^{-1}$. For example, the scan rate $750\,mVs^{-1}$ (Figure 3(a)) gave cathodic peak potential at $0.98\,V$ and anodic peak potential at 0.99 ($I_{pc} : I_{pa} = 1$). The cobalt complex of this ligand showed reversible redoxes for 100 and $250\,mVs^{-1}$ (Figure 3(b)) scan rates at $1.13\,V(E_{pc})$ and $1.1\,V(E_{pa})$ and at $0.99\,V(E_{pc})$ and $1.1\,V(E_{pa})$, respectively. The manganese(II) complex of the same ligand gave reversible processes for 100 and $500\,mVs^{-1}$. The rest of the data can be seen in Table 1. The redox processes for all the complexes can be defined as the formation of M(II) or M(I) with a simple one-electron process [24, 25]. The possible redox process for the ligands can be shown as in Figure 4.

3.3. Thermal Analysis.
Thermal gravimetric analysis was used to explore the thermal stability of these newly synthesized compounds and to verify the status of water or solvent

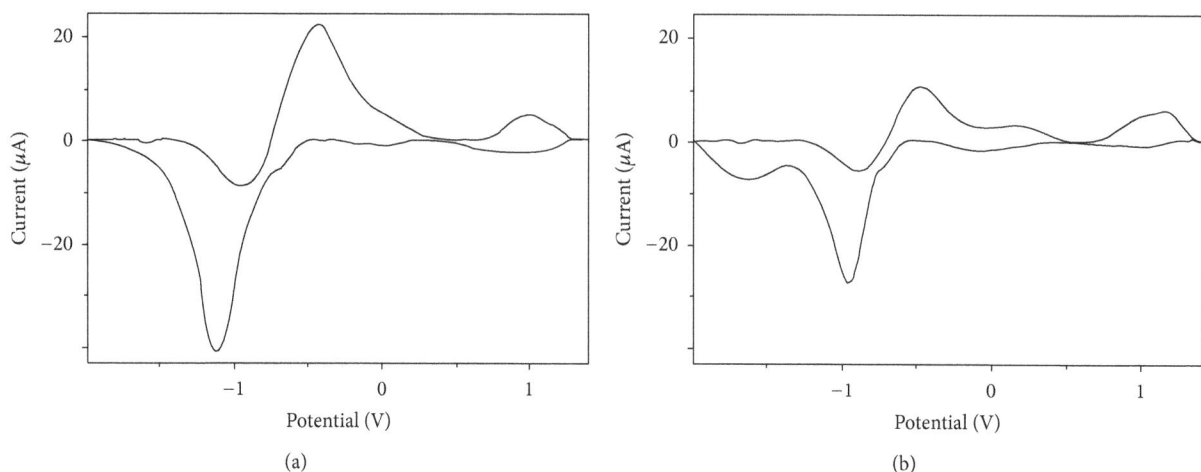

FIGURE 3: (a) Cyclic voltammogram of L^2 in the presence of 0.1 M NBu_4BF_4 in DMF solution at 1.10^{-3} M at $750\,mVs^{-1}$. (b) Cyclic voltammogram of L^2-Co in the presence of 0.1 M NBu_4BF_4 in DMF solution at 1.10^{-3} M at $250\,mVs^{-1}$.

TABLE 1: Electrochemical data of the title compounds.

Compound	Scan rate (mV/s)	E_{pc} (E_{pa}) (V)	$E_{1/2}$ (V)	ΔE_p (V)
L^1	100	0.87, −1.18 (−0.57, 0.88)	0.88	0.01
	250	0.87, −0.06, −1.24 (−0.48, 1.00)	—	0.13
	500	0.94, −0.08, −1.32 (−0.42, 1.04)	—	0.15
	750	0.90, −0.12, −1.40 (−0.38, 1.12)	—	0.22
	1000	0.90, −0.14, −1.42 (−0.35, 1.14)	—	0.24
$[Co(L^1)(H_2O)_2]2AcO$	100	0.72, −1.14 (−0.58, 0.80)	—	0.12
	250	0.76, −1.28 (−0.50, 0.93)	—	0.17
	500	0.93, −1.40 (−0.46, 1.03)	—	0.16
	750	0.97, −1.47 (−0.46, 1.07)	—	0.21
	1000	0.87, −1.54 (−0.42, 1.17)	—	0.30
$[Mn(L^1)(H_2O)_2]2AcO$	100	0.98, −1.11 (−0.52, 1.07)	—	0.09
	250	0.98, −1.13 (−0.54, 1.02)	—	0.04
	500	0.97, −1.28 (−0.40, 1.08)	—	0.13
	750	0.96, −1.41 (−0.33, 1.26)	—	0.30
	1000	0.97, −1.47 (−0.26, 1.22)	—	0.25
L^2	100	0.09, −0.90 (−0.56, −0.01, 0.82)	—	−0.10
	250	0.91, 0.03, −1.01 (−0.53, 0.93)	0.92	0.02
	500	0.94, −0.02, −1.09 (−0.50, 0.95)	0.95	0.01
	750	0.98, −1.12 (−0.44, 0.99)	0.99	0.01
	1000	0.96, −1.31 (−0.35, 0.99)	0.98	0.03
$[Co(L^2)(H_2O)_2]2AcO$	100	1.13, −0.03, −0.91 (−0.53, 0.14, 1.25)	—	0.17
	250	0.99, −0.07, −0.96 (−0.47, 0.20, 1.10)	0.72	0.12
	500	0.96, −0.08, −1.02 (−0.50, 1.13)	0.76	0.17
	750	0.95, −1.19 (−0.41, 1.10)	—	0.15
	1000	0.97, −1.42 (−0.27, 1.10)	—	0.13
$[Mn(L^2)(H_2O)_2]2AcO$	100	1.10, −0.05, −0.92 (−0.54, 1.19)	—	0.20
	250	0.96, 0.01, −1.00 (−0.49, 1.18)	0.75	0.22
	500	0.97, 0.13, −1.12 (−0.42, 1.07)	—	0.12
	750	0.96, −1.22 (−0.30, 1.11)	—	0.15
	1000	0.93, −1.43 (−0.24, 1.15)	—	0.22

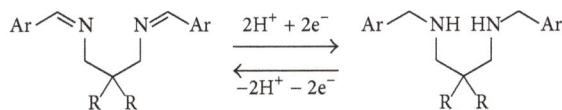

FIGURE 4: Reversible reduction oxidation processes of the title compounds L^1 and L^2 in DMF (1.10^{-3} M) solution.

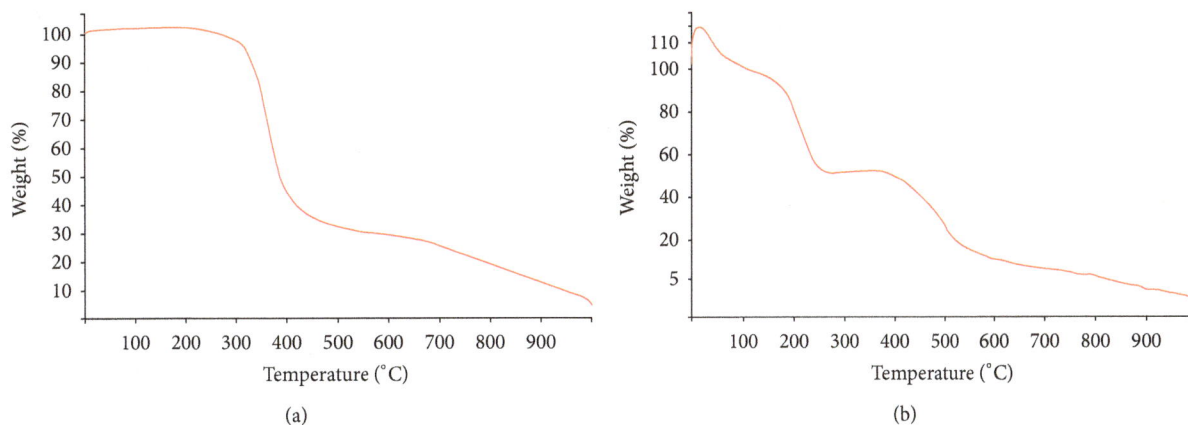

(a)

(b)

FIGURE 5: (a) TG plot of L^1 recorded under nitrogen atmosphere between the temperature ranges 30°C and 988°C at a heating rate of 10°C/min. (b) TG plot of L^2-Mn recorded under nitrogen atmosphere between the temperature ranges 30°C and 988°C at a heating rate of 10°C/min.

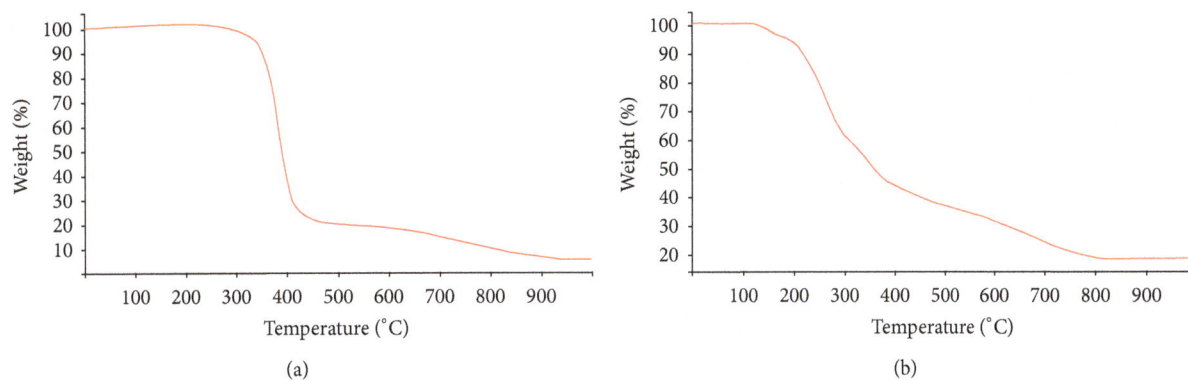

(a)

(b)

FIGURE 6: (a) TG plot of L^2 recorded under nitrogen atmosphere between the temperature ranges 30°C and 988°C at a heating rate of 10°C/min. (b) TG plot of L^1-Co recorded under nitrogen atmosphere between the temperature ranges 30°C and 988°C at a heating rate of 10°C/min.

molecules whether to be inside or outside of the coordination sphere of the complex compounds [26, 50]. The results of thermal analysis of the title compounds are in good agreement with the theoretical formulae as suggested from spectral analyses. The thermogram of L^1 (Figure 5(a)) shows decomposition starting around 220°C with the loss of two ethyl groups adjacent to nitrogen in the carbazole units with 12%(Calc.11.97%). The biggest loss was observed between the temperatures 340°C and 520°C with a percentage of 68.23(Calc.68%), which is assigned to the two carbazole units. The residual part was found as 19.87%(Calc.20%). One could also say that the 8.51%(Calc.8.67%) of the residual part may belong to the propyl unit which plays a bridge role between the two imine groups. TG plot of [CoL^1(H$_2$O)$_2$]2AcO complex gives somewhat distorted three decomposition steps (Figure 6(b)), the first of which is

observed between the temperature ranges 150°C and 220°C with a 9.2%(Calc.10%) loss belonging to the two ethyl groups on the nitrogen atoms. The second decomposition belongs to two carbazole units with a loss of 58.22%(Calc.57.35%). The third decomposition step, however, was assigned as the loss of $C_5H_8N_2$. The residual part gave percentages for metallic Co and the two water molecules. Mn(II) complex of this ligand starts to decompose at around 90°C. Between the temperatures 200°C and 430°C, our complex shows the highest percentage loss with 57.11%(Calc.57.75%) belonging to the carbazole groups. Our second ligand L^2 decomposes between the temperatures 300°C and 500°C with a total loss of 76.32%(Calc.76.17%) corresponding to the two carbazoles in our structures (Figure 6(a)). TG graphics of Mn(II) complex of L^2 (Figure 5(b)) shows three decomposition steps. The first one at 110°C corresponds to the loss of ethyl groups on the

TABLE 2: Thermal analysis of the ligands and their complexes.

Compound	M.W.	$T/^{\circ}C$	Mass loss/% found (calculated)	Assignment	Residual/% found (calculated)
L^1	484.26	220–330	12.00 (11.97)	C_2H_5	19.87 (20)
		340–520	68.23 (68)	$C_{12}H_8N$	
$[Co_2L^1(H_2O)_2]2AcO$	579.60	150–220	9.2 (10)	C_2H_5	10.80 (10) (Co)
		220–500	58.22 (57.35)	$C_{12}H_8N$	6.54 (6.15) (H_2O)
		500–700	16.17 (16.5)	$C_5H_8N_2$	
$[Mn_2L^1(H_2O)_2]2AcO$	575.60	90–200	10.3 (10)	C_2H_5	9.61 (9.5) (Mn)
		200–430	57.11 (57.75)	$C_{12}H_8N$	5.94 (6.2) (H_2O)
		430–590	16.8 (16.6)	$C_5H_8N_2$	
L^2	512.69	300–500	76.32 (76.17)	$C_{14}H_{13}N$	24.04 (23.83)
$[Co_2L^2(H_2O)_2]2AcO$	607.65	200–500	54.91 (54.7)	$C_{12}H_8N$	9.48 (9.69) (Co)
		830–900	20.48 (20.73)	$C_7H_{14}N_2$	5.89 (5.92) (H_2O)
					9.24 (8.96) (undetermined)
$[Mn_2L^2(H_2O)_2]2AcO$	603.66	90–150	10.16 (9.6)	C_2H_5	8.48 (9.1) (Mn)
		150–320	55.23 (55)	$C_{12}H_8N$	5.23 (5.43) (H_2O)
		400–490	20.9 (20.87)	$C_7H_{14}N_2$	

All thermal analyses were done under nitrogen atmosphere between the temperature ranges 30°C and 988°C at a heating rate of 10°C/min.

TABLE 3: Oxidation of styrene and cyclohexene.

Entry	Catalyst	Styrene conversion (%)	Selectivity (%)		
			Styreneoxide	Benzaldehyde	Others
1	Blank	11.5	4.5	10.0	85.5
2	$[CoL^1(H_2O)_2]2AcO$	76.7	21.8	44.5	33.7
3	$[CoL^2(H_2O)_2]2AcO$	81.3	17.4	56.1	26.5
4	$[MnL^1(H_2O)_2]2AcO$	85.3	28.2	38.9	32.9
5	$[MnL^2(H_2O)_2]2AcO$	77.8	31.8	40.5	27.7
Entry	Catalyst	Cyclohexene conversion (%)	Selectivity (%)		
			Cyclohexeneoxide	2-Cyclohexen-1-one	Others
1	Blank	17.7	7	14.3	78.7
2	$[CoL^1(H_2O)_2]2AcO$	65	14.2	61.2	24.6
3	$[CoL^2(H_2O)_2]2AcO$	69.7	18.7	58.3	23
4	$[MnL^1(H_2O)_2]2AcO$	71.2	9.3	52.4	38.3
5	$[MnL^2(H_2O)_2]2AcO$	80.9	11.2	64	24.8

Reaction conditions: alkene (10 mmol), H_2O_2 (20 mmol), catalyst (1 mmol), CH_3CN (20 mL), and nitrogen atmosphere at 90°C.

carbazoles. The second stage at 150°C corresponds to the loss of carbazole groups without the ethyls. The residual part gives a total mass loss of 13.71%(Calc.14.53%). The cobalt complex of the same ligand, however, gives the biggest mass loss of 54.91%(Calc.54.7%) between 200°C and 500°C belonging to the carbazole units. The residual part for this complex gives a total mass loss of 24.61%(Calc.24.57%), 9.24%(Calc.8.96%) of which can be assigned as the ethyl groups adjacent nitrogen on carbazole rings. The thermal data can be examined in Table 2.

3.4. Catalytic Activity. Results of the oxidation reactions (Table 3) show the catalytic activity of synthesized complexes. Comparison between neat and complex catalysed reactions proves that the oxidation reactions using catalysts give higher conversions than their corresponding neat reactions. Examination of Table 3 reveals that the coordination compounds were most effective towards the oxidation of styrene; in

particular, the manganese(II) complex gave 28.2% and 31.8% selectivities for styreneoxide. Oxidation of cyclohexene, on the other hand, gave poorer results compared to those of styrene oxidation. Higher selectivities (14.2% and 18.7%) were obtained for cyclohexeneoxide by using cobalt(II) complexes as catalyst. Comparing our results to those reported previously, we can say that our compounds and their catalytic activity results, although not whole, have similarities with salen and salophane type Schiff bases reported previously [8, 12, 18, 51]. The oxidation schemes for both styrene and cyclohexene can be seen in Figure 7.

3.5. Antimicrobial Activity and Minimal Inhibitory Concentration (MIC) Analysis. The title compounds were evaluated for antimicrobial activity against four gram negative (*K. pneumoniae*, *E. aerogenes*, *E. faecium*, and *E. coli*) and four gram positive bacteria (*B. subtilis*, *S. aureus*, *S. S. feacalis*, and *B. megaterium*) and fungi (*C. albicans*, *C. utilis*, and *S.*

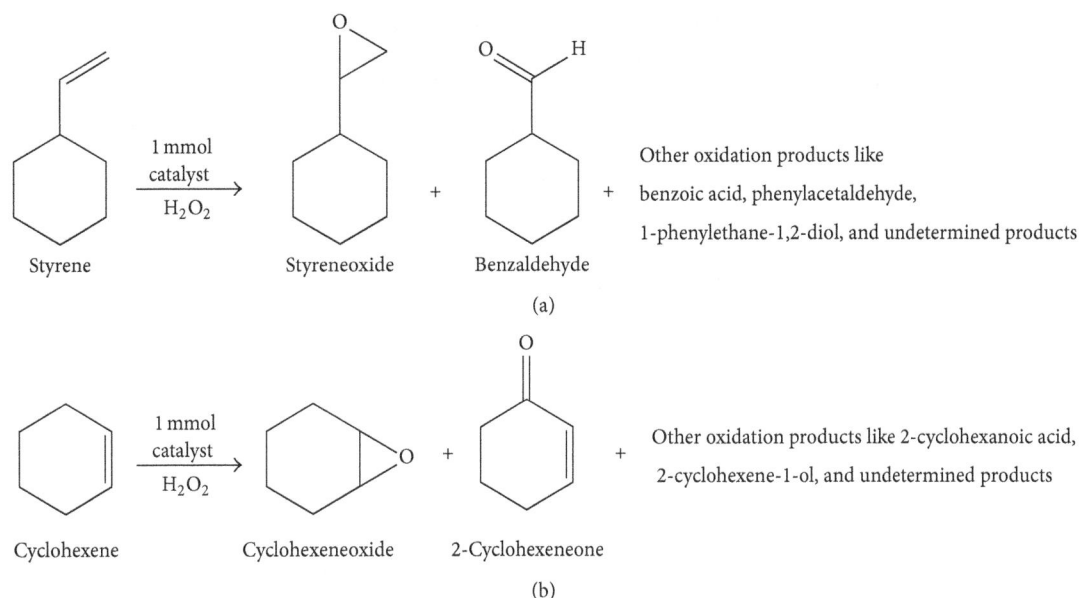

FIGURE 7: (a) Oxidation reaction of styrene. Catalyst : substrate : oxidant ratio: 1 : 10 : 20, under nitrogen atmosphere at 90°C. (b) Oxidation reaction of styrene. Catalyst : substrate : oxidant ratio: 1 : 10 : 20, under nitrogen atmosphere at 90°C.

TABLE 4: Antimicrobial data of synthesized compounds.

| Microorganisms | Compounds | | | | | |
	L^1	L^2	L^1–Co	L^2–Co	L^1–Mn	L^2–Mn
Gram (−)						
K. pneumoniae	18[a]	15	7	11	15	17
E. aerogenes	22	17	8	6	12	11
E. faecium	10	7	12	15	19	21
E. coli	12	11	10	9	14	16
Gram (+)						
B. subtilis	9	7	—[b]	—	8	10
S. aureus	24	21	11	9	13	12
S. faecalis	22	19	9	7	11	12
B. megaterium	15	12	—	7	11	9
Fungi						
C. albicans	11	12	—	7	9	8
C. utilis	27	25	9	10	16	17
S. cerevisiae	16	14	11	9	12	15

[a]Inhibition zone, mm.
[b]Undetermined inhibition zone.

cerevisiae). All antimicrobial and antifungal tests were performed in Mueller Hinton broth. Test tubes were incubated under normal atmospheric conditions at 37°C for 24 h for bacteria and at 30°C for 48 h for the yeasts and the microbial growth was determined by turbidimetric methods. All the synthesized compounds were effective for almost all of the microorganisms; in particular, for the microorganisms *K. pneumoniae, E. aerogenes, S. faecalis,* and *S. aureus* the title ligands brought about bigger inhibition zones. Generally,

they were most effective, both antimicrobial and antifungal. Among the fungi, they gave the biggest inhibition zones for *C. utilis* with 27 mm and 25 mm zones, respectively. Among the coordination compounds, the higher values, that are both antimicrobial and antifungal, were recorded for the manganese(II) complex compounds (Table 4). Minimal inhibition zone experiments revealed that the microorganisms *K. pneumoniae, E. aerogenes, S. faecalis, S. aureus,* and *C. utilis* were the most sensitive microorganisms with their MIC

TABLE 5: MIC values of the synthesized compounds.

Microorganisms[a]	Compounds					
	L^1	L^2	L^1–Co	L^2–Co	L^1–Mn	L^2–Mn
Gram (−)						
K. pneumoniae	500	500	1000	1000	500	500
E. aerogenes	500	500	1000	1000	500	500
E. faecium	750	1000	750	750	500	500
E. coli	1000	1000	1000	1000	1000	750
Gram (+)						
B. subtilis	1000	1000	>2500	>2500	1000	1000
S. aureus	500	500	750	1000	500	500
S. faecalis	500	500	1000	1000	500	500
B. megaterium	750	1000	>2500	1000	750	1000
Fungi[b]						
C. albicans	1000	1000	>2500	1000	1000	1000
C. utilis	500	500	1000	1000	500	500
S. cerevisiae	750	750	1000	1000	750	750

[a]All microorganisms tests were performed in Mueller Hinton broth (MHB).
[b]All fungi tests were performed in Sabouraud dextrose broth (SDB).

values of 500 μg/mL to our ligands and their manganese complexes. The other microorganisms were moderately resistant to the synthesized compounds (Table 5).

4. Conclusion

With this work, carbazoles containing efficient ligands and catalysts have been synthesized, characterized, and used for the oxidation reactions of styrene and cyclohexene. Catalytic activity results show the highest selectivities for styrene oxide formation and moderate results have been obtained for the oxidation of cyclohexene. Thermal analysis results are in agreement with the proposed structures of the ligands and their coordination compounds. Thermally most stable compound is the ligand L^2 with the decomposition temperature starting at 300°C. Following this is the ligand L^1. Among the coordination compounds, cobalt complexes seem to be more resistant to temperature than manganese ones. The biological activity results reveal that both ligands and their Mn(II) complexes are effective as being antimicrobial and antifungal. The cobalt(II) complexes, however, show moderate activities. Finally, the electronic features of these compounds have also been reported.

Conflict of Interests

The authors declare that there is no conflict of interests regarding the publication of this paper.

Acknowledgment

The authors would like to thank Kahramanmaras Sutcu Imam University Research Projects Coordination Unit for the financial support.

References

[1] J. Rudolph, K. L. Reddy, J. P. Chiang, and B. K. Sharpless, "Highly efficient epoxidation of olefins using aqueous H_2O_2 and catalytic methyltrioxorhenium/pyridine: pyridine-mediated ligand acceleration," *Journal of the American Chemical Society*, vol. 119, no. 26, pp. 6189–6190, 1997.

[2] K. Sato, M. Aoki, M. Ogawa, T. Hashimoto, and R. Noyori, "A practical method for epoxidation of terminal olefins with 30% hydrogen peroxide under halide-free conditions," *Journal of Organic Chemistry*, vol. 61, no. 23, pp. 8310–8311, 1996.

[3] C. Venturello and R. DAloisio, "Quaternary ammonium tetrakis(diperoxotungsto)phosphates(3-) as a new class of catalysts for efficient alkene epoxidation with hydrogen peroxide," *Journal of Organic Chemistry*, vol. 53, pp. 1553–1557, 1988.

[4] C. Coperet, H. Adolfson, and K. B. Sharpless, "A simple and efficient method for epoxidation of terminalalkenes," *Chemical Communications*, no. 16, pp. 1565–1566, 1997.

[5] D. E. de Vos, B. F. Sels, M. Reynaers, Y. V. Subba Rao, and P. A. Jacobs, "Epoxidation of terminal or electron-deficient olefins with H_2O_2, catalysed by Mn-trimethyltriazacyclonane complexes in the presence of an oxalate buffer," *Tetrahedron Letters*, vol. 39, no. 20, pp. 3221–3224, 1998.

[6] F. Heshmatpour, S. Rayati, M. Afghan Hajiabbas, P. Abdolalian, and B. Neumüller, "Copper(II) Schiff base complexes derived from 2,2′-dimethyl- propandiamine: Synthesis, characterization and catalytic performance in the oxidation of styrene and cyclooctene," *Polyhedron*, vol. 31, no. 1, pp. 443–450, 2012.

[7] W. Zeng, J. Li, and S. Qin, "The effect of aza crown ring bearing salicylaldimine Schiff bases Mn(III) complexes as catalysts in the presence of molecular oxygen on the catalytic oxidation of styrene," *Inorganic Chemistry Communications*, vol. 9, pp. 10–12, 2006.

[8] Y. Yang, Y. Zhang, S. Hao et al., "Heterogenization of functionalized Cu(II) and VO(IV) Schiff base complexes by direct immobilization onto amino-modified SBA-15: styrene oxidation catalysts with enhanced reactivity," *Applied Catalysis A: General*, vol. 381, pp. 274–281, 2010.

[9] G. Romanowski and J. Kira, "Oxidovanadium(V) complexes with chiral tridentate Schiff bases derived from R(-)-phenylglycinol: synthesis, spectroscopic characterization and catalytic activity in the oxidation of sulfides and styrene," *Polyhedron*, vol. 53, pp. 172–178, 2013.

[10] M. Silva, C. Freire, B. de Castro, and J. L. Figueiredo, "Styrene oxidation by manganese Schiff base complexes in zeolite structures," *Journal of Molecular Catalysis A: Chemical*, vol. 258, no. 1-2, pp. 327–333, 2006.

[11] S. Rayati and F. Ashouri, "Pronounced catalytic activity of oxo-vanadium(IV) Schiff base complexes in the oxidation of cyclooctene and styrene by tert-butyl hydroperoxide," *Comptes Rendus Chimie*, vol. 15, no. 8, pp. 679–687, 2012.

[12] S. Rayati, S. Zakavi, M. Koliaei, A. Wojtczak, and A. Kozakiewicz, "Electron-rich salen-type Schiff base complexes of Cu(II) as catalysts for oxidation of cyclooctene and styrene with tert-butylhydroperoxide: a comparison with electron-deficient ones," *Inorganic Chemistry Communications*, vol. 13, no. 1, pp. 203–207, 2010.

[13] Y. Yang, J. Guan, P. Qiu, and Q. Kan, "Enhanced catalytic performances by surface silylation of Cu(II) Schiff base-containing SBA-15 in epoxidation of styrene with H_2O_2," *Applied Surface Science*, vol. 256, no. 10, pp. 3346–3351, 2010.

[14] G. Romanowski, "Synthesis, characterization and catalytic activity in the oxidation of sulfides and styrene of vanadium(V) complexes with tridentate Schiff base ligands," *Journal of Molecular Catalysis A: Chemical*, vol. 368-369, pp. 137–144, 2013.

[15] S. Mukherjee, S. Samanta, B. C. Roy, and A. Bhaumik, "Efficient allylic oxidation of cyclohexene catalyzed by immobilized Schiff base complex using peroxides as oxidants," *Applied Catalysis A: General*, vol. 301, no. 1, pp. 79–88, 2006.

[16] Y. Chang, Y. Lv, F. Lu, F. Zha, and Z. Lei, "Efficient allylic oxidation of cyclohexene with oxygen catalyzed by chloromethylated polystyrene supported tridentate Schiff-base complexes," *Journal of Molecular Catalysis A: Chemical*, vol. 320, no. 1-2, pp. 56–61, 2010.

[17] M. Salavati-Niasari and H. Babazadeh-Arani, "Cyclohexene oxidation with tert-butylhydroperoxide and hydrogen peroxide catalyzed by new square-planar manganese(II), cobalt(II), nickel(II) and copper(II) bis(2-mercaptoanil)benzil complexes supported on alumina," *Journal of Molecular Catalysis A: Chemical*, vol. 274, no. 1-2, pp. 58–64, 2007.

[18] M. Salavati-Niasari, P. Salemi, and F. Davar, "Oxidation of cyclohexene with tert-butylhydroperoxide and hydrogen peroxide catalysted by Cu(II), Ni(II), Co(II) and Mn(II) complexes of N,N'-bis-(α-methylsalicylidene)-2,2-dimethylpropane-1,3-diamine, supported on alumina," *Journal of Molecular Catalysis A: Chemical*, vol. 238, no. 1-2, pp. 215–222, 2005.

[19] M. Salavati-Niasari, M. Hassani-Kabutarkhani, and F. Davar, "Alumina-supported Mn(II), Co(II), Ni(II) and Cu(II) N,N-bis(salicylidene)-2,2-dimethylpropane-1,3-diamine complexes: synthesis, characterization and catalytic oxidation of cyclohexene with tert-butylhydroperoxide and hydrogen peroxide," *Catalysis Communications*, vol. 7, pp. 955–962, 2006.

[20] D. Chatterjee, S. Mukherjee, and A. Mitra, "Epoxidation of olefins with sodium hypochloride catalysed by new Nickel_II/Schiff base complexes," *Journal of Molecular Catalysis A*, vol. 154, pp. 5–8, 2000.

[21] I. Cârlescu, G. Lisa, and D. Scutaru, "Thermal stability of some ferrocene containing schiff bases," *Journal of Thermal Analysis and Calorimetry*, vol. 91, no. 2, pp. 535–540, 2008.

[22] D. Apreutesei, G. Lisa, N. Hurduc, and D. Scutaru, "Thermal behavior of some cholesteric esters," *Journal of Thermal Analysis and Calorimetry*, vol. 83, no. 2, pp. 335–340, 2006.

[23] M. Tümer, D. Ekinci, F. Tümer, and A. Bulut, "Synthesis, characterization and properties of some divalent metal(II) complexes: their electrochemical, catalytic, thermal and antimicrobial activity studies," *Spectrochimica Acta A*, vol. 67, no. 3-4, pp. 916–929, 2007.

[24] G. Ceyhan, C. Celik, S. Urus, I. Demirtas, M. Elmastas, and M. Tumer, "Antioxidant, electrochemical, thermal, antimicrobial and alkane oxidation properties of tridentate Schiff base ligands and their metal complexes," *Spectrochimica Acta Part A*, vol. 81, pp. 184–198, 2011.

[25] M. Aslantas, E. Kendi, N. Demir, A. E. Sabik, M. Tumer, and M. Kertmen, "Synthesis, spectroscopic, structural characterization, electrochemical and antimicrobial activity studies of the Schiff base ligand and its transition metal complexes," *Spectrochimica Acta A: Molecular and Biomolecular Spectroscopy*, vol. 74, no. 3, pp. 617–624, 2009.

[26] M. Shebl, "Synthesis, spectroscopic characterization and antimicrobial activity of binuclear metal complexes of a new asymmetrical Schiff base ligand: DNA binding affinity of copper(II) complexes," *Spectrochimica Acta*, vol. 117, pp. 127–137, 2014.

[27] Y.-T. Liu, G.-D. Lian, D.-W. Yin, and B.-J. Su, "Synthesis, characterization and biological activity of ferrocene-based Schiff base ligands and their metal (II) complexes," *Spectrochimica Acta A*, vol. 100, pp. 131–137, 2013.

[28] T. A. Yousef, G. M. Abu El-Reash, O. A. El-Gammal, and R. A. Bedier, "Synthesis, characterization, optical band gap, in vitro antimicrobial activity and DNA cleavage studies of some metal complexes of pyridyl thiosemicarbazone," *Journal of Molecular Structure*, vol. 1035, pp. 307–317, 2013.

[29] D. Guo, P. Wu, H. Tan, L. Xia, and W. Zhou, "Synthesis and luminescence properties of novel 4-(N-carbazole methyl) benzoyl hydrazone Schiff bases," *Journal of Luminescence*, vol. 131, no. 7, pp. 1272–1276, 2011.

[30] R. Tang, W. Zang, Y. Luo, and J. Li, "Synthesis, fluorescence properties of Eu(III) complexes with novel carbazole functionalized β-diketone ligand," *Journal of Rare Earths*, vol. 27, no. 3, pp. 362–367, 2009.

[31] S. Zhao, X. Liu, W. Feng, X. Lü, W. Wong, and W. Wong, "Effective enhancement of near-infrared emission by carbazole modification in the Zn-Nd bimetallic Schiff-base complexes," *Inorganic Chemistry Communications*, vol. 20, pp. 41–45, 2012.

[32] L. Yang, W. Zhu, M. Fang, Q. Zhang, and C. Li, "A new carbazole-based Schiff-base as fluorescent chemosensor for selective detection of Fe^{3+} and Cu^{2+}," *Spectrochimica Acta A*, vol. 109, pp. 186–192, 2013.

[33] J. Liu and J.-S. Miao, "Blue electroluminescence of a novel Zn^{2+}-β-diketone complex with a carbazole moiety," *Chinese Chemical Letters*, vol. 25, no. 1, pp. 69–72, 2014.

[34] B. Ruan, Y. Tian, H. Zhou et al., "Synthesis, characterization and in vitro antitumor activity of three organotin(IV) complexes with carbazole ligand," *Inorganica Chimica Acta*, vol. 365, no. 1, pp. 302–308, 2011.

[35] F. B. Koyuncu, S. Koyuncu, and E. Ozdemir, "A novel donor-acceptor polymeric electrochromic material containing carbazole and 1,8-naphtalimide as subunit," *Electrochimica Acta*, vol. 55, no. 17, pp. 4935–4941, 2010.

[36] S. Koyuncu, B. Gultekin, C. Zafer et al., "Electrochemical and optical properties of biphenyl bridged-dicarbazole

oligomer films: electropolymerization and electrochromism," *Electrochimica Acta*, vol. 54, no. 24, pp. 5694–5702, 2009.

[37] S. Koyuncu, C. Zafer, E. Sefer et al., "A new conducting polymer of 2,5-bis(2-thienyl)-1*H*-(pyrrole) (SNS) containing carbazole subunit: electrochemical, optical and electrochromic properties," *Synthetic Metals*, vol. 159, no. 19-20, pp. 2013–2021, 2009.

[38] Y. Liu and M. Liu, "Langmuir-Blodgett film and acidichromism of a long chain carbazole-containing Schiff base," *Thin Solid Films*, vol. 415, no. 1-2, pp. 248–252, 2002.

[39] M. Grigoras and N. Antonoaia, "Synthesis and characterization of some carbazole-based imine polymers," *European Polymer Journal*, vol. 41, no. 5, pp. 1079–1089, 2005.

[40] K. R. Yoon, S. Ko, S. M. Lee, and H. Lee, "Synthesis and characterization of carbazole derived nonlinear optical dyes," *Dyes and Pigments*, vol. 75, no. 3, pp. 567–573, 2007.

[41] NCCLS, *Performance Standards for Antimicrobial Susceptibility Testing*, M100-S9, International Supplement, Villanova, Pa, USA, 9th edition, 1999.

[42] NCCLS, *Performance Standarts for Antimicrobial Disks Susceptibilty Tests*, Approved Standart M2-A8, NCCLS, Wayne, Pa, USA, 8th edition, 2003.

[43] C. H. Collins, P. M. Lyne, and J. M. Grange, *Microbiological Methods*, Butterworths, London, UK, 1989.

[44] L. J. Bradshaw, *Laboratory Microbiology*, Saundes College Publishing, Fort Worth, Tex, USA, 4th edition, 1992.

[45] M. Tümer, C. Çelik, H. Köksal, and S. Serin, "Transition metal complexes of bidentate Schiff base ligands," *Transition Metal Chemistry*, vol. 24, pp. 525–532, 1999.

[46] E. Ispir, "The synthesis, characterization, electrochemical character, catalytic and antimicrobial activity of novel, azo-containing Schiff bases and their metal complexes," *Dyes and Pigments*, vol. 82, no. 1, pp. 13–19, 2009.

[47] M. Tümer, N. Deligönül, A. Gölcü et al., "Mixed-ligand copper(II) complexes: investigation of their spectroscopic, catalysis, antimicrobial and potentiometric properties," *Transition Metal Chemistry*, vol. 31, pp. 1–12, 2006.

[48] L. P. Nitha, R. Aswathy, N. E. Mathews, B. S. Kumari, and K. Mohanan, "Synthesis, spectroscopic characterisation, DNA cleavage, superoxidase dismutase activity and antibacterial properties of some transition metal complexes of a novel bidentate Schiff base derived from isatin and 2-aminopyrimidine," *Spectrochimica Acta A: Molecular and Biomolecular Spectroscopy*, vol. 118, pp. 154–161, 2014.

[49] N. K. Singh and S. B. Singh, "Complexes of 1-isonicotinoyl-4-benzoyl-3-thiosemicarbazide with manganese(II), iron(III), chromium(III), cobalt(II), nickel(II), copper(II) and zinc(II)," *Transition Metal Chemistry*, vol. 26, no. 4-5, pp. 487–495, 2001.

[50] H. P. Ebrahimi, J. S. Hadi, Z. A. Abdulnabi, and Z. Bolandnazar, "Spectroscopic, thermal analysis and DFT computational studies of salen-type Schiff base complexes," *Spectrochim Acta A*, vol. 117, pp. 485–492, 2014.

[51] V. Mirkhani, M. Moghadam, S. Tangestaninejad, I. Mohammadpoor-Baltork, and N. Rasouli, "Catalytic oxidation of olefins with hydrogen peroxide catalyzed by [Fe(III)(salen)Cl] complex covalently linked to polyoxometalate," *Inorganic Chemistry Communications*, vol. 10, no. 12, pp. 1537–1540, 2007.

5

Ionic Liquids: Synthesis and Applications in Catalysis

Rajni Ratti[1,2]

[1] *Maitreyi College, New Delhi 110021, India*
[2] *Miranda House, University of Delhi, New Delhi 110007, India*

Correspondence should be addressed to Rajni Ratti; rajniratti@gmail.com

Academic Editor: Hideaki Shirota

Ionic liquids have emerged as an environmentally friendly alternative to the volatile organic solvents. Being designer solvents, they can be modulated to suit the reaction conditions, therefore earning the name "task specific ionic liquids." Though primarily used as solvents, they are now finding applications in various fields like catalysis, electrochemistry, spectroscopy, and material science to mention a few. The present review is aimed at exploring the applications of ionic liquids in catalysis as acid, base, and organocatalysts and as soluble supports for catalysts.

1. Introduction

One of the twelve principles of green chemistry is that the use of auxiliary substances such as solvents and separation agents should be made unnecessary and if used should be innocuous [1]. The toxic and hazardous properties of many solvents particularly chlorinated hydrocarbons pose crucial environmental concerns such as atmospheric emissions and contamination of water effluents. It is recognized that employing the use of nonconventional solvents as alternatives for environmentally unfriendly traditional solvents can reduce waste solvent production and hence reduce the negative impact on environment to a great extent [2]. The most prevalent of these new solvent systems includes, but not exclusively, water, supercritical fluids (like supercritical CO_2), ionic liquids, solventless processes, and fluorous techniques [3].

Of all the above mentioned nonconventional solvents of interest, ionic liquids have emerged as a promising alternative [4]. Ionic liquid is defined as a salt with melting point below the boiling point of water [5]. Ionic liquids are known by several different names like neoteric solvents, designer solvents, ionic fluids, and molten salts. Most of the ionic liquids are composed of organic cation and inorganic anions. In order to be liquid at room temperature, the cation should preferably be unsymmetrical; that is, the alkyl groups should be different. Polarity and hydrophilicity/hydrophobicity of

ionic liquids can be tuned by suitable combination of cation and anion. It is this property of ionic liquids which has earned them the accolade "designer solvents."

As solvents, ionic liquids have found applications in a number of reactions [6–16]. Dupont et al. extensively reviewed the application of ionic liquids as catalytic phase in various organometallic reactions [17]. Catalytic applications of metal nanoparticles have been explored in ionic liquid media by Migowski and Dupont [18, 19].

Besides the use of ionic liquids as alternate solvents, lately further work has led to the progress in designing functional ionic liquids also referred to as "task specific ionic liquids" (TSIL) [20]. The term task specific ionic liquids or functionalized ionic liquids actually indicates an attempt to capitalize on the potential "design" capacity of ionic liquids and make them true working systems rather than just reaction media.

2. Synthesis of Ionic Liquids

The first room temperature ionic liquid [EtNH$_3$][NO$_3$] (m.p. 12°C) was discovered in 1914 [21], but interest did not develop until the discovery of binary ionic liquids made from mixtures of aluminum(III) chloride and N-alkylpyridinium or 1,3-dialkylimidazolium chloride [22, 23].

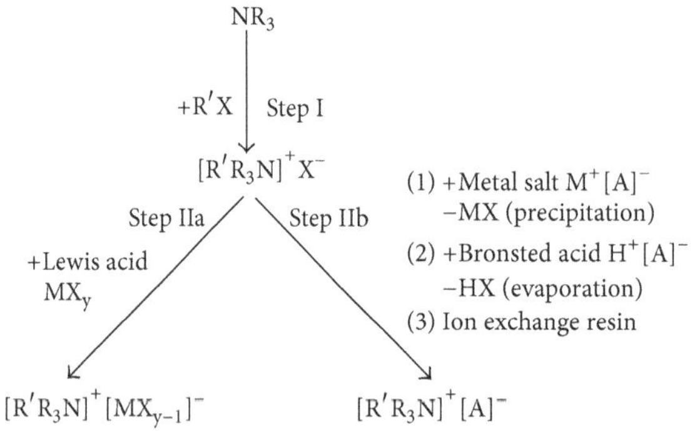

FIGURE 1: Synthesis path for the preparation of ionic liquids [24].

$$[\text{emim}]^+\text{Cl}^- + \text{AlCl}_3 \rightleftharpoons [\text{emim}]^+[\text{AlCl}_4]^- \qquad (1)$$

$$[\text{emim}]^+[\text{AlCl}_4]^- + \text{AlCl}_3 \rightleftharpoons [\text{emim}]^+[\text{Al}_2\text{Cl}_7]^- \qquad (2)$$

$$[\text{emim}]^+[\text{Al}_2\text{Cl}_7]^- + \text{AlCl}_3 \rightleftharpoons [\text{emim}]^+[\text{Al}_3\text{Cl}_{10}]^-. \qquad (3)$$

SCHEME 1: Series of equilibria in the reaction between [emim]Cl and AlCl$_3$.

Ionic liquids come in two main categories, namely, simple salts (made of a single anion and cation) and binary ionic liquids (salts where equilibrium is involved). For example, [EtNH$_3$][NO$_3$] is a simple salt whereas mixtures of aluminum(III) chloride and 1,3-dialkylimidazolium chlorides (a binary ionic liquid system) contain several different ionic species, and their melting point and properties depend upon the mole fractions of aluminum(III) chloride and 1,3-dialkylimidazolium chloride present.

The synthesis of ionic liquids can be described in two steps (Figure 1).

(1) The Formation of the Desired Cation. The desired cation can be synthesized either by the protonation of the amine by an acid or through quaternization reactions of amine with a haloalkane and heating the mixture.

(2) Anion Exchange. Anion exchange reactions can be carried out by treatment of halide salts with Lewis acids to form Lewis acid-based ionic liquids or by anion metathesis.

The most extensively studied and used Lewis acid based ionic liquids are AlCl$_3$ based salts [25–27]. Such salts involve simple mixing of the Lewis acid and the halide salt which results in the formation of more than one anionic species depending upon the ratio of quaternary halide salt Q$^+$X$^-$ and Lewis acid MX$_n$ as illustrated by the reaction between [emim]Cl and AlCl$_3$ in Scheme 1.

When [emim]Cl is present in molar excess over AlCl$_3$ the ionic liquid formed is basic (1); however, the molar excess of AlCl$_3$ leads to the formation of an acidic ionic liquid (3). When both [emim]Cl and AlCl$_3$ are present in equimolar quantities, it results in the formation of neutral ionic liquids. Apart from AlCl$_3$, other Lewis acids used are AlEtCl$_2$ [28], BCl$_3$ [29], CuCl [30], and InCl$_3$ [31] to mention a few.

Anion metathesis is the methodology of choice for the preparation of water and air stable ionic liquids based upon

TABLE 1: Examples of ionic liquids prepared by anion metathesis.

Salt	Anion source	References
[Cation][PF$_6$]	HPF$_6$	[32–34]
[Cation][BF$_4$]	HBF$_4$, NH$_4$BF$_4$, NaBF$_4$	[33–37]
[Cation][(CF$_3$SO$_2$)$_2$N]	Li[(CF$_3$SO$_2$)$_2$N]	[34, 38]
[Cation][CF$_3$SO$_3$]	CF$_3$SO$_3$CH$_3$, NH$_4$[CF$_3$SO$_3$]	[38]
[Cation][CH$_3$CO$_2$]	Ag[CH$_3$CO$_2$]	[35]
[Cation][CF$_3$CO$_2$]	Ag[CF$_3$CO$_2$]	[35]
[Cation][CF$_3$(CF$_3$)$_3$CO$_2$]	K[CF$_3$(CF$_3$)$_3$CO$_2$]	[38]
[Cation][NO$_3$]	AgNO$_3$, NaNO$_3$	[34, 38]
[Cation][N(CN)$_2$]	Ag[N(CN)$_2$]	[39]
[Cation][CB$_{11}$H$_{12}$]	Ag[CB$_{11}$H$_{12}$]	[40]
[Cation][AuCl$_4$]	HAuCl$_4$	[41]

1,3-dialkylimidazolium cations. This method involves the treatment of the halide salt with the silver/sodium/potassium salts of NO$_2$$^-$, NO$_3$$^-$, BF$_4$$^-$, SO$_4$$^{2-}$, and CO$_2CH_3$$^-$ or with the free acid of the appropriate anion. Table 1 gives examples of the few ionic liquids prepared by anion metathesis.

It is clear from the above discussion that large number of ionic liquids can be envisioned by simple combination of different cations and anions. The estimated number of single ILs is 10^{18} which further increases if we include binary and ternary ionic liquids. Because of their "tailor-made" nature the ionic liquids find applications as storage media for toxic gases, catalysts/solvents in organic syntheses, performance additives in pigments, and matrices [42–44].

Several new and improved methodologies using nonconventional techniques, such as irradiation with microwaves (MW) and power ultrasound (US), whether used alone or in combination, have considerably improved the synthesis of ILs, cutting down reaction times and improving yields [45–47]. The recent introduction of efficient, solventless, one-pot synthetic protocols should make ILs cheaper and thus encourage a wider use of these neoteric solvents [48–50].

3. Task Specific Ionic Liquids (TSILs)

In 1999, Davis Jr. and Forrester demonstrated the concept of designing ionic liquid to interact with a solute in a specific manner by using a thiazolium based IL as a solvent-catalyst for the benzoin condensation and introduced the term "task specific ionic liquid" for such ILs in which functional group

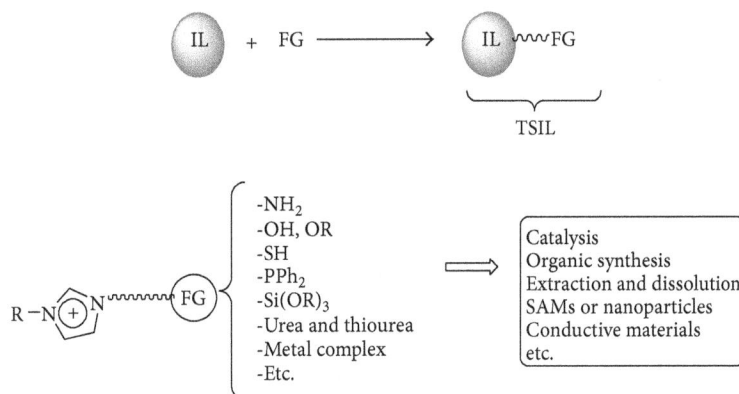

FIGURE 2

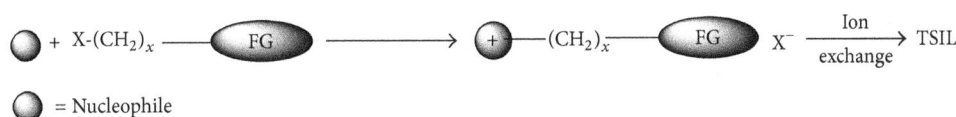

= Nucleophile

FIGURE 3

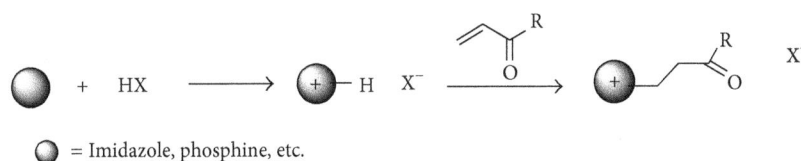

= Imidazole, phosphine, etc.

FIGURE 4

is incorporated as a part of the cation and/or anion structure [51, 52]. The covalent attachment of some functional group to cation/anion or both of an ordinary ionic liquid imparts it the capacity to behave not only as solvent but also as reagent and/or catalyst, catalyst in the chemical reactions [53, 54] (see Figures 2 and 3).

(i) For example, safe to handle Bronsted acidic ionic liquids containing sulphonic acid groups were used as solvent and/or catalyst for esterification and other acid catalyzed reactions [55].

(ii) Ionic liquids bearing appended amines can separate carbon dioxide from gas streams [56].

(iii) Ionic liquids with large aromatic head groups show enhanced activity for extraction of aromatics in aqueous biphasic systems [57].

(iv) Ionic liquids with a tethered hydroxyl group (–OH) have been used as phase transfer catalyst in the synthesis of ethoxybenzene [58].

(v) Ionic liquids containing metal ligating group find use in the extraction of metal ions from aqueous solution [59].

(vi) Ionic liquids with appended carboxylate groups have been used as supports for "IL-phase" synthesis which is a versatile extension of the solid phase synthesis concept [60].

A TSIL can be any of the following two types [61]:

(i) A room temperature ionic liquid, having covalently attached functional group, behaves not just as reaction media but also as reagent/catalyst.

(ii) A binary system of some functionalized salt, which may be solid at room temperature, dissolved in conventional ionic liquid.

3.1. Synthesis of TSILs. Conventional method used for synthesizing a TSIL involves the displacement of halide from an organic by a parent imidazole, phosphine, and so forth whereby the organic halide already incorporates a desired functional group. The displacement reaction is followed by anion exchange (see Figure 3).

This method is suitable for the synthesis of all ionic liquids which are stable towards bases; however, because of the strong basicity of imidazole, elimination of hydrogen halide or Hoffmann elimination occurs in some cases [62]. Generally the functional groups have been introduced directly to the imidazolium moiety using the direct quaternization route. For example, imidazolium cation with hydroxyl groups [58], carboxyl groups [60], thiol groups [63], alkyne groups [64], allyl groups [65], and fluorous chains [66] were successfully prepared.

Wasserscheid and coworkers introduced a new methodology to synthesize TSILs by making use of Michael reaction.

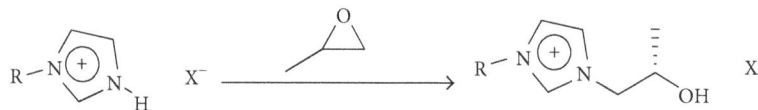

SCHEME 2

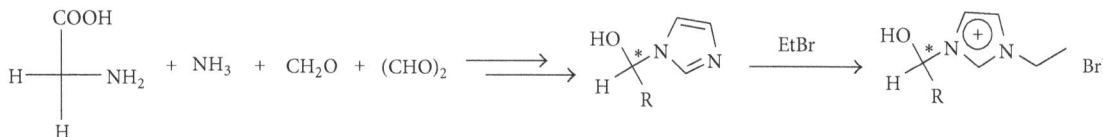

SCHEME 3

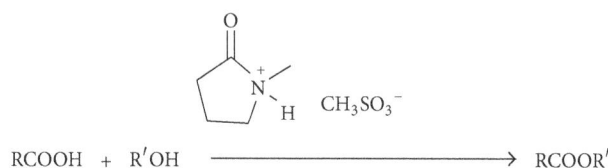

SCHEME 4

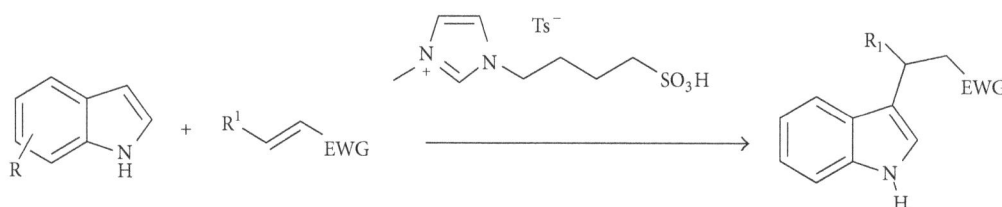

SCHEME 5

In this approach the nucleophile is protonated using the acid form of anion which will eventually be incorporated into the ionic liquid [67] (see Figure 4).

To synthesize –OH group containing TSILs two procedures have been reported as discussed below.

Holbrey et al. have described a simple, high yielding one pot method for the synthesis of alcohol-appended imidazolium TSIL, an ionic liquid type which was previously difficult to prepare cleanly. Preformed imidazolium-H salt of TSIL anion is allowed to react with an epoxide leading to ring opening without further alcohol-epoxide oligomerization [68] (see Scheme 2).

Bao et al. synthesized an imidazolium ring by the four components condensation of amino acids, ammonia, formaldehyde, and glyoxal. The procedure yields an optically active TSIL [69] (see Scheme 3).

4. Ionic Liquids as Catalysts

Although ionic liquids were initially introduced as alternative green reaction media because of their unique physical and chemical properties, today they have marched far beyond this border, showing their significant role in controlling the reaction as catalysts [70–75]. Depending upon the functional group attached to the cation and/or anion, the ionic liquid may behave as an acidic, basic, or organocatalyst.

4.1. As Acid Catalysts. The application of acidic (Bronsted as well as Lewis) task specific ionic liquids (TSILs) as a catalytic material is growing rapidly in the field of catalysis [76, 77]. Combining the useful characteristics of solid acids and mineral acids, TSILs have been synthesized to replace the traditional mineral liquid acids, such as hydrochloric acid and sulphuric acid, in the chemical reactions. In view of green chemistry, the substitution of harmful liquid acids by reusable TSILs is one of the most promising catalytic systems in chemistry.

The acidic nature of Bronsted acidic ionic liquids as catalysts has been exploited for many organic transformations like Pechmann reaction, Koch carbonylation, asymmetric Aldol condensation, Aza-Michael reaction, Beckmann rearrangement, synthesis of chalcones, oxidation reactions and Prin's reaction, synthesis of furfural, biodiesel, Hantzsch reaction, and Mannich reaction to mention a few [78–89].

Esterification of alcohols by carboxylic acids has been carried out in a halogen-free Bronsted acidic ionic liquid, N-methyl-2-pyrrolidinium methyl sulphonate under mild conditions, and without additional solvent [90] (see Scheme 4).

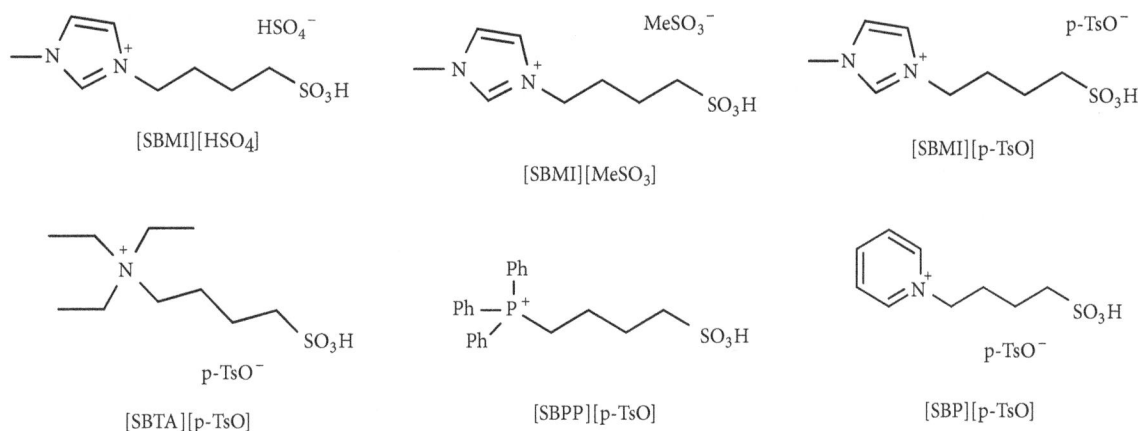

FIGURE 5

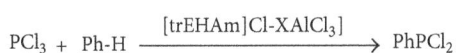

$$PCl_3 + Ph\text{-}H \xrightarrow{[trEHAm]Cl\text{-}XAlCl_3]} PhPCl_2$$

SCHEME 6

In a very recent report of TSIL, Das and coworkers have reported a sulfonic acid functionalized IL for efficient synthesis of indole derivatives [91]. The advantage of this IL is that it could be reused up to 10 cycles without any substantial loss of catalytic activity. The catalyst is versatile as it is also applicable to both aliphatic and aromatic amines and in the synthesis of bis(indolyl) methane (see Scheme 5).

Z.-W. Wang and L.-S. Wang reported the Friedel-Crafts reaction of PCl_3 and benzene in [trEHAm]Cl-XAlCl$_3$ ionic liquid for the clean synthesis of dichlorophenylphosphine (DCPP) [93] (see Scheme 6).

Compared with the classical methods this protocol allows the simple product isolation and lesser reusable catalyst consumption, which contributes to the greenness of the procedure.

Wang et al. screened various ionic liquids for Saucy-Marbet reaction between unsaturated alcohols and unsaturated ethers leading to corresponding unsaturated ketones [94]. It was observed that, with five ionic liquids bearing $[HSO_4^-]$ anion, the conversion decreases as the chain length of ionic liquid increases due to its lipophilic character. Among the various acidic ionic liquids $[Et_3NH]$ HSO_4 gave the best results in terms of conversion (88%) and selectivity (97%) for the model reaction involving dehydrolinalool and 2-ethoxypropene.

With neutral ionic liquids like $[bmim]BF_4$, the conversion was less than 10%. No reaction was observed when [bmim]Cl or $[bmim]PF_6$ was used as catalyst. This cost effective, solvent-free protocol has the advantages of easy work-up, recyclability with only slight decrease in activity, low toxicity of ammonium based ionic liquids, high activity, and selectivity (see Scheme 7).

Brandt et al. extensively reviewed the use of ionic liquids as deconstruction solvents for lignocellulosic biomass [95].

Ionic liquid disrupts the lignin and hemicellulosic network while decrystallizing the cellulose portion which further enhances the speed of saccharification.

The use of ionic liquids $[C_4C_1im][HSO_4]$ and $[C_4C_1im][MeSO_3]$ for lignocellulosic biomass treatment was successfully investigated even in the presence of significant quantities of water, thus eliminating the need for anhydrous conditions during pretreatment [96].

The use of acidic ionic liquids for the saccharification of cellulose and its subsequent conversion into important platform molecules like hydroxymethylfurfural, furfural, and levulinic acid has been well explored [97–103].

A green solvent-free, metal-free, mild, and efficient protocol for the synthesis of 3-vinyl indoles starting from indoles and ketones has been developed using a sulfonyl containing ionic liquid as a recyclable catalyst [104]. The simultaneous presence of sulfonyl and sulphonic acid groups in the same ionic liquid leads to an augmented catalytic activity. Even the challenging substrates like bulky ketones or ortho-substituted ketones gave satisfactory yields (see Scheme 8).

Six different Bronsted acidic ionic liquids (BAILs) have been synthesized and used as recyclable reaction media as well as acid promoters for Pd-phosphine catalyzed methoxy carbonylation of ethylene to produce methyl propionate in excellent yields [105] (see Figure 5). The use of BAILs not only hampered the formation of undesirable palladium black but also leads to the formation of a biphasic reaction media with the product thereby facilitating the product as well as catalyst recovery. The catalytic system has been found to be recyclable up to fifteen cycles without any appreciable loss in activity (see Scheme 9).

Titze-Frech and coworkers developed an efficient and selective methodology for the alkylation of phenol and anisole using Bronsted acidic triflate ionic liquid [MIMBS] [OTf] as catalyst in a biphasic reaction medium [106]. This protocol is advantageous over the existing ones as it negates the need for the neutralization of excess acid formed as a by-product. Also, ionic liquid catalyst being less oxophilic as compared to mineral acids leads to greater selectivities (see Scheme 10).

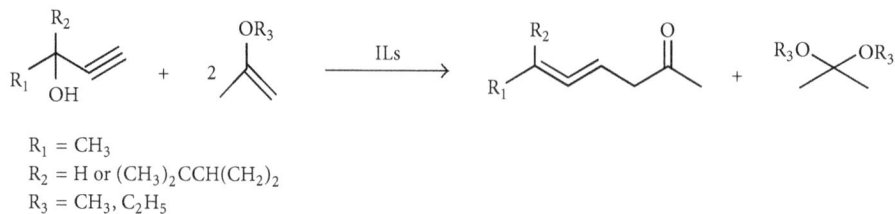

$R_1 = CH_3$
$R_2 = H$ or $(CH_3)_2CCH(CH_2)_2$
$R_3 = CH_3, C_2H_5$

SCHEME 7

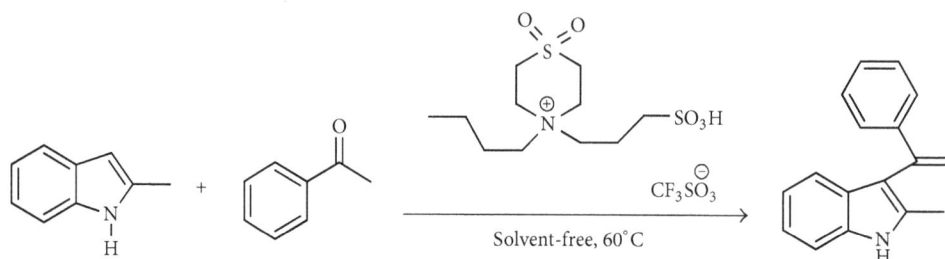

SCHEME 8

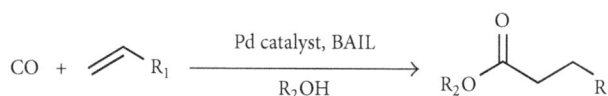

SCHEME 9

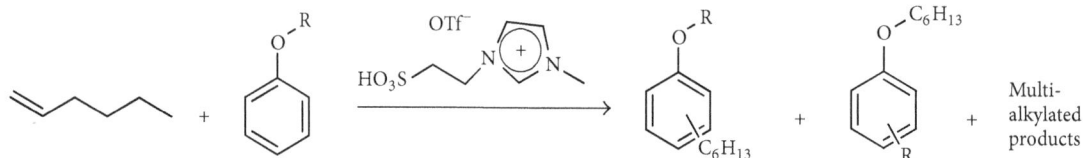

SCHEME 10

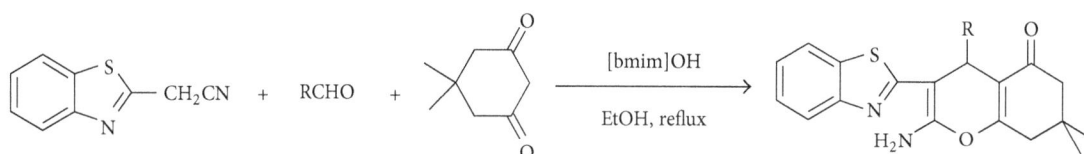

SCHEME 11

4.2. As Base Catalysts. Basic functionalized ionic liquids have aroused unprecedented interest because they showed more advantages, such as convenient recycling and higher catalytic efficiency than the mixture of inorganic base and ionic liquid for some base-catalyzed processes [107].

Basic ionic liquids have been used to catalyze a number of reactions like aza-Michael addition reaction, Michael addition of active methylene compounds, condensation reaction of aldehydes and ketones with hydroxylamine, synthesis of quinolines, pyrroles, and AGET ATRP of methyl methacrylate to mention a few [108–113].

A facile, mild, and quantitative procedure for the preparation of tetrahydrobenzo[b] pyran derivatives in the presence of an easily accessible basic ionic liquid [bmim]OH as catalyst

has been developed by Wen et al. [114]. The ionic liquid was used for at least nine times with consistent activity (see Scheme 11).

Xu et al. developed a green protocol for the Michael addition of N-heterocycles to α,β-unsaturated compounds at room temperature using a basic ionic liquid [bmim]OH as a catalyst and reaction medium [115] (see Scheme 12).

Wang and coworkers described synthesis and application of ethanolamine functionalized TSIL for the palladium-catalyzed Heck reaction [116] (see Scheme 13).

Here this IL performs a multifunctional role of base, ligand, and reaction media with added advantage of recyclability of the system. The catalyst system is very effective for a wide spectrum of substrates giving excellent yields.

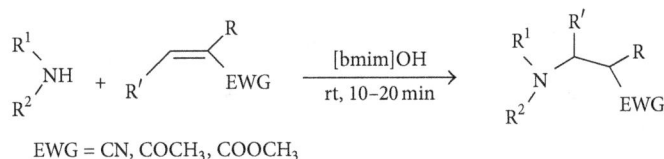

EWG = CN, COCH$_3$, COOCH$_3$

SCHEME 12

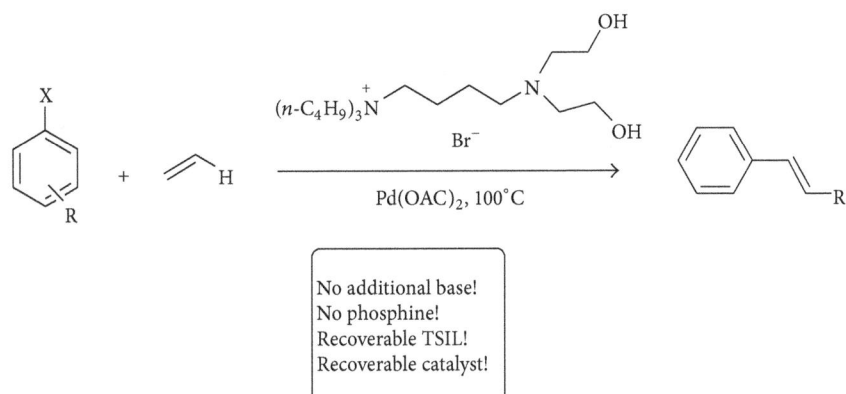

No additional base!
No phosphine!
Recoverable TSIL!
Recoverable catalyst!

SCHEME 13

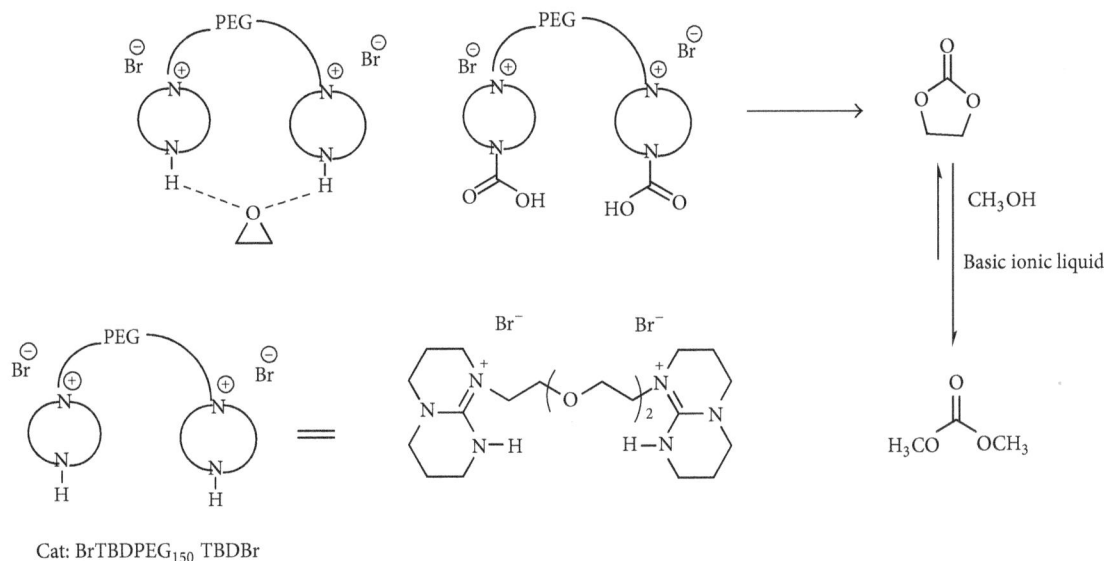

Cat: BrTBDPEG$_{150}$TBDBr

SCHEME 14

Yang et al. designed a series of PEG functionalized basic ionic liquids based on 1,2-dimethyl imidazole (DMIm); 1,8-diazabicyclo [5.4.0] undec-7-ene (DBU); 1,5-diazabicyclo [4.3.0] non-5-ene (DBN); 1-methyl imidazole (MIm) and 1,5,7-triazabicyclo [4.4.0] dec-5-ene (TBD); and tested them as catalysts, under identical conditions, for the conversion of carbon dioxide into useful organic carbonates [117]. Of all these catalysts BrTBDPEG$_{150}$ TBDBr has been found to be excellent recyclable catalyst under solvent-free conditions at low pressure (see Scheme 14).

The presence of both secondary and tertiary nitrogens in the cation of BrTBDPEG$_{150}$ TBDBr endows it with the ability to activate methanol leading to high activity for transesterification of ethylene carbonate with methanol. Therefore the use of basic ionic liquid BrTBDPEG$_{150}$ TBDBr as catalyst allows the integration of cycloaddition as well as transesterification as a single process.

Basic ionic liquids choline hydroxide [ChOH], choline methoxide (ChOMe), and choline imidazolium (ChIm) have been synthesized and checked for their catalytic activity for the production of biodiesel from soybean oil [118]. Of all the three ionic liquids, choline hydroxide was found to give the best results in terms of yield, efficiency, and recyclability. After studying various reaction parameters, 4 wt% catalyst

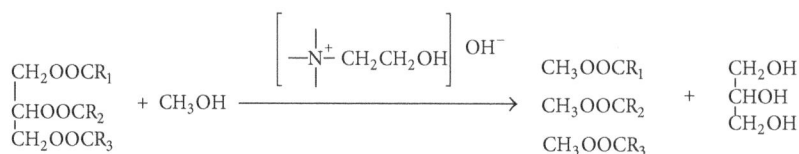

SCHEME 15

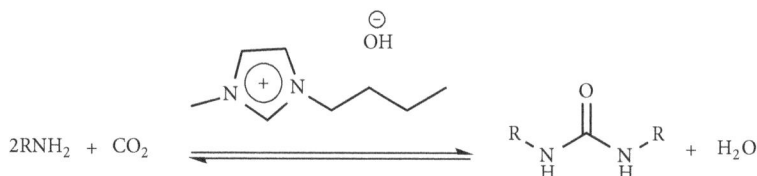

SCHEME 16

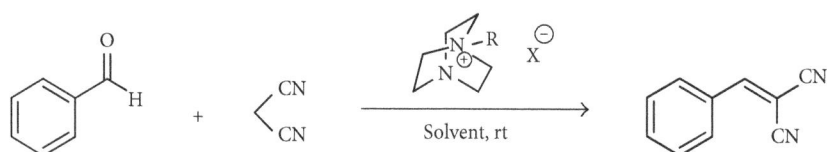

SCHEME 17

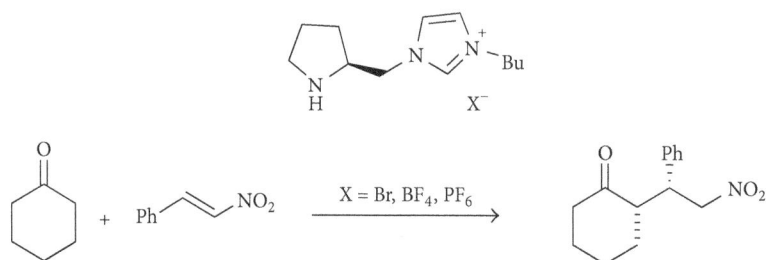

SCHEME 18

dosage at a temperature of 60°C was optimized to give the best results when the ratio of methanol to soybean oil was 9 : 1 (see Scheme 15).

Basic ionic liquid [bmim]OH has been successfully used as an efficient catalyst for the synthesis of substituted ureas starting from carbon dioxide and amines [119]. The main advantages of this methodology are solvent-free reaction conditions, no need of dehydrating agents to remove the water formed as a by-product, recyclability of catalyst, and operational simplicity. The developed protocol is quite general as aliphatic amines, cyclohexylamine, and benzylamine were converted to corresponding ureas efficiently and selectively (see Scheme 16).

Various [DABCO] based ionic liquids have been screened for executing Knoevengeal condensation reaction [120]. Of all these ionic liquids [C$_4$dabco] [BF$_4$] was found to give the best results. Using [C$_4$dabco] [BF$_4$] as catalyst in aqueous media various aromatic/aliphatic/heterocyclic/α,β unsaturated aldehydes and cyclic/acyclic ketones have been found to undergo efficient Knoevengeal condensation with active

methylene compounds. No product purification was required and the catalyst was found to be recyclable up to seven cycles without any decrease in activity. The reaction is highly stereoselective giving alkenes with E-geometry only (see Scheme 17).

4.3. As Organocatalysts. In the last few years a renewed interest in the use of organic compounds as catalysts has begun to emerge. Ionic liquids have the potential to have a huge impact in this area [121–124]. One of the promising approaches to organocatalysis is through hydrogen bonding interactions, and the reactions to which this has been most often applied are Diels-Alder cycloadditions and their derivatives.

Luo and coworkers used a functionalized chiral ionic liquid as an efficient reusable organocatalyst for asymmetric Michael addition of ketones/aldehydes with nitroalkenes [125] (see Scheme 18).

Pyrrolidine-based chiral ionic liquid has been developed by Ni and coworkers [126]. This chiral ionic liquid was found to catalyze the Michael addition reaction of aldehydes

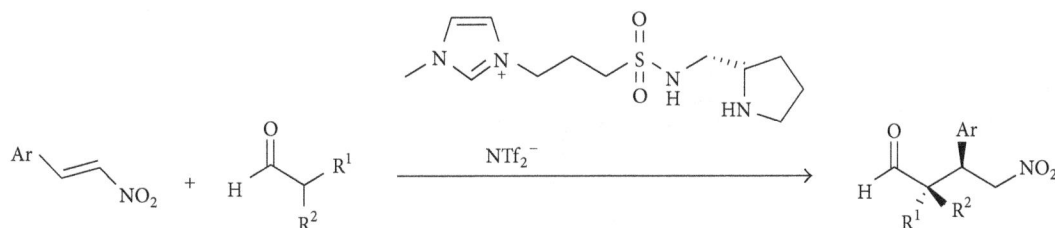

SCHEME 19

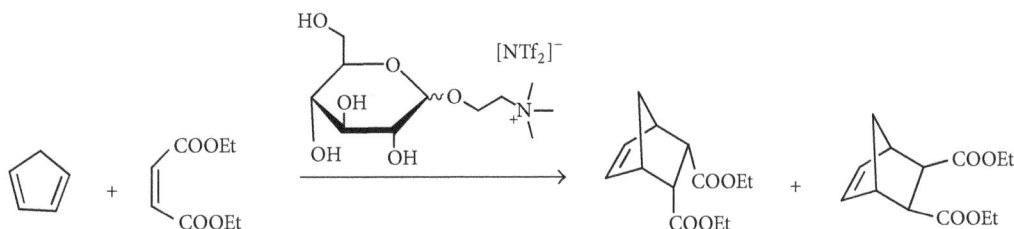

SCHEME 20

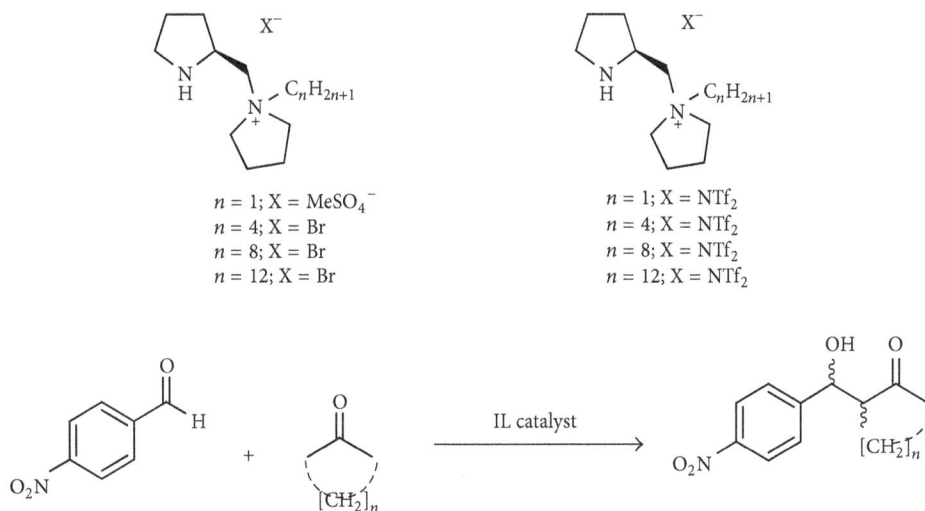

SCHEME 21

and nitrostyrenes to give moderate yields, good enantioselectives, high diastereoselectivities, and recyclability (see Scheme 19).

Though ionic liquids are green solvents, they are synthesized from the materials which use fossil fuels as their resource. Synthesizing ionic liquids from renewable raw materials will add to the green attributes of ionic liquids. Sugars are suitable, abundantly available raw material for the synthesis of ionic liquids. Also the presence of hydroxyl groups in the ionic liquids derived from sugars makes them highly coordinating solvents thus enabling them to be used in stereoselective and metal catalyzed reactions.

Erfut et al. synthesized novel hydrogen bond rich ionic liquids based on D-glucosopyranoside derivatives as cation precursor and low coordinating bistriflimide as anion [127]. Chloroalcohols have been utilized as source of hydroxyl

groups for the construction of ionic liquid cation. The synthesized ionic liquids were successfully used as organocatalysts (4 mol% with respect to dienophile) for Diels-Alder reaction of various dienes and dienophiles. Influence of number of hydroxyl groups on the reaction course has been thoroughly investigated. With all the ionic liquids the endo selectivity was found to be prevalent (see Scheme 20).

Starting from (S)-proline, several chiral ionic liquids have been synthesized by Vasiloiu and coworkers [128]. These ionic liquids were successfully used as organocatalysts to execute asymmetric aldol condensation giving good yields and selectivity up to 80% ee.

It has been observed that hydrophilic triflimide based chiral ionic liquids lead to greater yields and higher selectivities as compared to ionic liquids bearing methyl sulphate or bromide anions. This methodology not only negates

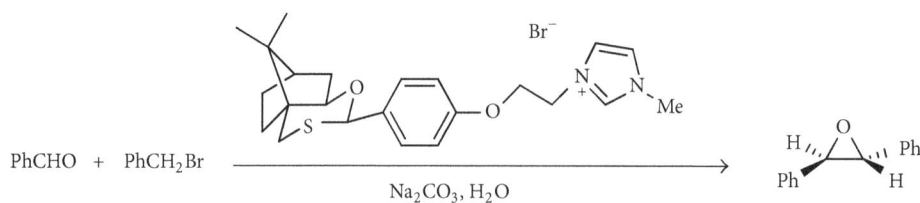

SCHEME 22

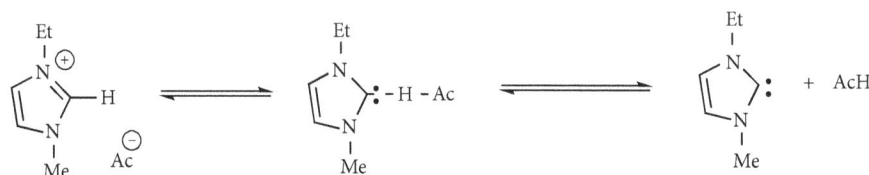

SCHEME 23

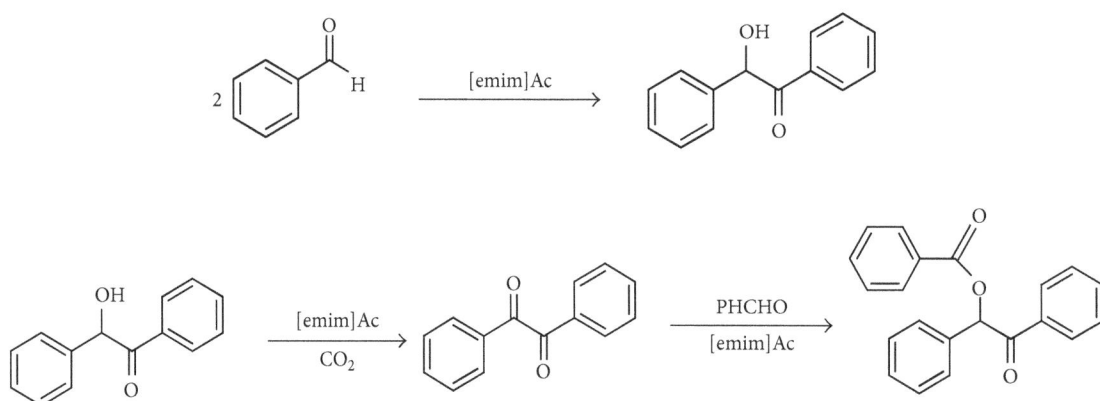

SCHEME 24

the requirement of corrosive trifluoroacetic acid but also widens the substrate scope for organocatalysis towards acid sensitive compounds (see Scheme 21).

Li et al. synthesized a sulphur functionalized chiral ionic liquid which has been used as an organocatalyst for epoxidation reaction of various aromatic aldehydes with benzyl bromide in water giving trans-epoxides with high diastereoselectivity and enantioselectivity up to 72% ee [129]. Sodium carbonate has been found to be the best base for this process. Work-up of this reaction is quite simple as the organocatalyst is insoluble in ether and soluble in water. The catalytic system is recyclable up to five cycles without any appreciable reduction in yields and enatioselectivities (see Scheme 22).

Imidazolium based ionic liquids can be used as pre-catalysts for N-heterocyclic carbene catalyzed reactions whereby the catalyst can be obtained by deprotonation (see Scheme 23).

Kelemen and coworkers successfully used imidazolium acetate as organocatalysts for benzoin condensation, hydroacylation and oxidation of alcohols using carbon-dioxide and air [130] (see Scheme 24).

5. Ionic Liquids as Soluble Supports

Due to their tunable solubility and practically nonvolatile nature, ionic liquids have been used as soluble supports for catalyst/reagent immobilization [92] (see Figure 6). Ionic liquid supported synthesis (ILSS) has been successfully applied for a number of organic reactions like 1,3-cycloadditions [131], Knoevengeal reaction [60], Suzuki coupling [132], synthesis of thiazolidinones [133], oligosaccharide synthesis [134], and Grieco's multicomponnet synthesis of tetrahydroquinolines [135].

Donga and coworkers described the synthesis of oligonucleotides in solution using a soluble ionic liquid support.

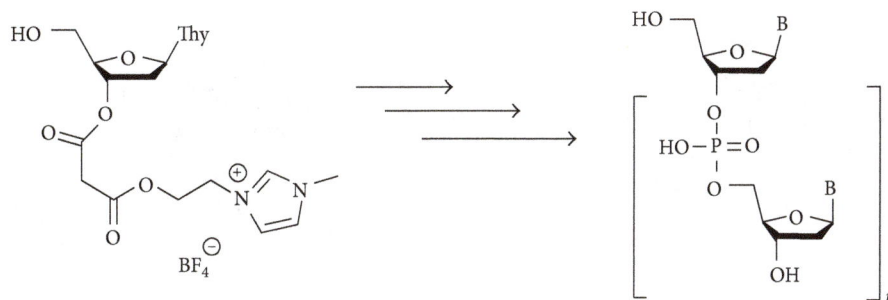

SCHEME 25

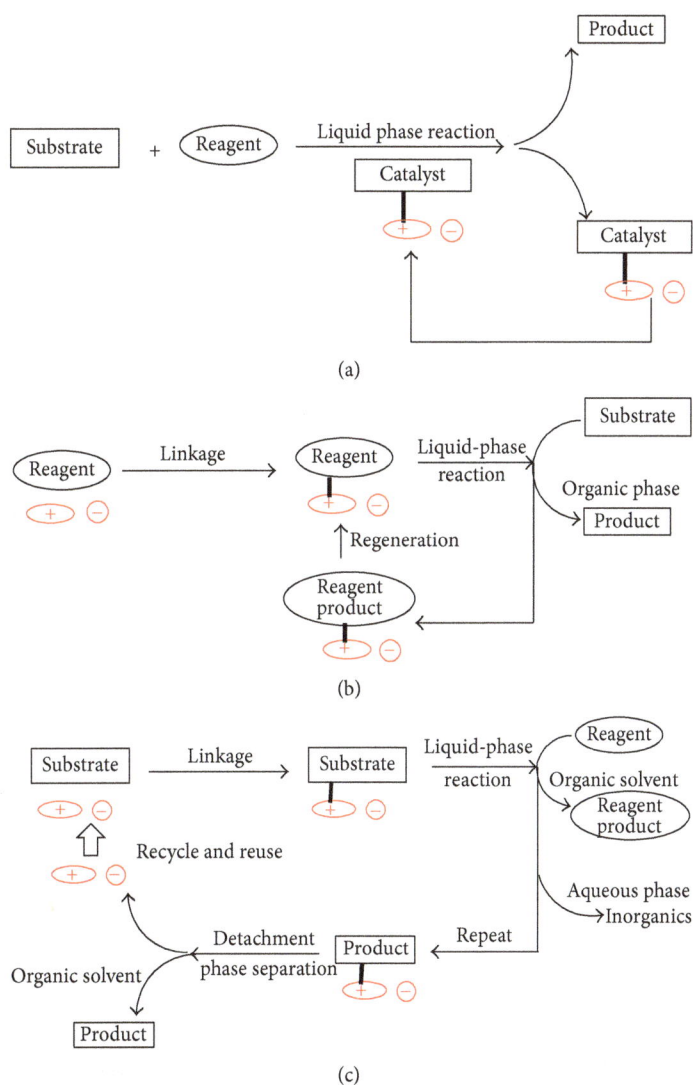

(a)

(b)

(c)

FIGURE 6: Ionic liquid supported synthesis (ILSS) (a) catalyst, (b) reagent, and (c) substrate [92].

Oligonucleotides up to tetrameric species have been synthesized and shown to be comparable to the products generated using standard automated DNA synthesis techniques [136] (see Scheme 25).

A novel and efficient route using ionic liquids as soluble supports has been reported for the synthesis of 1,4-benzodiazepine-2,5-dione by Xie et al. [137] (see Scheme 26).

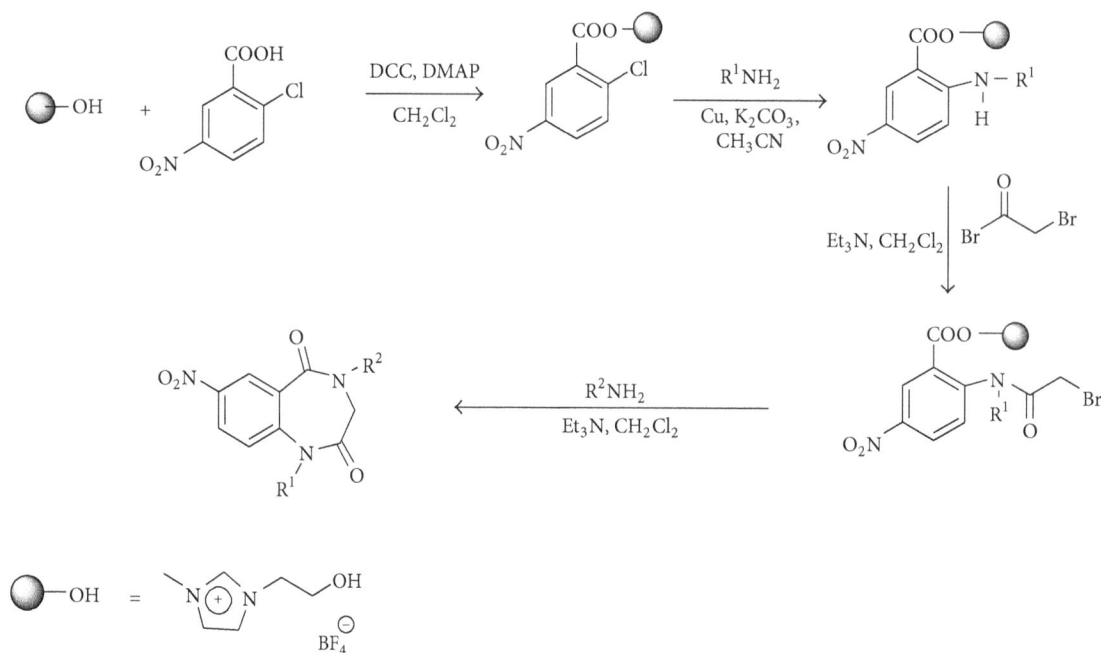

SCHEME 26

6. Conclusion

This review is focused on the synthesis, importance, and applications of ionic liquids. Not particularly as solvents, they are nowadays finding use as catalysts and catalytic supports in organic chemistry. Their scope has marched beyond academic research laboratories to industries where their practical applications have been leading to various sustainable technologies. Flexibility to modulate properties by changing design endows freedom to a chemist to design an ionic liquid according to one's own requirement. To conclude it can be said that the field of ionic liquid catalysis holds enormous possibilities to be explored.

Conflict of Interests

The author declares that there is no conflict of interests regarding the publication of this paper.

References

[1] P. T. Anastas and T. C. Williamson, *Green Chemistry: Frontiers in Benign Chemical Syntheses and Processes*, Oxford University Press, 1998.

[2] D. J. Adams, P. J. Dyson, and S. J. Tavener, *Chemistry in Alternative Reaction Media*, John Wiley & Sons, Chichester, UK, 2004.

[3] P. Pollet, E. A. Davey, E. E. Ureña-Benavides, C. A. Eckert, and C. L. Liotta, "Solvents for sustainable chemical processes," *Green Chemistry*, vol. 16, no. 3, pp. 1034–1055, 2014.

[4] P. Wasscheid and T. Welton, *Ionic Liquids in Synthesis*, Wiley-VCH, New York, NY, USA, 2nd edition, 2008.

[5] T. Welton, "Room-temperature ionic liquids: solvents for synthesis and catalysis," *Chemical Reviews*, vol. 99, no. 8, pp. 2071–2083, 1999.

[6] C. E. Song, "Enantioselective chemo- and bio-catalysis in ionic liquids," *Chemical Communications*, no. 9, pp. 1033–1043, 2004.

[7] Y. Xiao and S. V. Malhotra, "Diels-Alder reactions in pyridinium based ionic liquids," *Tetrahedron Letters*, vol. 45, no. 45, pp. 8339–8342, 2004.

[8] S. T. Handy, "Grignard reactions in imidazolium ionic liquids," *Journal of Organic Chemistry*, vol. 71, no. 12, pp. 4659–4662, 2006.

[9] S. Anjaiah, S. Chandrasekhar, and R. Gree, "Carbon-Ferrier rearrangements in ionic liquids using Yb(OTf)$_3$ as catalyst," *Journal of Molecular Catalysis A: Chemical*, vol. 214, pp. 133–136, 2004.

[10] P. Mastrorilli, C. F. Nobile, R. Paolillo, and G. P. Suranna, "Catalytic Pauson-Khand reaction in ionic liquids," *Journal of Molecular Catalysis A: Chemical*, vol. 214, no. 1, pp. 103–106, 2004.

[11] M. Lombardo, M. Chiarucci, and C. Trombini, "A recyclable triethylammonium ion-tagged diphenylphosphine palladium complex for the Suzuki-Miyaura reaction in ionic liquids," *Green Chemistry*, vol. 11, no. 4, pp. 574–579, 2009.

[12] Z. L. Shen, W. J. Zhou, Y. T. Liu, S. J. Ji, and T. P. Loh, "One-pot chemoenzymatic syntheses of enantiomerically-enriched O-acetyl cyanohydrins from aldehydes in ionic liquid," *Green Chemistry*, vol. 10, no. 3, pp. 283–286, 2008.

[13] L.-C. Feng, Y.-W. Sun, W.-J. Tang et al., "Highly efficient chemoselective construction of 2,2-dimethyl-6-substituted 4-piperidones via multi-component tandem Mannich reaction in ionic liquids," *Green Chemistry*, vol. 12, no. 6, pp. 949–952, 2010.

[14] D. Singh, S. Narayanaperumal, K. Gul, M. Godoi, O. E. D. Rodrigues, and A. L. Braga, "Efficient synthesis of selenoesters

from acyl chlorides mediated by CuO nanopowder in ionic liquid," *Green Chemistry*, vol. 12, no. 6, pp. 957–960, 2010.

[15] T. Fukuyama, T. Inouye, and I. Ryu, "Atom transfer carbonylation using ionic liquids as reaction media," *Journal of Organometallic Chemistry*, vol. 692, no. 1–3, pp. 685–690, 2007.

[16] A. Schenzel, A. Hufendiek, C. Barner-Kowollik, and M. A. R. Meier, "Catalytic transesterification of cellulose in ionic liquids: sustainable access to cellulose esters," *Green Chemistry*, vol. 16, no. 6, pp. 3266–3271, 2014.

[17] J. Dupont, R. F. de Souza, and P. A. Z. Suarez, "Ionic liquid (molten salt) phase organometallic catalysis," *Chemical Reviews*, vol. 102, no. 10, pp. 3667–3692, 2002.

[18] P. Migowski and J. Dupont, "Catalytic applications of metal nanoparticles in imidazolium ionic liquids," *Chemistry*, vol. 13, no. 1, pp. 32–39, 2007.

[19] J. Dupont, G. S. Fonseca, A. P. Umpierre, P. F. P. Fichtner, and S. R. Teixeira, "Transition-metal nanoparticles in imidazolium ionic liquids: recyclable catalysts for biphasic hydrogenation reactions," *Journal of the American Chemical Society*, vol. 124, no. 16, pp. 4228–4229, 2002.

[20] L. Zhen, Z. Yingwei, H. Feng et al., "Catalysis and applications of task-specific ionic liquids," *Scientia Sinica Chimica*, vol. 4, pp. 502–524, 2012.

[21] P. Walden, "Ueber die Molekulargrösse und elektrische Leitfähigkeit einiger geschmolzenen Salze," *Bulletin de l'Académie Impériale des Sciences de St.-Pétersbourg*, vol. 8, no. 6, pp. 405–422, 1914.

[22] H. L. Chum, V. R. Koch, L. L. Miller, and R. A. Osteryoung, "An electrochemical scrutiny of organometallic iron complexes and hexamethylbenzene in a room temperature molten salt," *Journal of the American Chemical Society*, vol. 97, no. 11, pp. 3264–3265, 1975.

[23] J. S. Wilkes, "A short history of ionic liquids - From molten salts to neoteric solvents," *Green Chemistry*, vol. 4, no. 2, pp. 73–80, 2002.

[24] P. Wasserscheid and W. Keim, "Ionic liquids—new "solutions" for transition metal catalysis," *Angewandte Chemie International Edition*, vol. 39, no. 21, pp. 3773–3789, 2000.

[25] F. H. Hurley and T. P. Wier, "Electrodeposition of metals from fused quaternary ammonium salts," *Journal of The Electrochemical Society*, vol. 98, no. 5, pp. 203–206, 1951.

[26] J. Robinson and R. A. Osteryoung, "An electrochemical and spectroscopic study of some aromatic hydrocarbons in the room temperature molten salt system aluminum chloride-n-butylpyridinium chloride," *Journal of the American Chemical Society*, vol. 101, no. 2, pp. 323–327, 1979.

[27] J. S. Wilkes, J. A. Levinsky, R. A. Wilson, and C. L. Hussey, "Dialkylimidazolium chloroaluminate melts: a new class of room-temperature ionic liquids for electrochemistry, spectroscopy and synthesis," *Inorganic Chemistry*, vol. 21, pp. 1263–1264, 1982.

[28] Y. Chauvin, S. Einloft, and H. Olivier, "Catalytic dimerization of propene by nickel-phosphine complexes in 1-butyl-3-methylimidazolium chloride/AlEtxCl3-x (x = 0, 1) ionic liquids," *Industrial and Engineering Chemistry Research*, vol. 34, no. 4, pp. 1149–1155, 1995.

[29] S. D. Williams, J. P. Schoebrechts, J. C. Selkirk, and G. Mamantov, "A new room temperature molten salt solvent system: organic cation tetrachloroborates," *Journal of the American Chemical Society*, vol. 109, no. 7, pp. 2218–2219, 1987.

[30] Y. Chauvin and H. O. Bourbigou, "Nonaqueous ionic liquids as reaction solvents," *Chemtech*, vol. 25, no. 9, pp. 26–30, 1995.

[31] K. R. Seddon, C. Hardacre, and B. J. McAuley, "Catalyst comprising indium salt and organic ionic liquid and process for friedel-crafts reactions," WO 2003028883, 2003.

[32] J. G. Huddleston and R. D. Rogers, "Room temperature ionic liquids as novel media for "clean" liquid—liquid extraction," *Chemical Communications*, no. 16, pp. 1765–1766, 1998.

[33] J. Fuller, R. T. Carlin, H. C. de Long, and D. Haworth, "Structure of 1-ethyl-3-methylimidazolium hexafluorophosphate: model for room temperature molten salts," *Journal of the Chemical Society, Chemical Communications*, no. 3, pp. 299–300, 1994.

[34] L. Cammarata, S. G. Kazarian, P. A. Salter, and T. Welton, "Molecular states of water in room temperature ionic liquids," *Physical Chemistry Chemical Physics*, vol. 3, no. 23, pp. 5192–5200, 2001.

[35] J. S. Wilkes and M. J. Zaworotko, "Air and water stable 1-ethyl-3-methylimidazolium based ionic liquids," *Journal of the Chemical Society, Chemical Communications*, no. 13, pp. 965–967, 1992.

[36] J. D. Holbrey and K. R. Seddon, "The phase behaviour of 1-alkyl-3-methylimidazolium tetrafluoroborates; ionic liquids and ionic liquid crystals," *Journal of the Chemical Society, Dalton Transactions*, no. 13, pp. 2133–2140, 1999.

[37] N. L. Lancaster, T. Welton, and G. B. Young, "A study of halide nucleophilicity in ionic liquids," *Journal of the Chemical Society, Perkin Transactions 2*, no. 12, pp. 2267–2270, 2001.

[38] P. Bonhote, A.-P. Dias, N. Papageorgiou, K. Kalyansundaram, and M. Gratzel, "Hydrophobic, highly conductive ambient-temperature molten salts," *Inorganic Chemistry*, vol. 35, no. 5, pp. 1168–1178, 1996.

[39] D. R. MacFarlane, S. A. Forsyth, J. Golding, and G. B. Deacon, "Ionic liquids based on imidazolium, ammonium and pyrrolidinium salts of the dicyanamide anion," *Green Chemistry*, vol. 4, no. 5, pp. 444–448, 2002.

[40] A. S. Larsen, J. D. Holbrey, F. S. Tham, and C. A. Reed, "Designing ionic liquids: Imidazolium melts with inert carborane anions," *Journal of the American Chemical Society*, vol. 122, no. 30, pp. 7264–7272, 2000.

[41] M. Hasan, I. V. Kozhevnikov, M. R. H. Siddiqui, A. Steiner, and N. Winterton, "Gold compounds as ionic liquids. Synthesis, structures, and thermal properties of N,N'-dialkylimidazolium tetrachloroaurate salts," *Inorganic Chemistry*, vol. 38, no. 25, pp. 5637–5641, 1999.

[42] N. D. Khupse and A. Kumar, "Ionic liquids: new materials with wide applications," *Indian Journal of Chemistry Section A: Inorganic, Physical, Theoretical and Analytical Chemistry*, vol. 49, no. 5-6, pp. 635–648, 2010.

[43] P. Domínguez de María, ""Nonsolvent" applications of ionic liquids in biotransformations and organocatalysis," *Angewandte Chemie International Edition*, vol. 47, no. 37, pp. 6960–6968, 2008.

[44] V. Blasucci, R. Hart, V. L. Mestre et al., "Single component, reversible ionic liquids for energy applications," *Fuel*, vol. 89, no. 6, pp. 1315–1319, 2010.

[45] M. Deetlefs and K. R. Seddon, "Improved preparations of ionic liquids using microwave irradiation," *Green Chemistry*, vol. 5, no. 2, pp. 181–186, 2003.

[46] J.-M. Lévêque, J.-L. Luche, C. Pétrier, R. Roux, and W. Bonrath, "An improved preparation of ionic liquids by ultrasound," *Green Chemistry*, vol. 4, no. 4, pp. 357–360, 2002.

[47] J.-M. Lévêque, S. Desset, J. Suptil et al., "A general ultrasound-assisted access to room-temperature ionic liquids," *Ultrasonics Sonochemistry*, vol. 13, no. 2, pp. 189–193, 2006.

[48] R. S. Varma and V. V. Namboodiri, "Solvent-free preparation of ionic liquids using a household microwave oven," *Pure and Applied Chemistry*, vol. 73, no. 8, pp. 1309–1313, 2001.

[49] R. S. Varma and V. V. Namboodiri, "An expeditious solvent-free route to ionic liquids using microwaves," *Chemical Communications*, no. 7, pp. 643–644, 2001.

[50] P. D. Vu, A. J. Boydston, and C. W. Bielawski, "Ionic liquids via efficient, solvent-free anion metathesis," *Green Chemistry*, vol. 9, no. 11, pp. 1158–1159, 2007.

[51] J. H. Davis Jr., K. J. Forrester, and T. Merrigan, "Novel organic ionic liquids (OILS) incorporating cations derived from the antifungal drug miconazole," *Tetrahedron Letters*, vol. 39, no. 49, pp. 8955–8958, 1998.

[52] J. H. Davis Jr. and K. J. Forrester, "Thiazolium-ion based organic ionic liquids (OILs). Novel oils which promote the benzoin condensation," *Tetrahedron Letters*, vol. 40, no. 9, pp. 1621–1622, 1999.

[53] J. H. Davis Jr., "Task-specific ionic liquids," *Chemistry Letters*, vol. 33, no. 9, pp. 1072–1077, 2004.

[54] A. D. Sawant, D. G. Raut, N. B. Darvatkar, and M. M. Salunkhe, "Recent developments of task-specific ionic liquids in organic synthesis," *Green Chemistry Letters and Reviews*, vol. 4, no. 1, pp. 41–54, 2011.

[55] H. Xing, T. Wang, Z. Zhou, and Y. Dai, "Novel Brønsted-acidic ionic liquids for esterifications," *Industrial and Engineering Chemistry Research*, vol. 44, no. 11, pp. 4147–4150, 2005.

[56] E. D. Bates, R. D. Mayton, I. Ntai, and J. H. Davis Jr., "CO_2 capture by a task-specific ionic liquid," *Journal of the American Chemical Society*, vol. 124, no. 6, pp. 926–927, 2002.

[57] A. E. Visser, J. D. Holbrey, and R. D. Rogers, "Hydrophobic ionic liquids incorporating N-alkylisoquinolinium cations and their utilization in liquid-liquid separations," *Chemical Communications*, no. 23, pp. 2484–2485, 2001.

[58] G. R. Feng, J. J. Peng, H. Y. Qiu, J. X. Jiang, L. Tao, and G. Q. Lai, "Synthesis of novel greener functionalized ionic liquids containing appended hydroxyl," *Synthetic Communications*, vol. 37, no. 16, pp. 2671–2675, 2007.

[59] A. E. Visser, R. P. Swatloski, W. M. Reichert et al., "Task-specific ionic liquids for the extraction of metal ions from aqueous solutions," *Chemical Communications*, no. 1, pp. 135–136, 2001.

[60] J. Fraga-Dubreuil and J. P. Bazureau, "Grafted ionic liquid-phase-supported synthesis of small organic molecules," *Tetrahedron Letters*, vol. 42, no. 35, pp. 6097–6100, 2001.

[61] S.-G. Li, "Functionalized imidazolium salts for task-specific ionic liquids and their applications," *Chemical Communications*, no. 10, pp. 1049–1063, 2006.

[62] A. Horvath, "Michael adducts in regioselective synthesis of N-substituted azoles," *Synthesis*, no. 9, pp. 1183–1189, 1995.

[63] H. Itoh, K. Naka, and Y. Chujo, "Synthesis of gold nanoparticles modified with ionic liquid based on the imidazolium cation," *Journal of the American Chemical Society*, vol. 126, no. 10, pp. 3026–3027, 2004.

[64] Z. Fei, D. Zhao, R. Scopelliti, and P. J. Dyson, "Organometallic complexes derived from alkyne-functionalized imidazolium salts," *Organometallics*, vol. 23, no. 7, pp. 1622–1628, 2004.

[65] D. Zhao, Z. Fei, T. J. Geldbach, R. Scopelliti, G. Laurenczy, and P. J. Dyson, "Allyl-functionalised ionic liquids: synthesis, characterisation, and reactivity," *Helvetica Chimica Acta*, vol. 88, no. 3, pp. 665–675, 2005.

[66] T. L. Merrigan, E. D. Bates, S. C. Dorman, and J. H. Davis Jr., "New fluorous ionic liquids function as surfactants in conventional room-temperature ionic liquids," *Chemical Communications*, no. 20, pp. 2051–2052, 2000.

[67] P. Wasscheid, B. Drießen-Hölscher, R. van Hal, H. C. Steffers, and J. Zimmermann, "New, functionalised ionic liquids from Michael-type reactions—a chance for combinatorial ionic liquid development," *Chemical Communications*, no. 16, pp. 2038–2039, 2003.

[68] J. D. Holbrey, M. B. Turner, W. M. Reichert, and R. D. Rogers, "New ionic liquids containing an appended hydroxyl functionality from the atom-efficient, one-pot reaction of 1-methylimidazole and acid with propylene oxide," *Green Chemistry*, vol. 5, no. 6, pp. 731–736, 2003.

[69] W. Bao, Z. Wang, and Y. Li, "Synthesis of chiral ionic liquids from natural amino acids," *The Journal of Organic Chemistry*, vol. 68, pp. 591–593, 2003.

[70] T. Welton, "Ionic liquids in catalysis," *Coordination Chemistry Reviews*, vol. 248, no. 21–24, pp. 2459–2477, 2004.

[71] A. K. Chakraborti and S. R. Roy, "On catalysis by ionic liquids," *Journal of the American Chemical Society*, vol. 131, no. 20, pp. 6902–6903, 2009.

[72] X. Li, D. Zhao, Z. Fei, and L. Wang, "Applications of functionalized ionic liquids," *Science in China, Series B: Chemistry*, vol. 49, no. 5, pp. 385–401, 2006.

[73] D. Zhao, M. Wu, Y. Kou, and E. Min, "Ionic liquids: applications in catalysis," *Catalysis Today*, vol. 74, no. 1-2, pp. 157–189, 2002.

[74] V. I. Pârvulescu and C. Hardacre, "Catalysis in ionic liquids," *Chemical Reviews*, vol. 107, no. 6, pp. 2615–2665, 2007.

[75] Z. Fei, T. J. Geldbach, D. Zhao, and P. J. Dyson, "From dysfunction to bis-function: on the design and applications of functionalised ionic liquids," *Chemistry A: European Journal*, vol. 12, no. 8, pp. 2122–2130, 2006.

[76] T. L. Greaves and C. J. Drummond, "Protic ionic liquids: properties and applications," *Chemical Reviews*, vol. 108, no. 1, pp. 206–237, 2008.

[77] L. He, G. H. Tao, W. S. Liu, W. Xiong, T. Wang, and Y. Kou, "One-pot synthesis of Lewis acidic ionic liquids for Friedel-Crafts alkylation," *Chinese Chemical Letters*, vol. 17, pp. 321–324, 2006.

[78] F. Dong, C. Jian, G. Kai, S. Qunrong, and L. Zuliang, "Synthesis of coumarins via pechmann reaction in water catalyzed by acyclic acidic ionic liquids," *Catalysis Letters*, vol. 121, no. 3-4, pp. 255–259, 2008.

[79] K. Qiao and C. Yokoyama, "Koch carbonylation of tertiary alcohols in the presence of acidic ionic liquids," *Catalysis Communications*, vol. 7, no. 7, pp. 450–453, 2006.

[80] G. Pousse, F. L. Cavelier, L. Humphreys, J. Rouden, and J. Blanchet, "Brønsted acid catalyzed asymmetric aldol reaction: a complementary approach to enamine catalysis," *Organic Letters*, vol. 12, no. 16, pp. 3582–3585, 2010.

[81] X. B. Liu, M. Lu, T. T. Lu, and G. L. Gu, "Functionalized ionic liquid promoted aza-michael addition of aromatic amines," *Journal of the Chinese Chemical Society*, vol. 57, no. 6, pp. 1221–1226, 2010.

[82] R. Turgis, J. Estager, M. Draye, V. Ragaini, W. Bonrath, and J.-M. Lévêque, "Reusable task-specific ionic liquids for a clean ε-Caprolactam synthesis under mild conditions," *ChemSusChem*, vol. 3, no. 12, pp. 1403–1408, 2010.

[83] J. Shen, H. Wang, H. Liu, Y. Sun, and Z. Liu, "Bronsted acidic ionic liquid as dual catalyst and solvent for environmentally friendly synthesis of Chalcone," *Journal of Molecular Catalysis A: Chemical*, vol. 280, no. 1-2, pp. 24–28, 2008.

[84] A. C. Chaskar, S. R. Bhandari, A. B. Patil, O. P. Sharma, and S. Mayeker, "Solvent-free oxidation of alcohols with potassium persulphate in the presence of bronsted acidic ionic liquids," *Synthetic Communications*, vol. 39, no. 2, pp. 366–370, 2009.

[85] W. Wang, L. Shao, W. Cheng, J. Yang, and M. He, "Koch carbonylation of tertiary alcohols in the presence of acidic ionic liquids," *Catalysis Communications*, vol. 7, no. 7, pp. 450–453, 2006.

[86] J. C. S. Ruiz, J. M. Campelo, M. Francavilla et al., "Efficient microwave-assisted production of furfural from C_5 sugars in aqueous media catalysed by Brönsted acidic ionic liquids," *Catalysis Science & Technology*, vol. 2, no. 9, pp. 1828–1832, 2012.

[87] L. Zhang, M. Xian, Y. He et al., "A Brønsted acidic ionic liquid as an efficient and environmentally benign catalyst for biodiesel synthesis from free fatty acids and alcohols," *Bioresource Technology*, vol. 100, no. 19, pp. 4368–4373, 2009.

[88] D. Patil, D. Chandam, A. Mulik et al., "Novel Brønsted acidic ionic liquid ([CMIM][CF_3COO]) prompted multicomponent hantzsch reaction for the eco-friendly synthesis of acridinediones: an efficient and recyclable catalyst," *Catalysis Letters*, vol. 144, no. 5, pp. 949–958, 2014.

[89] L. He, S. Qin, T. Chang, Y. Sun, and J. Zhao, "Geminal brønsted acid ionic liquids as catalysts for the mannich reaction in water," *International Journal of Molecular Sciences*, vol. 15, no. 5, pp. 8656–8666, 2014.

[90] H. Zhang, F. Xu, X. Zhou, G. Zhang, and C. Wang, "A Brønsted acidic ionic liquid as an efficient and reusable catalyst system for esterification," *Green Chemistry*, vol. 9, no. 11, pp. 1208–1211, 2007.

[91] S. Das, M. Rahman, D. Kundu, A. Majee, and A. Hajra, "Task-specific ionic-liquid-catalyzed efficient synthesis of indole derivatives under solvent-free conditions," *Canadian Journal of Chemistry*, vol. 88, no. 2, pp. 150–154, 2010.

[92] W. Miao and T. H. Chan, "Ionic-liquid-supported synthesis: a novel liquid-phase strategy for organic synthesisIonic-liquid-supported synthesis: a novel liquid-phase strategy for organic synthesis," *Accounts of Chemical Research*, vol. 39, no. 12, pp. 897–908, 2006.

[93] Z.-W. Wang and L.-S. Wang, "Friedel–Crafts phosphylation of benzene catalyzed by [trEHAm]Cl-XAlCl$_3$ ionic liquids," *Applied Catalysis A: General*, vol. 262, no. 1, pp. 101–104, 2004.

[94] C. Wang, W. Zhao, H. Li, and L. Guo, "Solvent-free synthesis of unsaturated ketones by the Saucy-Marbet reaction using simple ammonium ionic liquid as a catalyst," *Green Chemistry*, vol. 11, no. 6, pp. 843–847, 2009.

[95] A. Brandt, J. Gräsvik, J. P. Hallett, and T. Welton, "Deconstruction of lignocellulosic biomass with ionic liquids," *Green Chemistry*, vol. 15, no. 3, pp. 550–583, 2013.

[96] A. Brandt, M. J. Ray, T. Q. To, D. J. Leak, R. J. Murphy, and T. Welton, "Ionic liquid pretreatment of lignocellulosic biomass with ionic liquid-water mixtures," *Green Chemistry*, vol. 13, no. 9, pp. 2489–2499, 2011.

[97] A. S. Amarasekara and O. S. Owereh, "Hydrolysis and decomposition of cellulose in bronØsted acidic ionic liquids under mild conditions," *Industrial & Engineering Chemistry Research*, vol. 48, no. 22, pp. 10152–10155, 2009.

[98] C. Li and Z. K. Zhao, "Efficient acid-catalyzed hydrolysis of cellulose in ionic liquid," *Advanced Synthesis and Catalysis*, vol. 349, no. 11-12, pp. 1847–1850, 2007.

[99] M. E. Zakrzewska, E. B. Lukasik, and R. B. Lukasik, "Ionic liquid-mediated formation of 5-hydroxymethylfurfural—a promising biomass-derived building block," *Chemical Reviews*, vol. 111, pp. 397–417, 2011.

[100] T. Ståhlberg, W. Fu, J. M. Woodley, and A. Riisager, "Synthesis of 5-(hydroxymethyl)furfural in ionic liquids: paving the way to renewable chemicals," *ChemSusChem*, vol. 4, no. 4, pp. 451–458, 2011.

[101] J. C. S. Ruiz, J. M. Campelo, M. Francavilla et al., "Efficient microwave-assisted production of furfural from C_5 sugars in aqueous media catalysed by Brönsted acidic ionic liquids," *Catalysis Science and Technology*, vol. 2, pp. 1828–1832, 2012.

[102] Z. Sun, M. Cheng, H. Li et al., "One-pot depolymerization of cellulose into glucose and levulinic acid by heteropolyacid ionic liquid catalysis," *RSC Advances*, vol. 2, no. 24, pp. 9058–9065, 2012.

[103] H. Ren, Y. Zhou, and L. Liu, "Selective conversion of cellulose to levulinic acid via microwave-assisted synthesis in ionic liquids," *Bioresource Technology*, vol. 129, pp. 616–619, 2013.

[104] A. Taheri, C. Liu, B. Lai, C. Cheng, X. Pan, and Y. Gu, "Brønsted acid ionic liquid catalyzed facile synthesis of 3-vinylindoles through direct C3 alkenylation of indoles with simple ketones," *Green Chemistry*, vol. 16, no. 8, pp. 3715–3719, 2014.

[105] E. J. García-Suárez, S. G. Khokarale, O. N. van Buu, R. Fehrmann, and A. Riisager, "Pd-catalyzed ethylene methoxycarbonylation with Brønsted acid ionic liquids as promoter and phase-separable reaction media," *Green Chemistry*, vol. 16, no. 1, pp. 161–166, 2014.

[106] K. Titze-Frech, N. Ignatiev, M. Uerdingen, P. S. Schulz, and P. Wasserscheid, "Highly selective aromatic alkylation of phenol and anisole by using recyclable brønsted acidic ionic liquid systems," *European Journal of Organic Chemistry*, no. 30, pp. 6961–6966, 2013.

[107] A. R. Hajipour and F. Rafiee, "Basic ionic liquids: a short review," *Journal of the Iranian Chemical Society*, vol. 6, no. 4, pp. 647–678, 2009.

[108] B. C. Ranu and S. Banerjee, "Ionic liquid as catalyst and reaction medium. The dramatic influence of a task-specific ionic liquid, [bmIm]OH, in Michael addition of active methylene compounds to conjugated ketones, carboxylic esters, and nitriles," *Organic Letters*, vol. 7, no. 14, pp. 3049–3052, 2005.

[109] L. Yang, L.-W. Xu, W. Zhou, L. Li, and C.-G. Xia, "Highly efficient aza-Michael reactions of aromatic amines and N-heterocycles catalyzed by a basic ionic liquid under solvent-free conditions," *Tetrahedron Letters*, vol. 47, no. 44, pp. 7723–7726, 2006.

[110] H. Zang, M. Wang, B. W. Cheng, and J. Song, "Ultrasound-promoted synthesis of oximes catalyzed by a basic ionic liquid [bmIm]OH," *Ultrasonics Sonochemistry*, vol. 16, no. 3, pp. 301–303, 2009.

[111] E. Kowsari and M. Mallakmohammadi, "Ultrasound promoted synthesis of quinolines using basic ionic liquids in aqueous media as a green procedure," *Ultrasonics Sonochemistry*, vol. 18, pp. 447–454, 2011.

[112] I. Yavari and E. Kowsari, "Efficient and green synthesis of tetrasubstituted pyrroles promoted by task-specific basic ionic liquids as catalyst in aqueous media," *Molecular Diversity*, vol. 13, no. 4, pp. 519–528, 2009.

[113] Z. Deng, J. Guo, L. Qiu, Y. Zhou, L. Xia, and F. Yan, "Basic ionic liquids: a new type of ligand and catalyst for the AGET ATRP of methyl methacrylate," *Polymer Chemistry*, vol. 3, no. 9, pp. 2436–2443, 2012.

[114] L.-R. Wen, H.-Y. Xie, and M. Li, "A basic ionic liquid catalyzed reaction of benzothiazole, aldehydes, and 5,5-dimethyl-1,3-cyclohexanedione: efficient synthesis of tetrahydrobenzo[b] pyrans," *Journal of Heterocyclic Chemistry*, vol. 46, no. 5, pp. 954–959, 2009.

[115] J.-M. Xu, Q. Wu, Q.-Y. Zhang, F. Zhang, and X.-F. Lin, "Basic ionic liquid as catalyst and reaction medium: a rapid and facile protocol for Aza-Michael addition reactions," *European Journal of Organic Chemistry*, pp. 1798–1802, 2007.

[116] L. Wang, H. Li, and P. Li, "Task-specific ionic liquid as base, ligand and reaction medium for the palladium-catalyzed Heck reaction," *Tetrahedron*, vol. 65, no. 1, pp. 364–368, 2009.

[117] Z. Z. Yang, Y. N. Zhao, L. N. He, J. Gao, and Z. S. Yin, "Highly efficient conversion of carbon dioxide catalyzed by polyethylene glycol-functionalized basic ionic liquids," *Green Chemistry*, vol. 14, no. 2, pp. 519–527, 2012.

[118] M. Fan, J. Huang, J. Yang, and P. Zhang, "Biodiesel production by transesterification catalyzed by an efficient choline ionic liquid catalyst," *Applied Energy*, vol. 108, pp. 333–339, 2013.

[119] T. Jiang, X. Ma, Y. Zhou, S. Liang, J. Zhang, and B. Han, "Solvent-free synthesis of substituted ureas from CO_2 and amines with a functional ionic liquid as the catalyst," *Green Chemistry*, vol. 10, no. 4, pp. 465–469, 2008.

[120] D.-Z. Xu, Y. Liu, S. Shi, and Y. Wang, "A simple, efficient and green procedure for Knoevenagel condensation catalyzed by [C_4dabco][BF_4] ionic liquid in water," *Green Chemistry*, vol. 12, no. 3, pp. 514–517, 2010.

[121] F.-L. Yu, R.-L. Zhang, C.-X. Xie, and S.-T. Yu, "Synthesis of thermoregulated phase-separable triazolium ionic liquids catalysts and application for Stetter reaction," *Tetrahedron*, vol. 66, no. 47, pp. 9145–9150, 2010.

[122] D. Z. Xu, Y. Liu, S. Shi, and Y. Wang, "Chiral quaternary alkylammonium ionic liquid [Pro-dabco][BF_4]: as a recyclable and highly efficient organocatalyst for asymmetric Michael addition reactions," *Tetrahedron Asymmetry*, vol. 21, no. 20, pp. 2530–2534, 2010.

[123] O. V. Maltsev, A. S. Kucherenko, A. L. Chimishkyan, and S. G. Zlotin, "α,α-Diarylprolinol-derived chiral ionic liquids: recoverable organocatalysts for the domino reaction between α,β-enals and N-protected hydroxylamines," *Tetrahedron Asymmetry*, vol. 21, no. 21-22, pp. 2659–2670, 2010.

[124] S. S. Khan, J. Shah, and J. Liebscher, "Ionic-liquid tagged prolines as recyclable organocatalysts for enantioselective α-aminoxylations of carbonyl compounds," *Tetrahedron*, vol. 67, no. 10, pp. 1812–1820, 2011.

[125] S. Luo, X. Mi, L. Zhang, S. Liu, H. Xu, and J.-P. Cheng, "Functionalized chiral ionic liquids as highly efficient asymmetric organocatalysts for michael addition to nitroolefins," *Angewandte Chemie International Edition*, vol. 45, pp. 3093–3097, 2006.

[126] B. Ni, Q. Zhang, and A. D. Headley, "Functionalized chiral ionic liquid as recyclable organocatalyst for asymmetric Michael addition to nitrostyrenes," *Green Chemistry*, vol. 9, no. 7, pp. 737–739, 2007.

[127] K. Erfut, I. Wandzik, K. Walczak, K. Matuszek, and A. Chrobok, "Hydrogen-bond-rich ionic liquids as effective organocatalysts for Diels–Alder reactions," *Green Chemistry*, vol. 16, no. 7, pp. 3508–3514, 2014.

[128] M. Vasiloiu, D. Rainer, P. Gaertner, C. Reichel, C. Schröder, and K. Bica, "Basic chiral ionic liquids: a novel strategy for acid-free organocatalysis," *Catalysis Today*, vol. 200, no. 1, pp. 80–86, 2013.

[129] J. Li, X.-K. Xie, F. Liu, and Z.-Z. Huang, "Synthesis of new functionalized chiral ionic liquid and its organocatalytic asymmetric epoxidation in water," *Catalysis Communications*, vol. 11, pp. 276–279, 2009.

[130] Z. Kelemen, O. Holloczki, J. Nagy, and L. Nyulaszi, "An organocatalytic ionic liquid," *Organic & Biomolecular Chemistry*, vol. 9, pp. 5362–5364, 2011.

[131] J. F. Dubreuil and J. P. Bazureau, "Rate accelerations of 1,3-dipolar cycloaddition reactions in ionic liquids," *Tetrahedron Letters*, vol. 41, no. 38, pp. 7351–7355, 2000.

[132] W. Miao and T. H. Chan, "Exploration of ionic liquids as soluble supports for organic synthesis. demonstration with a suzuki coupling reaction," *Organic Letters*, vol. 5, no. 26, pp. 5003–5005, 2003.

[133] J. Fraga-Dubreuil and J. P. Bazureau, "Efficient combination of task-specific ionic liquid and microwave dielectric heating applied to one-pot three component synthesis of a small library of 4-thiazolidinones," *Tetrahedron*, vol. 59, no. 32, pp. 6121–6130, 2003.

[134] J.-Y. Huang, M. Lei, and Y.-G. Wang, "A novel and efficient ionic liquid supported synthesis of oligosaccharides," *Tetrahedron Letters*, vol. 47, no. 18, pp. 3047–3050, 2006.

[135] F. Hassine, S. Gmouh, M. Pucheault, and M. Vaultier, "Task specific onium salts and ionic liquids as soluble supports in Grieco's multicomponent synthesis of tetrahydroquinolines," *Monatshefte fur Chemie*, vol. 138, no. 11, pp. 1167–1174, 2007.

[136] R. A. Donga, S. M. Khaliq-Uz-Zaman, T.-H. Chan, and M. J. Damha, "A novel approach to oligonucleotide synthesis using an imidazolium ion tag as a soluble support," *Journal of Organic Chemistry*, vol. 71, no. 20, pp. 7907–7910, 2006.

[137] H. Xie, C. Lu, G. Yang, and Z. Chen, "Synthesis of 1,4-Benzodiazepine-2,5-diones using an ionic liquid as a Soluble Support," *Synthesis*, vol. 2009, no. 2, pp. 205–210, 2009.

6

Pulsed Supersonic Beams from High Pressure Source: Simulation Results and Experimental Measurements

U. Even

Sackler School of Chemistry, Tel Aviv University, 69978 Tel Aviv, Israel

Correspondence should be addressed to U. Even; even@post.tau.ac.il

Academic Editor: Zhengcheng Zhang

Pulsed beams, originating from a high pressure, fast acting valve equipped with a shaped nozzle, can now be generated at high repetition rates and with moderate vacuum pumping speeds. The high intensity beams are discussed, together with the skimmer requirements that must be met in order to propagate the skimmed beams in a high-vacuum environment without significant disruption of the beam or substantial increases in beam temperature.

1. Introduction

Supersonic beams have proven to be an essential source of information on molecular properties and collision processes. The early stage of their development relied on the production of continuous beams and was limited by available pumping capacity. Beam propagation was limited by collisions with background gases, usually to a distance of only a few mm before the beam was dispersed and attenuated. Much effort was required to ensure that the jet was interrogated before it was degraded, and building such a supersonic beam instrument became an art [1–5].

The introduction of pulsed supersonic beams solved some of the problems of CW (continuous wave) machines and is widely spread today, covering too many research areas to enumerate concisely. Since its early years it was evident that using pulsed beams enables the generation of more intense and colder beams at reduced pumping capacity [1–4]. The short pulse length (<50 μsec.) enables us to disregard interaction with the walls of the vacuum system or residual gas. In fact only interactions within a radius of 20 mm from the beam are relevant to these time scales, and the pumps have all the time between pulses to reduce the background pressure to negligible values.

Many designs of pulsed beam sources are based on various actuating mechanisms (mechanical [6], Lorenz force [7–9], Piezo driver [10, 11], or electromagnetic [12–17]). Several

years ago we introduced a shaped nozzle, which combined with a high pressure fast-acting pulsed valve enabled us to reach high beam intensities and lower jet temperatures than what was previously available [18, 19]. The development of this valve and its operating characteristics are described here, along with the changes in skimmer design that are required in order to make use of the higher on-axis beam intensities. A comprehensive review of our pulsed valve was recently published [20]. Simulations of gas flow [21] as well as measurements of actual beam parameters are presented. Previous designs of molecular beam skimmers [5, 20, 22–24] as originally developed for the earlier low-intensity continuous beams cannot be copied for high intensity beams because skimmer interference (also known as skimmer clogging) is simply too severe with the more intense pulsed beams that are now available. In this paper, we describe recent results on nozzle design (Section 2), pulsed valve design (Section 3), molecular beam properties (Section 4), skimmer design (Section 5), and conclusions (Section 6). These topics are discussed in the designated sections, Section 2–Section 6.

2. Nozzle Design and Testing

The simplest nozzle is a circular hole in a thin plate also known as a sonic nozzle. The expansion from this simple geometry can be calculated analytically [3], but we choose

FIGURE 1: Number density contour for three nozzles.

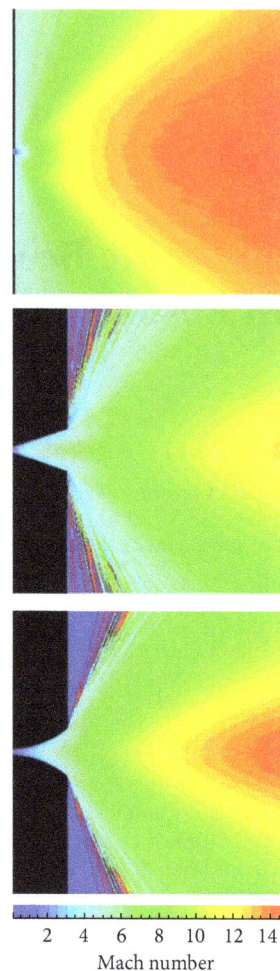

FIGURE 2: Mach number contours for three nozzles.

to show a simulation of this axisymmetric flow because this allows for more direct comparisons to the results for more complicated nozzle shapes. We use the freely available DSMC program [25], run on a modern personal computer for several hours for each run. The simulation run contains ~1 million atoms and becomes too long if we use a stagnation pressure larger than 0.3 bars expanding into a vacuum. The mean free path of the flow in the nozzle itself (and up to 50 nozzle diameters) is much smaller than the dimensions involved, leading us to believe that the results at this reduced pressure simulations (0.3 bars) can also represent the flow at the higher operating pressures (several tens of bars). No clustering is included in the simulation here. The simulation was run until no further changes are seen; that is, a steady state is achieved in the flow (50 ms of flow time). The simulation results are then compared to experimental results to test this assumption. We present results for the beam intensity (i.e., atomic number density) and Mach number contours for three nozzles: (I) sonic nozzle, (II) conical nozzle of 40 degrees full angle, and (III) parabolic nozzle shape. All nozzles have the same inlet diameters (0.2 mm). The last two nozzles have a length of 2 mm (10 nozzle diameters). The computation

extends to 10 mm distance from the nozzle (50 nozzle diameters). Figures 1 and 2 show the simulation results. Since the computations are performed at low stagnation pressures, we extrapolate that actual beam density in our experiments will be 100 times higher (at the applied experimental pressure of 30 bars). A central issue in the computation is how to model the interaction of individual atoms colliding with the nozzle surface. This is also known as the energy accommodation coefficients problem. Not much is known for our nozzle surface (polished stainless steel or more recently polished ruby). We chose a perpendicular accommodation coefficient of 0.4 and a tangential accommodation coefficient of 0.4 [26–29] in the simulations to simulate helium. These coefficients determine the boundary layer thickness in the flow (between the flowing gas and the stationary nozzle surface). We estimate that at the applied high pressures in our experiment this boundary layer is less than 10 micrometers thick.

Figure 1 displays the expected $\cos(\theta)$ angular dependence [30] for the sonic nozzle resulting in a wide beam with a rapid density fall-off and rapid cooling with distance (Mach number in Figure 2). The other two shaped nozzles show a slower expansion rate as the gas jet is confined by the nozzles walls.

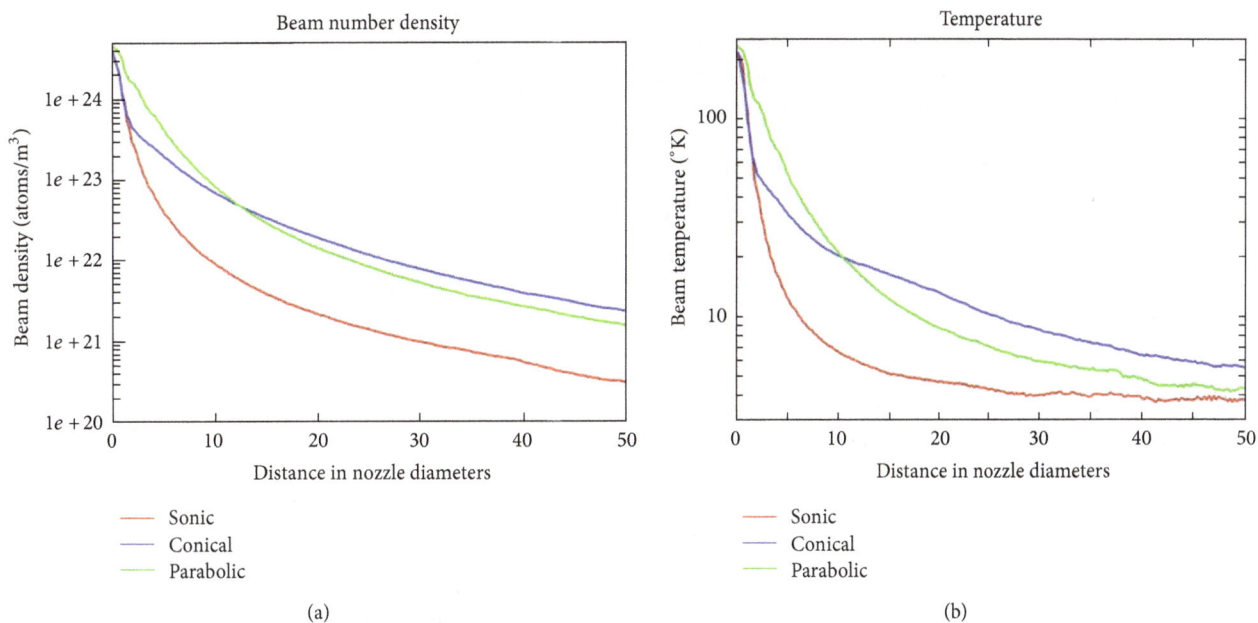

FIGURE 3: Number density and temperature as a function of the distance from the nozzle.

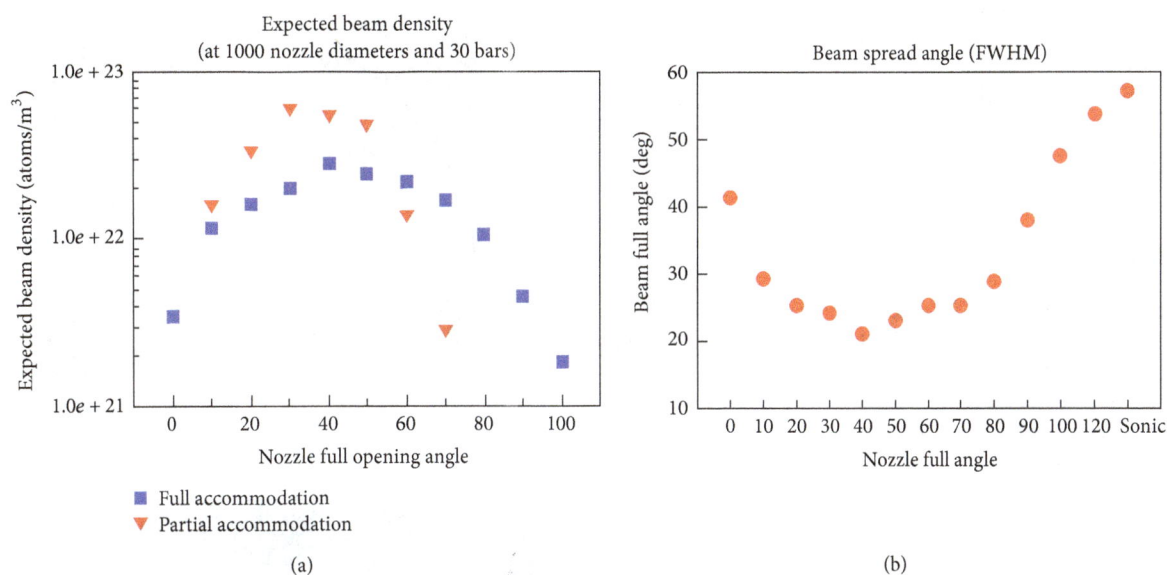

FIGURE 4: Beam density and angular spread (FWHM) as a function of the conical nozzle opening angle. Two accommodation models are shown in (a) (energy accommodation is 1 or 0.4); we calculated the expected beam density at a distance of 500 nozzle diameters (100 mm.) and a stagnation pressure of 30 bars. Admittedly this is a wild extrapolation of the simulated results which were performed at 0.3 bars only.

Figure 3 depicts the fall in number density and temperature with the distance from the nozzle. Both cone nozzle and parabolic nozzle show a slower expansion rate resulting in higher beam number density (i.e., beam intensity) at a given distance from the nozzle. This slower expansion results in a many more collisions before the beam density becomes too low (the so called sudden freeze model [5, 31–36]), enabling more efficient cooling due to two-body collisions, as well as encouraging cluster growth via three-body collisions. One can conclude from these simulations

that a conical or parabolic nozzle is much better than a sonic nozzle for obtaining high on-axis beam intensities. Either design provides an improvement of about a factor of five in beam intensity compared to the sonic nozzle. The parabolic nozzle is advantageous for the growth of weakly bound clusters because it avoids the prolonged period in which the expanding gas is held at 20–30 K that is found with the conical nozzle design.

Figure 4 shows the simulated on-axis beam atomic number density as a function of the opening angle of the cone nozzle. As can be seen, 40°–50° cones represent an optimum choice of opening angle. This is intuitively reasonable, as there is no difference between the limiting case of a low opening angle approaching 0° and a large opening angle approaching 180°.

We can compare these simulation results with experimental measurement of beam intensity and angular dependence in Figure 5.

These simulation results are compared to measurements of actual beam intensity as a function of angle in Figure 5, where results for sonic and 40° conical nozzles are presented for the same expansion conditions. The beam intensity was measured with a fast ion gauge as the nozzle was rotated. The conical nozzle produced an on-axis beam intensity that was more than a factor of 10 greater than the sonic nozzle. The greater increase in on-axis beam intensity compared to simulation results is probably due to the higher pressure used in the experiment, as compared to the simulation, which resulted in a thinner boundary layer near the nozzle wall.

Verification of the production of narrow, collimated beams from a shaped nozzle is demonstrated visually in Figure 6. Here a neon beam, excited by an electric discharge inside the conical nozzle [37], is photographed. The emerging excited neon atoms emit in the red, showing that the expansion diverges at an angle of approximately 10° (FWHM). The angular distribution of the emitting neon atoms is consistent with the measured angular distribution presented in Figure 5 and is a bit narrower than the simulated beam density of Figure 1.

3. Valve Design

The design of the pulsed valve is described below. It is actuated by a short (~30 μs), high current (~20 A), and low voltage (~30 V) pulse. This creates a strong pulsed magnetic field (~3 Tesla) that imparts momentum to the magnetic alloy stainless steel plunger. The coil design was simulated by solving the Maxwell equations using the properties of the actual materials. The resulting spatial distribution of the magnetic field is displayed in Figure 7.

The magnetic field was also measured using a small Faraday rotation crystal [38, 39] inserted inside the coil. The magnetic fields generated by the coil interact with the plunger to create an actuating force of about 10 N. The air gap and therefore the free movement of the plunger are limited to 50 μm only. The materials placed in the coil center are subjected to high fields and can easily reach deep saturation. In fact not enough data can be found in the literature to accurately predict material behavior under these extreme fields. However, the simulation does predict the measured response time quite accurately (opening time of ~20 μs.).

The assembled valve designed for room temperature operation is shown in Figure 9. Some modifications in the construction materials allow the valve to operate without adjustments from 10 K to 500 K.

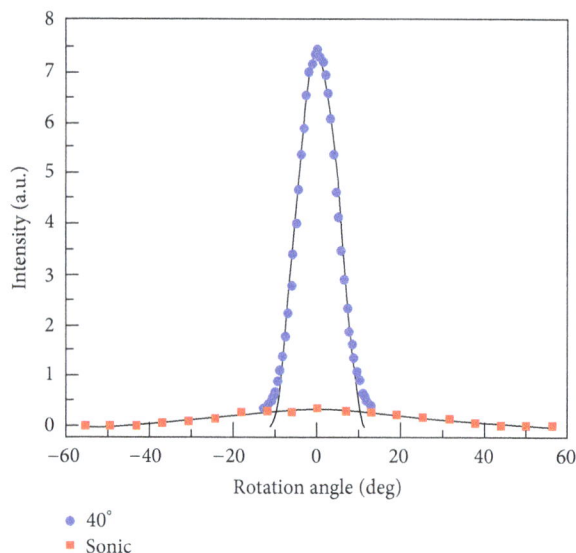

FIGURE 5: Experimental measurement of angular distribution of molecular beam intensity for sonic and 40° conical nozzles. Experimental conditions: 100 bars helium at 300 K. Nozzle length is 10 nozzle diameters. Beam intensity was measured by a fast ionization gauge [18].

FIGURE 6: Low divergence of an excited neon beam from a conical nozzle.

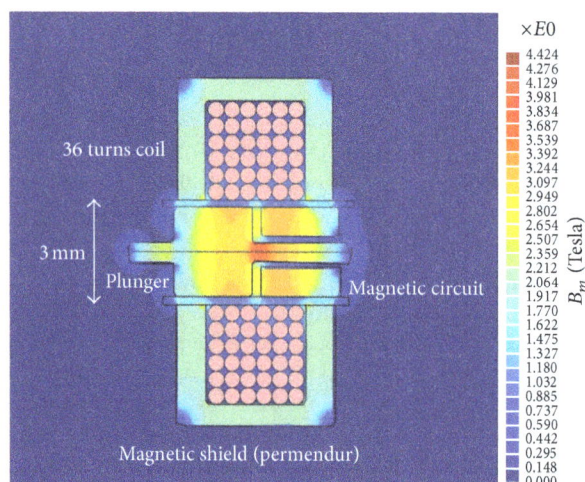

FIGURE 7: Coil and plunger forming the actuating magnetic circuit [14].

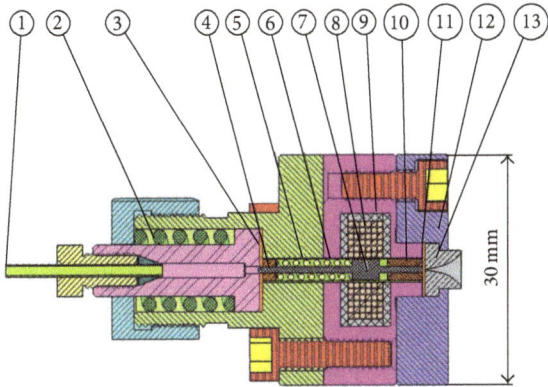

FIGURE 8: Assembled room temperature Even-Lavie valve [14].

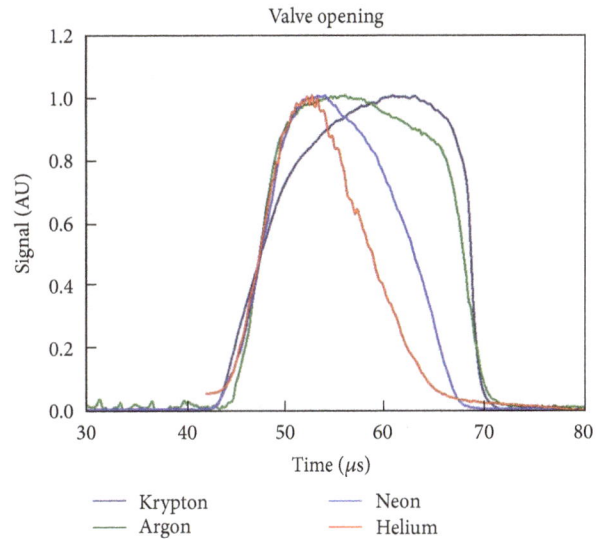

FIGURE 9: Gas pulse duration at the nozzle exit.

From Figure 8 we have the following pulsed valve components:

(1) stainless gas inlet tube (1/16″) gas feed;

(2) tightening spring (180 N) and pressure relief valve;

(3) Kapton foil gasket seals (rear, 0.125 mm. thick);

(4) ceramic (Zirconia or Ruby) rear guiding precision ferrule;

(5) return spring (stainless alloy);

(6) thin walled pressure vessel (Inconel or Zirconia ceramic);

(7) reciprocating plunger (magnetic stainless steel alloy);

(8) Kapton insulated copper coil;

(9) permendur magnetic shield and field concentrator;

(10) ceramic (Zirconia or Ruby) front guiding precision ferrule;

(11) Kapton foil gasket seal (front, 0.125 mm. thick);

(12) front flange and valve body (copper or stainless);

(13) conical (or parabolic) shape expansion nozzle (Zirconia ceramic or hardened stainless steel).

Some basic operating specifications of the valve are given below:

(1) spring pressure that maintains sealing under full temperature range 10–500 K;

(2) pressure range 0–100 bars;

(3) corrosion resistant materials (stainless steel, ceramics, and Kapton);

(4) generated gas pulse width 20–30 μs;

(5) repetition rate 0–600 Hz;

(6) energy consumption 6 mJ per pulse.

4. Beam Properties Generated by the Even-Lavie Valve

The time response, intensity, and the speed ratio of the supersonic jets, generated by this valve, were measured [40, 41]. The operating gases and stagnation conditions (pressure and temperature) were varied.

Time in Figure 9 is measured from the initiating current pulse rise. The gas pulse duration is not determined solely by the excitation current pulse duration. The plunger movement is determined by the initial momentum gained from the magnetic field pulse and the time to recoil from the end of its travel gap. There is a built in delay of ~45 μs, before the gas appears at the nozzle exit. Driving the valve with higher currents (or larger current pulse width) can eventually cause the plunger to bounce more than once, creating multiple pulses of gas. Note the effect of gas viscosity (and its molecular velocity) on the gas pulse duration and its delay. The heavier gases appear later and have longer pulse duration.

Once the gas pulse is formed (with a pulse duration of 20 μs corresponding to an axial length of 30 mm for helium at 300 K), it propagates unimpeded if the background pressure is sufficiently low. At pressures of 10–4 mbar, however, there is already severe beam attenuation over a distance of several cm. To experimentally investigate the gas pulse, we have monitored it using a fast ion gauge. One must remember that at the beginning and end of the plunger movement the valve is not fully opened, so the beam properties (such as pressure and speed) change during the gas pulse. To measure the properties of the fully developed expansion more accurately, it would be better to select only the central part of the pulse (say, at the peak opening of the valve) and measure the properties of the expanding gas in that part of the pulse. One way that this could be accomplished would be to use a short duration pulsed and focused electron beam, fired at the appropriate time to excite a narrow group of atoms in the central portion of the pulse. A fraction of the atoms will be

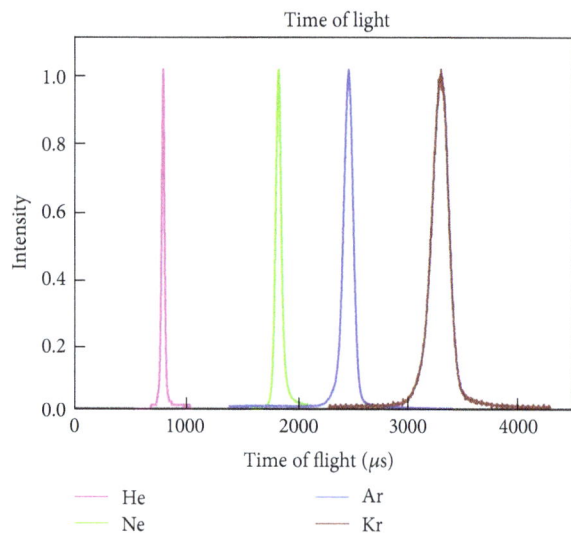

FIGURE 10: Time of flight for metastable atoms (1.4 m flight tube). Stagnation temperature is 300 K and pressure is 50 bars.

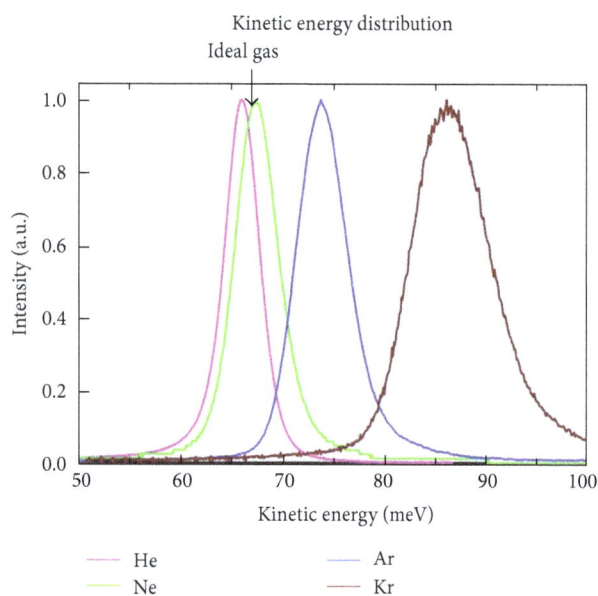

FIGURE 11: Kinetic energy distribution in the beam. Stagnation pressure and temperature as in Figure 10.

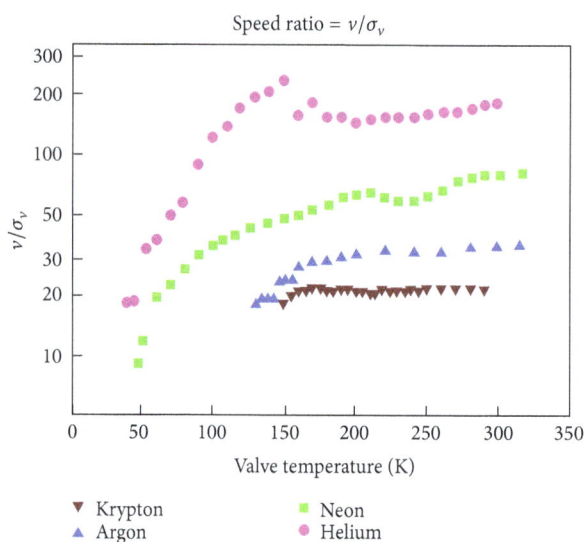

(a)

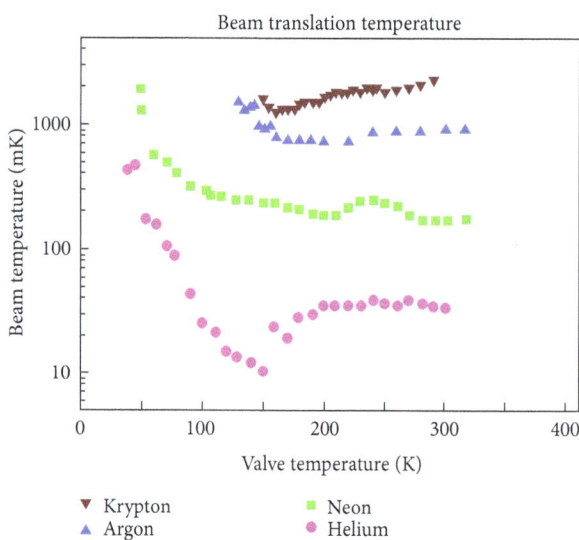

(b)

FIGURE 12: Speed ratio and attained temperature for various gases. Stagnation pressure is as in Figure 10. Errors in the measurements are 10%.

excited to neutral, metastable excited states (usually triplet states) that have long radiative lifetimes [41]. However, here we present results that were obtained by creating metastable atoms using a soft discharge mechanism inside the nozzle itself [37, 40]. This is a more efficient mechanism that creates large numbers of excited metastable atoms in the beam. The metastable atoms created by this mechanism contain sufficient electronic energy (10–20 eV) to allow them to be readily detected on a microchannel plate electron multiplier.

Figure 10 displays the pulse arrival time at a distance of 1 m downstream from the nozzle for expansions of the various rare gases. From these distributions, the average beam

velocity, v, and its standard deviation, σ_v, may be calculated. From these parameters, we can calculate the speed ratio, S:

$$S = \frac{v}{\sigma_v}. \qquad (1)$$

The speed ratio allows us to calculate the translational temperature assuming an ideal gas adiabatic expansion. It is evident from Figure 10 that helium produces the coldest beam and that the temperature deteriorates as the mass of the rare gas is increased. It is illuminating to compare the experimental results with those calculated for an ideal gas expansion. This is displayed in Figure 11, where we have replotted the data of Figure 10 after converting the time-of-flight distributions into kinetic energy distributions.

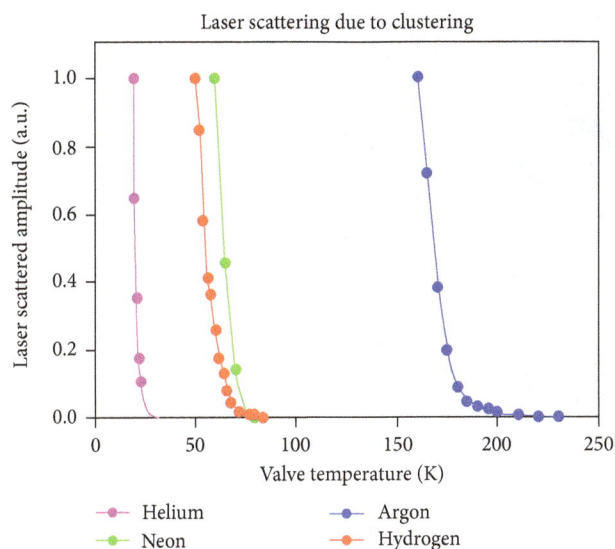

FIGURE 13: Scattered laser light, indicating extensive cluster growth below specific temperatures for each expansion gas.

(a)

(b)

FIGURE 14: Measured rotational contour of the band origin in aniline and tetracene. Comparing the rotational contours with calculation for aniline we can estimate the rotation temperature as indicated.

As can be seen in Figure 11, only helium and neon give kinetic energy distributions that are close to the calculated value for a monatomic ideal gas. Argon and even more so krypton deviate significantly from the results expected for an ideal monatomic gas. This is undoubtedly due to the formation of rare gas clusters for these more highly polarizable gases. It is the formation of clusters during the expansion, along with the consequent release of the heat of condensation that limits the cooling of the expanding jet [18]. This phenomenon can be better seen if we plot the velocity ratio, S, or the corresponding translational temperature, as the temperature of the nozzle is lowered (Figure 12).

Figure 12 indicates that a high speed ratio (and a low beam temperature) can be attained readily for helium and neon, with progressively worsening conditions for argon and krypton, due to their tendency to form clusters because of their higher polarizability. Extensive clustering in the molecular beam can be detected by the enhanced light scattering that occurs by clusters, as opposed to rare gas atoms. To investigate this effect, we measured the scattered light that is detected when the focused 4th harmonic radiation of a Nd-YAG laser (266 nm, 2 ns pulse duration) is used to irradiate the expanding gas near the nozzle. The experiment was conducted using He, Ne, H_2, and Ar as the expansion gas, as a function of the nozzle temperature. As displayed in Figure 13, it was found that for each gas there is a critical nozzle temperature below which clusters readily form.

We have shown that low temperatures (as low as 10 mK) can be achieved for pure helium expansions and 200 mK for neon, the question now is what limits can be achieved when we seed the beam with heavier molecules at low concentration? Obviously the limiting temperature in this case will be higher, mainly because even helium can condense on the polarizable molecules and raise the temperature by releasing the heat of condensation [19, 42]. Using LIF (Figure 14) for known rotational constants of some molecules

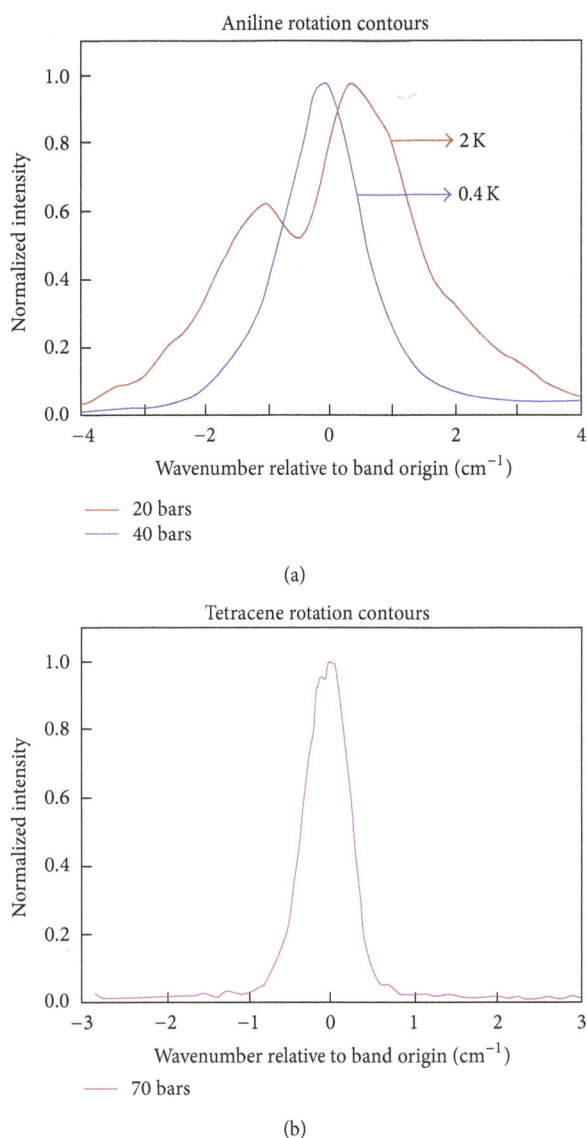

we can measure the rotational temperature quite accurately [19] and it does follow the translation temperatures closely [18]. Similar conclusions can be derived from laser alignment experiments [43–45] on seeded beams. Using the data from these experiments we conclude that temperatures as low as 0.4–0.6 K can be achieved, even for large molecules (containing 10–50 atoms), using helium or neon high pressure sources.

5. Skimmer Design

Because high pressure shaped nozzles can produce high on-axis beam intensities, it is not surprising that modifications of the accepted skimmer design are required. Since the early days of low intensity continuous beams, skimmers have been

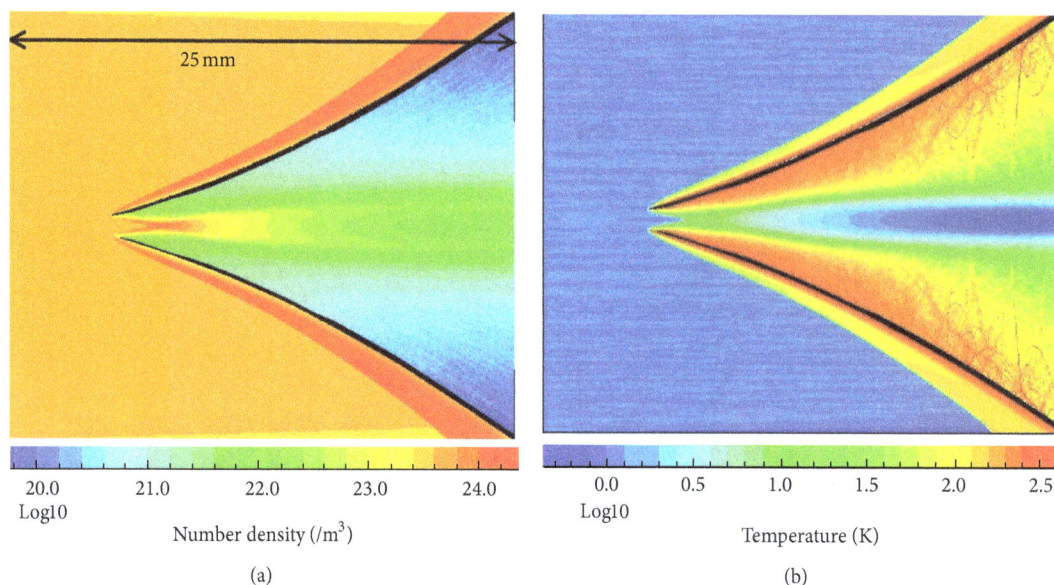

FIGURE 15: Density and temperature maps for a clogged skimmer flow at relatively high beam number density of 10^{23} atoms/m^3.

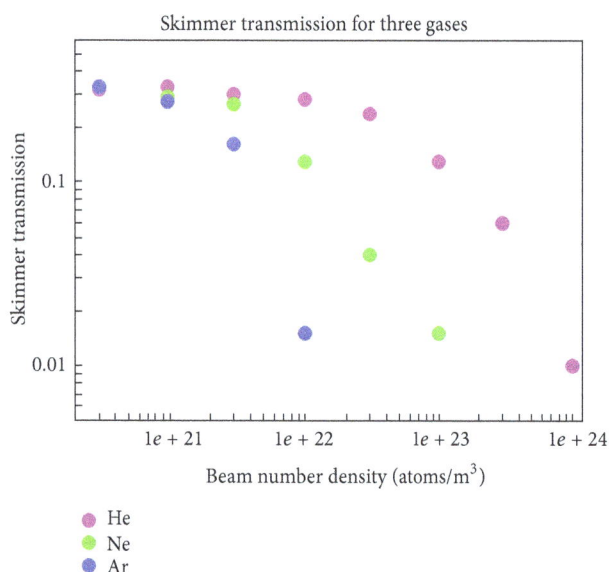

FIGURE 16: Skimmer transmission as a function of incoming beam gas density, for the skimmer depicted in Figure 15.

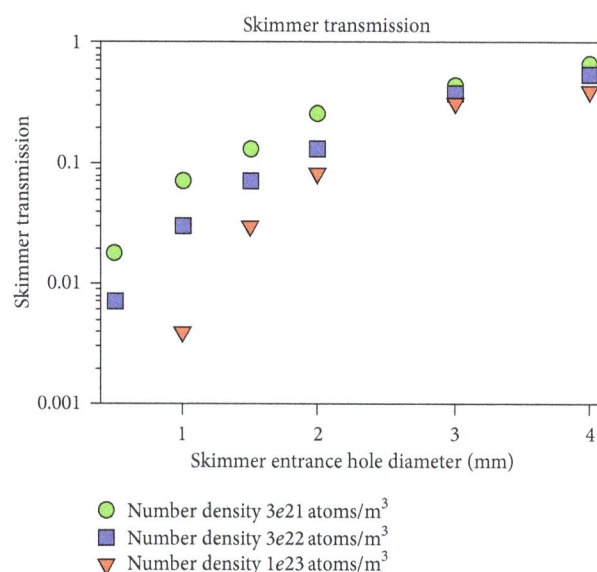

FIGURE 17: Skimmer transmission as a function of skimmer aperture (diameter) and beam density [40]. Simulations are for argon, where full energy accommodation is assumed.

recognized as more than a simple bystander that extracts the central portion of the beam. The effects of molecules scattering from the finite skimmer edges have long been recognized [22, 24, 31, 46–48], and the reduced transmission of the beam through the skimmer was termed "skimmer interference." At times the term "skimmer clogging" was used to describe the dramatic reduction in skimmer transmission. To study these effects, we employed the same DSMC gas flow simulation program [25] which is eminently suitable for the gas density encountered at the skimmer. Here we show how the skimmer distance, skimmer opening, and skimmer edge

sharpness influence the transmission of the beam through the skimmer.

We begin by examining a popular skimmer shape that was used extensively in low intensity continuous beams [49]. The skimmer opening is 1 mm. and its length is 22 mm. The included full angle at its tip is 25° and at its base 70°. The tip sharpness is determined by the simulation mesh size and is ~5 mm. The simulation represents the flow of a collimated and cold stream of neon at 1 K moving at 750 m/s to the right. We chose the neon beam number density ($3 * 10^{23}$ atoms/m^3) to represent the density of the gas from a

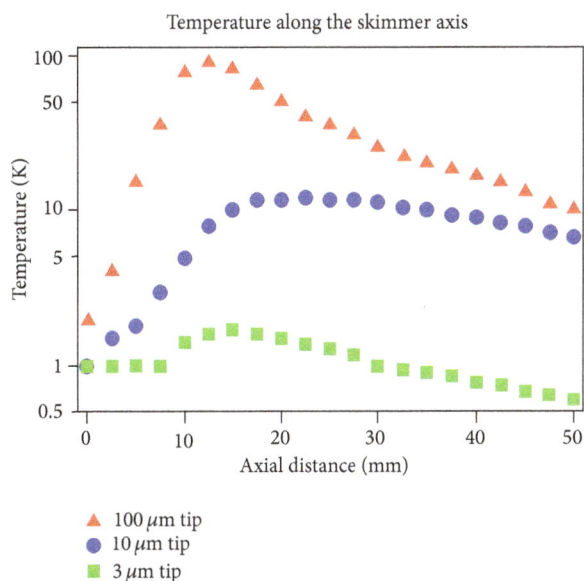

FIGURE 18: Temperature rise in a blunt skimmer [40]. The flowing gas is neon (energy accommodation coefficient 0.8) at 1 K and a number density of 10^{22} atoms/m^3. Skimmer entrance aperture is 1 mm. The temperature soak, lasting only few mS can destroy fragile clusters in the beam.

0.2 mm. conical nozzle at a stagnation pressure of 30 bars and at a distance of 10 mm from the nozzle (extrapolated from Figure 3). Figure 15 represents a clogged skimmer condition with a high density gas plug formed at the skimmer throat, serving as a secondary expansion point from this higher temperature plug (25 K). The transmitted beam intensity at the skimmer base is less than 2% of the incoming beam number density.

Skimmer clogging and therefore low beam transmission can occur quite suddenly as the gas beam density rises. This effect is illustrated in Figure 16, where skimmer transmission as a function of beam gas density is plotted for the skimmer described above and illustrated in Figure 16. In Figure 17, skimmer transmission is defined as the ratio of the gas density at the skimmer outlet to the density at the skimmer inlet. The simulation results are sensitive to the interaction parameters of the gas with the skimmer surface (energy accommodation occurring in the gas surface collision). The accommodation coefficient rises as the mass of the gas atoms approaches that of the surface atoms (nickel). We chose accommodation coefficients of 0.4 for He, 0.8 for Ne, and 1 (complete accommodation) for Ar [29].

A surprising result is that in order to allow for a reasonable beam transmission one has to move the skimmer to a larger distance from the nozzle, as large as 150 mm. (or 750 nozzle diameters!) for Ne or 55 mm. for helium. Lower beam transmission is found for Ar.

The next issue we address is the how does the skimmer entrance hole affect its transmission. The simulations show that smaller hole skimmer tends to clog up at lower beam densities than a larger entrance hole skimmers.

To achieve the best transmission results, it is necessary to employ a large diameter entrance aperture (2–4 mm) and to move the skimmer by as much as 1000 nozzle diameters downstream from the nozzle (100–200 mm) to avoid strong beam attenuation in the skimmer. This is illustrated in Figure 17, where the skimmer transmission is plotted as a function of skimmer entrance diameter for conical skimmers (20° full opening angle) subjected to gas densities ranging from 3×10^{21} to 1×10^{23} atoms/m^3. The necessity of using narrow conical skimmers with large diameter entrance apertures, positioned far downstream, is a very unorthodox result that will require the redesign of source chambers for a large number of currently employed high intensity supersonic sources.

The last question we wish to address is how sharp does the skimmer edge need to be? We can simulate the temperature rise of the cold beam due to molecules scattering from the room temperature skimmer edges. The results are shown in Figure 18. A long conical skimmer was used and the axial temperature along its axis is shown. Beam number density is 10^{22} of helium. A wall thickness of 3 μm causes little beam heating in the skimmer. A 10 μm tip thickness is already causing significant beam heating, and a 100 μm wall is simply bad. Sharply honed or electropolished edges can be machined on the skimmer tip to an edge that is smaller than 3 microns.

6. Summary and Conclusions

We have tested by simulation and experiments the flow characteristic of several nozzle shapes. A shaped nozzle can increase dramatically the on-axis beam number density at a given distance from the nozzle, while achieving high speed ratio. Our simulations do not include a clustering mechanism; thus we attribute the limited cooling achieved in experiments to cluster formation in the expanding jet. This is corroborated by following the cluster formation in the beam by laser light scattering. The trends in cluster formation below a certain source temperature (for each specific gas) support this assumption in general, but we could not follow this quantitatively. We tested the behavior of a popular skimmer shape to the increased gas load achieved in shaped nozzle sources. A clearer picture emerges as to what is happening when skimmer clogging occurs (i.e., skimmer interference effects). We show that skimmer interference is gas dependent (accommodation coefficient that is specific for each gas) and is also a function of the skimmer aperture and its edge sharpness. When using high pressure sources with a shaped nozzle the skimmer has to be placed some 500–1000 nozzle diameters away from the source. The required modifications in vacuum system design may prevent some common errors when replacing the source type to a high pressure source with a shaped nozzle.

Conflict of Interests

The author declares that there is no conflict of interests regarding the publication of this paper.

References

[1] R. C. Campargue, *Atomic and Molecular Beams: The State of the Art 2000*, Springer, Berlin, Germany, 2001.

[2] G. Scoles, *Atomic and Molecular Beam Methods*, Oxford University Press, Oxford, UK, 1989-1992.

[3] H. Pauly, *Atom, Molecule, and Cluster Beams I*, Springer, Berlin, Germany, 2000.

[4] H. Haberland, *Clusters of Atoms and Molecules*, Springer, Berlin, Germany, 1994.

[5] R. Campargue, A. Lebehot, and J. C. Lemonnier, in *Rarefied Gas Dynamics*, pp. 1033–1045, 1977.

[6] A. Amirav, U. Even, and J. Jortner, "Absorption spectroscopy of ultracold large molecules in planar supersonic expansions," *Chemical Physics Letters*, vol. 83, no. 1, pp. 1–4, 1981.

[7] W. R. Gentry and C. F. Giese, "Ten–microsecond pulsed molecular beam source and a fast ionization detector," *Review of Scientific Instruments*, vol. 49, p. 595, 1978.

[8] B. Yan, P. F. H. Claus, B. G. M. van Oorschot et al., "A new high intensity and short-pulse molecular beam valve," *Review of Scientific Instruments*, vol. 84, no. 2, Article ID 023102, 2013.

[9] M. D. Barry, N. P. Johnson, and P. A. Gorry, "A fast ($30\mu s$) pulsed supersonic nozzle beam source: application to the photodissociation of CS_2 at 193 nm," *Journal of Physics E: Scientific Instruments*, vol. 19, no. 10, pp. 815–819, 1986.

[10] D. Irimia, D. Dobrikov, R. Kortekaas et al., "A short pulse ($7 \mu s$ FWHM) and high repetition rate (dc-5kHz) cantilever piezovalve for pulsed atomic and molecular beams," *Review of Scientific Instruments*, vol. 80, no. 11, Article ID 113303, 2009.

[11] D. L. Proctor, D. R. Albert, and H. F. Davis, "Improved piezoelectric actuators for use in high-speed pulsed valves," *Review of Scientific Instruments*, vol. 81, Article ID 023106, 2010.

[12] M. R. Adriaens, W. Allison, and B. Feuerbacher, "A pulsed molecular beam source," *Journal of Physics E: Scientific Instruments*, vol. 14, no. 12, article no. 006, pp. 1375–1377, 1981.

[13] D. Bahat, O. Cheshnovsky, U. Even, N. Lavie, and Y. Magen, "Generation and detection of intense cluster beams," *Journal of Physical Chemistry*, vol. 91, no. 10, pp. 2460–2462, 1987.

[14] D. Pentlehner, R. Riechers, B. Dick et al., "Rapidly pulsed helium droplet source," *Review of Scientific Instruments*, vol. 80, no. 4, Article ID 043302, 2009.

[15] E. Hendell and U. Even, "Cluster-surface interaction at high kinetic energy. I. Electron emission," *The Journal of Chemical Physics*, vol. 103, no. 20, pp. 9045–9052, 1995.

[16] O. F. Hagena, "Pulsed valve for supersonic nozzle experiments at cryogenic temperatures," *Review of Scientific Instruments*, vol. 62, pp. 2038–2039, 1991.

[17] C. E. Otis and P. M. Johnson, "A simple pulsed valve for use in supersonic nozzle experiments," *Review of Scientific Instruments*, vol. 51, p. 1128, 1980.

[18] M. Hillenkamp, S. Keinan, and U. Even, "Condensation limited cooling in supersonic expansions," *Journal of Chemical Physics*, vol. 118, no. 19, pp. 8699–8705, 2003.

[19] U. Even, J. Jortner, D. Noy, N. Lavie, and C. Cossart-Magos, "Cooling of large molecules below 1 K and He clusters formation," *Journal of Chemical Physics*, vol. 112, no. 18, pp. 8068–8071, 2000.

[20] W. Christen, "Stationary flow conditions in pulsed supersonic beams," *The Journal of Chemical Physics*, vol. 139, Article ID 154202, 2013.

[21] G. A. Bird, *Molecular Gas Dynamics and the Direct Simulation of Gas Flows*, Oxford University Press, Oxford, UK, 1994.

[22] W. R. Gentry and C. F. Giese, "High-precision skimmers for supersonic molecular beams," *Review of Scientific Instruments*, vol. 46, no. 1, p. 104, 1975.

[23] G. A. Bird, "Transition regime behavior of supersonic beam skimmers," *Physics of Fluids*, vol. 19, no. 10, pp. 1486–1491, 1976.

[24] J. Braun, P. K. Day, J. P. Toennies, G. Witte, and E. Neher, "Micrometer-sized nozzles and skimmers for the production of supersonic He atom beams," *Review of Scientific Instruments*, vol. 68, no. 8, pp. 3001–3009, 1997.

[25] G. Bird, http://www.gab.com.au/.

[26] K. J. Daun, G. J. Smallwood, and F. Liu, "Molecular dynamics simulations of translational thermal accommodation coefficients for time-resolved LII," *Applied Physics B*, vol. 94, no. 1, pp. 39–49, 2009.

[27] W. F. N. Santos, "Gas-surface interaction effect on round leading edge aerothermodynamics," *Brazilian Journal of Physics*, vol. 37, no. 2, pp. 337–348, 2007.

[28] J. F. Padilla and I. D. Boyd, "Assessment of gas-surface interaction models for computation of rarefied hypersonic flow," *Journal of Thermophysics and Heat Transfer*, vol. 23, no. 1, pp. 96–105, 2009.

[29] D. J. Rader, W. M. Trott, J. R. Torczynski, J. N. Castañeda, and T. W. Grasser, Sandia National Laboratories SAND2005-6084, 2005, http://prod.sandia.gov/techlib/access-control.cgi/2005/056084.pdf.

[30] H. Ashkenas and S. F. Sherman, *Proceedings of the 4th International Symposium on Rarefied Gas Dynamics: Experimental Methods in Rarefied Gas Dynamics*, Academic Press, New York, NY, USA, 1966.

[31] R. Campargue, "Characteristics of supersonic beams applicable in collision or semi-collision experiments," *Journal de Chimie Physique et de Physico-Chimie Biologique*, vol. 77, no. 3, p. R15, 1980.

[32] R. Campargue, "Progress in overexpanded supersonic jets and skimmed molecular beams in free-jet zones of silence," *The Journal of Physical Chemistry*, vol. 88, no. 20, pp. 4466–4414, 1984.

[33] S. Semushin and V. Malka, "High density gas jet nozzle design for laser target production," *Review of Scientific Instruments*, vol. 72, no. 7, pp. 2961–2965, 2001.

[34] H. V. Tafreshi, G. Benedek, P. Piseri, S. Vinati, E. Barborini, and P. Milani, "A simple nozzle configuration for the production of low divergence supersonic cluster beam by aerodynamic focusing," *Aerosol Science and Technology*, vol. 36, no. 5, pp. 593–606, 2002.

[35] J. P. Toennies and K. Winkelman, "Theoretical studies of highly expanded free jets: Influence of quantum effects and a realistic intermolecular potential," *The Journal of Chemical Physics*, vol. 66, no. 9, p. 3965, 1977.

[36] J. T. McDaniels, R. E. Continetti, and D. R. Miller, "The effect of nozzle geometry on cluster formation in molecular beam sources," in *Proceedings of the 23rd international Symposium on Rarefied Gas Dynamics*, A. D. Ketsdever, Ed., American Institute of Physics, 2003.

[37] K. Luria, N. Lavie, and U. Even, "Dielectric barrier discharge source for supersonic beams," *Review of Scientific Instruments*, vol. 80, Article ID 104102, 2009.

[38] E. Narevicius, C. G. Parthey, A. Libson et al., "An atomic coilgun: Using pulsed magnetic fields to slow a supersonic beam," *New Journal of Physics*, vol. 9, article 358, 2007.

[39] E. Narevicius, A. Libson, M. F. Riedel et al., "Coherent slowing of a supersonic beam with an atomic paddle," *Physical Review Letters*, vol. 98, no. 10, Article ID 103201, 2007.

[40] K. Luria, W. Christen, and U. Even, "Generation and propagation of intense supersonic beams," *Journal of Physical Chemistry A*, vol. 115, no. 25, pp. 7362–7367, 2011.

[41] W. Christen, "Stationary flow conditions in pulsed supersonic beams," *The Journal of Chemical Physics*, vol. 139, no. 15, Article ID 154202, 2013.

[42] U. Even, I. Al-Hroub, and J. Jortner, "Small he clusters with aromatic molecules," *Journal of Chemical Physics*, vol. 115, no. 5, pp. 2069–2073, 2001.

[43] H. Stapelfeldt, "Laser aligned molecules: applications in physics and chemistry," *Physica Scripta*, vol. T110, p. 132, 2004.

[44] H. Stapelfeldt and T. Seideman, "Colloquium: Aligning molecules with strong laser pulses," *Reviews of Modern Physics*, vol. 75, no. 2, pp. 543–557, 2003.

[45] I. Nevo, L. Holmegaard, J. H. Nielsen et al., "Laser-induced 3D alignment and orientation of quantum state-selected molecules," *Physical Chemistry Chemical Physics*, vol. 11, no. 42, pp. 9912–9918, 2009.

[46] N. F. Ramsey, *Molecular Beams*, Oxford University Press, Oxford, UK, 1956.

[47] A. B. Bailey, R. Dawbarn, and M. R. Busby, "Effects of skimmer and endwall temperature of condensed molecular beams," *AIAA Journal*, vol. 14, no. 1, pp. 91–92, 1976.

[48] D. C. Jordan, R. Barling, and R. B. Doak, "Refractory graphite skimmers for supersonic free-jet, supersonic arc-jet, and plasma discharge applications," *Review of Scientific Instruments*, vol. 70, no. 3, pp. 1640–1648, 1999.

[49] Beam-Dynamics, http://www.beamdynamicsinc.com/skimmer_specs.htm.

7

A Simple and Advantageous Synthesis of the Privileged 1,4-Benzodiazepine Nucleus

Neetu Jain and Dharma Kishore

Department of Chemistry, Banasthali University, Banasthali, Rajasthan 304022, India

Correspondence should be addressed to Neetu Jain; to.neetu@yahoo.com

Academic Editor: Hideto Miyabe

A novel domino approach has been described for an easy access of the privileged nucleus of 5-carbomethoxy substituted 1,4-benzodiazepin-2-ones 4(a–i) from an in situ methanolic hydrolysis of an incipient species formed from the interaction of 1-chloroacetylisatin 2(a–i), hexamethyldisilazane, and n-butyl lithium. The reaction is believed to take place through a consecutive series of intramolecular reactions in a cascade to first generate a highly reactive carbene intermediate 3(a–i) from 1-chloroacetylisatin and n-butyl lithium which is simultaneously trapped by hexamethyldisilazane before undergoing its in situ hydrolysis with methanol to initiate its concomitant cyclocondensation to produce 4(a–i) in high yield and purity.

1. Introduction

Exploration of synthetic processes that lead to the development of small molecules of medicinal interest by telescoping the multicomponent operations into a single step or resorting to a process such as domino reactions is a rapidly emerging subject in medicinal chemistry. Ever since Koch et al. [1] carried out a quantitative analysis of physiologically active natural product scaffolds and showed that ones with two or three rings were most often found in bioactive natural products, the interest on various facets of the chemistry of small molecules has expanded exponentially thereafter. Benzodiazepines and their analogues have been recognized recently to belong to the class of privileged heterocyclic structures, [2–5] by virtue of their ability to form ligands to a number of functionally and structurally discrete biological receptors [6–13]. This property stimulated chemists to utilize their potential in the design and development of molecular probes for biological evaluations. Ubiquitous presence of this nucleus in the psychopharmacologically active agents and in molecules active against HIV infection, for example, TIBO (1) [14–17] and FDA approved dipyrido diazepine analogue nevirapine (2) [18–24] in Figure 1 provided an impetus for an enormous research effort to be directed towards the development of their structural analogues of medicinal importance [25, 26].

2. Results and Discussion

This communication reports the application of a novel domino process for an easy access of the privileged nucleus of 5-carbomethoxy substituted 1,4-benzodiazepin-2-ones 4(a–i) from the ring expansion of 1-chloroacetylisatins 2(a–i), initiated by hexamethyldisilazane under the influence of n-butyl lithium. N-Butyl lithium formed an obvious choice since it has been used as a catalyst in amination of active alkyl halides with hexamethyldisilazane. Amination of 1-chloroacetylisatin (2) formed the key step in allowing it to undergo ring expansion to give 4. Ogata and Matsumoto's [27] original procedure which employed Delepine reaction in the amination of 2 with methanolic solution of hexamine produced 4 in low yield. This called out to revisit this reaction to augment the scope of this reaction in view of the yield of this product.

It is believed that Delepine reaction proceeded with the formation of the hexaminium ion. We surmise that it was the bulky nature of hexamine which hindered its formation

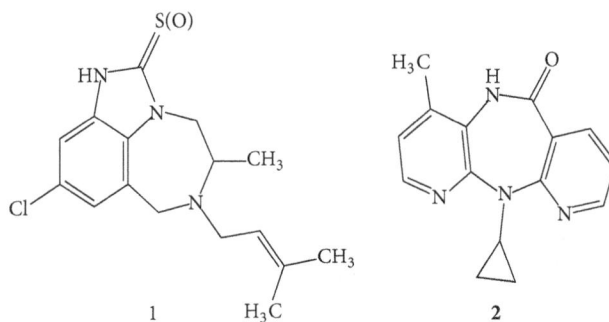

FIGURE 1

from **2**. We assumed that this problem could possibly be circumvented by carrying out the amination of **2** with less bulkier agents. In consideration of the potential of hexamethyldisilazane [28] in amination reaction, in our initial attempt, we replaced hexamine with this reagent. However, contrary to our expectation, **2** resisted its reaction with this reagent and caused it to be recovered unchanged from the reaction mixture. A search for possible use of a catalyst in this reaction revealed that n-butyl lithium has been used in aminations employing hexamethyldisilazane. This provided optimism for this reaction too, to succeed with the use of this catalyst. This expectation turned into a reality in all the runs using a wide variety of substituted 1-chloroacetylisatin derivatives to produce **4** in an exceptionally high yield and purity (Table 1).

We suggest that this reaction proceeds with the base catalysed dehydrochlorination of **2** to generate an acyl carbene intermediate **3** which is subsequently trapped by hexamethyldisilazane. The relatively small size of carbene precludes the steric factor in the reaction with hexamethyldisilazane. The hydrolysis of bis(trimethylsilyl) group from this with methanol sets the stage for it, to undergo ring expansion to give **4**.

In view of the extremely weak nucleophilic character of isatinylamide nitrogen of **3**, the possibility of the rearrangement of the acyl carbene intermediate had to be ruled out. (No rearranged product was, however, traceable from the reaction mixture.)

As the carbene was not likely to be trapped by the tertiary amine (hexamine) its reaction with hexamine was not examined.

3. Experiment

All the melting points were taken in open capillaries and are uncorrected. The purity of all the compounds were checked by TLC using the solvent systems (benzene : methanol, 9 : 1 v/v) and silica gel G as adsorbent. IR spectra were recorded on Shimadzu FTIR-8400 infrared spectrometer using KBr, ^{1}H NMR were recorded on Bruker AC 300F in CDCl$_3$ + DMSO-d$_6$ (2 : 1 v/v, TMS as internal reference and chemical shifts expressed in δ ppm), and mass spectra

were recorded on Jeol-JMS-D-D-300 mass spectrometer. Reagents, 5-fluoro-, 5-chloro-, 5-bromo-, 5-iodo-, 5-methyl-, 5-methoxy-, 5-nitro-, and 5,7-dimethylisatins were procured from commercial sources and used as such in the reaction without further purifications (see Scheme 1).

3.1. General Methods for the Preparation of **2(a–i)** *from* **1(a–i)**. 5-Fluoroisatin (**1b**, 0.068 mol) was vigorously refluxed with chloroacetyl chloride (0.090 mol) for 7 h. and the mixture was cooled overnight at 0–5°C. The crude product which settled was filtered, washed with 20 mL of ether, air-dried, and then recrystallised from ethyl acetate to give **2b**, yield: 88%, m.p. 165-166°C. Other compounds **2(a, c–i)** were prepared from **1(a, c–i)** using this procedure.

3.2. General Methods for the Preparation of **4(a–i)** *from* **2(a–i)**. 5-Fluoro-1-chloroacetylisatin (**2b**, 0.01 mol) was dissolved in dry THF (20 mL) and to this solution n-butyl lithium (0.01 mol) and hexamethyldisilazane (0.01 mol) were added. The reaction mixture was magnetically stirred for 2 h at room temperature. The progress of reaction was checked by TLC. Methanol (20 mL) was added to the reaction mass and the mixture was refluxed for 5 h. It was then poured on crushed ice, filtered, air-dried, and recrystallised from methanol to give **4b**, yield: 87%, m.p. 198–200°C. Other compounds **4(a, c–i)** were prepared from **2(a, c–i)** using the same procedure.

4. Conclusion

In summary, a high yielding n-butyl lithium catalysed one-pot domino process has been developed to the facile access of the privileged nucleus of methyl-1,3-dihydro-2H-1,4-benzodiazepin-2-one-5-carboxylates **4(a–i)** at ambient temperature from the ring expansion of the corresponding 1-chloroacetylisatins **2(a–i)** under the influence of hexamethyldisilazane.

Conflict of Interests

The authors declare that there is no conflict of interests regarding the publication of this paper.

TABLE 1: Physical and spectral data of the compounds **2(a–i)** and **4(a–i)**.

Entry	Molecular formula	M.W.	M.P. (°C)[a]	Yield (%)	IR (KBr) cm^{-1}	^{1}H NMR (CDCl$_3$-DMSO-d$_6$) δ ppm	Elemental analysis C (cald/exp.)	H (cald/exp.)	N (cald/exp.)
2a	$C_{10}H_6ClNO_3$	223.61	210–12	91	1735, 1680, 1675	7.56–7.19 (m, 4H, ArH), 4.32 (s, 2H, CH$_2$)	53.71/53.67	2.70/2.65	6.26/6.18
2b	$C_{10}H_5ClFNO_3$	241.60	165–66	88	1755, 1725, 1700	7.81–7.23 (m, 3H, ArH), 4.33 (s, 2H, CH$_2$)	49.71/49.67	2.49/2.40	5.80/5.69
2c	$C_{10}H_5Cl_2NO_3$	258.06	153–55	94	1785, 1740, 1720	7.81–7.53 (m, 3H, ArH), 4.32 (s, 2H, CH$_2$)	46.54/46.42	1.95/1.90	5.43/531
2d	$C_{10}H_5BrClNO_3$	302.51	178–79	90	1780, 1735, 1710	7.96–7.61 (m, 3H, ArH), 4.32 (s, 2H, CH$_2$)	39.70/39.62	1.67/1.56	4.63/4.59
2e	$C_{10}H_5ClINO_3$	349.51	206–07	89	1770, 1725, 1690	7.80–7.50 (m, 3H, ArH), 4.28 (s, 2H, CH$_2$)	28.95/28.89	1.21/1.17	3.38/3.27
2f	$C_{11}H_8ClNO_3$	237.64	185–86	92	1765, 1720, 1690	7.72–7.03 (m, 3H, ArH), 4.32 (s, 2H, CH$_2$), 2.85 (s, 3H, CH$_3$)	55.60/55.51	3.39/3.28	5.89/5.80
2g	$C_{11}H_8ClNO_4$	253.64	223–25	90	1780, 1740, 1705	7.72–7.03 (m, 3H, ArH), 4.32 (s, 2H, CH$_2$), 3.73 (s, 3H, OCH$_3$)	52.09/51.97	3.18/3.12	5.52/5.42
2h	$C_{10}H_5Cl_2NO_5$	268.61	130–32	91	1785, 1745, 1710	8.72–8.09 (m, 3H, ArH), 4.52 (s, 2H, CH$_2$)	44.71/44.56	1.88/1.71	10.43/10.37
2i	$C_{12}H_{10}ClNO_3$	251.67	168–70	92	1740, 1700, 1685	7.40–7.12 (m, 2H, ArH), 4.32 (s, 2H, CH$_2$), 2.35 (s, 6H, CH$_3$)	57.27/57.19	4.01/3.97	5.57/5.46
4a	$C_{11}H_{10}N_2O_3$	218.21	172–75	85	3190, 1725, 1675	8.0 (s, 1H, NH), 7.98–7.65 (m, 4H, ArH), 4.48 (s, 2H, CH$_2$), 3.65 (s, 3H, OCH$_3$) MS: m/z (%) 218.21 (M+, 100%), 210.20 (75%), 107.16 (35%) 8.2 (s, 1H, NH), 8.08 (d, 1H, ArH), 7.78 (d, 1H, ArH), 1H, ArH),	60.55/60.48	4.62/4.52	12.84/12.79
4b	$C_{11}H_9FN_2O_3$	236.20	198–200	87	3220, 1735, 1690	7.66 (s, 1H, ArH), 4.56 (s, 2H, CH$_2$), 3.78 (s, 3H, OCH$_3$) MS: m/z (%) 236.20 (M+ 75%), 205.17 (100%), 177.16 (35%) 8.1 (s, 1H, NH), 7.90 (s, 1H, ArH), 7.81 (d, 1H, ArH),	55.93/55.88	3.84/3.79	11.86/11.76
4c	$C_{11}H_9ClN_2O_3$	252.65	176–78	84	3215, 1728, 1684	7.60 (d, 1H, ArH), 4.48 (s, 2H, CH$_2$), 3.70 (s, 3H, OCH$_3$) MS: m/z (%) 252.65 (M+, 100%), 254.57 (M+2, 33%) 8.0 (s, 1H, NH), 7.90 (s, 1H, ArH), 7.75 (d, 1H, ArH),	52.29/52.23	3.59/3.51	11.09/11.00
4d	$C_{11}H_9BrN_2O_3$	297.98	243–45	86	3208, 1721, 1675	7.50 (d, 1H, ArH), 4.32 (s, 2H, CH$_2$), 3.12 (s, 3H, OCH$_3$) MS: m/z (%) 297.98 (M+, 100%), 299.90 (M+2, 100%), 177.16 (35%) 8.0 (s, 1H, NH), 7.85 (s, 1H, ArH), 7.66 (d, 1H, ArH),	44.47/44.41	3.05/2.99	9.43/9.36
4e	$C_{11}H_9IN_2O_3$	344.11	187–88	90	3200, 1710, 1670	7.40 (d, 1H, ArH), 4.32 (s, 2H, CH$_2$), 3.12 (s, 3H, OCH$_3$) MS: m/z (%) 344.11 (M+ 75%), 205.17 (100%), 147.16 (35%)	38.39/38.29	2.64/2.59	8.14/8.07

TABLE 1: Continued.

Entry	Molecular formula	M.W.	M.P. (°C)[a]	Yield (%)	IR (KBr) cm⁻¹	¹H NMR (CDCl₃-DMSO-d₆) δ ppm	Elemental analysis C (cald/exp.)	H (cald/exp.)	N (cald/exp.)
4f	$C_{12}H_{12}N_2O_3$	232.24	205–06	85	3188, 1723, 1672	8.0 (s, 1H, NH), 7.84 (s, 1H, ArH), 7.56 (d, 1H, ArH), 7.42 (d, 1H, ArH), 4.32 (s, 2H, CH₂), 3.12 (s, 3H, OCH₃), 2.34 (s, 3H, CH₃), MS: m/z (%) 232.24 (M+ 72%), 177.25 (25%), 149.17 (100%)	62.60/62.52	5.21/5.17	12.06/11.96
4g	$C_{12}H_{12}N_2O_4$	248.08	182–85	87	3202, 1728, 1677	8.0 (s, 1H, NH), 7.88 (d, 1H, ArH), 7.75 (d, 1H, ArH), 7.43 (s, 1H, ArH), 4.33 (s, 2H, CH₂), 3.12 (s, 3H, OCH₃), 3.63 (s, 3H, OCH₃), MS: m/z (%) 248.08 (M+ 68%), 240.11 (100%), 163.45 (36%)	58.06/57.98	4.87/4.79	11.29/11.19
4h	$C_{11}H_9N_3O_5$	263.05	202–05	83	3228, 1735, 1698	8.0 (s, 1H, NH), 8.12 (s, 1H, ArH), 7.98 (d, 1H, ArH), 7.66 (d, 1H, ArH), 4.55 (s, 2H, CH₂), 3.42 (s, 3H, OCH₃), MS: m/z (%) 263.05 (M+ 65%), 201.17 (100%), 170.80 (35%)	50.20/50.13	3.45/3.34	15.96/15.89
4i	$C_{13}H_{14}N_2O_3$	246.26	215–17	88	3185, 1720, 1670	8.0 (s, 1H, NH), 7.40 (s, 2H, ArH), 4.30 (s, 2H, CH₂), 3.10 (s, 3H, OCH₃), 2.34 (s, 3H, for two CH₃), MS: m/z (%) 246.26 (M+ 100%), 145.17 (86%), 104.16 (27%)	63.40/63.31	5.73/5.69	11.38/11.33

[a]All the melting points (M.P.) were found to be identical to the authentic samples prepared according to the procedure reported in the literature [22].
M.W.: Molecular Weight; cald./exp.: calculated/experimental.

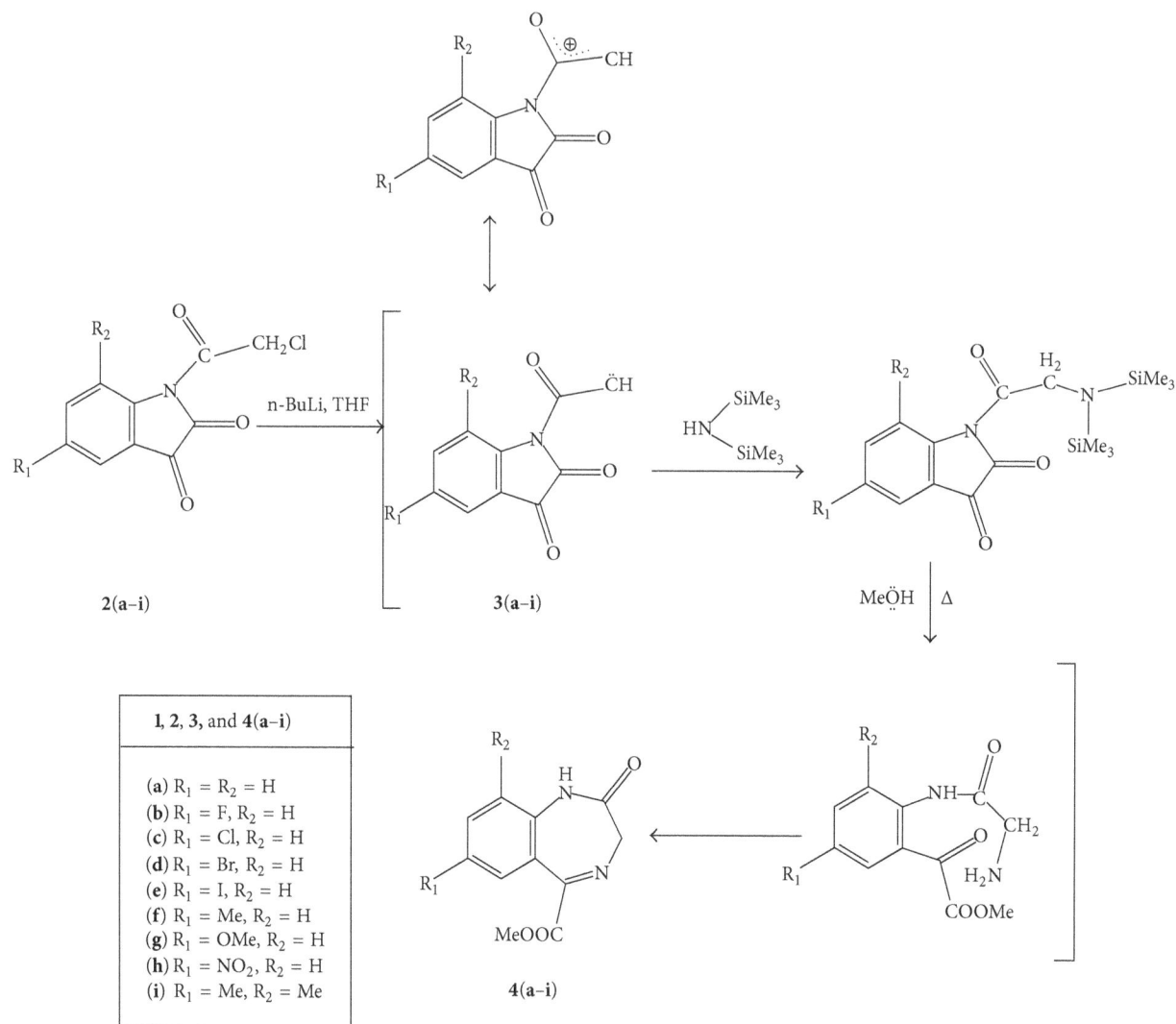

SCHEME 1: Schematic presentation of the synthesis of the compounds.

Acknowledgments

The authors are grateful to the Director of CDRI, Lucknow, India, for providing the spectral data of the compounds and to the Department of Science and Technology, DST, New Delhi, India, for providing the financial assistance to "Banasthali Centre for Education and Research in Basic Sciences," under their CURIE (Consolidation of University Research for Innovation and Excellence in Women Universities) programme.

References

[1] M. A. Koch, A. Schuffenhauer, M. Scheck et al., "Charting biologically relevant chemical space: a structural classification of natural products (SCONP)," *Proceedings of the National Academy of Sciences of the United States of America*, vol. 102, no. 48, pp. 17272–17277, 2005.

[2] W. Nawroca, B. Sztuba, A. Opolski, J. Wietrzk, M. W. Kowalska, and T. Glowiak, "Synthesis and antiproliferative activity in vitro of novel 1,5-benzodiazepines. Part II," *Archive der Pharmazie-Pharmaceutical and Medicinal Chemistry*, vol. 334, no. 1, pp. 3–10, 2000.

[3] B. E. Evans, K. E. Rittle, M. G. Bock et al., "Methods for drug discovery: development of potent, selective, orally effective cholecystokinin antagonists," *Journal of Medicinal Chemistry*, vol. 31, no. 12, pp. 2235–2246, 1988.

[4] R. Yarchoan, H. Mitsuya, R. V. Thomas et al., "In vivo activity against HIV and favorable toxicity profile of 2′,3′-dideoxyinosine," *Science*, vol. 245, no. 4916, pp. 412–415, 1989.

[5] J. R. Lokensgard, C. C. Chao, G. Gekker, S. Hu, and P. K. Peterson, "Benzodiazepines, glia, and HIV-1 neuropathogenesis," *Molecular Neurobiology*, vol. 18, no. 1, pp. 23–33, 1998.

[6] J. Poupaert, P. Carato, E. Colacino, and S. Yous, "2(3H)-benzoxazolone and bioisosters as "privileged scaffold" in the design of pharmacological probes," *Current Medicinal Chemistry*, vol. 12, no. 7, pp. 877–885, 2005.

[7] D. J. Triggle, "1,4-dihydropyridines as calcium channel ligands and privileged structures," *Cellular and Molecular Neurobiology*, vol. 23, no. 3, pp. 293–303, 2003.

[8] R. W. DeSimone, K. S. Currie, S. A. Mitchell, J. W. Darrow, and D. A. Pippin, "Privileged structures: applications in drug discovery," *Combinatorial Chemistry and High Throughput Screening*, vol. 7, no. 5, pp. 473–493, 2004.

[9] A. A. Patchett and R. P. Nargund, "Chapter 26. Privileged structures: an update," *Annual Reports in Medicinal Chemistry*, vol. 35, pp. 289–298, 2000.

[10] J. Yang, Q. Dang, J. Liu, Z. Wei, J. Wu, and X. Bai, "Preparation of a fully substituted purine library," *Journal of Combinatorial Chemistry*, vol. 7, no. 3, pp. 474–482, 2005.

[11] J. Liu, Q. Dang, Z. Wei, H. Zhang, and X. Bai, "Parallel solution-phase synthesis of a 2,6,8,9-tetrasubstituted purine library via a sulfur intermediate," *Journal of Combinatorial Chemistry*, vol. 7, no. 4, pp. 627–636, 2005.

[12] Q. Dang and J. E. Gomez-Galeno, "An efficient synthesis of pyrrolo[2,3-d]pyrimidines via inverse electron demand Diels-Alder reactions of 2-amino-4-cyanopyrroles with 1,3,5-triazines," *Journal of Organic Chemistry*, vol. 67, no. 24, pp. 8703–8705, 2002.

[13] P. Bhuyan, R. C. Boruah, and J. S. Sandhu, "Studies on uracils. 10. A facile one-pot synthesis of pyrido[2,3-d]- and pyrazolo[3,4-d]pyrimidines," *Journal of Organic Chemistry*, vol. 55, no. 2, pp. 568–571, 1990.

[14] R. H. Smith Jr., W. L. Jorgen, J. Tirado-Rives et al., "Prediction of binding affinities for TIBO inhibitors of HIV-1 reverse transcriptase using Monte Carlo simulations in a linear response method," *Journal of Medicinal Chemistry*, vol. 41, no. 26, pp. 5272–5286, 1998.

[15] B. A. Roberte, K. Andries, J. Desayter et al., "Potent and selective inhibition of HIV-1 replication in vitro by a novel series of TIBO derivatives," *Nature*, vol. 343, no. 6257, pp. 470–474, 1990.

[16] H. J. Breslin, M. J. Kukla, T. Kromis et al., "Synthesis and anti-HIV activity of 1,3,4,5-tetrahydro-2H-1,4- benzodiazepin-2-one (TBO) derivatives. Truncated 4,5,6,7-tetrahydro-5- methyl-imidazo[4,5,1-jk][1,4]benzodiazepin-2(1H)-ones (TIBO) analogues," *Bioorganic and Medicinal Chemistry*, vol. 7, no. 11, pp. 2427–2436, 1999.

[17] W. Ho, M. J. Kukla, H. J. Breslin et al., "Synthesis and anti-HIV-1 activity of 4,5,6,7-tetrahydro-5-methylimidazo-[4,5,1-jk][1,4]benzodiazepin-2(1H)-one (TIBO) derivatives. 4," *Journal of Medicinal Chemistry*, vol. 38, no. 5, pp. 794–802, 1995.

[18] B. D. Puodziunaite, R. Janciene, L. Kosychova, and Z. Stumbreviciute, "On the synthetic way to novel peri-annelated imidazo[1,5]benzodiazepinones as the potent non-nucleoside reverse transcriptase inhibitors," *Arkivoc*, vol. 2000, no. 4, pp. 512–522, 2000.

[19] M. D. Braccio, G. Grossi, G. Roma, L. Vargiu, M. Mura, and M. E. Marongiu, "1,5-Benzodiazepines. Part XII. Synthesis and biological evaluation of tricyclic and tetracyclic 1,5-benzodiazepine derivatives as nevirapine analogues," *European Journal of Medicinal Chemistry*, vol. 36, no. 11-12, pp. 935–949, 2001.

[20] A. Kamal, M. V. Rao, N. Laxman, G. Ramesh, and G. S. Reddy, "Recent developments in the design, synthesis and structure-activity relationship studies of pyrrolo[2,1-c][1,4]benzodiazepines as DNA-interactive antitumour antibiotics," *Current Medicinal Chemistry—Anti-Cancer Agents*, vol. 2, no. 2, pp. 215–254, 2002.

[21] P. Sharma, B. Vashistha, R. Tygai et al., "Application of microwave induced Delepine reaction to the facile one pot synthesis of 7-substituted 1,3-dihydro-2H-[1,4]-benzodiazepin-2-one-5-methyl carboxylates from the corresponding 1-chloroacetylisatins," *International Journal of Chemical Sciences*, vol. 8, no. 1, 2010.

[22] A. Singh, R. Sirohi, S. Shastri, and D. Kishore, "A facile one-pot synthesis of 7-substituted-5-methoxycarbonyl-1H-2, 3-dihydro-1, 4-benzodiazepin-2-ones from 5-substituted-N-chloroacetyl isatins," *Indian Journal of Chemistry B*, vol. 42B, no. 12, pp. 3124–3127, 2003.

[23] P. D. Popp, "The chemistry of isatin," in *Advances in Heterocyclic Chemistry*, vol. 18, pp. 1–58, 1975.

[24] M. Pal, N. K. Sharma, Priyanka, and K. K. Jha, "Synthetic and biological multiplicity of isatin: a review," *Journal of Advanced Scientific Research*, vol. 2, no. 2, pp. 35–44, 2011.

[25] N. Blazevic, D. Kolbah, B. Belin, V. Sunjic, and F. Kajfez, "Hexamethylenetetramine, "A versatile reagent in organic synthesis"," *Synthesis*, pp. 167–176, 1979.

[26] B. Rigoa, P. Caulieza, D. Fasseurb, and D. Couturierb, "Reaction of hexamethyldisilazane with diacylhydrazines: an easy 1,3,4-oxadiazole synthesis," *Synthetic Communications*, vol. 16, no. 13, pp. 1665–1669, 1986.

[27] M. Ogata and H. Matsumoto, "A convenient synthesis of 5-substituted 1,3-dihydro-2H-1,4-benzodiazepine-2-ones," *Chemistry and Industry*, p. 1067, 1976.

[28] H. Fujishima, H. Takeshita, S. Suzuki, M. Toyota, and M. Ihara, "Hexamethyldisilazanes mediated one-pot intramolecular Michael addition-olefination reactions leading to ejvo-olefinated bicyclo[6.4.0]dodecanes," *Journal of the Chemical Society, Perkin Transactions 1*, no. 18, pp. 2609–2616, 1999.

8

Synthesis and Antimicrobial Activity of Some Novel Heterocyclic Candidates via Michael Addition Involving 4-(4-Acetamidophenyl)-4-oxobut-2-enoic Acid

Maher A. EL-Hashash,[1] A. Essawy,[2] and Ahmed Sobhy Fawzy[2]

[1] *Chemistry Department, Faculty of Science, Ain Shams University, Cairo, Egypt*
[2] *Chemistry Department, Faculty of Science, Fayoum University, Fayoum, Egypt*

Correspondence should be addressed to Ahmed Sobhy Fawzy; asl868@fayoum.edu.eg

Academic Editor: Constantinos Pistos

This paper discusses the utility of 4-(4-acetamidophenyl)-4-oxobut-2-enoic acid as a key starting material for the preparation of a novel series of pyridazinones, thiazoles derivatives, and other heterocycles via interaction with nitrogen, sulfur, and carbon nucleophiles under Michael addition conditions and studies the antimicrobial activities of some of these compounds.

1. Introduction

β-Aroylacrylic acid derivatives showed high biological activity and exhibited a broad spectrum of physiological activities [1] (fungicidal, antitumor, hypotensive, hypolipidemic, and antibacterial). Also, β-aroylacrylic acids were considered as inhibitors for phospholipase [2, 3] and they have antiproliferative activity against human cervix carcinoma (Hela cells) [4, 5]. Besides that, β-aroylacrylic esters are important intermediates in field of medical science and agrochemicals [1]. Chemically, β-aroylacrylic acids are convenient polyelectrophilic reagents in the synthesis of heterocyclic compounds, for which the addition of nitrogen, sulfur, phosphorus, or carbon nucleophiles occurs exclusively at the α-carbon of the electrophilic center of the molecule [6–13]. On the other hand, aryl and heteroaryl substituted (E)-4-oxobut-2-enoic acids and their derivatives represent an important class of compounds with interesting pharmacological indications including antiulcer and cytoprotective properties [14] and kynurenine-3-hydroxylase [15] and human cytomegalovirus protease inhibiting activity [16]. Also several naturally occurring acylacrylic acids show notable antibiotic activity [17] and they are used as starting materials for the preparation of a novel series of pyridazinones and thiazoles, where many studies have been focused on pyridazinones which are characterized to possess good analgesic and anti-inflammatory activities; besides that, these studies have indicated that the heterocyclic ring substitutions at position six and the presence of acetamide side chain that is linked to the lactam nitrogen of pyridazinone ring at position two of the pyridazinone ring improve the analgesic and anti-inflammatory activities along with nil or very low ulcerogenicity [18].

The aim of this work is to study the behaviour of aza- and carba-Michael addition reactions involving 4-(4-acetamidophenyl)-4-oxobut-2-enoic acid and nitrogen and/or carbon nucleophiles which is considered as a first step to prepare pyridazinones, thiazoles, and other heterocyclic compounds.

2. Experimental

Melting points were determined on electrothermal apparatus using open capillary method and are uncorrected. Elemental analyses were carried out by the Micro Analytical Center at Cairo University. The IR spectra were recorded on FT/IR-300E Jasco spectrophotometer as potassium bromide discs. The mass spectra were run by a Shimadzu-GC-MS-QP 1000 EX apparatus at 70 eV. (^{1}H & ^{13}C) NMR spectra were recorded on Varian Mercury 300 MHz spectrometer using TMS as internal standard.

2.1. 4-(4-Acetamidophenyl)-4-oxobut-2-enoic Acid (1). 4-(4-Acetamidophenyl)-4-oxobut-2-enoic acid was prepared according to a published procedure [19].

2.2. General Procedure for the Synthesis of Aza-Michael Adduct (2). A mixture of β-aroylacrylic acid 1 (4 mmol) and nitrogen nucleophile (4 mmol) in dry benzene (20 mL) was left for 2 days. The resulting solid formed after concentration was filtered off, dried, and crystallized from ethanol and afforded the desired products.

2.2.1. 4-(4-Acetamidophenyl)-2-(benzylamino)-4-oxobutanoic Acid (2a). Yield 72%; m.p. 236°C; IR (KBr): 1610, 1685, 1710, 3246, 3361 and 3532 cm^{-1}; ^{1}H NMR (DMSO): δ ppm 2.08 (s, 3H, CH$_3$CO), 3.35 (s, 2H, methylene of benzyl moiety), 3.38 (m, 2H, diastereotopic methylene protons), 3.39, 3.59, 3.60, 3.95 (q, 1H stereogenic methine proton), 7.24, 7.26 (s, 2H, NH of amino and amide), 7.29 – 7.40 (m, 5H, ArH of benzyl amine moiety), 7.68, 7.72 (d, 4H, ArH), and 10.34 (s, 1H, COOH); ^{13}C NMR (DMSO): δ = 194.11, 179.7, 168.3, 139.95, 131.97, 128.91, 127.61, 126.89, 121.21, 64.13, 51.15, 45.42 and 23.91; MS m/z: 340 [M$^+$]; Anal. Calcd. for C$_{19}$H$_{20}$N$_2$O$_4$: C, 67.05; H, 5.92; N, 8.23%. Found: C, 66.92; H, 5.81; N, 8.32%.

2.2.2. 2-[3-(Imidazole-1-yl)propylamine]-4-(4-acetamido-phenyl)-4-oxobutanoic Acid (2b). Yield 58%; m.p. 215°C; IR (KBr): 1604, 1680, 2600, 3409 and 3255 cm^{-1}; ^{1}H NMR (DMSO): δ ppm 2.13 (s, 3H, CH$_3$CO), 1.75, 2.43, 4.15 (m, 6H, CH$_2$ of propyl group), 6.67, 7.35, 7.82 (m, 3H, CH of imidazole moiety), 7.12 (s, 1H, NH of amide), 7.97, 8.12 (m, 4H, ArH); ^{13}C NMR (DMSO): δ = 197.5, 180.9, 168.8, 147.8, 137.7, 132.4, 128.9, 128.2, 121.6, 120.5, 64.6, 47.4, 45.8, 44.2, 32.2, 23.9; MS m/z: 358 [M$^+$]; Anal. Calcd. for C$_{18}$H$_{22}$N$_4$O$_4$: C, 60.32; H, 6.19; N, 15.63%. Found: C, 60.15; H, 5.95; N, 15.73%.

2.2.3. 4-(4-Acetamidophenyl)-4-oxo-2-(pyridin-2-ylmethyl-amino)butanoic Acid (2c). Yield 62%; m.p. 216°C; IR (KBr): 1615, 1675, 1700, 3255, and 3412 cm^{-1}; ^{1}H NMR (DMSO): δ ppm 2.10 (s, 3H, CH$_3$CO), 3.31 (m, 2H, diastereotopic protons CH$_2$), 3.82 (s, 2H, CH$_2$ of picolyl moiety), 7.15, 7.57, 8.22 (m, 4H, CH of pyridine), 7.62, 7.83 (m, 4H, ArH); ^{13}C NMR (DMSO): δ = 197.3, 180.7, 168.8, 161.2, 148.5, 142.8, 139.7, 132.4, 129.1, 124.2, 121.4, 120.8, 64.2, 49.5, 45.6, 24.1; MS m/z: 341 [M$^+$]; Anal. Calcd. for C$_{18}$H$_{19}$N$_3$O$_4$: C, 63.33; H, 5.61; N, 12.31%. Found: C, 63.12; H, 5.49; N, 12.52%.

2.2.4. 4-(4-Acetamidophenyl)-4-oxo-2-(3-(trimethoxysil-yl)propylamino)butanoic Acid (2d). Yield 71%; m.p. > 360°C; IR (KBr): 1250, 1265, 1447, 1466, 1625, 1630, 1675, 1680, 3125 and 3300 cm^{-1}; ^{1}H NMR (DMSO): δ ppm 2.07 (s, 3H, CH$_3$CO), 3.45 (s, 9H, CH$_3$ of methoxy), 4.12 (t, 1H, methine), 7.15 (s, 1H, NH of amide), 7.79, 7.85 (m, 4H, ArH); MS m/z: 312 [M$^+$]; Anal. Calcd. for C$_{18}$H$_{28}$N$_2$O$_7$Si: C, 52.41; H, 6.84; N, 6.79%. Found: C, 52.14; H, 6.76; N, 6.92%.

2.2.5. 4-(4-Acetamidophenyl)-2-(4-methoxyphenylamino)-4-oxobutanoic Acid (2e). Yield 79%; m.p. 197°C; IR (KBr): 1633, 1655, 3101, 3250 cm^{-1}; ^{1}H NMR (DMSO): δ ppm 1.97 (s, 3H, CH$_3$CO), 2.75 (m, 2H, methylene), 3.59 (s, 3H, methoxy), 4.12 (m, 1H, methine), 6.51 (s, 4H, ArH), 7.43, 7.87 (d, 4H, ArH); ^{13}C NMR (DMSO): δ = 197.3, 174.8, 168.8, 151.6, 142.8, 139.8, 132.4, 129.1, 121.4, 115.7, 115.2, 63.7, 55.9, 45.4, 24.1; MS m/z: 356 [M$^+$]; Anal. Calcd. for C$_{19}$H$_{20}$N$_2$O$_5$: C, 64.04; H, 5.66; N, 7.86%. Found: C, 63.92; H, 5.45; N, 7.91%.

2.2.6. 4-(4-Acetamidophenyl)-2-(3-(dimethylamino)propyl-amino)-4-oxobutanoic Acid (2f). Yield 75%; m.p. 134°C; IR (KBr): 1655, 1685, 3177, 3250, 3300 cm^{-1}; ^{1}H NMR (DMSO): δ ppm 1.35, 2.31, 2.47 (m, 6H, propyl group), 2.21 (s, 3H, CH$_3$CO), 2.75 (s, 6H, N-methyl groups), 2.95 (m, 2H, methylene), 4.18 (m, 1H, methine), 6.93 (s, 1H, NH of amide), 7.62, 7.85 (m, 4H, ArH); MS m/z: 335 [M$^+$]; Anal. Calcd. for C$_{17}$H$_{25}$N$_3$O$_4$: C, 60.88; H, 7.51; N, 12.53%. Found: C, 60.75; H, 7.33; N, 12.64%.

2.2.7. 4-(4-Acetamidophenyl)-4-oxo-2-(3-(triethoxysilyl)pro-pylamino)butanoic Acid (2g). Yield 59%; m.p. > 360°C; IR (KBr): 1250, 1265, 1447, 1466, 1625, 1630, 1675, 1680, 2500, 3300 cm^{-1}; MS m/z: 454 [M$^+$]; Anal. Calcd. for C$_{21}$H$_{34}$N$_2$O$_7$Si: C, 55.48; H, 7.54; N, 6.16%. Found: C, 55.32; H, 7.29; N, 6.27%.

2.3. 4-Benzylamino-6(4-acetamidophenyl)-2,3,4,5-tetrahy-dro-3(2H) Pyridazinone (3). A mixture of aza-Michael adduct 2a (1.3 g, 0.003 mol) and hydrazine hydrate 80% (0.5 mL) in ethanol (20 mL) was refluxed for 2 hours. The reaction mixture was allowed to cool and the separated product was filtered off, dried, and crystallized from ethanol and afforded compound 3. Yield 56% m.p. 160°C; IR (KBr): 1663, 3213 and 3357 cm^{-1}; ^{1}H NMR (DMSO): δ ppm 1.75 (m, 2H, methylene) 2.11 (s, 3H, CH$_3$CO), 3.49 (S, 2H, PhCH$_2$), 7.35 (m, 5H, ArH of benzyl moiety), 7.85, 7.93 (m, 4H, ArH); ^{13}C NMR (DMSO): δ = 168.8, 163, 146.4, 140.9, 140.3, 132.1, 129.3, 128.5, 127.9, 127.1, 121.6, 67.1, 52.2, 33.8, 23.9; MS m/z: 336 [M$^+$]; Anal. Calcd. for C$_{19}$H$_{20}$N$_4$O$_2$: C, 67.84; H, 5.99; N, 16.66%. Found: C, 67.72; H, 5.86; N, 16.75%.

2.4. (E)-3-(4-Acetamidobenzoyl)-2-(benzylamino)-4-phenyl-but-3-enoic Acid (4a). A mixture of aza-Michael adduct 2a (1.3 g, 0.003 mol) and benzaldehyde (0.32 g, 0.003 mol) in ethanol (20 mL) was treated with few drops of triethylamine and refluxed for 2 hours. The reaction mixture was allowed to cool and the separated product was filtered off, dried, and crystallized from toluene and afforded compound 4a. Yield 74%; m.p. 183°C; IR (KBr): 1604, 1651, 1686, 3176, 3252 and 3302 cm^{-1}; ^{1}H NMR (DMSO): δ ppm 2.09 (s, 3H, CH$_3$CO), 3.59 (s, 2H, methylene of benzyl moiety), 4.21 (s, 1H, methine), 7.25 (s, 1H, NH of amide), 7.21, 7.45 (m, 5H, ArH of benzylamine moiety), 7.49, 7.91 (m, 9H, phenyl protons), 7.74 (s, 1H, CH of methine), 10.93 (s, 1H, COOH); ^{13}C NMR (DMSO): δ = 190.32, 173.91, 168.73, 144.12, 139.92, 138.19, 134.95, 134.13, 131.18, 128.23, 126.93, 121.89, 60.17, 51.52 and

24.12; MS *m/z*: 428 [M$^+$]; Anal. Calcd. for C$_{26}$H$_{24}$N$_2$O$_4$: C, 72.88; H, 5.65; N, 6.54%. Found: C, 72.76; H, 5.43; N, 6.62%.

2.5. (E)-3-(4-Acetamidobenzoyl)-2-(benzylamino)-4-(pyridin-2-yl)but-3-enoic Acid (4b). A mixture of aza-Michael adduct **2a** (1.3 g, 0.003 mol) and 2-pyridinecarboxaldehyde (0.32 g, 0.003 mol) in ethanol (20 mL) was treated with few drops of triethylamine and refluxed for 2 hours. The reaction mixture was allowed to cool and the separated product was filtered off, dried, and crystallized from toluene and afforded compound **4b**. Yield 69%; m.p. 182°C; IR (KBr): 1608, 1655, 1685, 3175 and 3251 cm^{-1}; ^{1}H NMR (DMSO): δ ppm 2.13 (s, 3H, CH$_3$CO), 3.61 (s, 2H, methylene of benzyl moiety), 7.28 (s, 1H, NH of amide), 7.22, 7.48 (m, 5H, ArH of benzylamine moiety), 7.93 (m, 9H, phenyl protons), 8.12 (s, 1H, of ethylene), 10.95 (s, 1H, COOH); ^{13}C NMR (DMSO): δ = 190.4, 174.2, 168.8, 154.6, 148.7, 144.2, 140.3, 139.9, 137.1, 135.2, 131.3, 128.4, 127.8, 124.2, 122.6, 122, 60.4, 51.6, 24.1; MS *m/z*: 429 [M$^+$]; Anal. Calcd. for C$_{25}$H$_{23}$N$_3$O$_4$: C, 69.92; H, 5.40; N, 9.78%. Found: C, 69.58; H, 5.32; N, 9.91%.

2.6. General Procedure for the Synthesis of Pyridazinone Derivatives 5(b, c, e, f). A mixture of aza-Michael adducts **2(b, c, e, f)** (0.003 mol) and hydrazine hydrate 80% (0.5 mL) in ethanol (20 mL) was refluxed for 2 hours. The reaction mixture was allowed to cool and the separated product was filtered off, dried, and crystallized from ethanol and afforded compound **5(b, c, e, f)**.

2.6.1. N-(4-(5-(3-(1H-Imidazole-1-yl)propylamino)-6-oxo-1,4,5,6-tetrahydropyridazin-3-yl)phenyl) Acetamide (5b). Yield 71%; m.p. 275°C; IR (KBr): 1621, 1650, 1666, 2962, 3286 and 3400 cm^{-1}; ^{1}H NMR (DMSO): δ ppm 2.18 (s, 3H, CH$_3$CO), 3.55 (m, 1H, methine), 1.65, 2.76, 4.27 (m, 6H, CH$_2$ of propyl group) 6.83, 7.18, 7.72 (m, 3H, CH of imidazole moiety), 7.15 (s, 1H, NH of amide), 7.51, 8.11 (m, 4H, ArH), 10.95 (s, 1H, COOH); MS *m/z*: 354 [M$^+$]; Anal. Calcd. for C$_{18}$H$_{22}$N$_6$O$_2$: C, 61.00; H, 6.26; N, 23.71%. Found: C, 60.91; H, 6.15; N, 23.85%.

2.6.2. N-(4-(6-Oxo-5-(pyridin-2-ylmethylamino)-1,4,5,6-tetrahydropyridazin-3-yl)phenyl) Acetamide (5c). Yield 73%; m.p. over 360°C; IR (KBr): 1616, 1670 and 3200 cm^{-1}; ^{1}H NMR (DMSO): δ ppm 2.19 (s, 3H, CH$_3$CO), 4.45 (d, 2H, methylene), 3.75 (m, 1H, methine), 7.34 (s, 1H, NH of acetamide), 7.52, 7.75 (m, 4H, ArH), 7.43, 7.89 and 8.57 (m, 4H, pyridine ring); ^{13}C NMR (DMSO): δ = 168.8, 163, 161.4, 148.5, 146.4, 140.9, 139.7, 132.1, 129.5, 124.2, 121.8, 120.8, 66.9, 50.3, 33.6, 23.9; MS *m/z*: 337 [M$^+$]; Anal. Calcd. for C$_{18}$H$_{19}$N$_5$O$_2$: C, 64.08; H, 5.68; N, 20.76%. Found: C, 63.92; H, 5.49; N, 20.89%.

2.6.3. N-(4-(5-(4-Methoxyphenylamino)-6-oxo-1,4,5,6-tetrahydropyridazin-3-yl)phenyl) Acetamide (5e). Yield 70%; m.p. 194°C; IR (KBr): 1600, 1668, 3189 and 3331 cm^{-1}; ^{1}H NMR (DMSO): δ ppm 1.79 (m, 2H, methylene of pyridazinone), 1.89 (s, 3H, CH$_3$CO), 3.95 (s, 3H, of methoxy), 6.85 (m, 4H, ArH), 7.41 (s, 1H, NH of acetamide), 7.65, 7.96 (m, 4H, ArH);

^{13}C NMR (DMSO): δ = 168.71, 162.82, 151.57, 146.32, 140.7, 139.7, 132.13, 127.7, 121.54, 115.61, 115.25, 69.34, 55.69, 33.17 and 23.95; MS *m/z*: 352 [M$^+$]; Anal. Calcd. for C$_{19}$H$_{20}$N$_4$O$_3$: C, 64.76; H, 5.72; N, 15.90%. Found: C, 64.53; H, 5.58; N, 16.14%.

2.6.4. N-(4-(5-(3-(Dimethylamino)propylamino)-6-oxo-1,4,5,6-tetrahydropyridazin-3-yl)phenyl) Acetamide (5f). Yield 57%; m.p. 286°C; IR (KBr): 1600, 1649, 3108, 3187, 3262 and 3304 cm^{-1}; ^{1}H NMR (DMSO): δ ppm 1.67 (m, 2H, methylene of pyridazinone), 2.17 (s, 3H, CH$_3$CO), 2.33 (s, 6H, N-methyl), 1.46, 2.38, and 2.69 (m, 6H, propyl group), 7.39 (s, 1H, NH of acetamide), 7.52, 7.76 (m, 4H, ArH); ^{13}C NMR (DMSO): δ = 169, 146.6, 140.8, 132.1, 129.3, 121.6, 67.4, 58.9, 47.1, 45.2, 26.8, 23.9; MS *m/z*: 331 [M$^+$]; Anal. Calcd. for C$_{17}$H$_{25}$N$_5$O$_2$: C, 61.61; H, 7.60; N, 21.13%. Found: C, 61.43; H, 7.39; N, 21.32%.

2.6.5. N-(4-(6-Oxo-5-(3-(trimethoxysilyl)propylamino)-1,4,5,6-tetrahydropyridazin-3-yl)phenyl) Acetamide (5d). A mixture of keto acid **2d** (1.2 g, 0.003 mol) and hydrazine hydrate 80% (0.5 mL) in dry benzene (20 mL) was heated gently for 30 minutes. The reaction mixture was allowed to cool and the separated product was filtered off, dried, and crystallized from ethanol and afforded compound **5d**. Yield 58%; m.p. over 360°C; IR (KBr): 1596, 1630, 1675, and 3180 cm^{-1}; ^{1}H NMR (DMSO): δ ppm 1.92 (m, 2H, methylene of pyridazinone), 2.33 (s, 3H, CH$_3$CO), 3.68 (s, 9H, of methoxy), 7.49 (s, 1H, NH of acetamide), 7.73, 7.91 (m, 4H, ArH); MS *m/z*: 408 [M$^+$]; Anal. Calcd. for C$_{18}$H$_{28}$N$_4$O$_5$Si: C, 52.92; H, 6.91; N, 13.71%. Found: C, 62.74; H, 6.82; N, 13.95%.

2.7. N-(4-(2-(3-Oxo-1,2,3,4-tetrahydroquinoxalin-2-yl)acetyl)phenyl) Acetamide (6). A mixture of β-aroylacrylic acid **1** (1 g, 0.004 mol) and o-phenylenediamine (0.43 g, 0.004 mol) in isopropyl alcohol (20 mL) was refluxed for 2 hours. The resulting solid formed after concentration was filtered off, dried, and crystallized from isopropyl alcohol and afforded compound **6**. Yield 74%; m.p. 220°C; IR (KBr): 1673, 3186 and 3373 cm^{-1}; ^{1}H NMR (DMSO): δ ppm 2.33 (s, 3H, CH$_3$CO), 3.45 (m, 2H, methylene), 4.22 (t, 1H methine), 5.47, 7.95 (s, 2H, NH of quinoxaline ring), 7.63, 7.84 (m, 4H, ArH), 6.83: 8.14 (m, 4H, ArH of quinoxaline); ^{13}C NMR (DMSO): δ = 197.5, 172.6, 168.8, 143, 139.9, 132.3, 129.1, 127.1, 125.2, 121.8, 121.4, 117.2, 114.7, 64.5, 45.8, 24.1; MS *m/z*: 323 [M$^+$]; Anal. Calcd. for C$_{18}$H$_{17}$N$_3$O$_3$: C, 66.86; H, 5.30; N, 13.00%. Found: C, 66.73; H, 5.45; N, 13.19%.

2.8. 4-(4-Acetamidophenyl)-2-(2-hydroxyphenylamino)-4-oxobutanoic Acid (7). A mixture of β-aroylacrylic acid **1** (1 g, 0.004 mol) and o-aminophenol (0.44 g, 0.004 mol) in isopropyl alcohol (20 mL) was refluxed for 2 hours. The resulting solid formed after concentration was filtered off, dried, and crystallized from isopropyl alcohol and afforded compound **7**. Yield 62%; m.p. 247°C; IR (KBr): 1616, 1675, 3255 and 3394 cm^{-1}; ^{1}H NMR (DMSO): δ ppm 2.18 (s, 3H, CH$_3$CO), 3.25 (m, 2H, methylene), 5.43 (s, 1H, ArOH), 4.15 (m, 2H, methine), 6.82, 6.91 (m, 4H, ArH), 7.18 (s, 1H, NH of acetamide), 7.84, 7.89 (m, 4H, ArH); MS *m/z*: 342 [M$^+$]; Anal.

Calcd. for $C_{18}H_{18}N_2O_5$: C, 63.15; H, 5.30; N, 8.18%. Found: C, 62.95; H, 5.19; N, 8.29%.

2.9. N-(4-(2-(3-Oxopiperazin-2-yl)acetyl)phenyl) Acetamide (8).

A mixture of β-aroylacrylic acid 1 (1 g, 0.004 mol) and ethylenediamine (0.26 g, 0.004 mol) in dry benzene (20 mL) was refluxed for 1 hour. The resulting solid formed after concentration was filtered off, dried, and crystallized from ethanol and afforded compound 8. Yield 69%; m.p. 204°C; IR (KBr): 1663, 3136, 3149 and 3294 cm^{-1}; ^{1}H NMR (DMSO): δ ppm 2.37 (s, 3H, CH$_3$CO), 2.93, 3.46 (m, 4H, methylene of piperazine moiety), 4.27 (t, 1H, methine), 7.14 (s, 1H, NH of acetamide), 7.83, 8.11 (m, 4H, ArH); ^{13}C NMR (DMSO): δ = 169.92, 168.89, 142.79, 132.19, 129.12, 121.45, 61.96, 46.22, 45.62, 36.43 and 24.17; MS m/z: 275 [M$^+$]; Anal. Calcd. for $C_{14}H_{17}N_3O_3$: C, 61.08; H, 6.22; N, 15.26%. Found: C, 61.92; H, 6.13; N, 15.35%.

2.10. N-(4-(2-(2-Amino-5-oxo-4,5-dihydrothiazol-4-yl)acetyl)phenyl) Acetamide (9).

A mixture of β-aroylacrylic acid 1 (1 g, 0.004 mol) and thiourea (0.3 g, 0.004 mol) in ethanol (20 mL) and few drops of glacial acetic acid were refluxed for 2 hours. The resulting solid formed after concentration was filtered off, dried, and crystallized from ethanol and afforded compound 9. Yield 77%; m.p. 238°C; IR (KBr): 1602, 1650, 1681, 3300 and 3347 cm^{-1}; ^{1}H NMR (DMSO): δ ppm 2.23 (s, 3H, CH$_3$CO), 2.91 (m, 2H, methylene), 7.18 (s, 1H, NH of acetamide), 7.79, 7.88 (m, 4H, ArH), 8.62 (s, 2H, amino); MS m/z: 291 [M$^+$]; Anal. Calcd. for $C_{13}H_{13}N_3O_3S$: C, 53.60; H, 4.50; N, 14.42%. Found: C, 53.43; H, 4.39; N, 14.64%.

2.11. N-(4-(6-Aminothiazolo[5,4-c]pyridazin-3-yl)phenyl) Acetamide (10).

A mixture of compound 9 (0.87 g, 0.003 mol) and hydrazine hydrate 80% (0.5 mL) in ethanol (20 mL) was refluxed for 2 hours. The reaction mixture was allowed to cool and the separated product was filtered off, dried, and crystallized from ethanol and afforded compound 10. Yield 55%; m.p. 244°C; IR (KBr): 1604, 1650, 3308 and 3417 cm^{-1}; ^{1}H NMR (DMSO): δ ppm 2.13 (s, 3H, CH$_3$CO), 6.78 (s, 2H, amino), 7.18 (s, 1H, NH of acetamide), 7.64 (s, 1H, of pyridazinone), 7.82, 7.93 (m, 4H, ArH); ^{13}C NMR (DMSO): δ = 168.67, 161.25, 142.38, 138.41, 128.57, 127.59, 125.18, 119.59, 108.14 and 24.18; MS m/z: 285 [M$^+$]; Anal. Calcd. for $C_{13}H_{11}N_5S$: C, 54.72; H, 3.89; N, 24.55%. Found: C, 54.63; H, 3.75; N, 24.81%.

2.12. N-(4-(6-Amino-4H-thiazolo[4,5-e][1,2]oxazin-3-yl)phenyl) Acetamide (11).

A mixture of compound 9 (0.87 g, 0.003 mol) and hydroxylamine hydrochloride (0.2 g, 0.003 mol) and few drops of sodium hydroxide 30% in ethanol (20 mL) was refluxed for 1 hour. The reaction mixture was allowed to cool and the separated product was filtered off, dried, and crystallized from ethanol and afforded compound 11. Yield 53%; m.p. 223°C; IR (KBr): 1620, 1660, 3200 and 3350 cm^{-1}; ^{1}H NMR (DMSO): δ ppm 2.24 (s, 3H, CH$_3$CO), 4.15 (s, 2H, oxazine ring), 6.85 (s, 2H, amino), 7.21 (s, 1H, NH of acetamide), 7.55 : 8.22 (m, 4H, ArH); MS m/z: 288 [M$^+$];

Anal. Calcd. for $C_{13}H_{12}N_4O_2S$: C, 54.15; H, 4.20; N, 19.43%. Found: C, 53.96; H, 4.11; N, 19.62%.

2.13. 2-(2-(4-Acetamidophenyl)-2-oxoethyl)-3-(ethoxycarbonyl)-4-oxopentanoic Acid (12).

A mixture of β-aroylacrylic acid 1 (1 g, 0.004 mol) and ethyl acetoacetate (0.52 g, 0.004 mol) and few drops of sodium hydroxide 30% in ethanol (20 mL) was left for 3 days and then acidified by HCl and the resulting solid formed after concentration was filtered off, dried, and crystallized from toluene and afforded compound 12. Yield 73%; m.p. over 360°C; IR (KBr): 1666, 1700, 3200 and 3332 cm^{-1}; ^{1}H NMR (DMSO): δ ppm 1.43 (t, 3H, CH$_3$ of ester) 2.26 (s, 3H, CH$_3$CO of acetamide), 2.45 (s, 3H, CH$_3$CO), 3.11 (d, 2H, methylene), 3.45, 3.84 (m, 2H, of 2 methine groups), 4.35 (q, 2H, methylene of ester), 7.17 (s, 1H, NH of acetamide), 7.63, 7.82 (m, 4H, ArH); MS m/z: 363 [M$^+$]; Anal. Calcd. for $C_{18}H_{21}NO_7$: C, 59.50; H, 5.83; N, 3.85%. Found: C, 59.34; H, 5.72; N, 3.94%.

2.14. 2-(2-(4-Acetamidophenyl)-2-oxoethyl)-3-benzoyl-4-ethoxy-4-oxobutanoic Acid (13).

A mixture of β-aroylacrylic acid 1 (1 g, 0.004 mol) and ethyl benzoyl acetate (0.77 g, 0.004 mol) and few drops of sodium hydroxide 30% in ethanol (20 mL) was left for 2 days and then acidified by HCl and the resulting solid formed after concentration was filtered off, dried, and crystallized from toluene and afforded compound 13. Yield 72%; m.p. over 360°C; IR (KBr): 1640, 1676, 1700, 1740, 3200 and 3415 cm^{-1}; ^{1}H NMR (DMSO): δ ppm 1.39 (t, 3H, CH$_3$ of ester) 2.17 (s, 3H, CH$_3$CO), 3.33 (d, 2H, methylene), 3.95, 4.23 (m, 2H, of 2 methine groups), 4.42 (q, 2H, methylene of ester), 7.19 (s, 1H, NH of acetamide), 7.68, 7.89 (m, 4H, ArH), 7.58, 766, 7.99 (m, 5H, ArH); ^{13}C NMR (DMSO): δ = 197.68, 195.32, 178.42, 169.82, 168.78, 142.71, 135.35, 132.87, 132.11, 128.46, 121.29, 61.16, 51.58, 36.13, 30.27, 23.94 and 14.25; MS m/z: 425 [M$^+$]; Anal. Calcd. for $C_{23}H_{23}NO_7$: C, 64.93; H, 5.45; N, 3.29%. Found: C, 64.81; H, 5.39; N, 3.34%.

2.15. 2-(2-(4-Acetamidophenyl)-2-oxoethyl)-3-acetyl-4-oxopentanoic Acid (14).

A mixture of β-aroylacrylic acid 1 (1 g, 0.004 mol) and acetylacetone (0.4 g, 0.004 mol) and few drops of sodium hydroxide 30% in ethanol (20 mL) was left for 2 days and then acidified by HCl and the resulting solid formed after concentration was filtered off, dried, and crystallized from toluene and afforded compound 14. Yield 74%; m.p. over 360°C; IR (KBr): 1630, 1661, 1690, 1710, 3200 and 3428 cm^{-1}; ^{1}H NMR (DMSO): δ ppm 2.27 (s, 3H, CH$_3$CO of acetamide), 2.45 (s, 6H, CH$_3$CO), 3.23 (m, 2H, methylene), 3.44, 3.63 (m, 2H, of 2 methine groups), 7.31 (s, 1H, NH of acetamide), 7.59, 7.78 (m, 4H, ArH); MS m/z: 333 [M$^+$]; Anal. Calcd. for $C_{17}H_{19}NO_6$: C, 61.25; H, 5.75; N, 4.20%. Found: C, 61.16; H, 5.59; N, 4.32%.

2.16. N-(4-(1,4-Dioxo-5-phenyl-2,3,4,4a,9,9a-hexahydro-1H-pyridazino[4,5-d][1,2]diazepin-8-yl)phenyl) Acetamide (15).

A mixture of compound 13 (1.28 g, 0.003 mol) and hydrazine hydrate (80% 0.5 mL) in ethanol (20 mL) was refluxed for 3 hours. The reaction mixture was allowed to cool and the

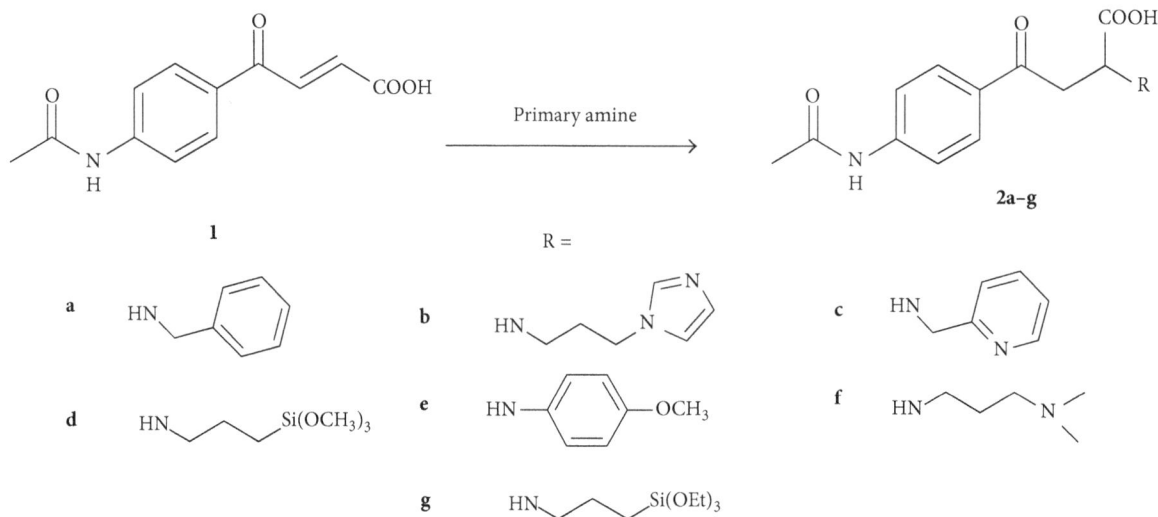

SCHEME 1: Reaction of β-aroylacrylic acid **1** with primary amines.

separated product was filtered off, dried, and crystallized from ethanol and afforded compound **15**. Yield 45%; m.p. over 360°C; IR (KBr): 1645, 1673, 3241, 3297 and 3321 cm^{-1}; ^{1}H NMR (DMSO): δ ppm 1.53 (m, 2H, methylene of diazepine), 2.12 (s, 3H, CH$_3$CO), 2.67, 3.24 (m, 2H of 2 methine of diazepine), 7.14 (s, 1H, NH of acetamide), 7.43, 7.85 (m, 5H, ArH), 7.62, 7.74 (m, 4H, ArH), 8.13 (d, 2H, 2 NH of pyridazinone); ^{13}C NMR (DMSO): δ = 177.1, 174.2, 168.9, 157.6, 146.5, 140.7, 134.1, 132.2, 130.9, 129.3, 128.7, 128.3, 121.6, 38.7, 26.6, 26.4, 23.9; MS *m/z*: 389 [M$^+$]; Anal. Calcd. for C$_{21}$H$_{19}$N$_5$O$_3$: C, 64.77; H, 4.92; N, 17.98%. Found: C, 64.61; H, 4.86; N, 18.09%.

3. Results and Discussion

The authors aimed through this research to study the behaviour of 4-(4-acetamidophenyl)-4-oxobut-2-enoic acid **1** towards some nitrogen nucleophiles. Thus the aza-Michael reaction of compound **1** with nitrogen nucleo-philes, namely, benzylamine, 3-(1-H-imidazole-1-yl)-propylamine, 2-pic-olylamine, 3-(trimethoxysilyl)propylamine, p-anisidine, 3-(N,N-dimethyl)aminopropylamine, and 3-(triethoxysilyl) propylamine, in dry benzene afforded the aza-Michael adducts' compounds (**2a–g**), respectively (Scheme 1). The structures of compounds (**2a–g**) were confirmed by elemental analysis and spectral data; EIMS of compound **2a** exhibits m/e 340 (M$^+$) and ^{1}H NMR of compound **2a** in DMSO shows signals at (δ ppm) 2.08 (s, 3H, CH$_3$CO), 3.35 (s, 2H, methylene of benzyl group), 3.38 (m, 2H, diastereotopic methylene protons), 3.39–3.95 (q, 1H, stereogenic methine proton), 7.24 and 7.26 (s, 2H, NH of amido group, exchangeable), 7.29–7.40 (m, 5H, ArH of benzylamine moiety), 7.68 and 7.72 (2d, 4H, phenyl protons), and 10.34 (s, 1H, COOH, exchangeable).

Furthermore the structure of compound **2a** was established chemically from the reaction with aromatic aldehydes (Scheme 2); when compound **2a** was allowed to react with

benzaldehyde and/or pyridine-2-carboxyaldehyde in boiling ethanol in the presence of triethylamine (TEA) as a base it afforded the arylidene derivatives' compounds (**4a, b**). The structures of compounds (**4a, b**) were confirmed by elemental analysis and spectral data; EIMS of compound **4a** exhibits the molecular ion peak m/e 428 (M$^+$) and ^{1}H NMR of **4a** reveals signals at (δ ppm) 2.09 (s, 3H, CH$_3$CO), 3.59 (s, 2H, methylene of benzyl moiety), 7.25 (s, 1H, NH of amide), and 10.93 (s, 1H, COOH).

In the recent years a substantial number of 3-(2*H*)-pyridazinones have been reported to possess antimicrobial [20, 21], potent analgesic [22], anti-inflammatory [22–26], antifeedant [27], herbicidal [28], antihypertensive [29–31] and antiplatelet, [32–34], anticancer [35], and other anticipated biological [9] and pharmacological properties [36, 37]. From the previous facts, authors planned to synthesize pyridazinones' derivatives through reacting acids **2a–f** with hydrazine hydrate (Schemes 2 and 3).

Thus when acids **2a–f** were allowed to react with hydrazine hydrate in boiling ethanol they afforded interesting pyridazinone derivatives **3** (Scheme 2) and (**5b–f**) (Scheme 3). The structures of compounds (**5b–f**) were ascertained by elemental analysis and spectral data; EIMS of compound **5b** exhibits m/e 354 (M$^+$) and ^{1}H NMR of compound **5b** in DMSO shows signals at (δ ppm) 2.18 (s, 3H, CH$_3$CO), 3.55 (m, 1H, methine), 1.65, 2.76, and 4.27 (m, 6H, CH$_2$ of propyl group), 6.83, 7.18, and 7.72 (m, 3H, CH of imidazole moiety), 7.15 (s, 1H, NH of amide), 7.51 and 8.11 (m, 4H, ArH), and 10.95 (s, 1H, COOH).

However the reactions of α and β unsaturated carbonyl compounds with binucleophiles provide a convenient route to interesting heterocycles (Scheme 4).

Recently [38] it was reported that β-aroylacrylic acids react with o-phenylenediamine to give quinoxalin-2-ones. Thus when compound **1** was allowed to react with o-phenylenediamine in isopropyl alcohol it yielded the quinoxaline derivative compound **6**; the reaction takes place via

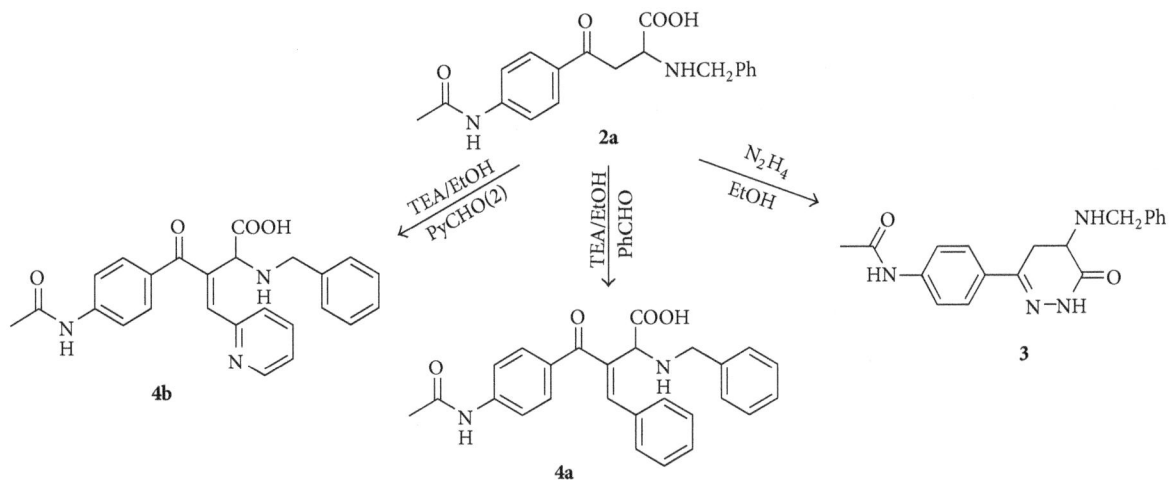

SCHEME 2: Reactions of aza-Michael adduct **2a**.

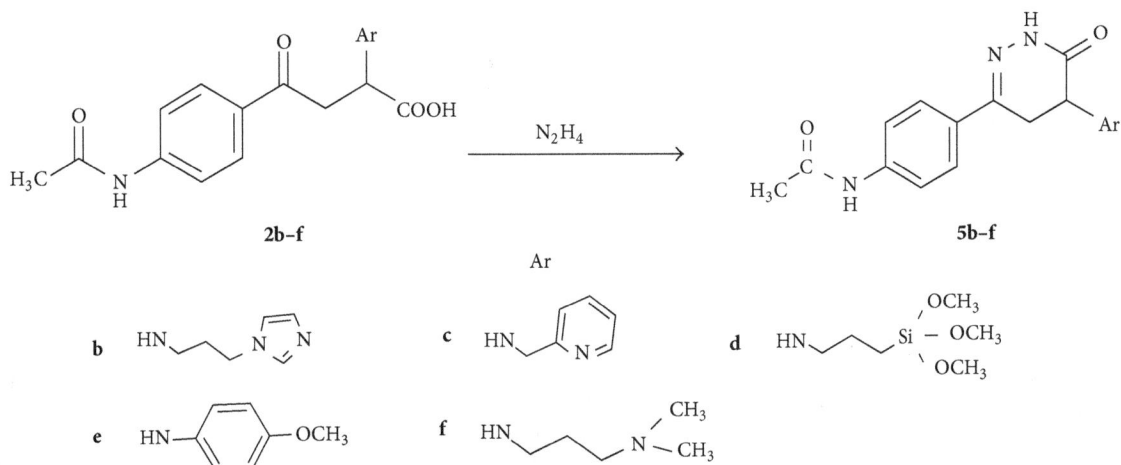

SCHEME 3: Synthesis of pyridazinone derivatives.

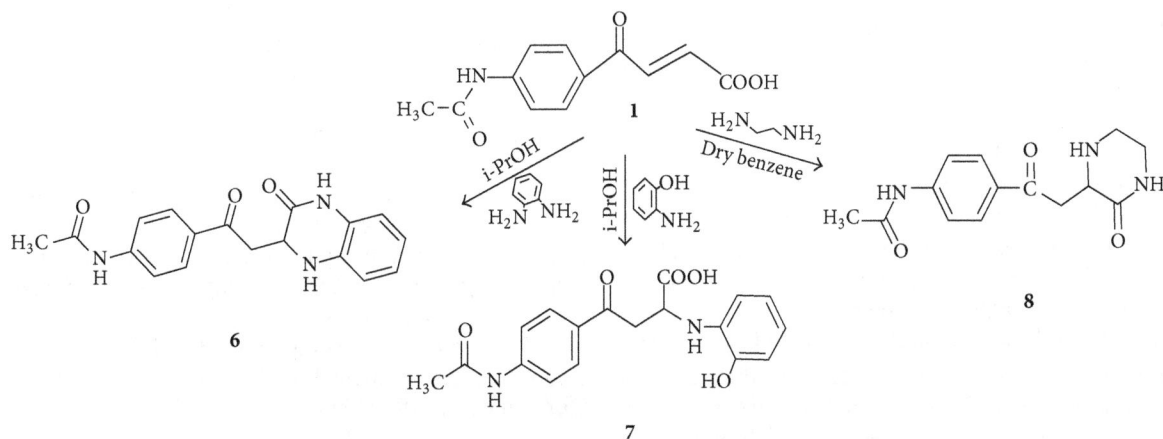

SCHEME 4: Reaction of β-aroylacrylic acid **1** with binucleophiles.

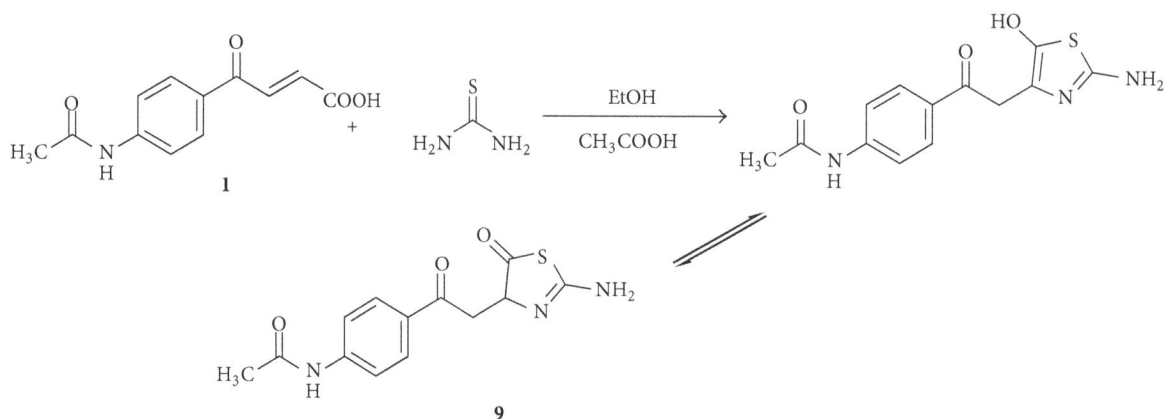

Scheme 5: Reaction of β-aroylacrylic acid **1** with thiourea.

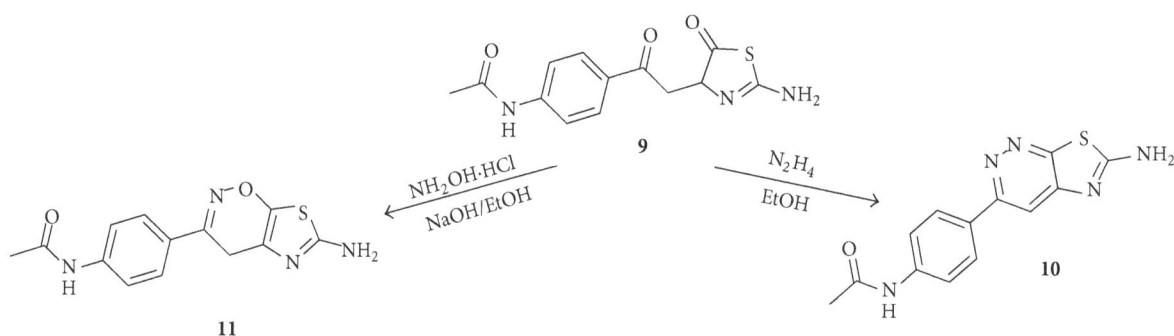

Scheme 6: Reaction of thiazole derivative **9** with hydroxyl amine and hydrazine hydrate.

aza-Michael addition followed by dehydration leading to the desired product; the structure of compound **6** was confirmed by elemental analysis and spectral data. IR spectrum of compound **6** revealed strong absorption bands at 1673, 3107, 3186, 3260, and 3373 cm^{-1} attributable to $v_{C=O}$ and v_{NH} bonded and nonbonded, respectively.

While the reaction of compound **1** with o-aminophenol in isopropyl alcohol afforded the aza-Michael adduct compound **7**, the structure of compound **7** was confirmed by elemental analysis and spectral data. Another binucleophile, namely, ethylenediamine, was reacted with 4-(4-acetamidophenyl)-4-oxobut-2-enoic acid **1** to give the piperazine derivative compound **8** via the addition of amino group to the α carbon of the activated double bound followed by ring closure; the structure of N-(4-(2-(3-oxopiperazin-2-yl)acetyl)phenyl)acetamide **8** was confirmed by elemental analysis and spectral data; ^{1}H NMR of compound **8** in DMSO exhibits signals at (δ ppm) 2.37 (s, 3H, CH$_3$CO), 2.93 and 3.46 (m, 4H, methylene of piperazin moiety), 4.27 (t, 1H, methine), 7.14 (s, 1H, NH of acetamide), and 7.83 and 8.11 (m, 4H, ArH).

Previously [39], it has been reported that thiourea reacted with 4-(4-chloro-3-methyl)phenyl-4-oxobut-2-enoic acid and yielded 2-amino-4-hydroxy-5-(4′-chloro-3′-methyl)benzoyl methyl thiazole. This prompted us to extend the study of the behaviour of the activated olefinic double bond in 4-(4-acetamido)phenyl-4-oxobut-2-enoic acid **1**

towards the same reagent. Thus when acid **1** was allowed to react with thiourea in boiling ethanol in the presence of few drops of glacial acetic acid, it afforded N-(4-(2-(2-amino-5-hydroxy-thiazol-4-yl)acetyl)phenyl)acetamide **9** (Scheme 5). The structure of compound **9** was confirmed by elemental analysis and spectral data; IR spectrum of **9** shows well-defined absorption bands attributable to v_{OH} and v_{NH} groups at 3347 and 3300 cm^{-1}, carbonyl group bands at 1681 and 1650 cm^{-1}, and $v_{C=N}$ of thiazole at 1602 cm^{-1}.

Furthermore, the reaction of thiazole derivative **9** with hydrazine hydrate and hydroxyl amine was investigated. In this way polynuclear systems containing a thiazole ring fused with another heterocyclic ring are usually formed (Scheme 6). Condensation of **9** with hydrazine hydrate in boiling ethanol yielded N-(4-(6-aminothiazolo[5,4-c]pyridazin-3-yl)phenyl)acetamide **10** (Scheme 6). The reaction of the thiazole **9** with hydroxylamine hydrochloride in alcoholic sodium hydroxide affected condensation with carbonyl group and subsequent ring closure, yielding oxazine derivative compound **11**.

The structures of compound **10** and compound **11** were confirmed by elemental analysis and spectral data; EIMS of compound **10** exhibits m/e 285 (M$^+$), while 1H NMR of compound **11** in DMSO shows signals at (δ ppm) 2.24 (s, 3H, CH$_3$CO), 4.15 (s, 2H, CH$_2$ of oxazine ring), 6.85 (s, 2H, amino), 7.21 (s, 1H, NH of acetamide), and 7.61–7.85 (m, 4H, phenyl protons).

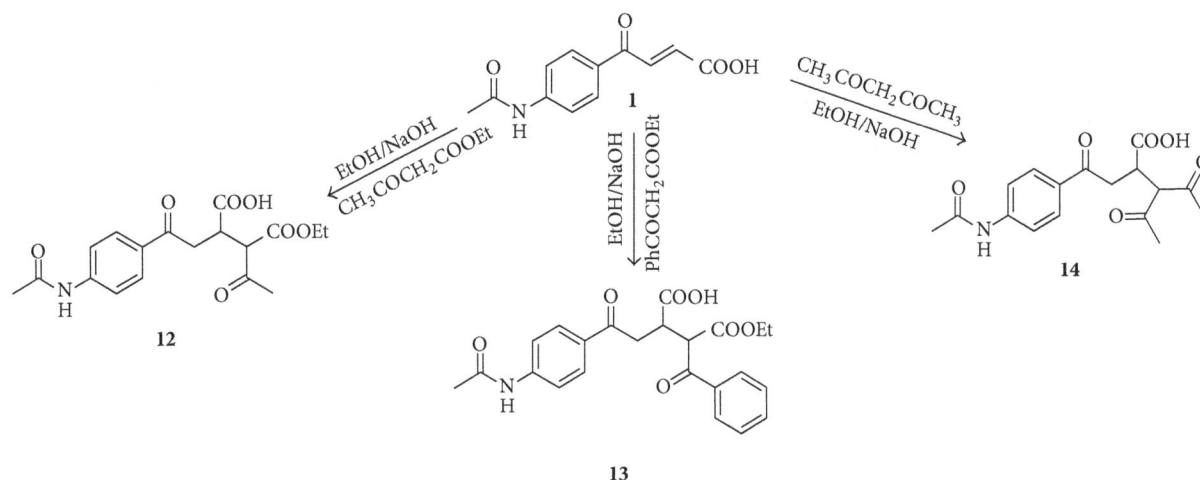

Scheme 7: Reaction of β-aroylacrylic acid **1** with carbon nucleophiles.

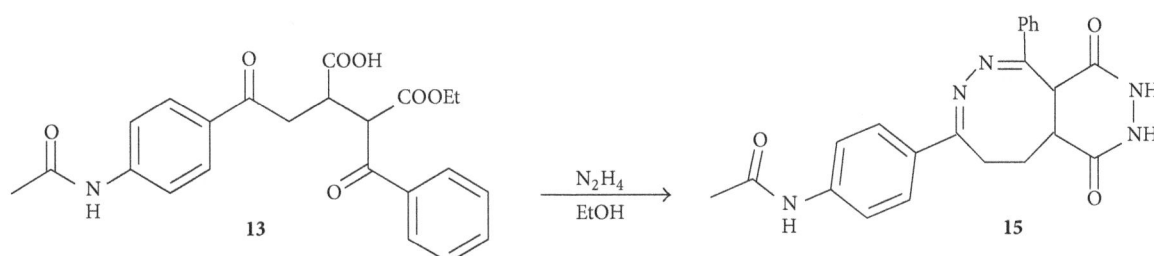

Scheme 8: Synthesis of diazepine derivative.

On the other hand, the authors studied the behavior of 4-(4-acetamidophenyl)-4-oxobut-2-enoic acid **1** towards some carbon nucleophiles under Michael reaction conditions. So when acid **1** was allowed to react with carbon nucleophiles, namely, ethyl acetoacetate, ethyl benzoylacetate, and acetylacetone, it yielded the Michael adducts **12**, **13**, and **14**, respectively (Scheme 7).

Furthermore the interaction of Michael adduct compound **13** with hydrazine hydrate gave the diazepine derivative compound **15** (Scheme 8); the structures of compounds **12–15** were confirmed by elemental analysis and spectral data.

4. Antimicrobial Activity

β-Aroylacrylic acid and its derivatives represent one of the most active classes of compounds that possess a wide spectrum of biological activity. Many of these compounds have been used for the treatment of various diseases and exhibit antibacterial activity and a broad spectrum of physiological (fungicidal, antitumor, hypotensive, hypolipidemic, etc.) activities [1]. In the present work, synthesis of some β-aroylacrylic acids and their derivatives was reported. Some of the new synthesized compounds have been tested for their antimicrobial activity evaluation. Antimicrobial activity of the tested samples was determined using a modified Kirby-Bauer disc diffusion method [40]. Briefly, 100 μl of the test bacteria was grown in 10 mL of fresh media (Mueller-Hinton agar) until they reached a count of approximately 108 cells/mL

for bacteria [41]. 100 μl of microbial suspension was spread onto agar plates corresponding to the broth in which they were maintained. The tested organisms were the gram +ve bacteria (*Staphylococcus aureus* ATCC 25923 and *Bacillus subtilis* MTCC 121) and the gram −ve bacteria (*Escherichia coli* ATCC 25922 and *Pseudomonas aeruginosa* ATCC 27853), by using sterile Whatman-No1 filter paper disks (8.0 mm diameter). Each compound was dissolved in DMSO. Filter paper disks were loaded with certain amount of the tested material (30 μg/disk) and then left with care under hot air to complete dryness. The disks were deposited on the surface of agar plates and the disks were incubated at 5°C for 1 h, to permit good diffusion. All the plates were then incubated for 24 h at 37°C. The diameter of inhibition zones was measured in mm. Table 1 represents the antibacterial activity of some new synthesized compounds.

5. Conclusions

Novel pyridazinone, thiazole, diazepine, and other heterocyclic compounds were successfully synthesized through simple methods. The structures for the new synthesized compounds were confirmed by elemental analysis, FTIR, NMR, and mass spectra. These compounds were evaluated for *in vitro* antibacterial activities against some strains of bacteria. And some of them showed significant activities for both gram positive and gram negative bacteria, where it was

TABLE 1

Sample	Inhibition zone diameter (mm/mg sample)			
	*Escherichia coli*ATCC 25922 (G−)	*Pseudomonas aeruginosa* ATCC 27853 (G−)	*Staphylococcus aureus* ATCC 25922 (G+)	*Bacillus subtilis* MTCC 121 (G+)
2b	++	+	+	++
2e	+	+	+	−
3	+	−	+	+
6	++	++	+	+
8	+	+	+	−
10	+++	+	+++	++
11	++	++	++	−
15	+++	−	+++	++

−: no activity, +: weak activity (diameter 5 : 10 mm), ++: moderate activity (diameter 10 : 15 mm), and +++: strong activity (diameter > 15 mm).

found that compounds **2b**, **2e**, and **3** showed moderate to weak activity against gram +ve and gram −ve bacteria which may be due to the pyridazinone moiety, while compounds **6** and **8** showed moderate to weak activity against gram +ve and gram −ve bacteria because of quinoxaline and pyrazine moiety, and compounds **10** and **11** showed strong to moderate activity against gram +ve and gram −ve bacteria which may be due to thiazole ring; finally compound **15** showed strong activity against gram +ve and gram −ve bacteria because of diazepine moiety.

Conflict of Interests

The authors declare that there is no conflict of interests regarding the publication of this paper.

References

[1] K.-I. Onoue, T. Shintou, C. S. Zhang, and I. Itoh, "An efficient synthesis of β-Aroylacrylic acid ethyl ester by the Friedel-Crafts reaction in the presence of diethyl sulfate," *Chemistry Letters*, vol. 35, no. 1, pp. 22–23, 2006.

[2] T. Köhler, G. Friedrich, and P. Nuhn, "Phospholipase A2 inhibition by alkylbenzoylacrylic acids," *Agents and Actions*, vol. 32, no. 1-2, pp. 70–72, 1991.

[3] T. Köhler, M. Heinisch, M. Kirchner, G. Peinhardt, R. Hirschelmann, and P. Nuhn, "Phospholipase A2 inhibition by alkylbenzoylacrylic acids," *Biochemical Pharmacology*, vol. 44, no. 4, pp. 805–813, 1992.

[4] Z. Juranic, L. J. Stevovi, B. Drakuli, T. Stanojkovi, S. Radulaovi, and I. Juranic, "Substituted (E)-b-(benzoyl)acrylic acids suppressed survival of neoplastic human HeLa cells," *Journal of the Serbian Chemical Society*, vol. 64, pp. 505–512, 1999.

[5] B. J. Drakulic, T. P. Stanojkovic, Z. S. Zizak, and M. M. Dabovic, "Antiproliferative activity of aroylacrylic acids. Structure-activity study based on molecular interaction fields," *European Journal of Medicinal Chemistry*, vol. 46, no. 8, pp. 3265–3273, 2011.

[6] A. Sammour and M. Elhashash, "Alkylation of aromatic hydrocarbons with β-aroylacrylic acids," *Journal für Praktische Chemie*, vol. 314, no. 5-6, pp. 906–914, 1972.

[7] M. Elhashash, Y. M. Elkady, and M. M. Mohamed, "Reactions & synthesis of 2-Aryl-3-(4-bromobenzoyl)propionic acids via Friedel Craft's alkylation of aromatic hydrocarbons with 3-(4-bromobenzoyl)acrylic acid," *Indian Journal of Chemistry B*, vol. 18, p. 136, 1979.

[8] S. A. Risk, M. Elhashash, and K. K. Mostafa, "Utility of β-aroyl acrylic acid in heterocyclic synthesis," *Egyptian Journal of Chemistry*, vol. 51, p. 611, 2008.

[9] A. S. A. Youssef, M. I. Marzouk, H. M. F. Madkour, A. M. A. El-Soll, and M. A. El-Hashash, "Synthesis of some heterocyclic systems of anticipated biological activities via 6-aryl-4-pyrazol-1-yl-pyridazin-3-one," *Canadian Journal of Chemistry*, vol. 83, no. 3, pp. 251–259, 2005.

[10] A. S. A. Youssef, H. M. F. Madkour, M. I. Marzouk, A. M. A. El-Soll, and M. A. El-Hashash, "Utility of 3-aroylprop-2-enoic acid in heterocyclic synthesis," *Afinidad*, vol. 61, no. 512, pp. 304–316, 2004.

[11] M. Umpreti, S. Pant, A. Dandia, and U. C. Pant, "Synthesis of 8-substituted-2-carboxy-4-(4-fluorophenyl)-2,3-dihydro-1,5-benzothiazepines," *Phosphorus, Sulfur, and Silicon and the Related Elements*, vol. 113, no. 1–4, pp. 165–171, 1996.

[12] A. N. Nesmeyanov, M. I. Rybinskaya, and A. I. Rybin, "Orientation of the nucleophilic attack on the activated double bond," *Uspekhi Khimii*, vol. 36, article 1089, 1967.

[13] R. D. Khachikyan, N. V. Karamyan, G. A. Panosyan, and M. G. Indzhikyan, "Reactions of β-aroylacrylic acids with N-nucleophiles," *Akademiia nauk SSSR. Izvestiia. Seriia khimicheskaia*, p. 1923, 2005.

[14] M. Bianchi, "Gastric anti-secretory, anti-ulcer and cytoprotective properties of substituted (E)-4-phenyl- and heteroaryl-4-oxo-2-butenoic acids," *European Journal of Medicinal Chemistry*, vol. 23, no. 1, pp. 45–52, 1988.

[15] A. Giordani, "4-phenyl-4-oxo-2-butenoic acid derivatives with kynurenine-3-hydroxylase inhibiting activity," United States Patent 6048896, 2000.

[16] I. L. Pinto, R. L. Jarvest, B. Clarke et al., "Inhibition of human cytomegalovirus protease by enedione derivatives of thieno[2,3-d]oxazinones through a novel dual acylation/alkylation mechanism," *Bioorganic and Medicinal Chemistry Letters*, vol. 9, no. 3, pp. 449–452, 1999.

[17] C. Pfefferle, C. Kempter, J. W. Metzger, and H.-P. Fiedler, "(E)-4-oxonon-2-enoic acid, an antibiotically active fatty acid produced by *Streptomyces olivaceus* Tü 4018," *Journal of Antibiotics*, vol. 49, no. 8, pp. 826–828, 1996.

[18] V. K. Chintakunta, V. Akella, M. S. Vedula et al., "3-O-Substituted benzyl pyridazinone derivatives as COX inhibitors," *European Journal of Medicinal Chemistry*, vol. 37, no. 4, pp. 339–347, 2002.

[19] D. Papa, E. Schwenk, F. Villani, and E. Klingsberg, "β-Aroylacrylic acids," *Journal of the American Chemical Society*, vol. 70, no. 10, pp. 3356–3360, 1948.

[20] G. H. Sayed, M. A. Sayed, M. R. Mahmoud, and S. S. Shaaban, "Synthesis and reactions of new pyridazinone derivatives of expected antimicrobial activities," *Egyptian Journal of Chemistry*, vol. 45, pp. 767–776, 2002.

[21] A. Katrusiak, A. Katrusiak, and S. Bałoniak, "Reactivity of 6-chloro-4- and 5-hydrazino-2-phenyl-3(2H)-pyridazinones with Vilsmeier reagent," *Tetrahedron*, vol. 50, no. 45, pp. 12933–12940, 1994.

[22] B. Okcelik, S. Unlu, E. Banoglu, E. Kupeli, E. Yesilada, and M. F. Sahin, "Investigations of new pyridazinone derivatives for the synthesis of potent analgesic and anti-inflammatory compounds with cyclooxygenase inhibitory activity," *Archiv der Pharmazie*, vol. 336, no. 9, pp. 406–412, 2003.

[23] D. S. Dogruer, M. F. Sahin, E. Kupeli, and E. Yesilada, "Synthesis and analgesic and anti-inflammatory activity of new pyridazinones," *Turkish Journal of Chemistry*, vol. 27, pp. 727–738, 2003.

[24] E. B. Frolov, F. J. Lakner, A. V. Khvat, and A. V. Ivachtchenko, "An efficient synthesis of novel 1,3-oxazolo[4,5-d]pyridazinones," *Tetrahedron Letters*, vol. 45, no. 24, pp. 4693–4696, 2004.

[25] E. Banoglu, C. Akoglu, S. Unlu, E. Kupeli, E. Yesilada, and M. F. Sahin, "Amide derivatives of [6-(5-methyl-3-phenylpyrazole-1-yl)-3(2H)-pyridazinone-2-yl]acetic acids as potential analgesic and anti-inflammatory compounds," *Archiv der Pharmazie*, vol. 337, no. 1, pp. 7–14, 2004.

[26] M. Gokçe, D. Dogruer, and M. F. Sahin, "Synthesis and antinociceptive activity of 6-substituted-3-pyridazinone derivatives," *Il Farmaco*, vol. 56, no. 3, pp. 233–237, 2001.

[27] S. Cao, X. Qian, G. Song, B. Chai, and Z. Jiang, "Synthesis and antifeedant activity of new oxadiazolyl 3(2H)-pyridazinones," *Journal of Agricultural and Food Chemistry*, vol. 51, no. 1, pp. 152–155, 2003.

[28] V. Dal Piaz, G. Ciciani, and M. P. Giovannoni, "5-Acetyl-2-methyl-4-nitro-6-phenyl-3(2H)-pyridazinone: versatile precursor to hetero-condensed pyridazinones," *Synthesis*, no. 7, pp. 669–671, 1994.

[29] C. Öğretir, S. Yarligan, and Ş. Demirayak, "Spectroscopic determination of acid dissociation constants of some biologically active 6-phenyl-4,5-dihydro-3(2H)-pyridazinone derivatives," *Journal of Chemical and Engineering Data*, vol. 47, no. 6, pp. 1396–1400, 2002.

[30] R. Barbaro, L. Betti, M. Botta et al., "Synthesis, biological evaluation, and pharmacophore generation of new pyridazinone derivatives with affinity toward α1- and α2-adrenoceptors," *Journal of Medicinal Chemistry*, vol. 44, no. 13, pp. 2118–2132, 2001.

[31] I. Sircar, "Synthesis of new 1,2,4-triazolo[4,3-b]pyridazines and related compounds," *Journal of Heterocyclic Chemistry*, vol. 22, no. 4, pp. 1045–1048, 1985.

[32] A. Coelho, E. Sotelo, N. Fraiz et al., "Pyridazines. Part 36: Synthesis and antiplatelet activity of 5-substituted-6-phenyl-3(2H)–pyridazinones," *Bioorganic & Medicinal Chemistry Letters*, vol. 14, pp. 321–324, 2004.

[33] E. Sotelo, N. B. Centeno, J. Rodrigo, and E. Ravina, "Pyridazine derivatives. Part 27: A joint theoretical and experimental approach to the synthesis of 6-phenyl-4,5-disubstituted-3(2H)-pyridazinones," *Tetrahedron Letters*, vol. 58, pp. 2389–2395, 2002.

[34] E. Sotelo, N. Fraiz, M. Yáez et al., "Pyridazines. Part XXIX: synthesis and platelet aggregation inhibition activity of 5-substituted-6-phenyl-3(2H)-pyridazinones. Novel aspects of their biological actions," *Bioorganic and Medicinal Chemistry*, vol. 10, no. 9, pp. 2873–2882, 2002.

[35] W. Malinka, A. Redzicka, and O. Lozach, "New derivatives of pyrrolo[3,4-d]pyridazinone and their anticancer effects," *Il Farmaco*, vol. 59, pp. 457–462, 2004.

[36] E. Sotelo, B. Pita, and E. Raviña, "Pyridazines. Part 22: highly efficient synthesis of pharmacologically useful 4-cyano-6-phenyl-5-substituted-3(2H)-pyridazinones," *Tetrahedron Letters*, vol. 41, no. 16, pp. 2863–2866, 2000.

[37] E. Sotelo, A. Coelho, and E. Ravina, "Pyridazine derivatives 32: stille-based approaches in the synthesis of 5-substituted-6-phenyl-3(2H)-pyridazinones," *Chemical and Pharmaceutical Bulletin*, vol. 51, no. 4, pp. 427–430, 2003.

[38] N. N. Kolos, L. Y. Kovalenko, S. V. Shishkina, O. V. Shishkin, and I. S. Konovalova, "Interaction of esters of β-aroylacrylic acids with o-phenylenediamines and 1,2-diamino-4-phenyl-imidazole," *Chemistry of Heterocyclic Compounds*, vol. 43, no. 11, pp. 1397–1405, 2007.

[39] M. El-Kady, M. A. El-Hashash, and M. A. Sayed, "Action of Hydrazine,Amine and Thiourea upon 3-(4-chloro-3-methyl)benzoyl acrylic acid," *Revue Roumaine de Chimie*, vol. 26, no. 8, pp. 1161–1167, 1981.

[40] A. W. Bauer, W. M. Kirby, J. C. Sherris, and M. Turck, "Antibiotic susceptibility testing by a standardized single disk method," *The American Journal of Clinical Pathology*, vol. 45, no. 4, pp. 493–496, 1966.

[41] M. A. Pfaller, L. Burmeister, M. S. Bartlett, and M. G. Rinaldi, "Multicenter evaluation of four methods of yeast inoculum preparation," *Journal of Clinical Microbiology*, vol. 26, no. 8, pp. 1437–1441, 1988.

9

A Conformational Model for MTPA Esters of Chiral N-(2-Hydroxyalkyl)acrylamides

Eduardo M. Rustoy,[1] Alicia Baldessari,[1] and Leandro N. Monsalve[1,2]

[1] *Laboratorio de Biocatálisis, Departamento de Química Orgánica y UMYMFOR, Facultad de Ciencias Exactas y Naturales,*
Universidad de Buenos Aires, Pabellón 2, Piso 3, Ciudad Universitaria, C1428EGA Buenos Aires, Argentina
[2] *INTI-CONICET, Avenida Gral. Paz 5445, Ed. 42, San Martín, B1650JKA Buenos Aires, Argentina*

Correspondence should be addressed to Leandro N. Monsalve; monsalve@inti.gob.ar

Academic Editor: Daniel Glossman-Mitnik

The absolute stereochemistry of novel chiral N-(2-hydroxylalkyl)acrylamides prepared by a lipase-catalyzed resolution was successfully determined by ^{1}H NMR of their MTPA esters. The method was validated for this particular case by computational experiments.

1. Introduction

It is well known that chiral acrylamides are useful compounds in organic synthesis [1–3].

The absolute stereochemistry of chiral compounds is determined by using several methods [4]. Among the methods available, NMR spectroscopy using chiral derivatizing agents such as 2-methoxy-2-(trifluoromethyl)phenylacetic acid (MTPA) has been widely used in the determination of configuration of stereogenic centers bearing either hydroxyl or amine groups. More recently, this methodology has been applied for the derivatization of chiral primary alcohols thus determining the absolute configuration of stereogenic centers at a two- and three-bond distance from the hydroxyl group [5–8]. Moreover, the modified Mosher's method has been used to determine the absolute configuration of primary alcohols with chiral methyl groups at C-2 [9].

However, it should be noted that this modification of Mosher's method seemed to be unsuccessful for determining their stereochemistry of MTPA esters of some simple C-2 branched primary alcohols with conjugated groups or a consecutive chiral centre at C-3. According to Tsuda et al. the difference in chemical shift between diastereotopic oxymethylene protons is similar for both diastereomers [9].

We have previously performed a lipase-catalyzed synthesis of nonchiral and chiral N-(2-hydroxyalkyl)acrylamides (Scheme 1) [10, 11].

We employed the modified Mosher's method for the determination of %ee. The absolute configuration of products was found to be (S). However, a report by Puertas et al. [12] describes the enantioselective behavior of *Candida antarctica* lipase B (CAL B) affording the (R)-acrylamides instead of (S)-acrylamides as products, using racemic amines as starting materials.

Considering the absence of a conformational model for MTPA esters of β-chiral primary alcohols and the results of our experiments showing opposite selectivity to previous experience, we decided to test the reliability of the method for our case. The aim of this work was to validate the results by conformational analysis of MTPA esters and thus obtain a reasonable explanation for NMR data on a molecular basis.

2. Results and Discussion

We performed the synthesis of chiral N-(2-hydroxyalkyl)acrylamides **2a–d** and their corresponding MTPA esters **3a–d** and **4a–d** as described in order to assign their absolute stereochemistry and the degree of stereoselectivity achieved in each case (Scheme 2).

After purification, the MTPA esters were analyzed by ^{1}H NMR spectroscopy in CDCl$_3$. The results observed for compounds (S)-**3b** and (S)-**4b** were taken as example and showed significant differences in the chemical shift for many

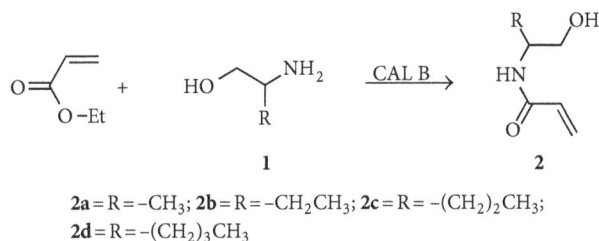

$2a = R = -CH_3$; $2b = R = -CH_2CH_3$; $2c = R = -(CH_2)_2CH_3$;
$2d = R = -(CH_2)_3CH_3$

Scheme 1: Lipase-catalyzed synthesis of 2a–d.

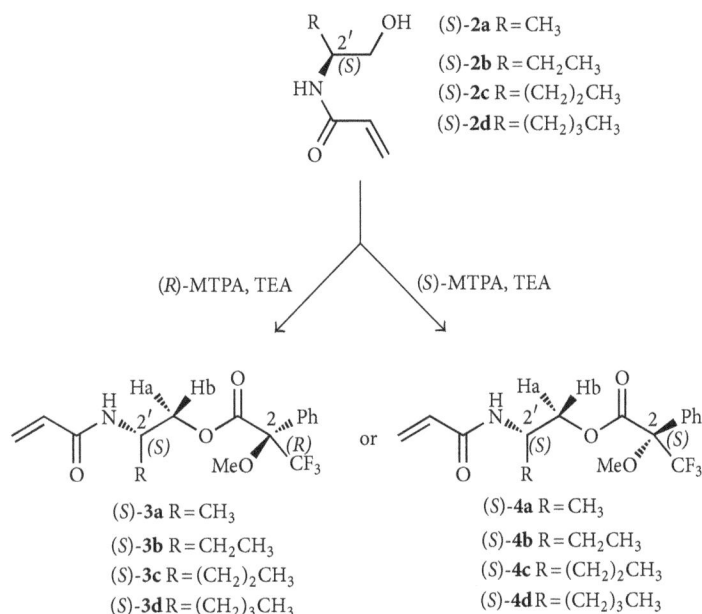

Scheme 2: MTPA esters (S)-3a–d and (S)-4a–d.

signals. Particularly, signals for diastereotopic protons H1'a and H1'b of the (R)-MTPA amido ester (S)-3b were for the major isomer at δ 4.36 and 4.37 ppm and for the minor isomer at δ 4.33 and 4.44 ppm (Figure 3, Spectrum A). For the (S)-MTPA ester (S)-4b these signals are reversed, so that those corresponding to the major isomer are observed at δ 4.33 and 4.44 ppm and the signals for the minor isomer are observed at δ 4.36 and 4.37 ppm (Figure 1, Spectrum B). The integration of the above-mentioned signals in both cases gave the same 35% ee value for 2b.

The same procedure was applied to study the enantiomeric purity of the other three products 2a, 2c, and 2d. Table 1 (columns 4 and 5) shows these results.

This pattern for the oxymethylene protons of chiral primary alcohols esterified with MTPA is according to previously published results on the determination of the absolute stereochemistry of a series of chiral primary alcohols with a methyl group at C-2 position [9] and chiral 1,3-dihidroxyketones [7]. In these works the absolute configuration of stereogenic centers was assigned by considering the $\Delta\delta$ between oxymethylene protons. Larger $\Delta\delta$ values of (R)-MTPA derivative were diagnosis for R stereochemistry on the carbon vicinal to oxymethylene whereas smaller $\Delta\delta$ values of (R)-MTPA derivative were diagnosis for the opposite

stereochemistry (S). Accordingly, (S)-MTPA derivatives of (R)-primary alcohols showed smaller $\Delta\delta$ values for their oxymethylene proton signals than those prepared from (S)-primary alcohols.

For (R)-MTPA derivatives of chiral N-(2-hydroxyalkyl)acrylamides here reported, we found $\Delta\delta$ = 0.01 ppm for the major isomer and $\Delta\delta$ = 0.11 ppm for the minor isomer. This fact should indicate that the absolute stereochemistry of the major isomer is $(2R,2'S)$ and therefore (S) is the absolute configuration of the products 2a–d.

On the other hand the stereoselective behavior of lipases towards hydrolysis of esters of chiral secondary alcohols has been extensively studied. The Kazlauskas rule is intended to predict the behavior of lipases in such cases and, according to this rule, lipases tend to hydrolyze esters of chiral secondary alcohols having absolute configuration (R) faster than their enantiomer [13]. Lipases also showed the same stereochemical preference in transesterification and aminolysis reactions [14, 15].

This model takes into consideration that the substituent seniority can be correlated with substituent size, which is not always the case. The enormous diversity of substrates accepted by lipases and reaction conditions applied showed that this rule is not met in some circumstances [16, 17].

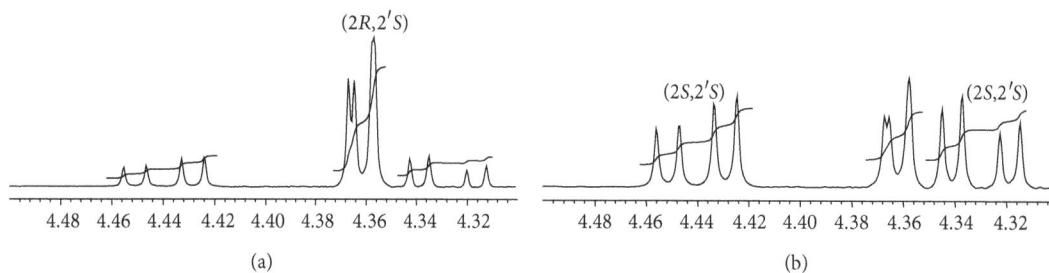

FIGURE 1: ^{1}H NMR signals for diastereotopic protons H1$'$a and H1$'$b in the (R)-MTPA amido ester (S)-**3b** (Spectrum A) and (S)-MTPA amido ester (S)-**4b** (Spectrum B).

TABLE 1: Conformers of both diastereomers of **3b**.

Compound	Diastereomer	Tor1 (°)	Tor2 (°)	$\triangle E$ (Kcal/mol)	%P
(R)-**3b**	(2R,2$'$R)	178	1	1.54	4.2
		−66	2	0.54	22.6
		−70	5	0.00	56.5
		66	2	0.72	16.7
(S)-**3b**	(2R,2$'$S)	73	−5	0.59	17.4
		−49	−4	1.70	2.6
		69	1	0.34	26.6
		−76	4	1.88	2.0
		−177	7	0.00	47.3
		168	4	1.60	3.1
		165	0	2.44	0.8
		162	−2	3.02	0.3
		−54	1	1.64	2.9
		156	−108	2.99	0.3
		−62	−140	1.72	2.6

For this reason, we considered that this primary conclusion should be submitted to further analysis, since this result seems difficult to be interpreted in terms of the Kazlauskas rule that predicted (R)-configuration. Therefore the stereochemical outcome of the enzymatic aminolysis of ethyl acrylate was not confirmed. Previous results on the enzyme-catalyzed transesterification of acrylates [18] derivatives of such compounds showed enantioselectivity towards (R)-enantiomer.

Furthermore, some authors reported that enzyme-catalyzed hydrolysis and transesterification of alcohols [14] and hydrolysis of amide [19] derivatives of such compounds showed enantioselectivity towards (R)- or (S)-enantiomer depending on the catalyst and the reaction conditions.

Moreover, we could not establish precisely if the reasoning from previously published results could also be applied in the determination of absolute stereochemistry of the chiral primary alcohols belonging to products **2a–d**, by means of ^{1}H NMR analysis of their MTPA derivatives. Besides the experimental fact that the configuration of the chiral products was known, no clear evidence was found that could be interpreted or generalized on a molecular basis (i.e., evidence of conformational restrictions). For this reason, we decided to perform some experiments *in silico* in order to provide an independent explanation for experimental results.

Conformational search experiments were performed with MTPA derivatives (S)-**3b** (2S,2$'$S) and (R)-**3b** (2R,2$'$S). In these experiments the torsion angles O–C1$'$–C2$'$–C3$'$ (Tor1) and MeO–C2–C1=O (Tor2) (Figure 2) were varied in order to obtain the most stable conformers of both compounds and thus interpret the chemical shift differences observed between their oxymethylene protons and H2$'$ in ^{1}H NMR experiments.

As it can be observed in Table 1, up to nine stable conformers were found for (S)-**3b** within 3 Kcal/mol above the global minimum. Most conformers had their methoxyl group synperiplanar to their carbonyl ester, which is according to previous calculations on other MTPA esters by DFT methods. However, many rotamers along O–C1$'$ and C1$'$–C2$'$ bond were found to be stable enough to be important contributions to conformer population. This fact explains the broader signals and lower $\Delta\delta$ values between oxymethylene protons for (S)-**3b** isomer. It was observed that signals at δ 4.36 and 4.37 ppm are broad (more than 6 Hz wide) and not completely resolved.

On the other hand, only four stable conformers were found for the diastereomer (R)-**3b**. All of them had their methoxyl group synperiplanar to their carbonyl ester. Major contributions to the conformational population were the H2$'$

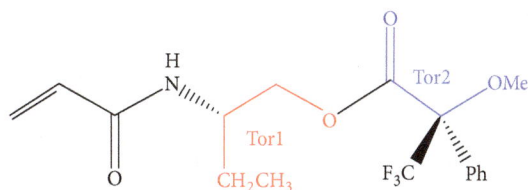

FIGURE 2: Torsion angles Tor1 (red and bold) and Tor2 (blue and bold) employed for conformational search of compound (S)-3b.

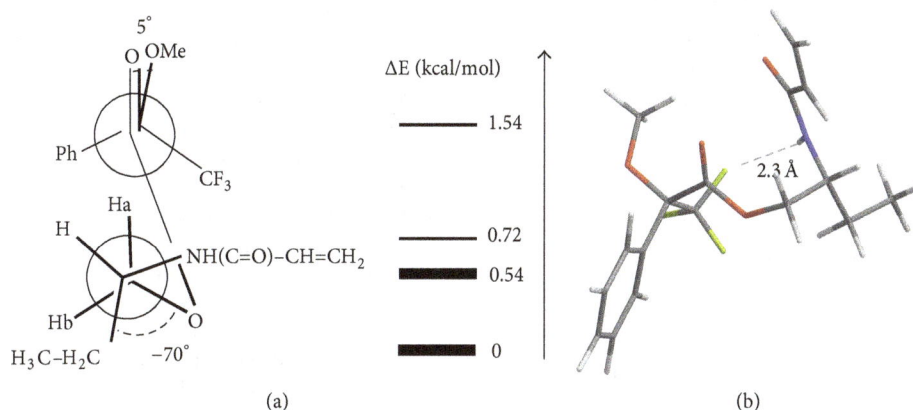

FIGURE 3: (a) Newman projections along the axes C1–C2 (up) and C1′–C2′ (down) for the most stable conformer of (R)-3b. Torsion angles Tor1 and Tor2 are indicated. Major contributions to this type of conformation are highlighted in the energy level diagram on the right. (b) Picture of a 3D molecular render of the same conformer showing intramolecular hydrogen bonding between amide hydrogen and a fluorine atom (H–F distance: 2.3 angstrom).

antiperiplanar to oxygen attached to C1′ (Figure 3). Two stable conformers have this conformation and they contribute to 79.1% of the total conformer population. Moreover, these conformers show one oxymethylene hydrogen (H1′b) synperiplanar to the aromatic ring and the other (H1′a) synperiplanar to the C=O bond. These observations could explain the large difference of chemical shifts between both oxymethylene protons (0.11 ppm) and the coupling constants observed for H1′a-H2′ and H1′b-H2′ (4.1 Hz). The occurrence of intramolecular hydrogen bonding between the amide hydrogen and a fluorine atom was also confirmed for both conformers. We assume that this hydrogen bond plays a key role in conformer stability.

These results are difficult to assimilate to those predicted in Kazlauskas rule based on the size of the substituents in the stereocenter and designed to predict which enantiomer of a secondary alcohol reacts faster in enzyme catalyzed reactions. The rule is reliable with substrates having substituents which differ significantly in size.

In the acylation reaction of every chiral alkanolamine used in this work the enzyme showed an enantioselective behavior opposite to that described by Kazlauskas rule, preferably getting the product by reaction of alkanolamine with (S) configuration instead of the (R) predicted by the rule. In 1a the fact could be explained considering that the hydroxymethyl group is clearly larger than the methyl group, but in 1b ethyl and hydroxymethyl substituents are about the same size and compounds 1c and 1d have their alkyl substituent larger than their hydroxyalkyl substituent. In these

two compounds the Kazlauskas rule could be applied in terms of substituent size.

The opposite configuration observed between the experimental results and that predicted by the rule could be attributed more likely to electronic effects than to steric hindrance. For instance, Maraite et al. showed that Pseudomonas stutzeri lipase stereoselectivity for benzoin acylation could be explained by hydrogen bonding between Tyr-54 hydroxyl group and carbonyl oxygen of the substrate [20]. A similar effect could be attained by hydrogen bonding between hydroxyl moiety of alkanolamine and the threonine-rich loop of residues 39–44 of CAL B. Therefore it could be suggested that the chemo- and enantioselectivity of the lipase-catalyzed N-acylation of chiral alkanolamines must be driven by the presence of the hydroxyl group in one substituent rather than the difference in the substituent size.

3. Conclusions

In this work we have proposed a conformational model for MTPA esters of chiral N-(2-hydroxyalkyl)acrylamides. The in silico conformational analysis of the corresponding diastereomers (R)-3b and (S)-3b showed that the differences in conformational restrictions, and consequently ^{1}H NMR signal splitting for oxymethylene protons, arose from specific intramolecular interactions rather than nonspecific steric hindrance. This analysis was also useful for explaining the chemical shift differences for both diastereomers and served for the validation of the stereochemical determination of chiral *N*-(2-hydroxyalkyl)acrylamides obtained by lipase-catalyzed resolution.

4. Experimental

4.1. General Remarks. ^{1}H NMR spectra were recorded in CDCl$_3$ as solvent using a Bruker Avance II 500 spectrometer operating at 500 MHz. Chemical shifts are reported in δ units (ppm) relative to tetramethylsilane (TMS) set at 0 ppm, and coupling constants are given in hertz. The synthesis of *N*-(2-hydroxyalkyl)acrylamides **2a–d** and their corresponding MTPA derivatives **3a–d** and **4a–d** were performed as described previously [11].

4.2. Molecular Modeling. The structures of (2*R*,2′*R*)-2′-(acryloylamino)butyl 3,3,3-trifluoro-2-methoxy-2-phenylpropanoate (*R*)-**3b** and (2*R*,2′*S*)-2′-(acryloylamino)butyl 3,3,3-trifluoro-2-methoxy-2-phenylpropanoate (*S*)-**3b** were submitted to conformational search using the molecular mechanics method MM+ and a conformational search algorithm integrated on HyperChem 8.0. The selected torsions to vary were O–C1′–C2′–C3′ (Tor1) and MeO–C2–C1=O (Tor2) and conformations of energies below 6 Kcal/mol over the global minimum were submitted to geometry optimization using Gamess [21]. Energies were minimized employing DFT method B3LYP using 6–31 g basis function. Solvent effect was simulated using the PCM method with the standard parameters for chloroform included in the software package. Optimized structures for each isomer were visualized using ChemBio3D Ultra 11.0 and overlaid in order to discard repeated structures.

Conflict of Interests

The authors declare that there is no conflict of interests regarding the publication of this paper.

Acknowledgments

The authors thank UBA X010, CONICET PIP 112-200801-00801/09, and ANPCyT PICT 2005-32735 for partial financial support. Eduardo M. Rustoy, Alicia Baldessari, and Leandro N. Monsalve are research members of CONICET.

References

[1] M. Nyerges, D. Bendell, A. Arany et al., "Silver acetate-catalysed asymmetric 1,3-dipolar cycloadditions of imines and chiral acrylamides," *Tetrahedron*, vol. 61, no. 15, pp. 3745–3753, 2005.

[2] Y. Tian, W. Lu, Y. Che, L. B. Shen, L. M. Jiang, and Z. Q. Shen, "Synthesis and characterization of macroporous silica modified with optically active poly[N-(oxazolinylphenyl)acrylamide] derivatives for potential application as chiral stationary phases," *Journal of Applied Polymer Science*, vol. 115, no. 2, pp. 999–1007, 2010.

[3] J. Tobis, Y. Thomann, and J. C. Tiller, "Synthesis and characterization of chiral and thermo responsive amphiphilic conetworks," *Polymer*, vol. 51, no. 1, pp. 35–45, 2010.

[4] J. M. Seco, E. Quiñoá, and R. Riguera, "The assignment of absolute configuration by NMR," *Chemical Reviews*, vol. 104, no. 1, pp. 17–118, 2004.

[5] T. Pehk, E. Lippmaam, M. Lopp, A. Pajú, B. C. Borer, and R. J. K. Taylor, "Determination of the absolute configuration of chiral secondary alcohols; new advances using ^{13}C- and 2D-NMR spectroscopy," *Tetrahedron Asymm*, vol. 4, no. 7, pp. 1527–1532, 1993.

[6] K. Akiyama, S. Kawamoto, H. Fujimoto, and M. Ishibashi, "Absolute stereochemistry of TT-1 (rasfonin), an α-pyrone-containing natural product from a fungus, Trichurus terrophilus," *Tetrahedron Letters*, vol. 44, no. 46, pp. 8427–8431, 2003.

[7] J. L. Galman and H. C. Hailes, "Application of a modified Mosher's method for the determination of enantiomeric ratio and absolute configuration at C-3 of chiral 1,3-dihydroxy ketones," *Tetrahedron Asymmetry*, vol. 20, no. 15, pp. 1828–1831, 2009.

[8] L. V. Parfenova, T. V. Berestova, T. V. Tyumkina et al., "Enantioelectivity of chiral zirconocenes as catalysts in alkene hydro-, carbo- and cycloalumination reactions," *Tetrahedron Asymmetry*, vol. 21, no. 3, pp. 299–310, 2010.

[9] M. Tsuda, Y. Toriyabe, T. Endo, and J. Kobayashi, "Application of modified Mosher's method for primary alcohols with a methyl group at C2 position," *Chemical & Pharmaceutical Bulletin*, vol. 51, no. 4, pp. 448–451, 2003.

[10] E. M. Rustoy and A. Baldessari, "Chemoselective enzymatic preparation of N-hydroxyalkylacrylamides, monomers for hydrophilic polymer matrices," *Journal of Molecular Catalysis B: Enzymatic*, vol. 39, no. 1–4, pp. 50–54, 2006.

[11] L. N. Monsalve, E. M. Rustoy, and A. Baldessari, "Biocatalytic synthesis of chiral N-(2-hydroxyalkyl)-acrylamides," *Biocatalysis and Biotransformation*, vol. 29, no. 2-3, pp. 87–95, 2011.

[12] S. Puertas, R. Brieva, F. Rebolledo, and V. Gotor, "Lipase catalyzed aminolysis of ethyl propiolate and acrylic esters. Synthesis of chiral acrylamides," *Tetrahedron*, vol. 49, no. 19, pp. 4007–4014, 1993.

[13] R. J. Kazlauskas, A. N. E. Weissfloch, A. T. Rappaport, and L. A. Cuccia, "A rule to predict which enantiomer of a secondary alcohol reacts faster in reactions catalyzed by cholesterol esterase, lipase from Pseudomonas cepacia, and lipase from Candida rugosa," *Journal of Organic Chemistry*, vol. 56, no. 8, pp. 2656–2665, 1991.

[14] F. Francalanci, P. Cesti, W. Cabri, D. Bianchi, T. Martinengo, and M. Foà, "Lipase-catalyzed resolution of chiral 2-amino 1-alcohols," *Journal of Organic Chemistry*, vol. 52, no. 23, pp. 5079–5082, 1987.

[15] J. González-Sabín, V. Gotor, and F. Rebolledo, "Enantioselective acylation of *rac*-2-phenylcycloalkanamines catalyzed by lipases," *Tetrahedron: Asymmetry*, vol. 16, no. 18, pp. 3070–3076, 2005.

[16] X. Xia, Y.-H. Wang, B. Yang, and X. Wang, "Wheat germ lipase catalyzed kinetic resolution of secondary alcohols in nonaqueous media," *Biotechnology Letters*, vol. 31, no. 1, pp. 83–87, 2009.

[17] P. Hoyos, V. Pace, J. V. Sinisterra, and A. R. Alcántara, "Chemoenzymatic synthesis of chiral unsymmetrical benzoin esters," *Tetrahedron*, vol. 67, no. 38, pp. 7321–7329, 2011.

[18] S. Akai, T. Naka, S. Omura et al., "Lipase-catalyzed domino kinetic resolution/intramolecular Diels-Alder reaction: one-pot synthesis of optically active 7-oxabicyclo[2.2.1]heptenes from furfuryl alcohols and β-substituted acrylic acids," *Chemistry: A European Journal*, vol. 8, pp. 4255–4264, 2002.

[19] N. W. Fadnavis, M. Sharfuddin, and S. K. Vadivel, "Resolution of racemic 2-amino-1-butanol with immobilised penicillin G acylase," *Tetrahedron Asymmetry*, vol. 10, no. 23, pp. 4495–4500, 1999.

[20] A. Maraite, P. Hoyos, J. D. Carballeira, A. C. Cabrera, M. B. Ansorge-Schumacher, and A. R. Alcántara, "Lipase from *Pseudomonas stutzeri*: purification, homology modelling and rational explanation of the substrate binding mode," *Journal of Molecular Catalysis B: Enzymatic*, vol. 87, pp. 88–98, 2013.

[21] M. W. Schmidt, K. K. Baldridge, J. A. Boatz et al., "General atomic and molecular electronic structure system," *Journal of Computational Chemistry*, vol. 14, no. 11, pp. 1347–1363, 1993.

10

Synthesis of Highly Stable Cobalt Nanomaterial Using Gallic Acid and Its Application in Catalysis

Saba Naz,[1,2,3] **Abdul Rauf Khaskheli,**[4,5] **Abdalaziz Aljabour,**[5] **Huseyin Kara,**[2,6]
Farah Naz Talpur,[3] **Syed Tufail Hussain Sherazi,**[2,3,6] **Abid Ali Khaskheli,**[3] **and Sana Jawaid**[3]

[1] Dr. M. A. Kazi Institute of Chemistry, University of Sindh, Jamshoro 76080, Pakistan
[2] Department of Chemistry, Faculty of Science, Selcuk University, 42075 Konya, Turkey
[3] National Centre of Excellence in Analytical Chemistry, University of Sindh, Jamshoro, Pakistan
[4] Department of Pharmacy, Shaheed Mohtarma Benazir Bhutto Medical University, Larkana 77150, Pakistan
[5] Advanced Technology Research and Application Center, Selcuk University, 42075 Konya, Turkey
[6] Department of Biotechnology, Faculty of Science, Necmettin Erbakan University, 42090 Konya, Turkey

Correspondence should be addressed to Saba Naz; saba0208@gmail.com

Academic Editor: Young-Seok Shon

We report the room temperature (25–30°C) green synthesis of cobalt nanomaterial (CoNM) in an aqueous medium using gallic acid as a reducing and stabilizing agent. pH 9.5 was found to favour the formation of well dispersed flower shaped CoNM. The optimization of various parameters in preparation of nanoscale was studied. The AFM, SEM, EDX, and XRD characterization studies provide detailed information about synthesized CoNM which were of 4–9 nm in dimensions. The highly stable CoNM were used to study their catalytic activity for removal of azo dyes by selecting methyl orange as a model compound. The results revealed that 0.4 mg of CoNM has shown 100% removal of dye from 50 μM aqueous solution of methyl orange. The synthesized CoNM can be easily recovered and recycled several times without decrease in their efficiency.

1. Introduction

Metal nanoparticles have attracted much attention in nanoscale science and engineering technology over the past decades due to their unusual chemical and physical properties, such as catalytic activity, novel electronic, and optical and magnetic properties. Their main application areas include catalysts, absorbents, chemical and biological sensors, optoelectronics, information storage, and photonic and electronic devices [1]. Cobalt nanomaterial (CoNM) exhibits high resistance to oxidation, corrosion, and wear. CoNM have been prepared by several synthetic methods including solvothermal process [2], thermal decomposition method [3], hydrothermal microemulsion process [4], high temperature solution phase method [5], and reduction by $NaBH_4$ at room temperature [6]. Among all, the wet chemical reduction method has the advantage over the others in easy control

of the reaction process. However, most of the wet chemical reduction methods reported to date rely strongly on the use of environmentally and biologically hazardous organic solvents and reducing agents (i.e., hydrazine, sodium borohydride, dimethyl formamide, formaldehyde, sodium hypophosphite, or hydroxylamine hydrochloride, etc.) [1].

Recently, there is an increased emphasis on the subject of green chemistry, to avoid the problems related to toxic chemicals and solvents. Nanomaterials prepared by green rout are environmentally benevolent. Raveendran et al. [7] prepared silver nanoparticles using water as a solvent, β-D-glucose as a reducing agent, and starch as a protecting agent. Liu et al. [8] synthesized gold nanocrystals using β-D-glucose as both the reducing and stabilizing agent. In addition Xiong et al. [1], worked on the synthesis of highly stable nanosized copper particles with an average particle size of less than 2 nm using a nontoxic L-ascorbic acid as a reducing and capping agent;

precursor in aqueous medium was studied by Xiong et al. [1]. Moreover Martinez-Castanon et al. [9] worked on the synthesis and antibacterial activity of silver nanoparticles with gallic acid in an aqueous chemical reduction method. Gallic acid is a natural poly-phenolic compound that can be used as a reducing agent [10]. To our best knowledge, the nanomaterials with urchinlike and flowerlike architectures can be used in catalysis because of their high specific surface area [11].

Here, we tried firstly one-step method for the synthesis of biocompatible highly stable flower shape CoNM, by using natural gallic acid as reductant and stabilizing agent at the room temperature without any additional protecting reagents. We also monitor the catalytic activity of synthesized CoNM by selecting methyl orange (MO) azo dyes for their removal by adsorption. Regarding the toxic effect of dyes, studies have been carried out that show the extensive release of a number of azo dyes (such as MO) from effluents of the textile industry and many other sources which cause aquatic environmental pollution leading to severe health problems in aquatic life [12]. Numerous approaches have been introduced to purify the waters including biodegradation, ion exchange, and adsorption [13]. However, these processes are insufficient to control the pollution because they basically perform transformation of the hazards from one phase to another and need additional costs for treatments like incineration or land filling to terminate the end product [14]. In view of the risky effects of the dye, we have fabricated small size CoNM via greener route that possess marvelous potential as catalytic materials to completely degrade MO azo dye, which can be taken as a model for other degradable dyes.

Because of the magnetism of CoNM, they can be recovered by a magnet after adsorption. Hence, it can be applied to use in the field of catalysis and wastewater treatment, which may play important roles in the future of industrial effluents. The effect of gallic acid moles to cobalt metal on the size and shape of CoNM as well as their characteristic was also investigated. UV-Vis and IR spectroscopy, scanning electron microscopy (SEM), and X-ray diffraction (XRD) were employed in the characterization of the prepared CoNM [10].

2. Experimental

2.1. Chemicals and Reagents.
$CoCl_2 \cdot 6H_2O$ (99.9%), gallic acid (99%), MO dye (99%), and sodium hydroxide (98%) were purchased from Sigma-Aldrich, ACS reagent. All chemicals were of analytical grade and were used as received without further purification. Ultrapure water was used for preparation of CoNM.

2.2. Preparation of Stock Solutions.
Stock solutions of 0.1 M $CoCl_2 \cdot 6H_2O$ and 0.1 M gallic acid were prepared in 100 mL volumetric flasks using the required quantities of each and diluted to the mark with ultrapure CoNM in aqueous medium. Further, 1 M NaOH solution was prepared to maintain desired pH. 0.001 M stock solution of MO dye was

prepared in ultrapure water to carry out the catalytic activity of CoMN for reduction/degradation of dye.

2.3. Procedure for Fabrication of CoNM.
The synthesis of CoNM was carried out at room temperature by using gallic acid as reductant and stabilizer. Typically, a 10 mL aqueous solution containing 1×10^{-3} M $CoCl_2 \cdot 6H_2O$ solution was taken. To this solution 1.0 mL of 1×10^{-3} M solution of gallic acid is added under continuous stirring and then the resulting solution was adjusted to pH value 9.5 by addition of 1 M NaOH solution. The reaction solution was left for 10–15 min to confirm the completion of reduction reaction. The solution color was observed to change from light pink to blue and then brown after adjusting pH 9.5. After 10 min, no further change in color took place, indicating that the reactions were complete. As-prepared gallic acid derived CoNM obtained in an aqueous medium were separated by magnet or simple decantation of remaining aqueous phase. The collected CoNM were then dried for 24 hours in air. The solution was analyzed by UV-Vis spectroscopy and it was found that the as-prepared sample of CoNM was stable up to 60 days.

2.4. Instrumentation.
The UV-Vis spectroscopic absorbance was taken using UV Probe 2.35 spectrophotometer (shimadzu). Fourier transform infrared (FTIR) spectra of standard gallic acid and CoNM were recorded using a Vertex 70 (Bruker, Germany) with Platinum ATR Diamond. Atomic force microscopy (AFM) images were recorded using a NT-MDT, NTEGRA (Russia) AFM, MFM, and Nanoscope IV controller. Scanning electron microscopy (SEM) EVO LS 10, AEISS (England), images were taken on aluminum sample holder. Energy dispersive X-ray analysis EDX was completed with Bruker 123 eV (Germany). The X-ray diffraction (XRD) pattern of synthesized nanoparticles was recorded using a Bruker Advance D8 XRD instrument, equipped with Cu Kα source (wavelength = 1.5406). The XRD pattern was obtained in powder mode.

2.5. Sample Preparation for AFM Studies.
In a typical AFM analysis, 10–20 μL volumes of dispersed solution of CoNM were put through drop casting method on glass cover slip and heated at 60°C for 30 min followed by air drying up to 5 min to ensure binding of CoNM with the glass surface and thereby loss of water molecules.

2.6. Sample Preparation for SEM/EDX Studies.
Small quantities of dispersed solution of CoNM in aqueous medium were mounted on aluminum sample holder (Gold sputter quarter 7 nm), by a dip coating method and forwarded for vacuum drying in a Cressington Sputter Coater, Auto 108, in the presence of Argon gas for 2 min to clean the surface and ensure that solvent had been removed. The sample prepared in this way was also used for EDX studies.

2.7. Sample Preparation for XRD and FTIR Studies.
Solid state CoNM capped with gallic acid was dried and poured into

a glass tube. As-prepared Co nanoparticles were utilized for XRD and FTIR analysis without any additional pretreatment.

2.8. Catalytic Test for Reduction of Dye.

2.8. Catalytic Test for Reduction of Dye. The catalytic activity of CoNM was inspected for reduction/degradation of MO dye by putting 0.4 mg of CoNM in a solution having 50 μM concentrations of dye and diagnosed with the help of UV-Vis spectroscopy. Several experiments like dosage of catalyst, concentration of dye solution, and time study were performed to find out the catalytic performance of CoNM.

3. Results and Discussions

3.1. Co Nanomaterial Synthesized with Gallic Acid. In this work we present a simple green aqueous method for the fabrication of small size CoNM using gallic acid as reducing and stabilizing agent in the absence of any other capping reagents. The phenol group in the gallic acid molecule is responsible for the reduction of metal ions by providing electron through redox reaction. Gallic acid has two pKa values, first is 4.1 for the carboxylic group and second is 8.38 for the hydroxyl group [15], and that is why the stabilization of CoNM at higher pH value could come through the complex formation between oxygen of hydroxyl group and the CoNM. Basically, pH shows a great role on the size of nanoparticles; therefore, at high pH value, it could be possible that the complexation of Co (II) ions by NaOH decreases its reactivity and slows down the rate of nucleation and growth resulting in a smaller particle size [16]. UV-Vis absorbance spectroscopy is a very useful technique for studying metal nanoparticles because the shapes and positions of peak are sensitive to size of particle. The influence of gallic acid concentration on the UV-Vis absorbance of as-prepared CoNM is shown in Figure 1. As we can see that the surface plasmon band of CoNM has shown a continuous blue shift with the variation in the concentration of gallic acid which indicates the presence of very small size Co nanoparticles. Regarding the above results, the high amount of gallic acid leads to the formation of small size nanoparticles due to its enhanced capping ability.

3.2. XRD Analysis. Figure 2(a) shows the diffraction pattern obtained for the 4–9 nm CoNM; this analysis was made to confirm the identity of the products. The diffractogram shows broad prominent peaks around $2\theta = 47.01°$ corresponding to the (111) plane of fcc (face centered cube) cobalt which have a good match with the standard diffraction pattern (JSPDC no. 05-0727) [17]. The broad XRD peak indicates a nanocrystalline nature. Owing to the noisy XRD pattern the other weaker peaks that correspond to the fcc cobalt nanomaterial was not visibly seen in a diffraction pattern [18]. The average crystallite size is determined through X-ray diffraction line broadening by the Debye-Scherrer formula. Consider the following:

$$D = \frac{K\lambda}{\beta \cos\theta}. \tag{1}$$

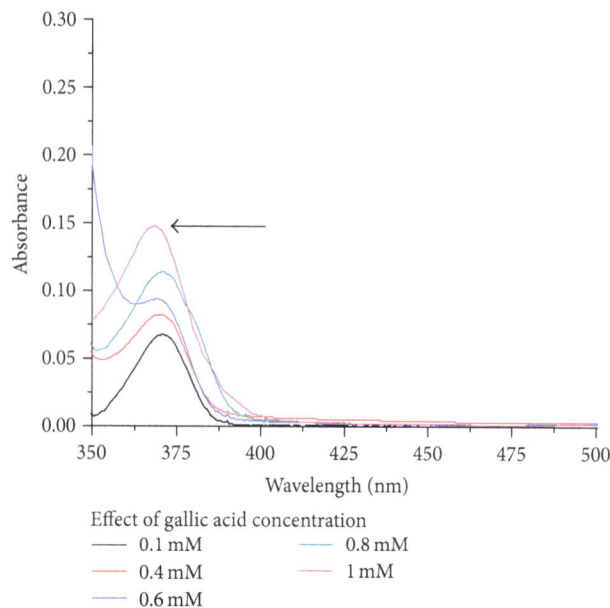

FIGURE 1: The absorption spectra of CoNM synthesized by various amounts of gallic acid to metal at room temperature.

In (1), D shows the average crystallite size, $K = 0.89$ is the Scherrer constant, $\lambda = 1.5406$ A° is the wavelength of X-ray (Cu Kα_1 radiation), θ is the diffraction angle of the peak, and β represents the full width at half maximum of the peaks [19]. The average crystallite size of CoNM was found to be 4–9 nm which is also confirmed by SAXS pattern as shown in Figure 2(b).

3.3. Fourier Transform Infrared (FTIR) Spectroscopy. IR spectrum was also measured for the further investigation to identify the possible molecular response for efficient stabilization of CoNM. The dried CoNM were used for IR analysis. Figures 3(a) and 3(b) show the infrared spectra of gallic acid standard and CoNM obtained by gallic acid reduction of CoCl$_2$·6H$_2$O. In Figure 3(a), the strong and broad band between 3600 and 2500 cm^{-1} and the strong and narrow peak at 1702 cm^{-1} could be assigned to the stretching vibration of OH group and carbonyl group, which indicated the existence of carboxyl group in the gallic acid. Three peaks observed at 1616, 1541, and 1450 cm^{-1} are typical stretching vibrations of C–C bonds in an aromatic ring. There are several peaks in 1300–1000 cm^{-1} region that could be assigned to the stretching vibration of C–O bond and bending vibration of O–H bond of gallic acid. In Figure 3(b), the strong and broad band in the 3650–2700 cm^{-1} region was considered to be stretching vibration of OH group, which basically covered C–H bond stretching vibration at about 3100 cm^{-1}. Stretching vibration of C–O bond and bending vibration of O–H bond in 1300–1000 cm^{-1} region was still retained, but the intensity obviously decreased. In contrast with that of gallic acid shown in Figure 3(a), the IR absorption spectrum of CoNM observed from Figure 3(b) indicated that the stretching vibration peak of carbonyl group shifted from

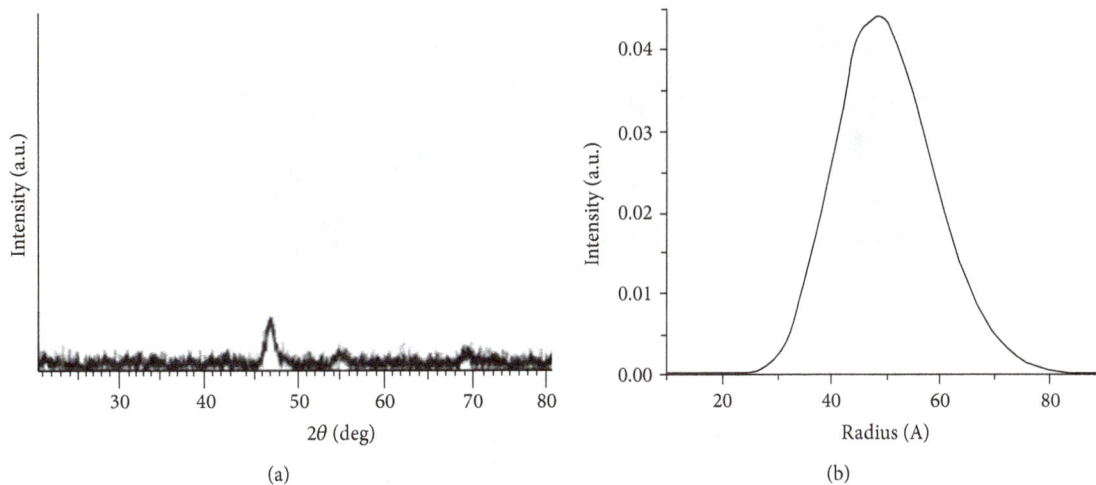

FIGURE 2: (a) The XRD diffraction pattern for the 4–9 nm CoNM (b) with SAXS profile.

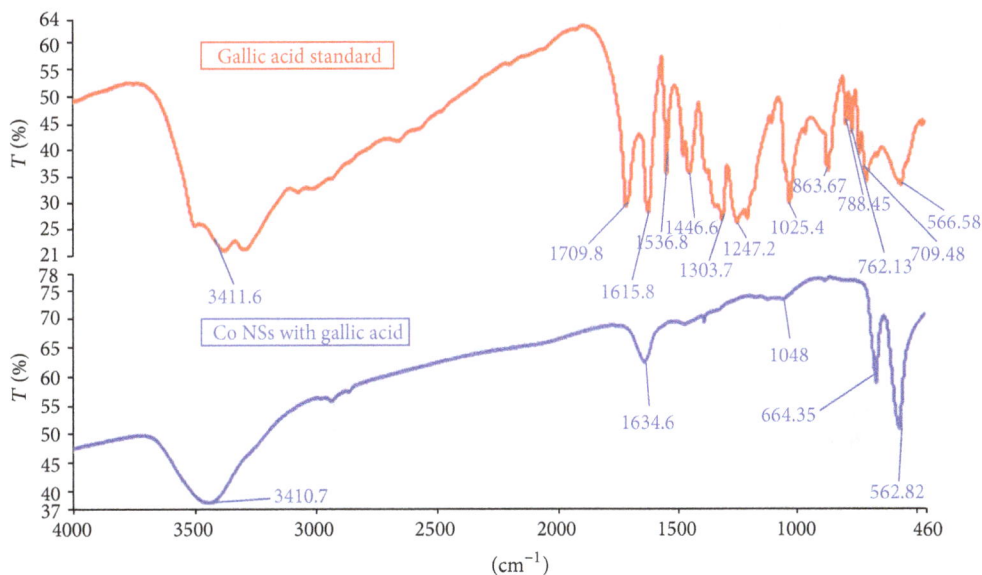

FIGURE 3: (a) FTIR spectra for gallic acid standard and (b) CoNM capped with gallic acid molecule.

1709 to 1634 cm^{-1}, and the stretching vibration of C–C bond at 1616 and 1537 cm^{-1} was covered with a broad and middle intensity band at around 1634 cm^{-1} [10]. Since phenolic compounds are easily oxidized to form quinones, it was speculated that the product of gallic acid reduction of $CoCl_2 \cdot 6H_2O$ might be a quinoid compound. The results indicated that quinoid compound with keto-enol system might be produced by gallic acid reduction of $CoCl_2 \cdot 6H_2O$ and absorbed on the surface of CoNM. Usually, when molecule absorbs on the nano scale metal island, surface-enhanced Raman and infrared spectra can be observed [20].

3.4. Morphology of CoNM by SEM and AFM.
Scanning electron microscopy was used to investigate the surface morphology, structure, and particle size of CoNM. Figures 4(a)

and 4(b) show the SEM images of the synthesized CoNM with two different magnifications. It clearly reveals that uniform flower-like microspheres can be synthesized successfully by a simple reduction of gallic acid. Atomic force micrographs (AFM) present highly distributed CoNM grown after capping with gallic acid molecules on glass cover slips as shown in Figure 4(c). One of the most beneficial features of atomic force microscopy is its ability to quantitatively measure the spectral dimensions of different surface features and high resolution images. The flower images obtained from SEM were relatively similar to AFM images which confirm the formation of nanoflowers shape CoNM.

3.5. Stability Study of CoNM.
Stability of synthesized CoNM was assessed for 60 days of storage at ambient room temperature 25 ± 2°C by UV-Vis analysis. There was no change

(a)

(b)

(c)

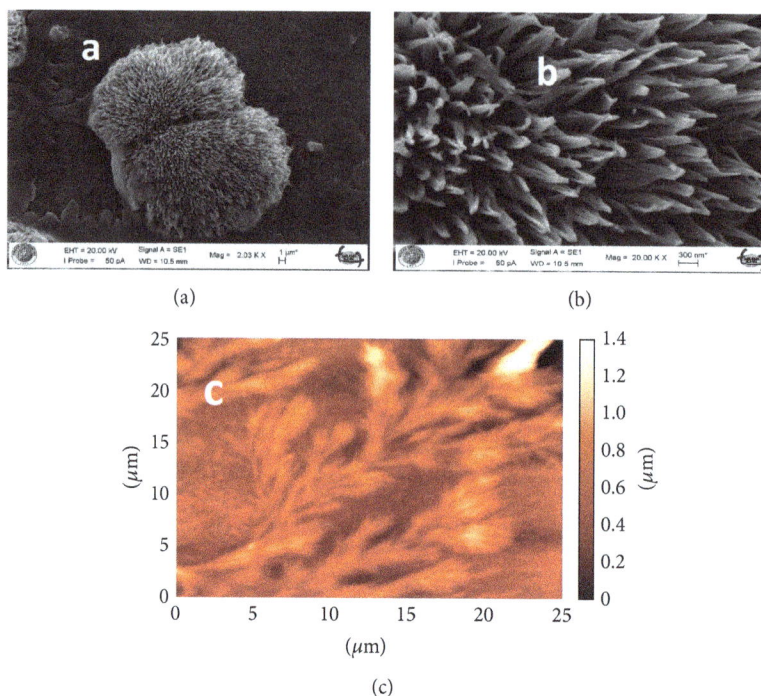

FIGURE 4: (a) SEM micrographs of CoNM at low magnification, (b) at high magnification, and (c) AFM image of CoNM.

observed in the appearance or color intensity during the entire period of storage time.

3.6. Catalytic Performance of Methyl Orange (MO) Dye. Methyl orange ($C_{14}H_{14}N_3NaO_3S$, MW = 327.33 Da) is an intensely colored compound used in dyeing and printing textiles. It is made from sodium nitrite, dimethylaniline, and sulfanilic acid through diazotization process. It is considered to be a harmful and carcinogenic pollutant causing various diseases and disorders in living organisms [6]. UV-Vis spectroscopy was used to measure the concentration of MO dye in an aqueous solution. The color of MO solution changes from orange to colorless after the addition of CoNM within 50 seconds. Figure 5 shows the decrease of MO dye absorbance with respect to time which suggests the 100% removal of MO dye within 50 sec. The as-prepared CoNM showed high adsorption capacity and a faster reaction rate and thus can be potentially used to remove harmful dyes from aqueous solution within a small time and easily separated from aqueous solution.

3.7. Recovery of CoNM Catalyst after Reuse. The newly synthesized CoNM can be used several times without loss of their catalytic efficiency. In addition the CoNM can be easily recovered after the removal of MO dye then washed with sufficient amount of deionized water followed by drying. Then, the dried CoNM is kept at room temperature. The high catalytic efficiency of CoNM was observed during recycling experiments the results shows that 0.4 mg CoNM can be even eight times reuse for removal of 50 μM solution of MO dye within 50 sec each time.

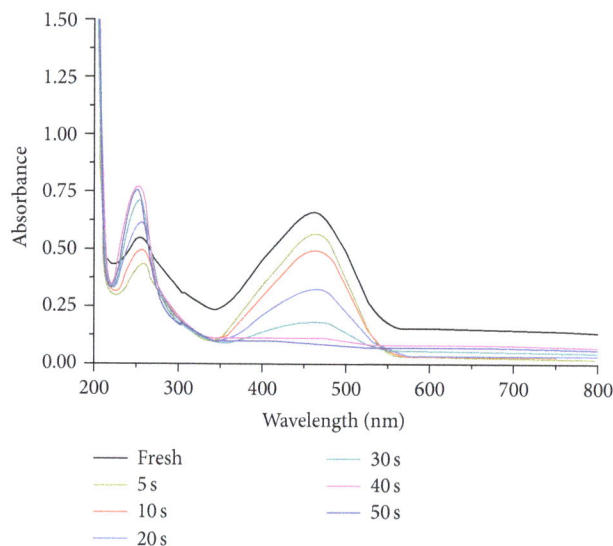

FIGURE 5: UV-Visible spectra of methyl orange in the presence of CoNM.

4. Conclusions

In conclusion, we have demonstrated a facile green method to synthesize low cost CoNM 4–9 nm in size on average with a narrow size distribution and a uniform shape by employing gallic acid as both the reducing and capping agent. The prepared dispersions of CoNM are highly stable and do not show any sign of sedimentation even after storage for 60 days. Since the reagents used in the reaction medium are completely nontoxic and environmentally friendly, this green method can be readily used for biomedical applications. Moreover, the highly stable freshly prepared CoNM shows highly catalytic performance for MO azo dyes removal within 50 sec at room temperature. The CoNM found to be recovered easily and recycle several times without loss of their catalytic activity.

Conflict of Interests

The authors declare that they do not have any conflict of interests.

Acknowledgment

The authors would like to thank TUBITAK (The Scientific and Technological Research Council of Turkey) for the finance support (Program no. 2216).

References

[1] J. Xiong, Y. Wang, Q. Xue, and X. Wu, "Synthesis of highly stable dispersions of nanosized copper particles using l-ascorbic acid," *Green Chemistry*, vol. 13, no. 4, pp. 900–904, 2011.

[2] L. P. Zhu, W. D. Zhang, H. M. Xiao, Y. Yang, and S. Y. Fu, "Facile synthesis of metallic Co hierarchical nanostructured microspheres by a simple solvothermal process," *The Journal of Physical Chemistry C*, vol. 112, no. 27, pp. 10073–10078, 2008.

[3] N. Matoussevitch, A. Gorschinski, W. Habicht et al., "Surface modification of metallic Co nanoparticles," *Journal of Magnetism and Magnetic Materials*, vol. 311, no. 1, pp. 92–96, 2007.

[4] W. Liu, W. Zhong, X. Wu, N. Tang, and Y. Du, "Hydrothermal microemulsion synthesis of cobalt nanorods and self-assembly into square-shaped nanostructures," *Journal of Crystal Growth*, vol. 284, no. 3-4, pp. 446–452, 2005.

[5] Y. Su, X. OuYang, and J. Tang, "Spectra study and size control of cobalt nanoparticles passivated with oleic acid and triphenylphosphine," *Applied Surface Science*, vol. 256, no. 8, pp. 2353–2356, 2010.

[6] X. Liang and L. Zhao, "Room-temperature synthesis of air-stable cobalt nanoparticles and their highly efficient adsorption ability for Congo red," *RSC Advances*, vol. 2, no. 13, pp. 5485–5487, 2012.

[7] P. Raveendran, J. Fu, and S. L. Wallen, "Completely "green" synthesis and stabilization of metal nanoparticles," *Journal of the American Chemical Society*, vol. 125, no. 46, pp. 13940–13941, 2003.

[8] J. Liu, G. Qin, P. Raveendran, and Y. Ikushima, "Facile "green" synthesis, characterization, and catalytic function of β-D-glucose-stabilized Au nanocrystals," *Chemistry*, vol. 12, no. 8, pp. 2131–2138, 2006.

[9] G. A. Martinez-Castanon, N. Nino-Martinez, F. Martinez-Gutierrez, J. R. Martinez-Mendoza, and F. Ruiz, "Synthesis and antibacterial activity of silver nanoparticles with different sizes," *Journal of Nanoparticle Research*, vol. 10, no. 8, pp. 1343–1348, 2008.

[10] W. Wang, Q. Chen, C. Jiang, D. Yang, X. Liu, and S. Xu, "One-step synthesis of biocompatible gold nanoparticles using gallic acid in the presence of poly-(N-vinyl-2-pyrrolidone)," *Colloids and Surfaces A: Physicochemical and Engineering Aspects*, vol. 301, no. 1-3, pp. 73–79, 2007.

[11] Y. Chen, L. Hu, M. Wang, Y. Min, and Y. Zhang, "Self-assembled Co_3O_4 porous nanostructures and their photocatalytic activity," *Colloids and Surfaces A: Physicochemical and Engineering Aspects*, vol. 336, no. 1-3, pp. 64–68, 2009.

[12] K. Zhang and W. C. Oh, "The photocatalytic decomposition of different organic dyes under UV irradiation with and without H_2O_2 on Fe-ACF/TiO$_2$ photocatalysts," *Journal of the Korean Ceramic Society*, vol. 46, no. 6, pp. 561–567, 2009.

[13] C. Umpuch and S. Sakaew, "Removal of methyl orange from synthetic wastewater onto chitosan-coated-montmorillonite clay in fixed-beds," *GMSARN International Journal*, vol. 6, pp. 175–180, 2012.

[14] D. Kamel, A. Sihem, C. Halima, and S. Tahar, "Decolourization process of an azoïque dye (Congo red) by photochemical methods in homogeneous medium," *Desalination*, vol. 247, no. 1-3, pp. 412–422, 2009.

[15] A. E. Fazary, M. Taha, and Y. H. Ju, "Iron complexation studies of gallic acid," *Journal of Chemical & Engineering Data*, vol. 54, no. 1, pp. 35–42, 2008.

[16] K. W. Huang, C. J. Yu, and W. L. Tseng, "Sensitivity enhancement in the colorimetric detection of lead(II) ion using gallic acid-capped gold nanoparticles: improving size distribution and minimizing interparticle repulsion," *Biosensors and Bioelectronics*, vol. 25, no. 5, pp. 984–989, 2010.

[17] V. V. Matveev, D. A. Baranov, G. Y. Yurkov, N. G. Akatiev, I. P. Dotsenko, and S. P. Gubin, "Cobalt nanoparticles with preferential hcp structure: a confirmation by X-ray diffraction and NMR," *Chemical Physics Letters*, vol. 422, no. 4-6, pp. 402–405, 2006.

[18] N. S. Gajbhiye, S. Sharma, A. K. Nigam, and R. S. Ningthoujam, "Tuning of single to multi-domain behavior for monodispersed ferromagnetic cobalt nanoparticles," *Chemical Physics Letters*, vol. 466, no. 4-6, pp. 181–185, 2008.

[19] M. Alagiri, C. Muthamizhchelvan, and S. Hamid, "Synthesis of superparamagnetic cobalt nanoparticles through solvothermal process," *Journal of Materials Science: Materials in Electronics*, vol. 24, no. 11, pp. 4157–4160, 2013.

[20] Q. Song, X. Ai, D. Wang et al., "Preparation of gold/triblock copolymer composite nanoparticles," *Journal of Nanoparticle Research*, vol. 2, no. 4, pp. 381–385, 2000.

11

Mycotoxin Analysis: New Proposals for Sample Treatment

Natalia Arroyo-Manzanares, José F. Huertas-Pérez,
Ana M. García-Campaña, and Laura Gámiz-Gracia

Department of Analytical Chemistry, Faculty of Sciences, University of Granada, Campus Fuentenueva s/n, E-18071 Granada, Spain

Correspondence should be addressed to Laura Gámiz-Gracia; lgamiz@ugr.es

Academic Editor: Brijesh Tiwari

Mycotoxins are toxic secondary metabolites produced by different fungi, with different chemical structures. Mycotoxins contaminate food, feed, or raw materials used in their production and cause diseases and disorders in humans and livestock. Because of their great variety of toxic effects and their extreme heat resistance, the presence of mycotoxins in food and feed is considered a high risk to human and animal health. In order to ensure food quality and health consumers, European legislation has set maximum contents of some mycotoxins in different matrices. However, there are still some food commodities susceptible to fungal contamination, which were not contemplated in this legislation. In this context, we have developed new analytical techniques for the multiclass determination of mycotoxins in a great variety of food commodities (some of them scarcely studied), such as cereals, pseudocereals, cereal syrups, nuts, edible seeds, and botanicals. Considering the latest technical developments, ultrahigh performance liquid chromatography coupled to tandem mass spectrometry has been chosen as an efficient, fast, and selective powerful analytical technique. In addition, alternative sample treatments based on emerging methodologies, such as dispersive liquid-liquid microextraction and QuEChERS, have been developed, which allow an increased efficiency and sample throughput, as well as reducing contaminant waste.

1. Introduction

Mycotoxins are toxic natural secondary metabolites produced by several species of fungi (as *Fusarium*, *Aspergillus*, and *Penicillium* genera) on agricultural commodities. The presence of mycotoxins in food and feed may affect human and animal health, as they may cause many different adverse effects such as estrogenic, gastrointestinal, and kidney disorders, induction of cancer, and mutagenicity. Furthermore, some mycotoxins are also immunosuppressive and reduce resistance to infectious diseases [1, 2]. Mycotoxins grow under a wide range of climatic conditions and the Food and Agriculture Organization (FAO) has estimated that they affect 25% of the world crops. On the other hand, mycotoxins are the hazard category with the highest number of border rejections reported by the Rapid Alert System for Food and Feed (RASFF) [3]; therefore their impact on economy is evident.

Hundreds of mycotoxins have been recognized with diverse chemical structures, different toxicity, and biological effects. The most relevant groups of mycotoxins found in food are aflatoxins (aflatoxin B1 (AFB_1, included in group 1 of carcinogenic to humans by the International Agency for Research on Cancer (IARC) [4]), aflatoxin B2 (AFB_2), aflatoxin G1 (AFG_1), aflatoxin G2 (AFG_2), and aflatoxin M1 (AFM_1, metabolite of AFB_1, excreted in the milk of mammals)); ochratoxin A (OTA); trichothecenes (HT-2 and T-2 toxin and deoxynivalenol (DON)); zearalenone (ZEN); fumonisins B1 and B2 (FB_1 and FB_2); citrinin (CIT); patulin (PAT) and ergot alkaloids [2, 5]. Figure 1 shows the structures of some common mycotoxins and Table 1 includes the most important toxins, their main producing fungi, and typical food commodities that may be contaminated by them.

In order to protect consumer health, international institutions and organizations have proposed regulatory limits for some mycotoxins. Thus, the European Commission (EC) establishes maximum permitted levels for most mycotoxins in foods by means of the Commission Regulation (EC) number 1881/2006 [6] (or recommended levels for other mycotoxins [7]), as well as methods of sampling and analysis for

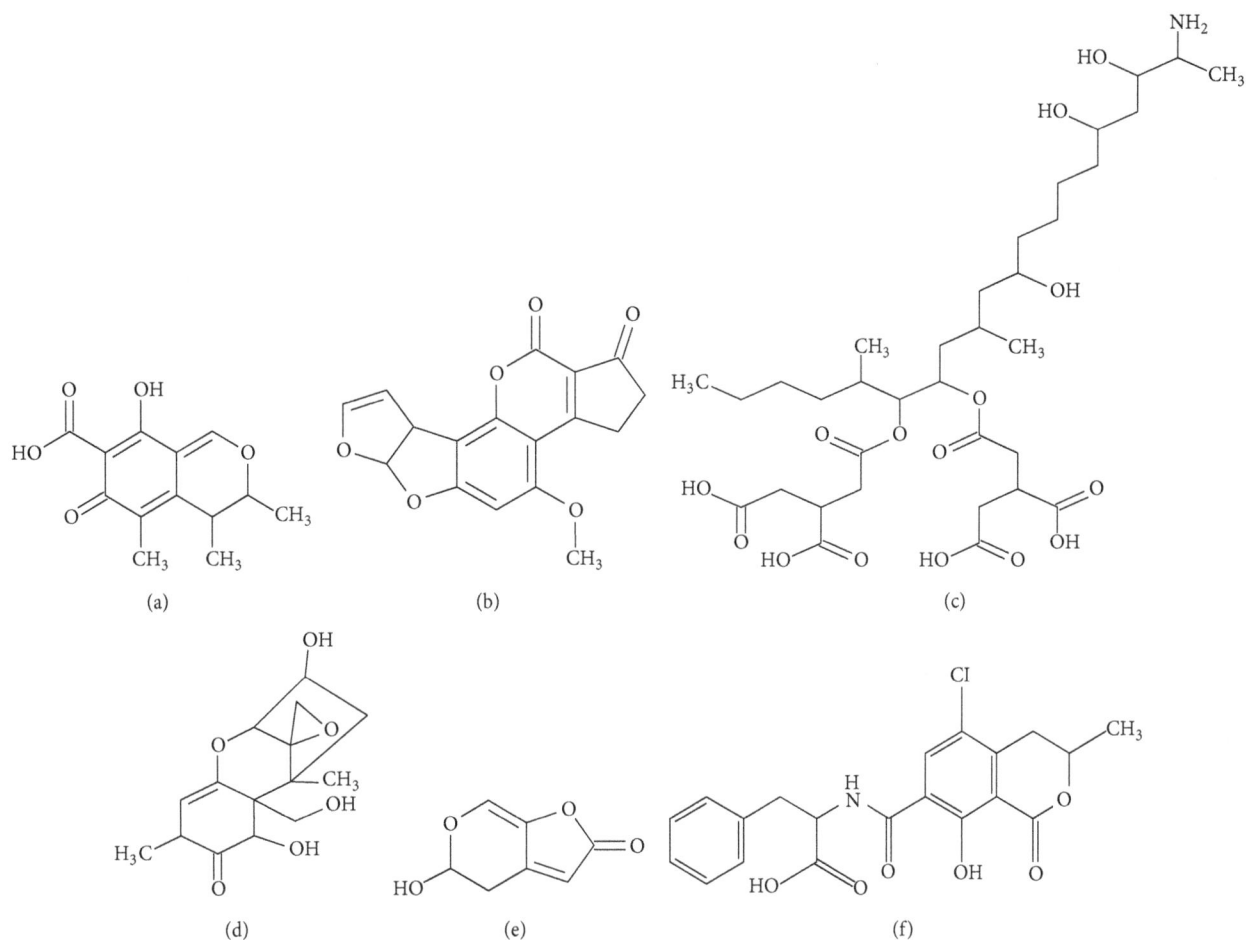

FIGURE 1: Structures of some common mycotoxins: (a) citrinin; (b) aflatoxin B_1; (c) fumonisin B_1; (d) deoxynivalenol; (e) patulin; (f) ochratoxin A.

their control by Commission Regulation (EC) number 401/2006 [8], which have been subsequently amended. Regulations are also established by the US Food and Drug Administration (FDA). FDA mycotoxin compliance programs provide introductory information about mycotoxins, products prone to contamination, and analytical methods [9].

In this context, the use of robust analytical methodologies for sampling, sample treatment, and identification/quantification of mycotoxins in food and feed is mandatory, in order to protect the consumer health [10–12].

2. Current Analytical Methods for Determination of Mycotoxins in Food

Different analytical methods have been proposed for mycotoxin determination in food, such as thin layer chromatography (TLC) [13], ELISA [14], gas chromatography (GC) [15] or capillary electrophoresis (CE) [16]. However, the most popular technique is high performance liquid chromatography (HPLC) with UV/Vis, fluorescence (FL) [17–19], or mass spectrometry (MS) detection [20–22]. Recently, ultrahigh performance liquid chromatography (UHPLC) coupled with tandem mass spectrometry (MS/MS) has become very popular, especially for multiclass determination of mycotoxins and

for multiresidue determination with other contaminants [23–26].

Because of the complexity of food matrices, an extraction and clean-up purification step is usually required before analysis. Different approaches have been proposed. The most common methodology implies solid-liquid extraction (SLE) followed by solid phase extraction (SPE) with immunoaffinity columns (IACs), which contain specific antibodies to the analyte of interest [10]. Several reviews present an overview of the different methodologies proposed for the determination of mycotoxins in food including the most frequent sample treatments [12, 27–30].

However, IACs are expensive and complex purification systems which suffer from low recoveries for some mycotoxins and their use in multiclass analysis is limited because of their high selectivity. As a consequence, simpler, more efficient, multiclass, and environmentally friendly extraction systems are demanded. Among the different proposals, the so-called QuEChERS (quick, easy, cheap, effective, rugged, and safe) and dispersive liquid-liquid microextraction (DLLME) are becoming increasingly popular treatments.

DLLME is based on the use of a ternary component solvent system; an appropriate mixture of a few microliters of an organic extraction solvent and a small volume of

TABLE 1: Most common mycotoxins, main producing fungal species, and food commodities frequently contaminated.

Fungal species	Mycotoxin	Food commodity	Maximum permitted levels[a] ($\mu g\,Kg^{-1}$)
Aspergillus parasiticus	Aflatoxins B1, B2, G1, G2	Maize, wheat, rice, sorghum, ground nuts, tree nuts, and figs	B1: 0.10–12 B1 + B2 + G1 + G2: 4–15
Aspergillus flavus	Aflatoxins B1 y B2	Idem	
Metabolite of aflatoxin B1 in mammals	Aflatoxin M1	Milk and milk products	0.025–0.050
Fusarium sporotrichioides	Toxins T-2 and HT-2	Cereals and cereal products	T-2 + HT-2: 15–2000[b]
Fusarium graminearum	Deoxynivalenol zearalenone	Cereals and cereal products	DON: 200–1750 ZEN: 20–400
Fusarium moniliforme (F. verticillioides)	Fumonisins B1, B2	Maize, maize products, sorghum, and asparagus	B1 + B2: 200–4000
Penicillium verrucosum Aspergillus ochraceus	Ochratoxin A	Cereals, wine, fruits, coffee, and spices	0.50–80
Penicillium expansum	Patulin	Apples, apple juice, and apple products	25–50
Aspergillus, Penicillium, and Monascus	Citrinin	Cereals, red rice, fruits, and cheese	nr
Aspergillus versicolor	Sterigmatocystin	Cereals, coffee, ham, pepper, and cheese	nr
Hypocreales (Claviceps purpúrea), Eurotiales	Ergot alkaloids	Cereals	nr

[a] Range of maximum permitted levels in the EU, depending on the food commodity [6].
[b] Recommended level [7].
nr: not regulated.

a disperser solvent (miscible with the extraction solvent and with water) is rapidly injected into an aqueous medium, with the result of a stable emulsion. The organic analytes present in the aqueous medium rapidly migrate to the extraction solvent, because of the large contact surface between the organic and the aqueous phases. After phases separation, the organic phase with the analytes of interest is collected and analysed by an appropriate technique [31–33]. DLLME has been applied for the determination of OTA in wine by HPLC-MS [34], cereals by HPLC-FL [35], and patulin in apple juices by CE-UV [36]. Also, we have developed two DLLME methods (one of them using an ionic liquid as extraction solvent) for the determination of OTA in wine by capillary-HPLC with laser induced fluorescence detection (LIF) with excellent results [18, 19].

On the other hand, QuEChERS is a fast and inexpensive method widely used in the last years, mainly for the extraction of pesticides and presents some advantages such as its simplicity, minimum steps, and effectiveness for cleaning-up complex samples [37, 38]. It comprises two steps: (i) an extraction based on partitioning via salting-out, involving the equilibrium between an aqueous and an organic layer; (ii) a dispersive SPE (dSPE) for further clean-up using combinations of $MgSO_4$ and different sorbents, such as C_{18} or primary and secondary amine (PSA). QuEChERS-based methods have been recently reported for the extraction of different mycotoxins in cereal products [39–41], bread [42], eggs [43], or spices [44] and in the multiresidue extraction of different contaminants (including mycotoxins) in foods [45, 46]. Also, we proposed this methodology for the determination of OTA in wine samples by capillary HPLC-LIF [18].

3. New Proposals for Determination of Mycotoxins in Different Foods

Considering the above described advances in sample treatments, we have proposed a multiclass method for the determination of mycotoxins in different food commodities. Taking advantage of UHPLC-MS/MS characteristics, we optimised a separation method that allows the determination of 15 mycotoxins in only four minutes. The studied mycotoxins are included in Regulation (EC) number 1881/2006 or considered as dangerous by the IARC [6, 47].

Moreover, in order to propose alternative methods for multiclass mycotoxins determination in scarcely investigated matrices and considering previous results obtained for the determination of OTA, we have tried to explore the advantages of the above mentioned sample treatments (DLLME and QuEChERS) as green, easy, and simple alternatives to other well-established methodologies, as IAC or SPE.

Below, we will explain the UHPLC-MS/MS conditions (common to all the methods) and then we will focus on the studied samples: cereals and pseudocereals, cereal syrups, edible nuts and seeds, and milk thistle. In all cases, a validation was performed in order to assess the compliance with the current requirements for mycotoxin determination in foods [8]. The validation included matrix effect study, establishment of matrix-matched calibrations, limits of detection (LODs) and quantification (LOQs), and intraday and intermediate precision. Moreover, recovery studies at three different concentration levels were carried out by comparison of the signal obtained for a sample spiked with a known concentration of mycotoxins before the sample treatment with the signal of

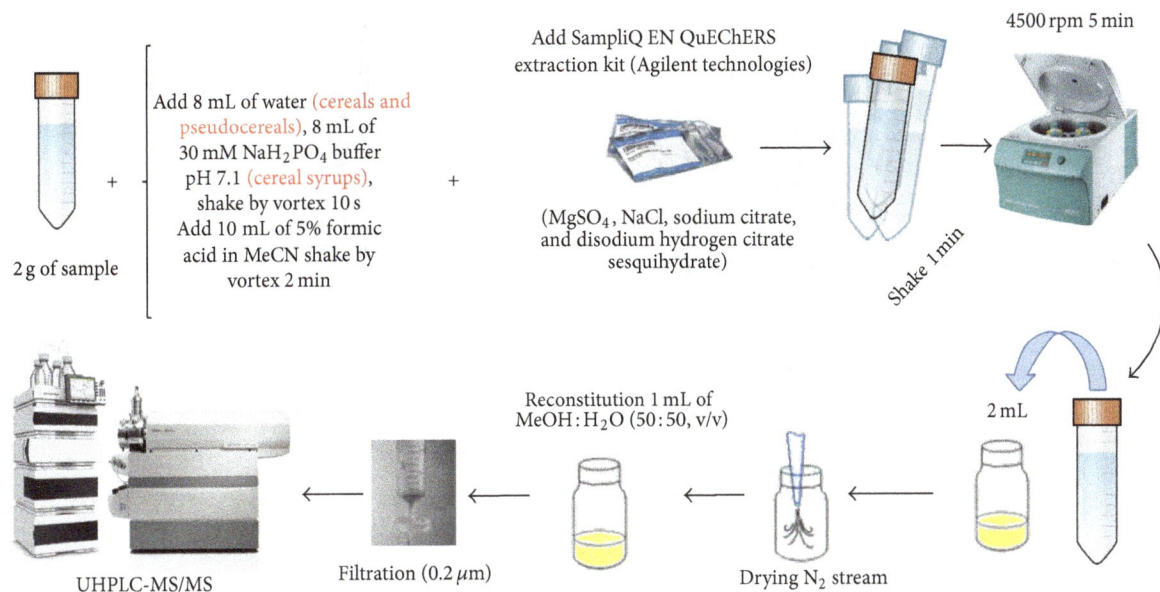

Figure 2: Sample treatment for the determination of mycotoxins in cereals and pseudocereals (MeCN: acetonitrile; MeOH: methanol).

an spiked extract obtained after the sample treatment. As a summary, the most significant analytical characteristics of the developed methods are shown in Table 2.

3.1. Chromatographic and MS Conditions.
UHPLC separations were performed on an Agilent 1290 Infinity LC under the conditions summarised in Table 2.

The triple quadrupole mass spectrometer API 3200 (AB Sciex) worked with electrospray ionization in positive mode (ESI+), under multiple reaction monitoring (MRM) conditions shown in Table 3. The ionization source parameters were source temperature 500°C; curtain gas (nitrogen) 30 psi; ion spray voltage 5000 V; GAS 1 and GAS 2 (both of them nitrogen) 50 psi. Moreover, a precursor ion and two product ions (the most abundant for quantification and the other one for confirmation) were selected, obtaining four identification points, fulfilling the requirements established by European Union (EU) for confirmation of contaminants in foodstuff [48].

3.2. Analysis of Cereals and Pseudocereals.
Cereals are a commodity of great interest, highly prone to microbial contamination because of their chemical composition. Rice is one of the most consumed cereals in the world. Moreover, brown rice and red rice (obtained by the fermentation of rice with *Monascus* fungi [49], that can produce CIT [50]) are increasingly chosen by customers because of their health benefits. Other cereal of interest is spelt. Its nutritional properties, high resistance in unfavourable environments, and low fertilization requirements made it increasingly valuable for food product manufacturers and consumers [51]. Matrices of concern are also pseudocereals, such as amaranth, quinoa, and buckwheat. Though botanically they are not true cereal grains, they produce starch-rich seeds consumed like cereals. Pseudocereals are also susceptible to fungal growth and

therefore to mycotoxin contamination. However, this issue has received little attention in literature.

Taking into account the interest and the scarce data about the determination of mycotoxins in some of the above mentioned matrices, we developed and validated an analytical method for the simultaneous identification and quantification of 15 mycotoxins (AFB$_1$, AFB$_2$, AFG$_1$, AFG$_2$, OTA, FB$_1$, FB$_2$, T-2, HT-2, CIT, STE, F-X, NIV, DON, and ZEN) in pseudocereals, spelt, and white, red, and brown rice. As a sample treatment we proposed a simple salting out assisted solid-liquid extraction (i.e., a QuEChERS-based extraction, see Figure 2). No further clean-up was required, although matrix effect was higher than |20%| for some mycotoxins (aflatoxins, DON, and NIV). Thus matrix-matched calibration was applied. A typical chromatogram corresponding to a spiked white rice sample submitted to the proposed method is shown in Figure 3(a). This methodology has proved to be a suitable and efficient choice for multiclass mycotoxin determination in these matrices, with LOQs below the contents currently regulated. It provides good recoveries (between 60.0% and 103.5%) and precision (RSD lower than 12% in all cases), allows extraction time reduction, and is environmentally friendly. Among all the samples analysed, a red rice sample was positive for AFB$_1$ (Figure 4(a)). The result was confirmed by comparison with a standard method [52].

3.3. Analysis of Cereal Syrups.
Cereal syrups are obtained by isolation of starch after wet milling of grains, hydrolysis, and further purification and are widely used in food and pharmaceutical industry. Mycotoxins may also be found in cereal syrups as result of using contaminated raw material, or because of contamination of the final manufactured product by microorganisms during storage [53]. However, there are very few methods for the determination of mycotoxins in these matrices.

TABLE 2: Analytical characteristics of proposed methods.

Analytes	Matrix	Sample treatment	UHPLC chromatographic conditions	LOQs (μg Kg^{-1})	Recovery (%)	Intraday precision (%RSD, $n = 9$)	Intermediate precision (%RSD, $n = 15$)
AFB$_1$, AFB$_2$, AFG$_1$, AFG$_2$, OTA, FB$_1$, FB$_2$, T-2, HT-2, STE, CIT, DON, NIV, F-X, ZEN	Cereals and pseudocereals	QuEChERS-based extraction	**Mobile phase:** (A) H$_2$O with 0.3% formic acid and 5 mM ammonium formate (B) MeOH with 0.3% formic acid and 5 mM ammonium formate **Gradient profile:** 0 min: 5% B; 1 min: 50% B; 2 min: 72% B; 4 min: 80% B and 6 min: 90% B **Column:** Zorbax Eclipse Plus RRHD C$_{18}$ (50 mm × 2.1 mm, 1.8 μm) **Flow-rate:** 0.4 mL min^{-1} **Temperature:** 35°C **Injection volume:** 5 μL	From 0.23 (AFG$_1$) to 233 (NIV)	From 60.0 (NIV in red rice) to 103.5 (AFG$_2$ in white rice)	From 1.3 (FB$_2$) to 8.8 (ZEN)	From 6.4 (HT-2) to 11.9 (FB$_1$, FX)
OTA, FB$_1$, FB$_2$, T-2, HT-2, STE, CIT, DON, F-X, ZEN	Cereal syrups	QuEChERS-based extraction		From 0.45 (STE) to 75.2 (DON)	From 62.8 (STE in wheat syrup) to 100.6 (CIT in rice syrup)	From 1.3 (T-2, CIT) to 8.4 (DON)	From 2.6 (CIT) to 11.5 (OTA)
AFB$_1$, AFB$_2$, AFG$_1$, AFG$_2$, OTA, FB$_1$, FB$_2$, T-2, HT-2, STE, CIT, DON, F-X, ZEN	Edible nuts and seeds	QuEChERS-based extraction + DLLME		From 0.57 (OTA) to 150 (F-X)	From 61.7 (STE in almond) to 104.3 (T-2 in hazelnut)	From 0.6 (ZEN) to 8.9 (CIT)	From 6.0 (T-2) to 10.6 (DON)
AFB$_1$, AFB$_2$, AFG$_1$, AFG$_2$, OTA, FB$_1$, FB$_2$, T-2, HT-2, STE, CIT, DON, NIV, F-X, ZEN	Milk thistle	QuEChERS-based extraction + DLLME		From 1.50 (AFG$_1$) to 1530 (NIV)	From 60.0 (CIT in extract) to 98.9 (AFB$_2$ in extract)	From 4.4 (OTA) to 9.3 (AFG$_2$)	From 5.3 (FB$_1$) to 9.9 (T2)

TABLE 3: Monitored ions of target analytes and MS/MS parameters.

Analyte	Retention time (min)	Precursor ion (m/z)	Molecular ion	DP[a]	EP[a]	CEP[a]	Product ions[b]	CEn[a]	CXP[a]
NIV	1.36	313.1	$[M + H]^+$	41.0	5.0	14.0	174.9 (Q)	17.0	4.0
							128.0 (I)	75.0	6.0
DON	1.70	297.1	$[M + H]^+$	36.0	5.5	16.0	249.2 (Q)	17.0	4.0
							161.0 (I)	29.0	4.0
F-X	2.02	355.1	$[M + H]^+$	26.0	12.0	18.0	174.7 (Q)	23.0	4.0
							137.1 (I)	31.0	4.0
AFG$_2$	2.52	331.1	$[M + H]^+$	61.0	6.0	42.0	245.1 (Q)	39.0	4.0
							313.1 (I)	27.0	6.0
AFG$_1$	2.61	329.0	$[M + H]^+$	76.0	9.5	16.0	243.1 (Q)	39.0	6.0
							311.1 (I)	29.0	6.0
AFB$_2$	2.72	315.1	$[M + H]^+$	81.0	4.0	34.0	286.9 (Q)	33.0	6.0
							259.0 (I)	39.0	8.0
AFB$_1$	2.79	313.1	$[M + H]^+$	46.0	12.0	26.0	241.0 (Q)	41.0	4.0
							284.9 (I)	39.0	4.0
CIT	2.90	251.2	$[M + H]^+$	26.0	11.0	18.0	233.0 (Q)	23.0	23.0
							204.8 (I)	73.0	10.0
HT-2	3.18	442.0	$[M + NH_4]^+$	21.0	5.5	21.0	262.8 (Q)	22.0	8.0
							215.4 (I)	19.0	4.0
FB$_1$	3.24	722.2	$[M + H]^+$	71.0	10.0	30.0	334.2 (Q)	51.0	6.0
							352.2 (I)	47.0	6.0
T-2	3.44	484.0	$[M + NH_4]^+$	21.0	10.0	22.0	215.0 (Q)	22.0	4.0
							185.0 (I)	29.0	4.0
ZEN	3.71	319.0	$[M + H]^+$	26.0	8.0	20.0	282.9 (Q)	19.0	4.0
							301.0 (I)	15.0	10.0
OTA	3.76	404.0	$[M + H]^+$	41.0	7.5	16.0	238.9 (Q)	31.0	6.0
							102.1 (I)	91.0	6.0
FB$_2$	3.86	706.2	$[M + H]^+$	71.0	10.5	20.0	336.3 (Q)	43.0	14.0
							318.3 (I)	45.0	12.0
STE	3.88	325.1	$[M + H]^+$	66.0	3.5	26.0	281.0 (Q)	43.0	4.0
							310.0 (I)	37.0	4.0

[a](DP) Declustering potential, (EP) entrance potential, (CEP) collision cell entrance potential, (CXP) collision cell exit potential, and (CEn) collision energy. All expressed in voltage.
[b]Product ions: (Q) transition used for quantification and (I) transition employed to confirm the identification.

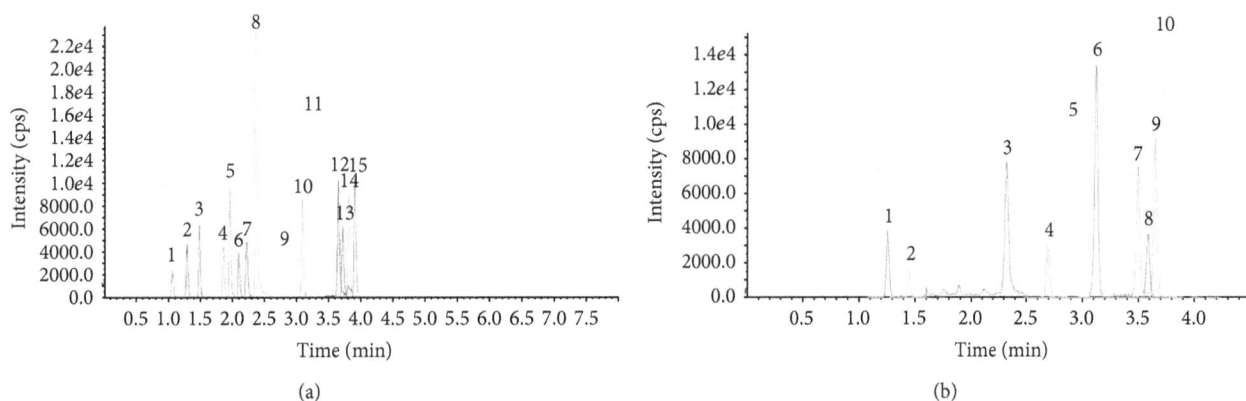

FIGURE 3: (a) Chromatogram of a spiked white rice sample; (b) chromatogram of a spiked barley syrup sample applying the proposed QuEChERS-UHPLC-MS/MS methodology. (Aflatoxins, OTA and STE: 25 μg kg^{-1}; CIT: 50 μg kg^{-1}; FB$_1$, FB$_2$, T-2, HT-2, and ZEN: 250 μg kg^{-1}; DON: 1000 μg kg^{-1}; F-X, NIV: 2500 μg kg^{-1}). 1: NIV; 2: DON; 3: F-X; 4: AFG$_2$; 5: AFG$_1$; 6: AFB$_2$; 7: AFB$_1$; 8: CIT; 9: HT-2; 10: FB$_1$; 11: T-2; 12: ZEN; 13: OTA; 14: FB$_2$; 15: STE.

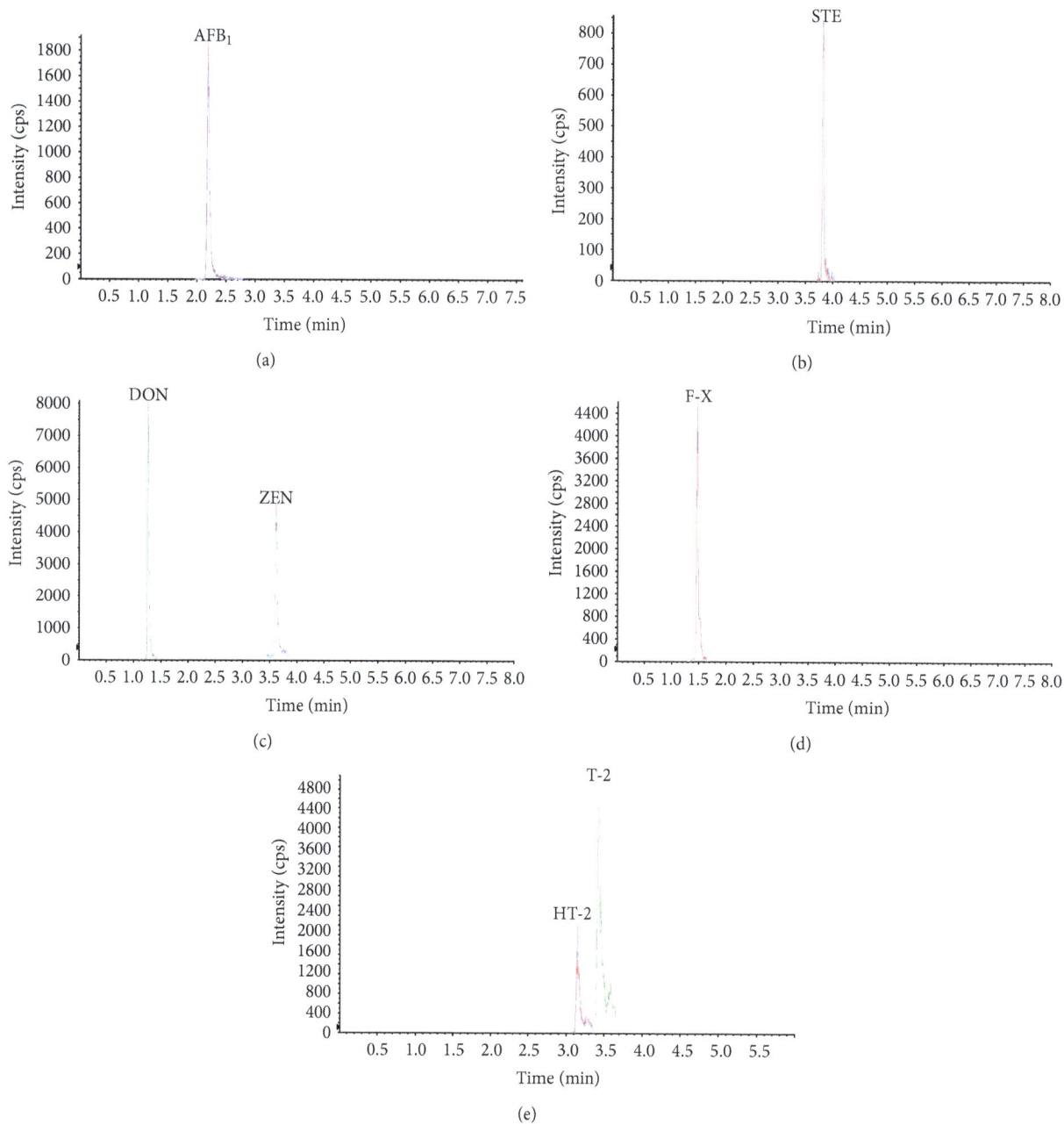

FIGURE 4: Extracted ion chromatogram of positive samples: (a) red rice ($8.3\,\mu\mathrm{g\,kg}^{-1}$ of AFB$_1$); (b) sunflower seed sample ($3.7\,\mu\mathrm{g\,kg}^{-1}$ of STE); (c) walnut sample ($222\,\mu\mathrm{g\,kg}^{-1}$ of ZEN and $346\,\mu\mathrm{g\,kg}^{-1}$ of DON); (d) macadamia nut sample ($2466\,\mu\mathrm{g\,kg}^{-1}$ of F-X); (e) milk thistle ($827\,\mu\mathrm{g\,kg}^{-1}$ of T-2 and $944\,\mu\mathrm{g\,kg}^{-1}$ of HT-2).

Trying to fill this gap, we modified the sample treatment previously described for cereals and pseudocereals (Figure 2) to make it suitable for the determination of 10 mycotoxins (OTA, FB$_1$, FB$_2$, T-2, HT-2, CIT, STE, F-X, ZEN, and DON) in wheat, barley, and rice syrup. Unfortunately, very low recoveries were obtained for aflatoxins and NIV and their quantification was impossible in these matrices. Good recoveries were obtained (between 62.8% and 100.6%), with RSD lower than 11.5% [54]. A chromatogram corresponding to a spiked barley syrup sample submitted to the proposed method is shown in Figure 3(b). It must be pointed out

that there is no specific legislation for this kind of matrices, although the low LOQs obtained allowed the determination of these mycotoxins at concentrations lower than the maximum contents usually established by current legislation in different foodstuffs.

3.4. Analysis of Edible Nuts and Seeds. It is well known that nuts and seeds are susceptible to mould growth and consequently to mycotoxin contamination [55–57]. Insect feeding-damage is the principal factor leading to preharvest fungal infection and subsequent mycotoxin contamination, but

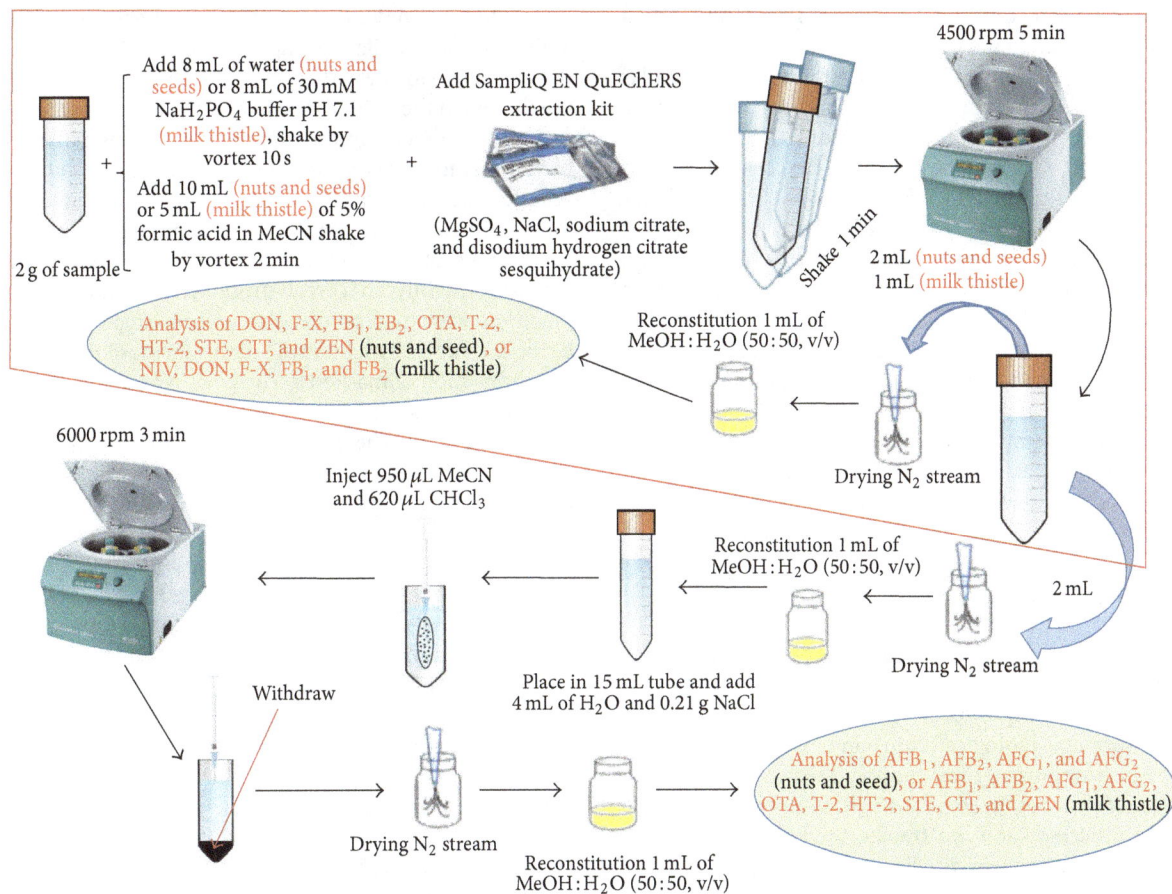

FIGURE 5: Sample treatment for the determination of mycotoxins in nuts and seeds and milk thistle (MeCN: acetonitrile; MeOH: methanol).

infection may also occur after harvesting and storage. Current EU food safety legislation only regulates the content of aflatoxins in these matrices, with maximum permitted levels which depend on the kind of nut or seed for direct human consumption [6, 58].

Considering the good results obtained with the previously described matrices, the same approach was attempted. In this case, the QuEChERS-based extraction allowed the determination of OTA, T-2, HT-2, STE, CIT, ZEN, FB_1, FB_2, DON, and F-X. However, a further purification of the extracts was required for subsequent determination of aflatoxins. Thus, an additional clean-up step based on DLLME was proposed in order to reduce matrix effect, allowing the determination of aflatoxins. A flow chart of the whole procedure is shown in Figure 5. This methodology was applied for the determination of 14 mycotoxins in different nuts and seeds (almonds, peanuts, sunflower seeds, pumpkin seeds, walnuts, macadamia nuts, pistachios, hazelnuts, and pine nuts). The low LOQs obtained for aflatoxins, the only mycotoxins regulated in nut and seed matrices, allowed their quantification at concentrations lower than the maximum level established by current legislation. RSD was lower than 11% and recoveries ranged from 60.7% to 104.3%. Among the samples analysed, a sunflower seed sample showed a high content of STE (Figure 4(b)), a walnut sample of ZEN and

DON (Figure 4(c)), and a macadamia nut sample of F-X (Figure 4(d)) [59].

3.5. Analysis of Herbal Products: Milk Thistle. The consumption of products with specific nutritional and/or functional characteristics has significantly increased during the last decade. Among them, there are many food supplements containing herbal products and/or their derivatives as ingredients. Previous studies have demonstrated that these materials can suffer from fungi and mycotoxin contamination [60, 61]. However, maximum levels for mycotoxins in food supplements are not established; only, the European Pharmacopoeia sets a maximum level for AFB_1 ($2\,\mu g\,kg^{-1}$) and for the sum of AFB_1, AFB_2, AFG_1, and AFG_2 ($4\,\mu g\,kg^{-1}$) for herbal products used as drug ingredients [62].

Most of the scarce methods proposed for the determination of mycotoxins in herbal products use SLE and IAC for clean-up, including the method recommended by the Pharmacopoeia for the determination of AFB_1 [62].

In this context, we proposed a method for the multiclass determination of mycotoxins in milk thistle (*Silybum marianum*), a botanical consumed as food supplement because of its protective effects on the liver. Moreover, although analytical methods for studying the occurrence of mycotoxins in other herbal products (as tea, ginseng, or ginger) have been previously reported, milk thistle has been scarcely studied and only for aflatoxin content [63].

The sample treatment optimized for this matrix is similar to that proposed for nuts and seeds, with some modifications (Figure 5). After QuEChERS-based extraction, FB_1, FB_2, NIV, DON, and F-X were quantified. However, a second clean-up step based on DLLME was needed for the determination of AFB_1, AFB_2, AFG_1, AFG_2, OTA, T-2, HT-2, STE, CIT, and ZEN. The method allowed the quantification of aflatoxins at concentrations lower than their maximum level established in botanicals by Pharmacopoeia. The rest of mycotoxins were determined at concentrations lower than their usual established limits in different foodstuff. Good recoveries were obtained (between 62.3% and 98.9%, except for ZEN in seed samples and CIT in extract). Among the different commercial samples of milk thistle (seeds and natural extract), two were contaminated with T-2 and HT-2 (Figure 4(e)) and ZEN was detected in one of them [64].

4. Future Trends

Currently, mycotoxins analysis presents several challenges that still need to be addressed and overcome. On the first place, carryover of mycotoxins from contaminated feeds to animal tissues and biological fluids and eventually to products intended for human consumption (meat, milk, and eggs) is a matter of concern. In some animals, mycotoxins undergo metabolic processes and are transformed into other compounds with different toxicity. For instance, most mammals metabolize AFB_1 into AFM_1 that is transferred to the milk, while poultries metabolize AFB_1 into toxic hydroxylated metabolites that may pass to eggs [65]. Animal feed is the first link in food chain and, therefore, the production of safe food depends not only on the manufacturers compliance with current legislation, but also on the use of safe feed by the farmers. As a solution, the use of detoxifying agents (a new group of feed additives) has been proposed in order to decrease the effect of mycotoxin contamination in feeds. These products are mainly adsorbent agents (i.e., activated charcoal, clays, silicates, and some synthetic polymers) that reduce the absortion of mycotoxins in the gastrointestinal tract of animal. Also, some enzymes and microorganisms are used as agents capable of transforming the mycotoxins by modification of their structure, although their use is more limited [66]. However, because of the different properties of mycotoxins, an adsorbent may be effective against a mycotoxin while ineffective against others. Recently, some published studies deal with various natural extracts, such as derivatives of honey [67] and organosulfur compounds derived from allium, like garlic [68], which could be natural alternatives to the usual binders, opening a very interesting field of research.

Another issue of great importance is the evaluation of occurrence of the so-called "masked mycotoxins" in food and feed. The metabolism of some plants (which have natural detoxification mechanisms) can generate conjugated compounds (masked mycotoxins), with different chemical behaviors than the mycotoxins of origin. Thus, some plants are able to transform the relatively nonpolar trichothecenes and ZEN into more polar derivatives by conjugation with sugars, amino acids, or sulfate groups, which are then isolated into the vacuoles. However, these forms can be hydrolyzed to their precursors in the animal digestive tract, thus showing similar toxicity than free mycotoxins. Although this phenomenon has been studied mainly on *Fusarium* toxins (trichothecenes, zearalenone, and fumonisins), it has also been described for other mycotoxins. In addition, some technological food processings play an important role in the mechanisms of masking, mainly in cereal products, as they can induce reactions with macromolecules such as sugars, proteins, or lipids and inversely release the native forms of mycotoxins by decomposition of masked derivatives. Nowadays, toxicological data on masked mycotoxins are scarce, although several studies highlight the potential threat of these compounds for consumer safety. In particular, the possible hydrolysis of masked mycotoxin (generating the initial mycotoxin) during mammalian digestion would be a risk factor to be considered. For instance, products with an apparent low mycotoxin contamination have induced toxic effects due to the presence of "occult" fumonisins liberated upon hydrolysis and not detected in routine analysis. In this way, masked mycotoxins can quantitatively contribute to the total amount of mycotoxins, especially in cereals [69, 70]. Consequently, the development of analytical methods for multiclass analysis of mycotoxins including their transformation products is a challenge, in order to assess their real risk to the health of consumers.

Finally, the study of the so-called emerging mycotoxins derived from *Fusarium* fungi (as fusaproliferin, moniliformin, beauvericin, and enniatins), present prevalently in foods from northern Europe and Mediterranean countries, must also be highlighted. Unlike other better studied mycotoxins, permitted maximum levels have not yet been established for these mycotoxins. This is mainly because of the scarce data related to their presence in food, level of contamination, and toxicity. Although no cases of mycotoxicosis by the intake of these mycotoxins have been described, some studies (most *in vitro*) revealed the possible toxicity of these compounds, which could be increased as a result of the interaction of several mycotoxins present in food [71, 72]. For this reason, more studies are mandatory in order to evaluate the risk of these emerging toxins. Once again, reliable analytical methods are urgently required.

5. Conclusions

Alternative UHPLC-MS/MS analytical methods for multiclass determination of mycotoxins based on QuEChERS and DLLME for sample treatment have been developed. The proposed methods have been evaluated in diverse food commodities, most of them scarcely investigated. They showed as general advantages their efficacy, simplicity, versatility, and accuracy, as well as their low impact on the environment, shorter analysis time, and the relatively low-cost, compared with conventional IAC. Thus they fulfill the current requirements of analytical methods for the determination of contaminants. However, there are some aspects concerning mycotoxin determination that still are a challenge for the scientific community, as the development of new analytical methods including the determination of masked or emerging mycotoxins. The proposed methodologies could be applied also to these analytes, opening new perspectives that, combined with powerful analytical techniques, such as UHPLC-MS/MS, offer interesting perspectives in this field.

Conflict of Interests

The authors declare that there is no conflict of interests regarding the publication of this paper.

References

[1] European Food Safety Authority (EFSA), 2012, http://www.efsa.europa.eu/en/topics/topic/mycotoxins.htm.

[2] S. Marin, A. J. Ramos, G. Cano-Sancho, and V. Sanchis, "Mycotoxins: occurrence, toxicology, and exposure assessment," *Food and Chemical Toxicology*, vol. 60, pp. 218–237, 2013.

[3] http://ec.europa.eu/food/safety/rasff/index_en.htm.

[4] IARC, *Monographs on the Evaluation of Carcinogenic Risks to Humans*, vol. 56, IARC, Lyon, France, 1993.

[5] A. L. Capriotti, C. Caruso, C. Cavaliere, P. Foglia, R. Samperi, and A. Laganà, "Multiclass mycotoxin analysis in food, environmental and biological matrices with chromatography/mass spectrometry," *Mass Spectrometry Reviews*, vol. 31, no. 4, pp. 466–503, 2012.

[6] "Commission regulation (EC) No. 1881/2006 of 19 December 2006 setting maximum levels for certain contaminants in foodstuffs," *Official Journal of the European Union*, pp. L364/5–L364/24, 2006.

[7] "Commission recommendation 2013/165/EU on the presence of T-2 and HT-2 toxin in cereals and cereal products," *Official Journal of the European Union L*, vol. 91, p. 12, 2013.

[8] "Commission Regulation (EC) No. 401/2006 laying down the methods of sampling and analysis for the official control of the levels of mycotoxins in foodstuffs," *Official Journal of the European Union*, vol. L70, pp. 12–34, 2006.

[9] FDA, "Molecular biology and natural toxins. Mycotoxins in domestic and imported foods," in *Compliance Program Guidance Manual*, chapter 7, Food and Drug Administration, Washington, DC, USA, 2008, http://www.fda.gov/Food/ComplianceEnforcement/FoodCompliancePrograms/ucm071496.htm.

[10] G. S. Shephard, "Determination of mycotoxins in human foods," *Chemical Society Reviews*, vol. 37, pp. 2468–2477, 2008.

[11] S. de Saeger, Ed., *Determining Mycotoxins and Mycotoxigenic Fungi in Food and Feed*, Woodhead, Cambridge, UK, 2011.

[12] G. S. Shephard, F. Berthiller, P. A. Burdaspal et al., "Developments in mycotoxin analysis: an update for 2011-2012," *World Mycotoxin Journal*, vol. 6, no. 1, pp. 3–30, 2013.

[13] D. Heperkan, F. K. Güler, and H. I. Oktay, "Mycoflora and natural occurrence of aflatoxin, cyclopiazonic acid, fumonisin and ochratoxin A in dried figs," *Food Additives and Contaminants*, vol. 29, pp. 277–286, 2012.

[14] J. S. Dos Santos, C. R. Takabayashi, E. Y. S. Ono et al., "Immunoassay based on monoclonal antibodies versus LC-MS: deoxynivalenol in wheat and flour in Southern Brazil," *Food Additives & Contaminants A*, vol. 28, no. 8, pp. 1083–1090, 2011.

[15] S. C. Cunha and J. O. Fernandes, "Development and validation of a method based on a QuEChERS procedure and heart-cutting GC-MS for determination of five mycotoxins in cereal products," *Journal of Separation Science*, vol. 33, no. 4-5, pp. 600–609, 2010.

[16] N. Arroyo-Manzanares, L. Gámiz-Gracia, A. M. Gámiz-Gracia, J. J. Soto-Chinchilla, A. M. García-Campaña, and L. E. García-Ayuso, "On-line preconcentration for the determination of aflatoxins in rice samples by micellar electrokinetic capillary chromatography with laser-induced fluorescence detection," *Electrophoresis*, vol. 31, no. 13, pp. 2180–2185, 2010.

[17] L. Campone, A. L. Piccinelli, R. Celano, and L. Rastrelli, "Application of dispersive liquid-liquid microextraction for the determination of aflatoxins B1, B2, G1 and G2 in cereal products," *Journal of Chromatography A*, vol. 1218, no. 42, pp. 7648–7654, 2011.

[18] N. Arroyo-Manzanares, A. M. García-Campaña, and L. Gámiz-Gracia, "Comparison of different sample treatments for the analysis of ochratoxin A in wine by capillary HPLC with laser-induced fluorescence detection," *Analytical and Bioanalytical Chemistry*, vol. 401, p. 2987, 2011.

[19] N. Arroyo-Manzanares, L. Gámiz-Gracia, and A. M. García-Campaña, "Determination of ochratoxin A in wines by capillary liquid chromatography with laser induced fluorescence detection using dispersive liquid-liquid microextraction," *Food Chemistry*, vol. 135, no. 2, pp. 368–372, 2012.

[20] J. Rubert, C. Soler, and J. Mañes, "Application of an HPLC–MS/MS method for mycotoxin analysis in commercial baby foods," *Food Chemistry*, vol. 133, pp. 176–183, 2012.

[21] I. Sospedra, J. Blesa, J. M. Soriano, and J. Mañes, "Use of the modified quick easy cheap effective rugged and safe sample preparation approach for the simultaneous analysis of type A- and B-trichothecenes in wheat flour," *Journal of Chromatography A*, vol. 1217, no. 9, pp. 1437–1440, 2010.

[22] P. Li, Z. Zhang, X. Hu, and Q. Zhang, "Advanced hyphenated chromatographic-mass spectrometry in mycotoxin determination: Current status and prospects," *Mass Spectrometry Reviews*, vol. 32, no. 6, pp. 420–452, 2013.

[23] M. Zachariasova, O. Lacina, A. Malachova et al., "Novel approaches in analysis of Fusarium mycotoxins in cereals employing ultra performance liquid chromatography coupled with high resolution mass spectrometry," *Analytica Chimica Acta*, vol. 662, no. 1, pp. 51–61, 2010.

[24] M. M. Aguilera-Luiz, P. Plaza-Bolaños, R. Romero-González, J. L. Martínez-Vidal, and A. Garrido-Frenich, "Comparison of the efficiency of different extraction methods for the simultaneous determination of mycotoxins and pesticides in milk samples by ultra high-performance liquid chromatography-tandem mass spectrometry," *Analytical and Bioanalytical Chemistry*, vol. 399, no. 8, pp. 2863–2875, 2011.

[25] P. Pérez-Ortega, B. Gilbert-López, J. F. García-Reyes, and A. Molina-Díaz, "Generic sample treatment method for simultaneous determination of multiclass pesticides and mycotoxins in wines by liquid chromatography-mass spectrometry," *Journal of Chromatography A*, vol. 1249, pp. 32–40, 2012.

[26] J. O'Mahony, L. Clarkea, M. Whelan et al., "The use of ultra-high pressure liquid chromatography with tandem mass spectrometric detection in the analysis of agrochemical residues and mycotoxins in food—challenges and applications," *Journal of Chromatography A*, vol. 1292, pp. 83–95, 2013.

[27] J. P. Meneely, F. Ricci, H. P. van Egmond, and C. T. Elliott, "Current methods of analysis for the determination of trichothecene

mycotoxins in food," *TrAC—Trends in Analytical Chemistry*, vol. 30, no. 2, pp. 192–203, 2011.

[28] A. Veršilovskis and S. de Saeger, "Sterigmatocystin: occurrence in foodstuffs and analytical methods-an overview," *Molecular Nutrition & Food Research*, vol. 54, no. 1, pp. 136–147, 2010.

[29] R. Köppen, M. Koch, D. Siegel, S. Merkel, R. Maul, and I. Nehls, "Determination of mycotoxins in foods: current state of analytical methods and limitations," *Applied Microbiology and Biotechnology*, vol. 86, no. 6, pp. 1595–1612, 2010.

[30] E. Reiter, J. Zentek, and E. Razzazi, "Review on sample preparation strategies and methods used for the analysis of aflatoxins in food and feed," *Molecular Nutrition and Food Research*, vol. 53, no. 4, pp. 508–524, 2009.

[31] A. Zgoła-Grzeskowiak and T. Grzeskowiak, "Dispersive liquid-liquid microextraction," *TrAC Trends in Analytical Chemistry*, vol. 30, no. 9, pp. 1382–1399, 2011.

[32] V. Andruch, I. S. Balogh, L. Kocúrová, and J. Sandrejová, "Five years of dispersive liquid–liquid microextraction," *Applied Spectroscopy Reviews*, vol. 48, no. 3, pp. 161–259, 2013.

[33] H. Yan, H. Wang, and J. Chromatogr, "Recent development and applications of dispersive liquid–liquid microextraction," *Journal of Chromatography A*, vol. 1295, pp. 1–15, 2013.

[34] L. Campone, A. L. Piccinelli, and L. Rastrelli, "Dispersive liquid-liquid microextraction combined with high-performance liquid chromatography-tandem mass spectrometry for the identification and the accurate quantification by isotope dilution assay of Ochratoxin A in wine samples," *Analytical and Bioanalytical Chemistry*, vol. 399, no. 3, pp. 1279–1286, 2011.

[35] L. Campone, A. L. Piccinelli, R. Celano, and L. Rastrelli, "PH-controlled dispersive liquid-liquid microextraction for the analysis of ionisable compounds in complex matrices: case study of ochratoxin A in cereals," *Analytica Chimica Acta*, vol. 754, pp. 61–66, 2012.

[36] M. D. Víctor-Ortega, F. J. Lara, A. M. García-Campaña, M. del Olmo-Iruela, and A. M. García-Campaña, "Evaluation of dispersive liquid–liquid microextraction for the determination of patulin in apple juices using micellar electrokinetic capillary chromatography," *Food Control*, vol. 31, no. 2, pp. 353–358, 2013.

[37] S. J. Lehotay, M. Anastassiades, and R. E. Majors, "QuEChERS, a sample preparation technique that is "catching on": an up-to-date interview with the inventors," *LC-GC North America*, vol. 28, no. 7, pp. 504–516, 2010.

[38] M. Anastassiades, S. J. Lehotay, D. Stajnbaher, and F. J. Schenck, "Fast and easy multiresidue method employing acetonitrile extraction/partitioning and "dispersive solid-phase extraction" for the determination of pesticide residues in produce," *Journal of AOAC International*, vol. 86, no. 2, pp. 412–431, 2003.

[39] A. Desmarchelier, J. M. Oberson, P. Tella, E. Gremaud, W. Seefelder, and P. Mottier, "Development and comparison of two multiresidue methods for the analysis of 17 mycotoxins in cereals by liquid chromatography electrospray ionization tandem mass spectrometry," *Journal of Agricultural and Food Chemistry*, vol. 58, no. 13, pp. 7510–7519, 1021.

[40] L. Vaclavik, M. Zachariasova, V. Hrbek, and J. Hajslova, "Analysis of multiple mycotoxins in cereals under ambient conditions using direct analysis in real time (DART) ionization coupled to high resolution mass spectrometry," *Talanta*, vol. 82, no. 5, pp. 1950–1957, 2010.

[41] U. Koesukwiwat, K. Sanguankaew, and N. Leepipatpiboon, "Evaluation of a modified QuEChERS method for analysis of mycotoxins in rice," *Food Chemistry*, vol. 153, pp. 44–51, 2014.

[42] P. Paíga, S. Morais, T. Oliva-Teles et al., "Extraction of ochratoxin A in bread samples by the QuEChERS methodology," *Food Chemistry*, vol. 135, no. 4, pp. 2522–2528, 2012.

[43] A. Garrido-Frenich, R. Romero-González, M. L. Gómez-Pérez, J. L. Martínez-Vidal, and J. Chromatogr, "Multi-mycotoxin analysis in eggs using a QuEChERS-based extraction procedure and ultra-high-pressure liquid chromatography coupled to triple quadrupole mass spectrometry," *Journal of Chromatography A*, vol. 1218, no. 28, pp. 4349–4356, 2011.

[44] P. Yogendrarajah, C. van Poucke, B. de Meulenaer, and S. de Saeger, "Development and validation of a QuEChERS based liquid chromatography tandem mass spectrometry method for the determination of multiple mycotoxins in spices," *Journal of Chromatography A*, vol. 1297, pp. 1–11, 2013.

[45] R. Romero-González, A. Garrido-Frenich, J. L. Martínez-Vidal, O. D. Prestes, and S. L. Grio, "Simultaneous determination of pesticides, biopesticides and mycotoxins in organic products applying a quick, easy, cheap, effective, rugged and safe extraction procedure and ultra-high performance liquid chromatography–tandem mass spectrometry," *Journal of Chromatography A*, vol. 1218, no. 11, pp. 1477–1485, 2011.

[46] J. M. Zhang, Y. L. Wu, and Y. B. Lu, "Simultaneous determination of carbamate insecticides and mycotoxins in cereals by reversed phase liquid chromatography tandem mass spectrometry using a quick, easy, cheap, effective, rugged and safe extraction procedure," *Journal of Chromatography B*, vol. 915-916, pp. 13–20, 2013.

[47] International Agency for Research on Cancer (IARC), http://www.iarc.fr.

[48] "Commission Decision of 12 August 2002 implementing Council Directive 96/23/EC concerning the performance of analytical methods and the interpretation of results (2002/657/EC)," *Official Journal of the European Communities L*, vol. 221, p. 8, 2002.

[49] T. Wang and T. Lin, "*Monascus* rice products," *Advances in Food and Nutrition Research*, vol. 53, pp. 123–159, 2007.

[50] N. I. Samsudin and N. Abdullah, "A preliminary survey on the occurrence of mycotoxigenic fungi and mycotoxins contaminating red rice at consumer level in Selangor, Malaysia," *Mycotoxin Research*, vol. 29, pp. 89–96, 2013.

[51] G. Bonafaccia, V. Galli, R. Francisci, V. Mair, V. Skrabanja, and I. Kreft, "Characteristics of spelt wheat products and nutritional value of spelt wheat-based bread," *Food Chemistry*, vol. 68, no. 4, pp. 437–441, 2000.

[52] N. Arroyo-Manzanares, J. F. Huertas-Pérez, A. M. García-Campaña, and L. Gámiz-Gracia, "Simple methodology for the determination of mycotoxins in pseudocereals, spelt and rice," *Food Control*, vol. 36, no. 1, pp. 94–101, 2014.

[53] C. M. Hazel and S. Patel, "Influence of processing on trichothecene levels," *Toxicology Letters*, vol. 153, no. 1, pp. 51–59, 2004.

[54] N. Arroyo-Manzanares, J. F. Huertas-Pérez, L. Gámiz-Gracia, and A. M. García-Campaña, *Food Chemistry*. In press.

[55] A. C. Baquiãoa, P. Zorzetea, T. A. Reisa, E. Assunçãoa, S. Vergueiro, and B. Correa, "Mycoflora and mycotoxins in field samples of Brazil nuts," *Food Control*, vol. 28, no. 2, pp. 224–229, 2012.

[56] C. N. Ezekiel, M. Sulyok, B. Warth, A. C. Odebode, and R. Krska, "Natural occurrence of mycotoxins in peanut cake from Nigeria," *Food Control*, vol. 27, no. 2, pp. 338–342, 2012.

[57] J. Rubert, C. Soler, and J. Mañes, "Occurrence of fourteen mycotoxins in tiger-nuts," *Food Control*, vol. 25, no. 1, pp. 374–379, 2012.

[58] "Commission Regulation (EU) No. 165/2010 amending Regulation (EC) No 1881/2006 setting maximum levels for certain contaminants in foodstuffs as regards aflatoxins," *Official Journal of the European Union L*, vol. 50, p. 8, 2010.

[59] N. Arroyo-Manzanares, J. F. Huertas-Pérez, L. Gámiz-Gracia, and A. M. García-Campaña, "A new approach in sample treatment combined with UHPLC-MS/MS for the determination of multiclass mycotoxins in edible nuts and seeds," *Talanta*, vol. 115, pp. 61–67, 2013.

[60] L. Santos, S. Marín, V. Sanchis, and A. J. Ramos, "Screening of mycotoxin multicontamination in medicinal and aromatic herbs sampled in Spain," *Journal of the Science of Food and Agriculture*, vol. 89, pp. 1802–1807, 2009.

[61] J. D. di Mavungu, S. Monbaliu, M.-L. Scippo et al., "LC-MS/MS multi-analyte method for mycotoxin determination in food supplements," *Food Additives and Contaminants A*, vol. 26, no. 6, pp. 885–895, 2009.

[62] European Pharmacopoeia 6.0, "01/2008:20818. Determination of aflatoxin B1 in herbal drugs," pp. 256-257.

[63] V. H. Tournasa, C. Sapp, and M. W. Trucksess, "Occurrence of aflatoxins in milk thistle herbal supplements," *Food Additives & Contaminants: Part A*, vol. 29, no. 6, p. 994, 2012.

[64] N. Arroyo-Manzanares, A. M. García-Campaña, and L. Gámiz-Gracia, "Multiclass mycotoxin analysis in *Silybum marianum* by ultra high performance liquid chromatography–tandem mass spectrometry using a procedure based on QuEChERS and dispersive liquid–liquid microextraction," *Journal of Chromatography A*, vol. 1282, pp. 11–19, 2013.

[65] L. Afsah-Hejri, S. Jinap, P. Hajeb, S. Radu, and S. Shakibazadeh, "A review on mycotoxins in food and feed: Malaysia case study," *Comprehensive Reviews in Food Science and Food Safety*, vol. 12, no. 6, pp. 629–651, 2013.

[66] G. Jard, T. Liboz, F. Mathieu, A. Guyonvarch, and A. Lebrihi, "Review of mycotoxin reduction in food and feed: from prevention in the field to detoxification by adsorption or transformation," *Food Additives & Contaminants A*, vol. 28, no. 11, pp. 1590–1609, 2011.

[67] C. Siddoo-Atwal and A. S. Atwal, "A possible role for honey bee products in the detoxification of mycotoxins," *Acta Horticulturae*, vol. 963, pp. 237–245, 2012.

[68] K. Mylona, *Fusarium species in grains: dry matter losses, mycotoxin contamination and control strategies using Ozone and chemical compounds [Ph.D. thesis]*, Cranfield University, 2012.

[69] G. Galaverna, C. Dallsta, M. A. Mangia, A. Dossena, and R. Marchelli, "Masked mycotoxins: an emerging issue for food safety," *Czech Journal of Food Sciences*, vol. 27, pp. S89–S92, 2009.

[70] F. Berthiller, C. Crews, C. Dall'Asta et al., "Masked mycotoxins: a review," *Molecular Nutrition and Food Research*, vol. 57, no. 1, pp. 165–186, 2013.

[71] M. Jestoi, "Emerging *Fusarium*-mycotoxins fusaproliferin, beauvericin, enniatins, and moniliformin—a review," *Critical Reviews in Food Science and Nutrition*, vol. 48, no. 1, pp. 21–49, 2008.

[72] A. Santini, G. Meca, S. Uhlig, and A. Ritieni, "Fusaproliferin, beauvericin and enniatins: occurrence in food: a review," *World Mycotoxin Journal*, vol. 5, no. 1, pp. 71–81, 2012.

12

Separation and Characterization of Synthetic Polyelectrolytes and Polysaccharides with Capillary Electrophoresis

Joel J. Thevarajah,[1,2] **Marianne Gaborieau,**[1,2] **and Patrice Castignolles**[1]

[1] *University of Western Sydney (UWS), School of Science and Health, Australian Centre for Research on Separation Sciences (ACROSS), Parramatta, NSW 2751, Australia*
[2] *University of Western Sydney (UWS), School of Science and Health, Molecular Medicine Research Group (MMRG), Parramatta, NSW 2751, Australia*

Correspondence should be addressed to Patrice Castignolles; p.castignolles@uws.edu.au

Academic Editor: Alejandro Sosnik

The development of macromolecular engineering and the need for renewable and sustainable polymer sources make polymeric materials progressively more sophisticated but also increasingly complex to characterize. Size-exclusion chromatography (SEC or GPC) has a monopoly in the separation and characterization of polymers, but it faces a number of proven, though regularly ignored, limitations for the characterization of a number of complex samples such as polyelectrolytes and polysaccharides. Free solution capillary electrophoresis (CE), or capillary zone electrophoresis, allows usually more robust separations than SEC due to the absence of a stationary phase. It is, for example, not necessary to filter the samples for analysis with CE. CE is mostly limited to polymers that are charged or can be charged, but in the case of polyelectrolytes it has similarities with liquid chromatography in the critical conditions: it does not separate a charged homopolymer by molar mass. It can thus characterize the topology of a branched polymer, such as poly(acrylic acid), or the purity or composition of copolymers, either natural ones such as pectin, chitosan, and gellan gum or synthetic ones.

1. Introduction to CE and Limitations of Size-Exclusion Chromatography (SEC/GPC)

Free solution capillary electrophoresis (CE), or capillary zone electrophoresis, is a robust polymer separation method. CE differs from the commonly known slab electrophoresis or capillary gel electrophoresis as the capillary does not contain any stationary phase: it is just filled with a buffer (also named background electrolyte). CE does not require tedious sample preparation, not even filtration (e.g., see later in Section 3.2.2). It has several advantages over traditional separation techniques for the characterization of polyelectrolytes which will be outlined in this review. The most commonly used method for the separation and characterization of polymers is size-exclusion chromatography (SEC, also known as GPC). SEC is relatively quick and affordable in obtaining data regarding the size or molar mass of a polymer with good repeatability [1]. Among

SEC's main limitations is its poor reproducibility in terms of molar mass analysis: round-robin tests often show poor accuracy of the values of the determined molar mass [2]. This is detailed in Berek's recent critical review [3]. The review linked the common accuracy issue to the difficulties in obtaining a pure size-exclusion separation: secondary retention mechanisms, side processes, parasitic processes, osmotic effects, secondary exclusion, concentration effects, preferential interactions, and SEC band broadening. For the ultrahigh molar masses, the sample is generally thought to be degraded by shear [4], although a change of conformation of the polymer chains may also take place, leading to a new separation mechanism [5]. In addition, even in ideal conditions (pure size-exclusion mechanism, no degradation), SEC separates by hydrodynamic volume not by molar mass [6]. Apparent molar masses determined by SEC (e.g., polystyrene-equivalent molar masses) thus have a variable and sometimes limited accuracy [7, 8]. Different topologies (branching) or compositions of the polymer sample influence the hydrodynamic volume and the separation is then incomplete in terms of molar mass when a range of branching structures or of compositions is present in a sample [9–11]. This can render the simple determination of molar mass using Mark-Houwink-Sakurada parameters inaccurate, like in the case of most poly(alkyl acrylates) [12, 13]. Up to 100% error in the determination of the molar mass of branched polymers has been measured using multiple detection SEC (light scattering and viscometry) [14, 15].

We recently discussed the SEC of branched polymers and polysaccharides in a review [16]. Composition of copolymers, branching, and purity are often overlooked in polymer characterization, since SEC has a quasi-monopoly and is not suited for these types of characterization. However, alternative methods are being developed, especially alternative liquid chromatography methods [17]. Liquid chromatography in critical conditions (or at the critical conditions) [18] is one of the most prominent alternative chromatography technique: the critical conditions for one homopolymer correspond to the absence of separation by molar mass for this homopolymer, allowing for separation solely by its topology if branched [19] or solely by its composition if copolymerized [20]. These critical conditions are, however, tedious to establish and low recoveries have been observed [21, 22]. CE offers an alternative and the objective of this review is to present and discuss the potential of CE for synthetic polymers and polysaccharides.

CE (defined here as free solution capillary electrophoresis) involves separation in a capillary filled with only buffer (no stationary phase) under high voltage [23]. The use of only a buffer and no stationary phase prevents the common problem of adsorption onto the stationary phase (and of degradation or deformation of the ultrahigh molar mass chains) commonly faced in SEC. The velocity of different analytes is proportional to the electric field: the proportional constant is named the electrophoretic mobility, μ_{ep}. The selectivity of CE separation relates to the difference in electrophoretic mobility of the analytes (see Figure 1 for the experimental determination of μ_{ep}). The electroosmotic flow (EOF) is created by the movement of the ions of the background electrolyte through the capillary under electric field. The EOF is contributing to the migration of all molecules, even neutral ones. At a high pH the silanol groups of the glass layer of the capillary are completely ionized. This generates a strong zeta potential and an electrical double layer of silanolate groups and positive ions from the background electrolyte. The higher the pH, the higher the density of the electrical double layer which increases the EOF [23].

Successful applications of CE to polymer characterization have been the object of a number of publications, especially by Cottet's group, and the earliest works have been reviewed [24]. Building on these advances, using CE, our group was able to reliably characterize several natural and synthetic polymers, especially polysaccharides and poly(acrylic acid) as discussed in this review.

2. Free Solution Capillary Electrophoresis (CE)

Characterization of polymers by CE can be divided into at least four categories: separation of monomer units after depolymerization (see Section 2.1), separation of oligoelectrolytes (see Section 2.2), and separation of longer polyelectrolytes (see Section 3). The fourth category is the separation of polymers bearing a single charge or no charge. For the latter category, the reader is referred to the pioneering work of the groups of Cottet [25, 26] and Cifuentes [24, 27].

2.1. Average Composition of Polysaccharides

2.1.1. Robust Separation of a Mixture of Monosaccharides. A number of polysaccharides, such as hemicellulose [28, 29], have highly complex chemical structures: they are composed of several different monomer units, mainly monosaccharides. The analysis of these polysaccharides is extremely difficult. The average composition can be determined after depolymerization (hydrolysis) and quantification of the different resulting monosaccharides. Currently high performance liquid chromatography (HPLC) is used to separate carbohydrates using different modes; however, this technique and the different modes used have limitations in regard to coelution [30], tedious sample preparation, and short column life [31]. The detection of monosaccharides is another difficulty. CE is most easily and classically applied to analytes that are charged and possess chromophores. The pKa of most mono- and disaccharides is around 12 [32, 33] and separation in CE was obtained at high pH [34] but initially indirect UV detection, conductivity detection, or derivatization was required for detection. Rovio et al. showed that different hemicelluloses can be characterized not only with CE but also with direct UV detection [32, 33]. This method was applied to plant fiber samples without any derivatization: CE achieved a high-resolution separation of the depolymerized fiber samples (Figure 1) and was compared to the various common HPLC methods and IC (HPAEC) [30, 32]. The CE separation can be optimized at minimal cost by changing the capillary length, buffer counter-ion, and/or the buffer concentration [35]. The main advantage of CE is the robustness of the technique, especially the minimum sample preparation that is required. The precision of the peak identification and the quantification were greatly improved with the use of an electroosmotic flow (EOF) marker and an internal standard [30]. Figure 1 shows electropherograms when raw data (migration time) are compared to corrected data (electrophoretic mobility). Figure 1 also gives the equation used to perform this transformation. In the equation, μ_{ep} is electrophoretic mobility, V is voltage, L_d is the length to the detector, L_t is the total length of the capillary, t_m is the time of migration, and t_{eo} is the migration of a neutral species. Using electrophoretic mobility (thus correcting for EOF variations) allows easy visual comparison of results, for example, to allow identification of trace sugars in ethanol fermentation [36].

CE was able to resolve and quantify mannose, galactose, and xylose. The CE quantification of these sugars results in larger amounts when compared to the HPLC results. This might indicate incomplete recovery in HPLC possibly due

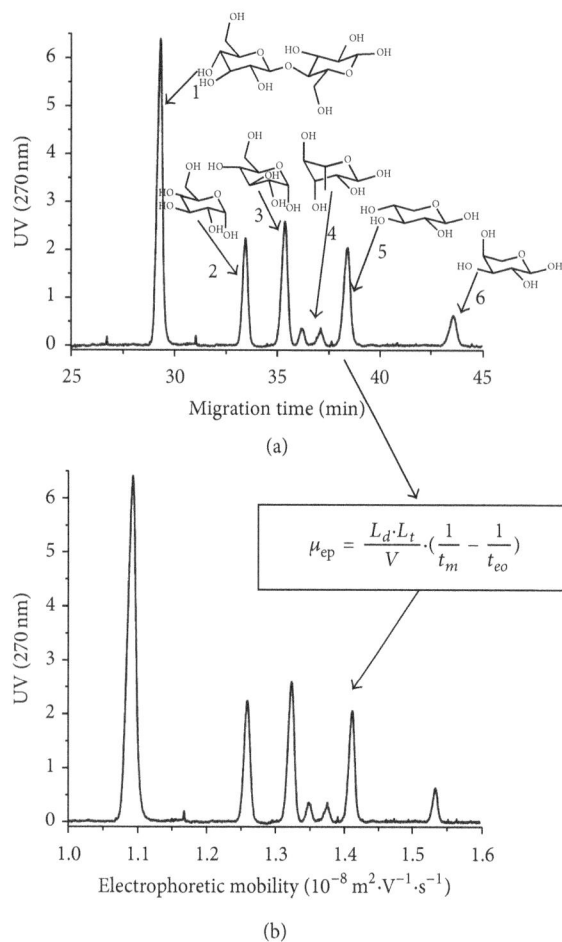

$$\mu_{ep} = \frac{L_d \cdot L_t}{V} \cdot \left(\frac{1}{t_m} - \frac{1}{t_{eo}} \right)$$

FIGURE 1: Separation by CE at high pH (12.6) and with direct UV detection of a depolymerized plant fiber sample plotted as a function of electrophoretic mobility (a) and of migration time (b). The sample contains (1) cellobiose, (2) galactose, (3) glucose, (4) rhamnose, (5) arabinose, and (6) xylose (the molecular structures are given for the sole purpose of identification, and they are in equilibrium with a number of linear and charged forms) [30].

to adsorption onto the stationary phase, which is a common problem associated with the HPLC of samples in complex matrices. There were weaknesses in the direct UV detection in CE which have been addressed recently (see Section 2.1.2).

2.1.2. Direct Detection due to the Photooxidation of Sugars.
Rovio et al. [33] showed that the detection of monosaccharides was possible at 270 nm at pH 12.6. Sarazin et al. [43] suggested the method of detection was due to a photooxidation reaction occurring at the detection window. We confirmed that the detection is by photooxidation using a combination of simulation, multidimensional CE migration, and NMR spectroscopy analysis [30, 37]. The detection occurs without the electric field (i.e., in pressure mobilization instead of CE) but the electric field enhances the sensitivity of the detection. The photooxidation is initiated either by hydroxyl radicals formed by minimal but sufficient water decomposition or by

direct decomposition of the carbohydrates under UV irradiation [35]. The diode-array detector (DAD) emits UV light down to 190 nm. These wavelengths are not leading to any known sample degradation, except for the photooxidation of carbohydrates at high pH. The photooxidation is a type of *in situ* derivatization. If one wishes to avoid the photooxidation reaction taking place, then the lowest wavelengths need to be filtered out: on commercial equipment this simply means using UV detection and not a DAD. Even by using a DAD, most of the sugar molecules are not photooxidized (in the timeframe of the detection). The UV-absorbing species are intermediates in the photooxidation process [37]. These intermediates are present at low concentration but have a high UV absorption coefficient. The final products do not absorb UV and are likely obtained after reacting with oxygen. NMR spectroscopy used as offline detection after CE migration allowed for the identification of a number of carboxylated compounds in the final products. This CE method is ideally suited for the separation of mono- and disaccharides in complex matrices. Direct detection has the advantage of simplicity and of using the most common detector in CE (diode-array detection). The detection of the CE was found to have a limit of detection 10–100x better than HPLC and a better selectivity of detection. Direct detection in CE is not as sensitive as more convoluted methods based on derivatization or pulsed-amperometric detection in ion-chromatography [36]. The method will thus need further improvement to be used for trace detection. Preliminary results showed that using a radical photoinitiator can increase the sensitivity of the direct UV detection [37].

The CE method (Figure 2) is useful not only to determine the average composition of complex polysaccharides, but also to monitor carbohydrates, for example, in a fermentation process. The latest developments showed that fermentation products such as ethanol can also be determined [35]. Ethanol is inhibiting the photooxidation process and this leads to indirect detection of ethanol in the presence of a sugar, such as sucrose. This indirect detection was successfully applied to monitoring lignocellulosic fiber fermentation in terms of both ethanol and sugars alcohols [36].

2.2. Oligoelectrolytes.
Cottet and Gareil have shown that oligo(styrene sulfonate)s can be separated by their molar mass up to a degree of polymerization of 9 [44]. Oligo(sodium acrylates)—oligoAAs—are used in the paint and coating industries to stabilize emulsions [45]. Controlled polymerization methods such as reversible addition-fragmentation chain transfer (RAFT) allow the controlled synthesis of oligoAA. CE can separate oligoAAs (Figure 3) at a higher resolution than that ever obtained with SEC (for oligomers) [38] even using optimal SEC conditions [46, 47].

CE was able to separate and quantify the residual RAFT agent used to obtain the oligoAAs as well as the species of degrees of polymerization (DP) of one, two, and three. The identification of these peaks was obtained by the online coupling of CE with ESI-MS-TOF (electrospray ionization-mass spectrometry-time of flight) [48]. MS analysis showed that the oligomers are separated not only according to their

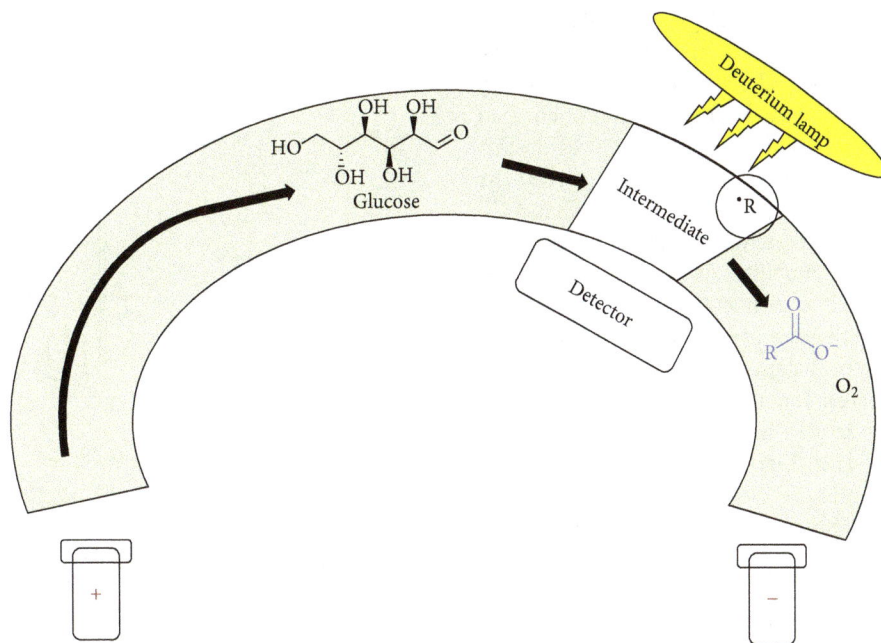

FIGURE 2: Mechanism of direct UV detection in CE of carbohydrates owing to a photooxidation reaction [37].

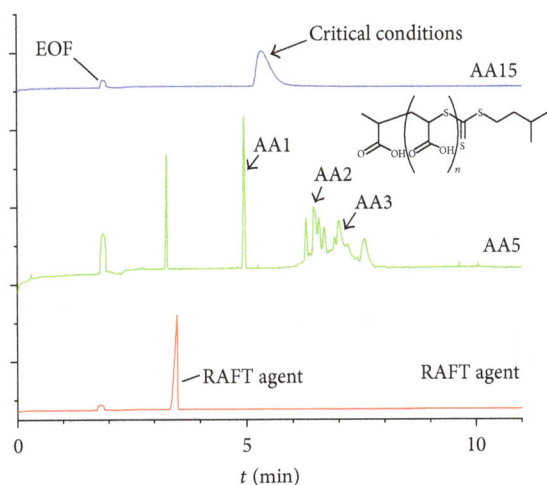

FIGURE 3: Electropherograms in lithium borate for two different oligoAAs, AA5 and AA15, produced by RAFT polymerization, where 5 and 15 correspond to the degree of polymerization obtained at the maximum of the mass spectrum from ESI-MS-TOF (electrospray ionization-mass spectrometry-time of flight) direct infusion (adapted from [38]). AA15 is not separated by molar mass; it is thus in the critical conditions. The bottom electropherogram is of the RAFT agent, that is, the control agent for the polymerization.

degree of polymerization and end-group, but also according to their tacticity. The shortest oligoAAs were shown to contain 50% of unreacted RAFT agent, while the direct infusion in ESI-MS estimated that the sample contained only 2% of unreacted RAFT agent. This large discrepancy is due to the known issue of the bias of the ionization towards low degrees of polymerization and hydrophilic species in MS analysis [49]. CE was shown to be a relevant and fast method

in the study of kinetics of polymerization of RAFT. It has also been used to shed light on the kinetics and mechanism of ring opening polymerization of either 2-oxazoline [50] or N-carboxyanhydrides [26].

Most importantly, the high-resolution separation of CE by molar mass is limited to oligoelectrolytes, with degrees of polymerization below about 10. For large polyelectrolytes, no separation by molar mass is obtained, which corresponds to the "critical conditions" described below.

3. CE in the Critical Conditions

3.1. Explanation of "Critical Conditions". The first example of analysis of synthetic polyelectrolytes by electrophoresis dealt with poly(4-vinyl-N-n-butylpyridinium bromide) more than half a century ago [51]. The authors concluded that "the electrophoretic behavior of polyvinylbutylpyridinium is not very sensitive to molecular weight." CE in the "critical conditions" differs from the CE undertaken in the separation and characterization of oligoelectrolytes such as oligoAA. Critical conditions refer to the conditions sought in liquid chromatography (LC) in which a homopolymer is not separated by molar mass (see Section 1). While these critical conditions are of no use to characterize simple (homo)polymers, most, if not all, polymers are not simple in the sense that they possess a distribution of molar masses as well as different end-groups, distribution(s) of compositions for copolymers, distribution of branch molar masses and of positions of branching points for branched polymers, and so forth. Polymeric samples are multidimensional: the critical conditions allow separation by one (or only a few) dimension at one time (since the distribution of molar masses does not influence the separation any more). The critical conditions thus enable the characterization of complex polymers through the simplification of a multidimensional problem. While a lot of research has been devoted to LC and critical conditions, the method remains tedious and plagued with low accuracy and recovery [21]. Applications to hydrophilic and/or charged polymers are very limited. CE is an alternative to LC in this specific, but important, case of complex polyelectrolytes. The molecular reasons behind critical conditions in LC and CE are completely different and are not widely accepted in any case. The electrophoretic mobility always depends on the charge to friction ratio. It does not depend on the ratio of the charge to the size in the case of polyelectrolytes, since the friction is not only hydrodynamic in this case. The critical conditions do not correspond to the free draining model as proposed by Flory, in which the solvent penetrates the polymer chains freely [52]. Electrostatic friction, however, screens the hydrodynamic friction [53, 54] and leads to the electrophoretic mobility having a very weak dependence on molar mass for a degree of polymerization generally above 15–20 [44, 55, 56]. Thus, CE leads to migration independent from molar mass for polyelectrolytes and this corresponds to the critical conditions sought in LC-CC. CE has been used in the critical conditions outside of our group in the separation of pectins [57] and carboxymethylcellulose [58] according to their composition. In our group using CE in the critical conditions (CE-CC) has allowed the investigation of the composition of natural polymers as well as of synthetic polymers. Further, we have also been able to look at the degree of branching of synthetic polymers.

3.2. Separation by Composition

3.2.1. Pectin and Carboxymethylcellulose (CMC). CE-CC effectively separates the polysaccharide pectin by composition. Several studies reported the separation of pectin by its degree of substitution (DS, which may include either esterification or methyl-esterification) [57, 59–62]. Within one sample, pectins macromolecules with different degrees of esterification (DE) could be separated. It was later hypothesized that the shape of the peaks could additionally be used to indicate a distribution of methyl esters of pectin within samples [60]. Guillotin et al. [62] established a protocol in which pectin's degree of amidification, degree of methyl-esterification, and subsequently the degree of substitution could be determined.

Other research involved the use of capillary electrophoresis to determine the DS of carboxymethylcellulose [58]. The study showed the possibility of not only determining the average DS but also determining the heterogeneity/distribution of the compositions of CMC.

3.2.2. Gellan Gum. Gellan gum is a natural polymer which is widely distributed in the environment. Due to its rheological properties it is viewed as a possible stabilizing agent in various industries [63]. Gellan gum's monomer unit structure contains D-glucuronic acid, D-glucose either with or without acyl substituents, and L-rhamnose. The proportion of acyl chains attached to the glucose and the distribution of these

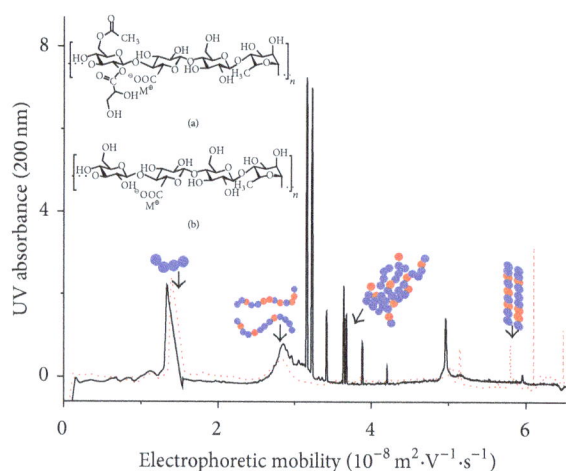

FIGURE 4: Electropherograms of a low acyl gellan gum (dotted red line) and a high-acyl gellan gum (solid black line) (in potassium borate, pH 9.2), after a few hours of dissolution [39].

acyl chains along the polysaccharide vary from sample to sample. Gellan gum is often characterized by its degree of acylation, which affects its desired properties. CE allowed some separation of gellan gum oligomers according to their degree of polymerization and separation of polymers by their degree of acylation (composition) [39]. CE gave a unique separation of gellan gums that could not be attained with any other existing separation methods. CE could characterize not only a low acyl gellan gum but also a high acyl gellan gum (Figure 4). The latter sample was a turbid dispersion: while obtaining a true solution was not possible, the characterization of this dispersion is relevant for its applications such as the stabilization of carbon nanotubes [64]. Characterization of the high acyl gellan gum showed the presence of aggregates forming during the dissolution of the gellan gum samples and appearing as very narrow peaks due to their very low diffusion coefficients. This illustrates the robustness of the method as it did not require sample filtration (while the background electrolyte still requires filtration). Filtration would have changed the nature of this colloidal sample and should thus not be performed for a meaningful characterization. Complementing the CE separation with a simple pressure mobilization analysis (a qualitative version of Taylor dispersion analysis [65]; see 3.4), the presence of oligomers of gellan gum was confirmed while they had never been identified previously in these gellan gums. Through the CE separation, the oligomers could be separated and quantified. Separation and characterization of this dispersion could only be obtained by CE or field flow fractionation (FFF). In the most common form of FFF, flow FFF, the oligomers would have been lost through the membrane. The low acyl samples contained more oligomers than the high acyl samples suggesting the occurrence of some degradation during the deacylation process. The electrophoretic mobility is also sensitive to the conformation of gellan gum and complementary to light scattering characterization. A high mobility peak, present in the high acyl sample and becoming more intense in the presence of potassium borate ions, suggests the possibility

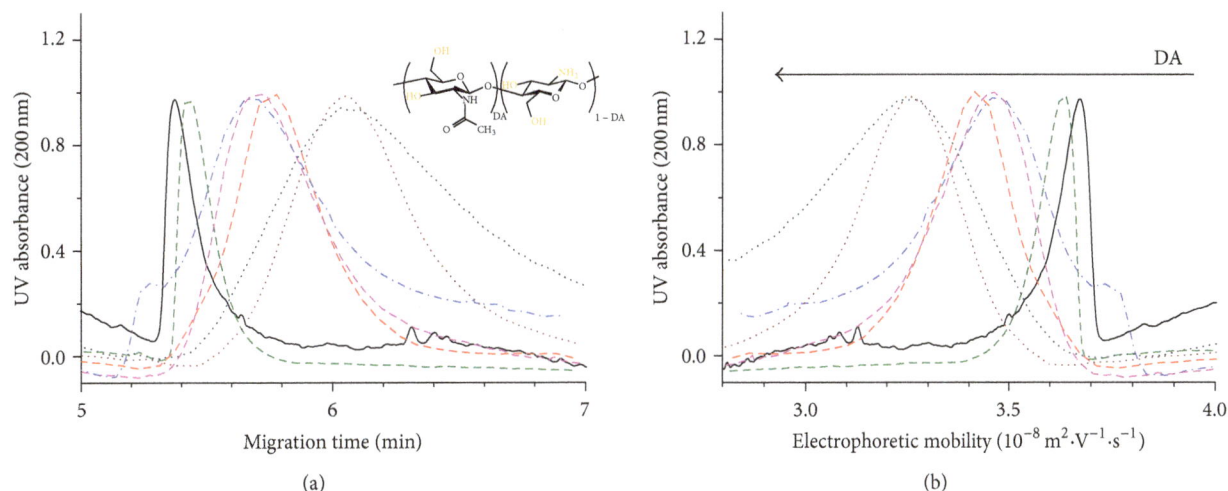

FIGURE 5: Separation of chitosan samples by their DA with CE (sodium phosphate, pH 3): electropherogram shown as a function of (a) migration time and (b) apparent electrophoretic mobility. Samples with different degrees of acetylation (weight-average DA determined by NMR spectroscopy) are shown: black solid line (4%), green dashed line (4.3%), blue dashed-dotted line (16.5%), red dashed line (18.7%), pink dashed line (19.8%), brown dotted line (22.4%), and black dotted line (23.6%) [40].

of a double-helix conformation. This peak differs in mobility and, thus, also suggests that the rest of the macromolecules are in a random coil conformation.

In order of increasing mobility, the electropherograms show the presence of gellan gum oligomers, then gellan gum polymer chains containing different degrees of acylation (random coil conformation), then aggregates of gellan gum chains, and finally gellan gum polymer chains in a helix conformation. The gellan gums studied in this work are copolymers containing both repeating units (not necessarily forming blocks). The top molecular structure represents fully acetylated gellan gum and the bottom one represents fully deacylated gellan gum. The monomer unit is constituted from left to right of one D-glucose with two acyl substituents (one acetyl and one glycerate bearing a diol or glycol), one D-glucuronic acid, one D-glucose without substituents, and one L-rhamnose.

The study undertaken on gellan gum is an example of the robustness of the CE technique. Whilst allowing the successful separation of complex samples by composition, it also provides information on the conformation of the polymer chains and the presence of aggregates.

3.2.3. Chitosan.

Chitosan is a polysaccharide produced from the N-deacetylation of chitin. Chitin is the main component of the shells of crabs and shrimps and can also be found in the cell wall of fungi. It is a renewable resource that it is a large waste product. Chitosan's structure contains N-acetyl-D-glucosamine as well as D-glucosamine units. The composition of this copolymer is quantified by the degree of acetylation (DA), which is the fraction of N-acetyl-D-glucosamine. Chitosan is receiving extensive research interest due to its inherent properties. It is biocompatible, antimicrobial, antifungal, biodegradable, and pH-responsive [66, 67]. However, one limitation of chitosan is in the incomplete characterization by current methods; while being a natural

product, there is a large variation among samples. Chitosan is often characterized by its average degree of acetylation (DA) [40]. Chitosan samples are, however, not composed of polymer chains with all the same DA, but they contain a distribution of DAs. The complexity and importance of the distribution of DAs have been revealed recently through a coupling of SEC with ^{1}H NMR spectroscopy [68]; however, it still has not been measured. ^{1}H NMR spectroscopy can determine number-average as well as weight-average DAs [40]. The measurements are accurate, but precise results are time-consuming and alternative methods are often considered. Chitosan is often characterized by only one of its average DAs, which is implicitly and incorrectly assuming the sample is homogeneous in terms of DA, that is, does not have a distribution of DAs.

CE-CC separates chitosan by its degree of acetylation (Figure 5) [40]. Chitosan macromolecules with a lower DA have a higher mobility (at low pH below the glucosamine monomer unit pKa) since they have a higher number of free amino groups which increase their charge and therefore their electrophoretic mobility. Another important attribute revealed in the CE separation is the broadness and shape of the peaks of the chitosan samples. The broadness corresponds to the distribution of DAs and some samples have broader distributions than others (Figure 5). These differences in distributions will likely affect functional properties such as adhesion, biodegradability, and bacteriostaticity [68]. With proper calibration, the CE separation will allow the determination of the distribution of DAs. The distribution can be calculated from the UV signal taking into account the nonlinear relation between electrophoretic mobility and migration, as it is done to calculate molar mass distributions in SEC [10].

We have also used CE to assist in the grafting of synthetic polymers, poly(sodium styrene sulfonate) and poly(methyl methacrylate-co-acrylonitrile), onto a chitosan backbone [41]

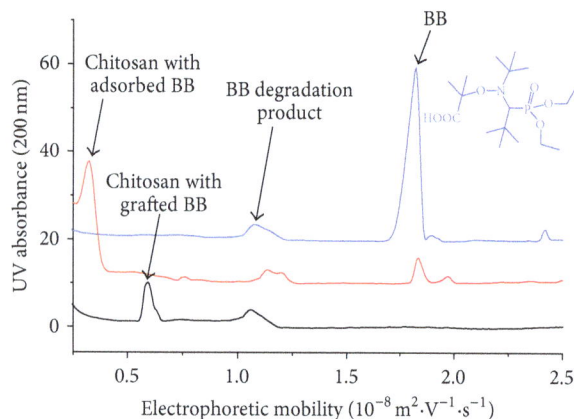

FIGURE 6: CE of pure BlocBuilder (BB, blue line), chitosan with adsorbed BB (red line), and chitosan with grafted BB (red line) in sodium borate buffer (pH 9.2) [41].

to address the variability of the mechanical properties of chitosan films [69]. CE could analyze samples produced along the synthetic pathway. This allowed the validation of the grafting process. Chitosan was first functionalized with the introduction of an acrylamide and/or acrylate function. This was followed by a radical addition of BlocBuilder (BB) alkoxyamine to allow a controlled (nitroxide-mediated) polymerization [70] of the grafted monomers. The limited solubility of chitosan, added to the hydrophobicity of the grafted compound, made the complete solubilization of the samples impossible. Despite incomplete dissolution, the robustness of CE allowed us to analyze these samples: chitosan functionalized with BlocBuilder was separated from pure BlocBuilder (Figure 6). This experiment was undertaken in a high pH buffer and therefore the negative charge expressed by BB results in a nonzero mobility in the CE electropherogram whilst the neutral chitosan has no mobility. Pure BB, chitosan covalently functionalized with BB, and chitosan physically mixed with BB (control) were then injected at the same pH. The chitosan grafted with BB encounters more hydrodynamic friction than BB alone which reduces the electrophoretic mobility of the chitosan grafted with BB in comparison to the pure BB. The method even allowed the discrimination of covalently grafted BB from BB adsorbed to chitosan since the latter has a lower electrophoretic mobility in comparison to the grafted sample. This type of separation, neutral polymer chain from slightly modified polymer chain, was also used by the group of Schoenmakers with nonaqueous CE to prove the presence of charged end-groups in part of their poly(2-oxazoline) samples [50].

Using both CE and CE-CC a more thorough analysis can be completed on both pure and modified chitosan samples. There is also a possibility of extending the research on chitosan with CE in terms of the determination of the distribution of the degrees of acetylation which has not been previously examined [40].

3.3. *Separation by Topology (Branching).* There is a large interest in the characterization of water-soluble polyacrylates

as their use has increased to include a range of applications from industrial protective coatings to food packaging. The poly(sodium acrylate)s, PNaA, studied by our group were produced by nitroxide-mediated polymerisation (NMP) [70]. The aim of the study was to characterize the branching in PAA using CE-CC. Different topologies of the PAA samples, linear, hyperbranched, and three-arm star, were separated within 15 min [42]. Figure 7 presents the CE results both as raw data, as a function of migration time and as EOF corrected data, as a function of electrophoretic mobility. This highlights the importance of converting the results to electrophoretic mobility plots as a trend is not seen in the migration time results due to the variation in EOF between injections. When the EOF correction is made, it can be seen that the hyperbranched polymer exhibits the lowest electrophoretic mobility followed by three-arm star and finally the linear ones. The differences in electrophoretic mobility can be explained by a decrease in the effective charge of the branched samples when compared to the linear samples. The results obtained were highly repeatable and reproducible with relative standard deviations (RSD) values below 1.6%. The separations were also successfully reproduced in a different buffer and whilst they produced different electrophoretic mobility values (as expected, due to the different counterions), the electrophoretic mobility remained lower for the more branched structures. This study highlights also the accuracy (related to the reproducibility of the separation) of the technique.

The CE results obtained also provided information regarding the homogeneity of the branching topology. These samples are expected to have a controlled molar mass owing to the reversible termination with the nitroxide but the broadness of the peaks obtained suggests some heterogeneity in the branching structure. The broad range of electrophoretic mobilities is attributed to a broad range of branching topologies produced in the polymerization.

The CE-CC separation was also shown to be influenced by end-groups. In Figure 8 the red solid line represents PAA with a SG1 [N-tert-butyl-N-(1-diethylphosphono-2,2-dimethylpropyl) nitroxide] moiety as the end-group. There is a marked difference between this sample and the sample with the PAA that has been heated in the presence of thiophenol to replace the SG1 end-groups with hydrogen. The electrophoretic mobility increases with the removal of the SG1 end-group, as expected, since the bulky SG1 molecule would contribute to the hydrodynamic friction of the PAA chains more than hydrogen (and neither contributes to its charge). The result also suggests heterogeneity of the sample in terms of branching. The thiophenol treatment was also applied to a PAA sample obtained from the hydrolysis of poly(t-butyl acrylate) and CE-CC showed that the hydrolysis of poly(t-butyl acrylate) did not just cleave the targeted t-butyl groups, but also likely lead to some degradation of the SG1 end-group. This led to a greater heterogeneity of the sample and is expressed in the broadness of the peak.

Through CE-CC, we were able to separate polymers by their topology (branching) as well as by their end-groups. The ability of CE-CC to separate based on the presence of SG1 (control agent used for nitroxide-mediated polymerization)

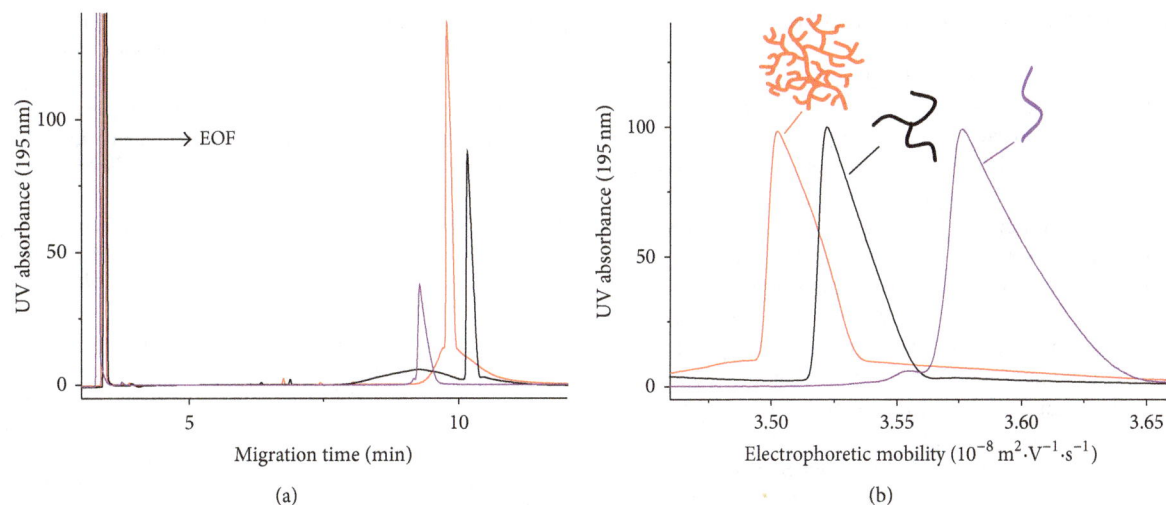

FIGURE 7: Separation of linear (purple line), three-arm star (black line), and hyperbranched (red line) poly(sodium acrylate) by capillary electrophoresis in sodium borate buffer (pH 9.2) shown as a function of (a) migration time and (b) electrophoretic mobility, which is a more reproducible quantity than the former [42].

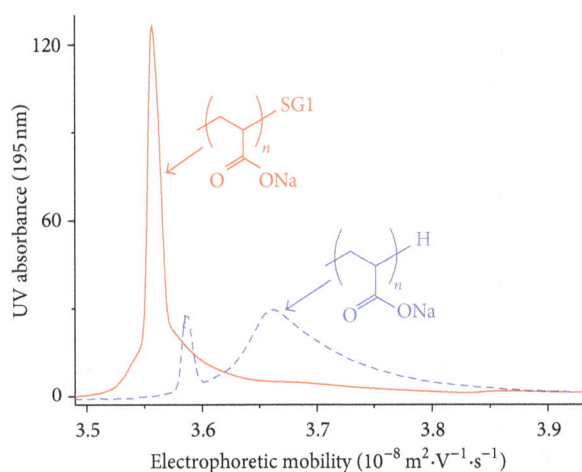

FIGURE 8: Electrophoretic mobility distributions of a PNaA obtained by nitroxide-mediated polymerization of acrylic acid initiated by the monofunctional initiator Monams (red solid line) followed by cleavage of the SG1 end-group by treatment with thiophenol (blue dashed line) in sodium borate (pH 9.2) [42].

reveals information regarding the "livingness" of the obtained sample (allowing continuing reacting through NMP) [42, 71]. This allows the optimization of the method used to produce the sample and provides information regarding further functionalization of the PAA.

3.4. Size Determination with TDA.
One limitation of CE for polymer characterization, compared to multiple detection SEC [16], is the limited number of detectors, especially the lack of molar mass sensitive detectors. The group of Cottet is, however, bridging this gap rapidly by demonstrating that the CE separation can be coupled to Taylor dispersion analysis (TDA). TDA is a method that does not involve

separation but allows the determination of the diffusion coefficient/hydrodynamic radius of a sample. TDA has been looked at previously to obtain diffusion coefficients in liquid systems [72, 73]. Further it has proven to be practical in the size characterization of macromolecules and particles of virtually any molar mass [65, 74]. TDA has several advantages including that it is an absolute method, meaning that no calibration is required. The group of Cottet has shown that a CE instrument is particularly well suited to carry out TDA [65, 75]. Le Saux and Cottet [76] coupled CE and TDA. A copolymer mixture of 2-acrylamido-2-methylpropanesulfonate/acrylamide and DNA was injected into a fused silica capillary. The mixture was separated in CE conditions (with an electric field) and then pressure was used to push the samples to the detection window where the analysis took place. The experiment proved that the coupling of CE and TDA allowed not only a complete separation of the copolymer from the DNA in the mixture, but also the successful determination of the diffusion coefficient of both the copolymer and the DNA. A successful coupling of CE with TDA allowed a combination of a high performance and throughput method with an absolute method for the calculation of diffusion coefficients [77]. The diffusion coefficient can then be related to the hydrodynamic volume of the macromolecule as it is classically done in light scattering by the following equation:

$$D = kT6\pi\eta r, \tag{1}$$

where k is the Boltzmann constant, T is the temperature, η is the viscosity of the solvent, and r is the hydrodynamic radius of the macromolecule [78].

The relation of the hydrodynamic radius to the molar mass is complex and is influenced by branching and copolymer composition as discussed in Section 1 for SEC [16].

4. Conclusions and Future Directions

Characterization by capillary electrophoresis involves both separation and characterization of complex polymers. This review outlined the broad range of samples that can be analyzed with CE. The characterization of complex polymers is significant for various industries including food, biomedical, energy/fuel, and materials (such as paint and bioplastics) industries. The robustness of the method, especially the minimal sample preparation, is one of the main strengths of the method (shared with field flow fractionation). The continual development of the methods in CE and their coupling with other techniques such as TDA widens the scope and depth of the possible characterization and meets the ever-growing needs of progressively increasing complex macromolecular structures for increasingly advanced applications. CE in the critical conditions (CE-CC) has the most potential and can be applied to a wide variety of charged polymers to characterize their topology, composition, or end-groups. The method is complementary to SEC.

Future directions will look into further characterizing complex polyelectrolytes, for example, in terms of determination of the molar mass distribution of one block in block copolymers, or in terms of sugars quantification in food samples. The different distributions related to branching will also be studied. CE coupled with TDA is a very useful and simple technique that will definitely be examined further as it is able to provide extremely valuable information regarding the size and shape of sample molecules by the calculation of their diffusion coefficient. Its simplicity and being an absolute method mean that it can be applied to a variety of samples investigated by our research group and other polymer research groups.

Conflict of Interests

The authors declare that there is no conflict of interests regarding the publication of this paper.

Acknowledgments

Patrice Castignolles and Marianne Gaborieau would like to thank Professor Herve Cottet (University of Montpellier II) and Professor Emily Hilder (ACROSS, University of Tasmania) for discussions over the last 10 years. The authors thank Alison Maniego, Adam Sutton, James Oliver, Danielle Taylor, and their entire macromolecular characterization team.

References

[1] A. M. Striegel, J. J. Kirkland, W. W. Yau, and D. D. Bly, *Modern Size Exclusion Chromatography*, John Wiley & Sons, Hoboken, NJ, USA, 2009.

[2] R. J. Bruessau, "Experiences with interlaboratory GPC experiments," *Macromolecular Symposia*, vol. 110, pp. 15–32, 1996.

[3] D. Berek, "Size exclusion chromatography—a blessing and a curse of science and technology of synthetic polymers," *Journal of Separation Science*, vol. 33, no. 3, pp. 315–335, 2010.

[4] H. G. Barth and F. J. Carlin Jr., "A review of polymer shear degradation in size-exclusion chromatography," *Journal of Liquid Chromatography*, vol. 7, no. 9, pp. 1717–1738, 1984.

[5] E. Uliyanchenko, S. van der Wal, and P. J. Schoenmakers, "Deformation and degradation of polymers in ultra-high-pressure liquid chromatography," *Journal of Chromatography A*, vol. 1218, no. 39, pp. 6930–6942, 2011.

[6] L. K. Kostanski, D. M. Keller, and A. E. Hamielec, "Size-exclusion chromatography—A review of calibration methodologies," *Journal of Biochemical and Biophysical Methods*, vol. 58, no. 2, pp. 159–186, 2004.

[7] M. Netopilík and P. Kratochvíl, "Polystyrene-equivalent molecular weight versus true molecular weight in size-exclusion chromatography," *Polymer*, vol. 44, no. 12, pp. 3431–3436, 2003.

[8] Y. Guillaneuf and P. Castignolles, "Using apparent molecular weight from SEC in controlled/living polymerization and kinetics of polymerization," *Journal of Polymer Science A: Polymer Chemistry*, vol. 46, no. 3, pp. 897–911, 2008.

[9] A. E. Hamielec and A. C. Ouano, "Generalized universal molecular-weight calibration parameter in GPC," *Journal of Liquid Chromatography*, vol. 1, pp. 111–120, 1978.

[10] M. Gaborieau, R. G. Gilbert, A. Gray-Weale, J. M. Hernandez, and P. Castignolles, "Theory of multiple-detection size-exclusion chromatography of complex branched polymers," *Macromolecular Theory and Simulations*, vol. 16, no. 1, pp. 13–28, 2007.

[11] M. Gaborieau, J. Nicolas, M. Save et al., "Separation of complex branched polymers by size-exclusion chromatography probed with multiple detection," *Journal of Chromatography A*, vol. 1190, no. 1-2, pp. 215–223, 2008.

[12] L. Couvreur, G. Piteau, P. Castignolles et al., "Pulsed-laser radical polymerization and propagation kinetic parameters of some alkyl acrylates," *Macromolecular Symposia*, vol. 174, pp. 197–207, 2001.

[13] P. Castignolles, R. Graf, M. Parkinson, M. Wilhelm, and M. Gaborieau, "Detection and quantification of branching in polyacrylates by size-exclusion chromatography (SEC) and melt-state 13C NMR spectroscopy," *Polymer*, vol. 50, no. 11, pp. 2373–2383, 2009.

[14] T. Junkers, M. Schneider-Baumann, S. S. P. Koo, P. Castignolles, and C. Barner-Kowollik, "Determination of propagation rate coefficients for methyl and 2-ethylhexyl acrylate via high frequency PLP-SEC under consideration of the impact of chain branching," *Macromolecules*, vol. 43, no. 24, pp. 10427–10434, 2010.

[15] P. Castignolles, "Transfer to polymer and long-chain branching in PLP-SEC of acrylates," *Macromolecular Rapid Communications*, vol. 30, no. 23, pp. 1995–2001, 2009.

[16] M. Gaborieau and P. Castignolles, "Size-exclusion chromatography (SEC) of branched polymers and polysaccharides," *Analytical and Bioanalytical Chemistry*, vol. 399, no. 4, pp. 1413–1423, 2011.

[17] E. Uliyanchenko, P. J. C. H. Cools, S. van der Wal, and P. J. Schoenmakers, "Comprehensive two-dimensional ultrahigh-pressure liquid chromatography for separations of polymers," *Analytical Chemistry*, vol. 84, no. 18, pp. 7802–7809, 2012.

[18] M. Rollet, D. Glé, T. N. T. Phan, Y. Guillaneuf, D. Bertin, and D. Gigmes, "Characterization of functional poly(ethylene oxide)s and their corresponding polystyrene block copolymers by liquid chromatography under critical conditions in organic solvents," *Macromolecules*, vol. 45, no. 17, pp. 7171–7178, 2012.

[19] M. Al Samman, W. Radke, A. Khalyavina, and A. Lederer, "Retention behavior of linear, branched, and hyperbranched

polyesters in interaction liquid chromatography," *Macromolecules*, vol. 43, no. 7, pp. 3215–3220, 2010.

[20] W. Lee, D. Cho, T. Chang, K. J. Hanley, and T. P. Lodge, "Characterization of polystyrene-b-polyisoprene diblock copolymers by liquid chromatography at the chromatographic critical condition," *Macromolecules*, vol. 34, no. 7, pp. 2353–2358, 2001.

[21] A. Favier, C. Petit, E. Beaudoin, and D. Bertin, "Liquid chromatography at the critical adsorption point (LC-CAP) of high molecular weight polystyrene: pushing back the limits of reduced sample recovery," *E-Polymers*, pp. 1–15, 2009.

[22] E. Beaudoin, A. Favier, C. Galindo et al., "Reduced sample recovery in liquid chromatography at critical adsorption point of high molar mass polystyrene," *European Polymer Journal*, vol. 44, no. 2, pp. 514–522, 2008.

[23] R. Weinberger, *Practical Capillary Electrophoresis*, CA Academic Press, San Diego, Calif, USA, 2000.

[24] H. Cottet and P. Gareil, "Separation of synthetic (co)polymers by capillary electrophoresis techniques.," *Methods in Molecular Biology*, vol. 384, pp. 541–567, 2008.

[25] H. Miramon, F. Cavelier, J. Martinez, and H. Cottett, "Highly resolutive separations of hardly soluble synthetic polypeptides by capillary electrophoresis," *Analytical Chemistry*, vol. 82, no. 1, pp. 394–399, 2010.

[26] W. Vayaboury, O. Giani, H. Cottet, S. Bonaric, and F. Schué, "Mechanistic Study Of α-amino acid N-carboxyanhydride (NCA) polymerization by capillary electrophoresis," *Macromolecular Chemistry and Physics*, vol. 209, no. 15, pp. 1628–1637, 2008.

[27] M. R. Aguilar, A. Gallardo, J. San Román, and A. Cifuentes, "Micellar electrokinetic chromatography: a powerful analytical tool to study copolymerization reactions involving ionic species," *Macromolecules*, vol. 35, no. 22, pp. 8315–8322, 2002.

[28] I. Spiridon and V. I. Popa, "Hemicelluloses: major sources, properties and applications," in *Monomers, Polymers and Composites from Renewable Resources*, M. N. Belgacem and A. Gandini, Eds., chapter 13, Elsevier, Amsterdam, The Netherlands, 2008.

[29] S. Brudin and P. Schoenmakers, "Analytical methodology for sulfonated lignins," *Journal of Separation Science*, vol. 33, no. 3, pp. 439–452, 2010.

[30] J. D. Oliver, M. Gaborieau, E. F. Hilder, and P. Castignolles, "Simple and robust determination of monosaccharides in plant fibers in complex mixtures by capillary electrophoresis and high performance liquid chromatography," *Journal of Chromatography A*, vol. 1291, pp. 179–186, 2013.

[31] M. Verzele, G. Simoens, and F. van Damme, "A critical review of some liquid chromatography systems for the separation of sugars," *Chromatographia*, vol. 23, no. 4, pp. 292–300, 1987.

[32] S. Rovio, H. Simolin, K. Koljonen, and H. Sirén, "Determination of monosaccharide composition in plant fiber materials by capillary zone electrophoresis," *Journal of Chromatography A*, vol. 1185, no. 1, pp. 139–144, 2008.

[33] S. Rovio, J. Yli-Kauhaluoma, and H. Sirén, "Determination of neutral carbohydrates by CZE with direct UV detection," *Electrophoresis*, vol. 28, no. 17, pp. 3129–3135, 2007.

[34] A. E. Vorndran, P. J. Oefner, H. Scherz, and G. K. Bonn, "Indirect UV detection of carbohydrates in capillary zone electrophoresis," *Chromatographia*, vol. 33, no. 3-4, pp. 163–168, 1992.

[35] J. D. Oliver, M. Gaborieau, and P. Castignolles, "Ethanol determination using pressure mobilization and free solution capillary electrophoresis by photo-oxidation assisted UV detection," *Journal of Chromatography A*, vol. 1348, pp. 150–157, 2014.

[36] J. D. Oliver, A. T. Sutton, N. Karu et al., "Simple and robust monitoring of ethanol fermentations by capillary electrophoresis," *Biotechnology and Applied Biochemistry*, 2014.

[37] J. D. Oliver, A. A. Rosser, C. M. Fellows et al., "Understanding and improving direct UV detection of monosaccharides and disaccharides in free solution capillary electrophoresis," *Analytica Chimica Acta*, vol. 809, pp. 183–193, 2014.

[38] P. Castignolles, M. Gaborieau, E. F. Hilder, E. Sprang, C. J. Ferguson, and R. G. Gilbert, "High-resolution separation of oligo(acrylic acid) by capillary zone electrophoresis," *Macromolecular Rapid Communications*, vol. 27, no. 1, pp. 42–46, 2006.

[39] D. L. Taylor, C. J. Ferris, A. R. Maniego, P. Castignolles, M. In Het Panhuis, and M. Gaborieau, "Characterization of gellan gum by capillary electrophoresis," *Australian Journal of Chemistry*, vol. 65, no. 8, pp. 1156–1164, 2012.

[40] M. Mnatsakanyan, J. J. Thevarajah, R. S. Roi, A. Lauto, M. Gaborieau, and P. Castignolles, "Separation of chitosan by degree of acetylation using simple free solution capillary electrophoresis," *Analytical and Bioanalytical Chemistry*, vol. 405, no. 21, pp. 6873–6877, 2013.

[41] C. Lefay, Y. Guillaneuf, G. Moreira et al., "Heterogeneous modification of chitosan via nitroxide-mediated polymerization," *Polymer Chemistry*, vol. 4, no. 2, pp. 322–328, 2013.

[42] A. R. Maniego, D. Ang, Y. Guillaneuf et al., "Separation of poly(acrylic acid) salts according to topology using capillary electrophoresis in the critical conditions," *Analytical and Bioanalytical Chemistry*, vol. 405, no. 28, pp. 9009–9020, 2013.

[43] C. Sarazin, N. Delaunay, C. Costanza, V. Eudes, J. Mallet, and P. Gareil, "New avenue for mid-UV-range detection of underivatized carbohydrates and amino acids in capillary electrophoresis," *Analytical Chemistry*, vol. 83, no. 19, pp. 7381–7387, 2011.

[44] H. Cottet and P. Gareil, "From small charged molecules to oligomers: a semiempirical approach to the modeling of actual mobility in free solution," *Electrophoresis*, vol. 21, pp. 1493–1504, 2000.

[45] M. Siauw, B. S. Hawkett, and S. Perrier, "Short chain amphiphilic diblock co-oligomers via RAFT polymerization," *Journal of Polymer Science A: Polymer Chemistry*, vol. 50, no. 1, pp. 187–198, 2012.

[46] C. J. Ferguson, R. J. Hughes, D. Nguyen et al., "Ab initio emulsion polymerization by RAFT-controlled self-assembly," *Macromolecules*, vol. 38, no. 6, pp. 2191–2204, 2005.

[47] I. Lacík, M. Stach, P. Kasak et al., "SEC analysis of poly(acrylic acid) and poly(methacrylic acid)," *Macromolecular Chemistry and Physics*, 2014.

[48] M. Gaborieau, T. J. Causon, Y. Guillaneuf, E. F. Hilder, and P. Castignolles, "Molecular weight and tacticity of oligoacrylates by capillary electrophoresis-mass spectrometry," *Australian Journal of Chemistry*, vol. 63, no. 8, pp. 1219–1226, 2010.

[49] C. M. Guttman, K. M. Flynn, W. E. Wallace, and A. J. Kearsley, "Quantitative mass spectrometry and polydisperse materials: creation of an absolute molecular mass distribution polymer standard," *Macromolecules*, vol. 42, no. 5, pp. 1695–1702, 2009.

[50] A. Chojnacka, K. Kempe, H. C. van de Ven et al., "Molar mass, chemical-composition, and functionality-type distributions of poly(2-oxazoline)s revealed by a variety of separation techniques," *Journal of Chromatography A*, vol. 1265, pp. 123–132, 2012.

[51] E. B. Fitzgerald and R. M. Fuoss, "Polyelectrolytes .11. electrophoresis in solutions of poly-4-vinyl-N-N-butylpyridinium bromide," *Journal of Polymer Science*, vol. 14, pp. 329–339, 1954.

[52] P. J. Flory, "Configurational and frictional properties of the polymer molecule in dilute solution," in *Principles of Polymer Chemistry*, Cornell University Press, Ithaca, NY, USA, 1st edition, 1953.

[53] M. Muthukumar, "Theory of electrophoretic mobility of a polyelectrolyte in semidilute solutions of neutral polymers," *Electrophoresis*, vol. 17, no. 6, pp. 1167–1172, 1996.

[54] J. L. Barrat and J. F. Joanny, "Theory of polyelectrolyte solutions," in *Advances in Chemical Physics, Vol Xciv*, John Wiley & Sons, New York, NY, USA, 1996.

[55] N. C. Stellwagen, C. Gelfi, and P. G. Righetti, "The Free Solution Mobility of DMA," *Biopolymers*, vol. 42, no. 6, pp. 687–703, 1997.

[56] H. Cottet, P. Gareil, O. Theodoly, and C. E. Williams, "A semi-empirical approach to the modeling of the electrophoretic mobility in free solution: application to polystyrenesulfonates of various sulfonation rates," *Electrophoresis*, vol. 21, no. 17, pp. 3529–3540, 2000.

[57] H. J. Zhong, M. A. K. Williams, R. D. Keenan, D. M. Goodall, and C. Rolin, "Separation and quantification of pectins using capillary electrophoresis: a preliminary study," *Carbohydrate Polymers*, vol. 32, no. 1, pp. 27–32, 1997.

[58] K. A. Oudhoff, F. A. Buijtenhuijs, P. H. Wijnen, P. J. Schoenmakers, and W. T. Kok, "Determination of the degree of substitution and its distribution of carboxymethylcelluloses by capillary zone electrophoresis," *Carbohydrate Research*, vol. 339, no. 11, pp. 1917–1924, 2004.

[59] C.-M. Jiang, M.-C. Wu, W.-H. Chan, and H.-M. Chang, "Determination of random- and blockwise-type de-esterified pectins by capillary zone electrophoresis," *Journal of Agricultural and Food Chemistry*, vol. 49, no. 11, pp. 5584–5588, 2001.

[60] M. A. K. Williams, T. J. Foster, and H. A. Schols, "Elucidation of pectin methylester distributions by capillary electrophoresis," *Journal of Agricultural and Food Chemistry*, vol. 51, no. 7, pp. 1777–1781, 2003.

[61] C.-M. Jiang, S.-C. Liu, M.-C. Wu, W.-H. Chang, and H.-M. Chang, "Determination of the degree of esterification of alkaline de-esterified pectins by capillary zone electrophoresis," *Food Chemistry*, vol. 91, no. 3, pp. 551–555, 2005.

[62] S. E. Guillotin, E. J. Bakx, P. Boulenguer, H. A. Schols, and A. G. J. Voragen, "Determination of the degree of substitution, degree of amidation and degree of blockiness of commercial pectins by using capillary electrophoresis," *Food Hydrocolloids*, vol. 21, no. 3, pp. 444–451, 2007.

[63] C. J. Ferris, K. J. Gilmore, G. G. Wallace, and M. in het Panhuis, "Modified gellan gum hydrogels for tissue engineering applications," *Soft Matter*, vol. 9, no. 14, pp. 3705–3711, 2013.

[64] C. John Ferrisa and M. in het Panhuism, "Conducting biomaterials based on gellan gum hydrogels," *Soft Matter*, vol. 5, no. 18, pp. 3430–3437, 2009.

[65] H. Cottet, J. Biron, and M. Martin, "Taylor dispersion analysis of mixtures," *Analytical Chemistry*, vol. 79, no. 23, pp. 9066–9073, 2007.

[66] A. Domard, "A perspective on 30 years research on chitin and chitosan," *Carbohydrate Polymers*, vol. 84, no. 2, pp. 696–703, 2011.

[67] I. Aranaz, M. Mengíbar, R. Harris et al., "Functional characterization of chitin and chitosan," *Current Chemical Biology*, vol. 3, no. 2, pp. 203–230, 2009.

[68] S. Nguyen, S. Hisiger, M. Jolicoeur, F. M. Winnik, and M. D. Buschmann, "Fractionation and characterization of chitosan by analytical SEC and 1H NMR after semi-preparative SEC," *Carbohydrate Polymers*, vol. 75, no. 4, pp. 636–645, 2009.

[69] C. Gartner, B. L. López, L. Sierra, R. Graf, H. W. Spiess, and M. Gaborieau, "Interplay between structure and dynamics in chitosan films investigated with solid-state NMR, dynamic mechanical analysis, and X-ray diffraction," *Biomacromolecules*, vol. 12, no. 4, pp. 1380–1386, 2011.

[70] J. Nicolas, Y. Guillaneuf, C. Lefay, D. Bertin, D. Gigmes, and B. Charleux, "Nitroxide-mediated polymerization," *Progress in Polymer Science*, vol. 38, no. 1, pp. 63–235, 2013.

[71] D. Gigmes, D. Bertin, C. Lefay, and Y. Guillaneuf, "Kinetic modeling of nitroxide-mediated polymerization: conditions for living and controlled polymerization," *Macromolecular Theory and Simulations*, vol. 18, no. 7-8, pp. 402–419, 2009.

[72] A. C. Ouano, "Diffusion in liquid systems. I. A simple and fast method of measuring diffusion constants," *Industrial and Engineering Chemistry Fundamentals*, vol. 11, no. 2, pp. 268–271, 1972.

[73] K. C. Pratt and W. A. Wakeham, "Mutual diffusion-coefficient of ethanol-water mixtures—determination by a rapid, new method," *Proceedings of the Royal Society of London A-Mathematical Physical and Engineering Sciences*, vol. 336, pp. 393–406, 1974.

[74] R. Callendar and D. G. Leaist, "Diffusion coefficients for binary, ternary, and polydisperse solutions from peak-width analysis of Taylor dispersion profiles," *Journal of Solution Chemistry*, vol. 35, no. 3, pp. 353–379, 2006.

[75] H. Cottet, M. Martin, A. Papillaud, E. Souaïd, H. Collet, and A. Commeyras, "Determination of dendrigraft poly-L-lysine diffusion coefficients by Taylor dispersion analysis," *Biomacromolecules*, vol. 8, no. 10, pp. 3235–3243, 2007.

[76] T. le Saux and H. Cottet, "Size-based characterization by the coupling of capillary electrophoresis to taylor dispersion analysis," *Analytical Chemistry*, vol. 80, no. 5, pp. 1829–1832, 2008.

[77] A. Ibrahim, R. Meyrueix, G. Pouliquen, Y. P. Chan, and H. Cottet, "Size and charge characterization of polymeric drug delivery systems by Taylor dispersion analysis and capillary electrophoresis," *Analytical and Bioanalytical Chemistry*, vol. 405, no. 16, pp. 5369–5379, 2013.

[78] J. R. Lakowicz, *Principles of Fluorescence Spectroscopy*, Springer, New York, NY, USA, 2007.

13

A Rapid Extractive Spectrophotometric Method for the Determination of Tin with 6-Chloro-3-hydroxy-2-(2′-thienyl)-4-oxo-4H-1-benzopyran

Ramesh Kataria[1] and Harish Kumar Sharma[2]

[1] *Department of Chemistry and Centre of Advanced studies in Chemistry, Panjab University, Chandigarh 160014, India*
[2] *Department of Chemistry, Kurukshetra University, Kurukshetra, Haryana 136119, India*

Correspondence should be addressed to Ramesh Kataria; rkataria@pu.ac.in

Academic Editor: Gerd-Uwe Flechsig

An extractive spectrophotometric method for the determination of the trace amounts of tin has been carried out by employing 6-chloro-3-hydroxy-2-(2′-thienyl)-4-oxo-4H-1-benzopyran (in acetone) (CHTB) for the complexation of the metal ion in HCl medium. The colored species thus produced is quantitatively extracted into dichloromethane and shows the maximum absorbance at 432–437 nm. The method obeys Beer's law in the range 0.0–1.3 μg mL^{-1} of tin with molar absorptivity and Sandell's sensitivity of 5.81×10^4 L mol^{-1} cm^{-1} and 0.0020 μg Sn cm^{-2}, respectively, at 435 nm. The method is highly selective and free from the interference of a large number of elements including platinum metals. The ratio of metal to ligand in the extracted species is 1 : 2. Utilizing this method, the analysis of various synthetic and technical samples including gun metal and tin can have been carried out satisfactorily.

1. Introduction

Tin does not occur free in nature and is found almost exclusively as tin oxide known as cassiterite or tin stone. Tin although a toxic metal, still it is being widely employed in manufacturing important alloys [1] and as solders for the joining of electronic components. The excess use of tin in daily life as fungicides in crops, in food packaging, and as stabilizer for polyvinyl chloride may introduce the inorganic tin {Sn(II) and Sn(IV)} in the environment. Out of these two, Sn(II) seems to be more toxic as compared to Sn(IV) [2]. In the literature, there are numerous analytical methods for the measurement of tin which are based on sophisticated instruments [3–10]. These methods are highly sensitive but generally tedious and prone to serious interferences from other elements. In contrast spectrophotometric methods are preferred due to their simplicity and speed in routine analysis. The reported studies have shown that a large number of reagents such as methyl orange [11], benzopyran derivatives [1, 12, 13], 2-(5-nitro-2-pyrilazo)-5-[N-n-propyl-N-(3-sulfopropyl)amino-phenoyl] [14], pyrocatechol violet [15–18], phenylfluorone [19, 20], dibromohydroxyphenylfluorone [21], arsenazo-M [22], isoamyl xanthate [23], diacetyl-monoxime-p-hydroxybenzoyl-hydrazine [24], bromopyrogallol red [25], potassium ethylxanthate [26], ferron [27], and 5,7-dichloro-8-quinolinol [28] have been used for the spectrophotometric determination of tin(II,IV) content. Among these many reagents [18, 21, 23, 26–28] are nonselective as they suffer from the interference, have low sensitivity [11, 12, 23, 24, 26–28], and some of them are time consuming, as they require time for full color development [14, 18, 20]. Some of the sensitive reagents [16, 17, 21, 22, 25] are reported, but these require the use of the surfactants, plasticizer, and critical pH adjustment. Thus in the view of the above facts it reveals that there is still a lot of scope for working out new methods and effecting amendments in the existing ones especially because of their lower sensitivity and selectivity. Keeping in mind the scope of the reported facts, a chromone derivative 6-chloro-3-hydroxy-2-(2′-thienyl)-4-oxo-4H-1-benzopyran(CHTB) has been used for complexation and spectrophotometric determination of trace amount of tin(II). The reagent 6-chloro-3-hydroxy-2-(2′-thienyl)-4-oxo-4H-1-benzopyran was found to give a sensitive reaction with Sn(II). In the present communication, optimization of conditions for the quantitative extraction of Sn(II)-CHTB complex was worked out apart from the studies involving stoichiometry and Beer's law range determination. The interference studies for diverse ions were also carried out and the extraction of Sn(II)-CHTB was made free from interference of large number of metal ions by using suitable masking agents. The extraction of the Sn(II)-CHTB complex into dichloromethane forms the basis of the proposed method, which provides the advantages particularly in respect of sensitivity, selectivity, and color development time to the existing methods. Some synthetic and technical samples including gun metal and tin can have been analyzed for tin contents with good agreement.

2. Experimental

2.1. Apparatus. A model-140-02, Shimadzu with 10 mm matched cells was used for the routine absorbance measurements and spectral studies.

2.2. Reagents and Solutions. The standard stock solution (250 mL) of Sn(II) containing $1 mg mL^{-1}$ of the metal ion was prepared by dissolving an accurately weighed amount (0.475 g) of $SnCl_2·2H_2O$ (RANBAXY) in 20 mL of concentrated hydrochloric acid, diluting with deionized water up to the mark and standardized by the SnO_2 method gravimetrically [29]. Lower concentration at $\mu g mL^{-1}$ level was prepared by suitable dilution of this solution containing $0.5 mol L^{-1}$ HCl final acidity. The containers of the tin solution were wrapped with carbon paper and kept in dark place. Stock solutions of other metal ions were prepared at $mg mL^{-1}$ level by dissolving their sodium or potassium salts in deionized water or dilute acid. They were suitably diluted to give $\mu g mL^{-1}$ level concentration of the metal ions.

6-Chloro-3-hydroxy-2-(2'-thienyl)-4-oxo-4H-1-benzo-pyran (CHTB; m.p. 200–202°C) was synthesized by the literature method [30] and dissolved in acetone to give 0.1% (m/v) solution. The chemical composition of CHTB is $C_{13}H_7O_3SCl$ and its structure is given in Figure 4.

Dichloromethane (Ranbaxy) was used for extraction as such.

2.3. The Samples. Synthetic samples were prepared by mixing tin solution with solutions of various metal ions in suitable proportions so as to give the composition as shown in Table 1.

2.4. Gun Metal. A weighed sample of gun metal (0.2 g) was dissolved in 10 mL of concentrated hydrochloric acid and 2–4 mL of concentrated nitric acid on heating and the volume was made up to 100 mL in a volumetric flask. 10 ML of this solution was diluted to 100 mL to get a working solution of low concentration. An aliquot (0.25 mL) of this solution was analyzed by the proposed method.

2.5. Tin Can. A weighed sample (0.6 g) of tin taken in a 10 mL beaker was heated gently with 5 mL of concentrated

TABLE 1: The analysis of various samples with the proposed method.

Matrix*	Sn added, μg	Sn found**, μg
Zn(0.02), Pb(0.01), Cu(0.001)[a]	10.0	10.03 ± 0.85
Cu(0.070), Co(0.014)[b]	8.0	7.89 ± 0.71
Co(1), Ba(2), U(0.01), Mo(0.020)[c]	7.0	6.83 ± 0.58
Cd(2), Fe(0.1), V(0.1)[d]	12.0	11.96 ± 0.71
Cr(0.1), Sr(1), Ag(0.5), Zr(0.01)[e]	12.0	11.99 ± 0.62
Pb (2), Nb(0.1), Th(0.05) [f]	5.0	5.13 ± 1.63
As(2), Se(3), Ti(0.1)[e]	8.0	7.92 ± 0.56
Re(0.01), Ta(0.05), Bi(1)[g]	5.0	5.03 ± 0.82
Be(2), Pt(0.01), Ir(0.01)	10.0	9.85 ± 0.41
Gun metal	4.9%[h]	4.48% ± 0.72
Tin can	—	0.15%[i]

*Amount of metal ion shown in parentheses is in mg. **Average of triplicate analyses; mean ± % RSD. [a,b]Correspond to kneiss metal and argental, respectively. [c]In presence of 0.5 mg dithionite. [d]In presence of 100 mg ascorbic acid. [e]In presence of 7 mg phosphate. [f]In presence of 4 mg oxalate. [g]In presence of 100 mg iodide. [h]Certified value. [i]Confirmed by SnO_2 method.

hydrochloric acid. The sample was dissolved completely by adding 5–10 mL of distilled water and heating until the volume was reduced to 2–5 mL. After cooling, the volume of the solution was made up to 25 mL and suitable portions of the sample solution were analyzed for tin content.

2.6. Procedure. To 1 mL aliquot of the sample solution containing ≤13 μg Sn(II) in $0.5 mol L^{-1}$ hydrochloric acid, were added 1 mL of 6-chloro-3-hydroxy-2-(2'-thienyl)-4-oxo-4H-1-benzopyran (0.1% in acetone) solution and distilled water to make the aqueous volume up to 10 mL in a short stemmed 125 mL separating funnel. The contents were mixed well and equilibrated with 10 mL of dichloromethane for 20 s. The two layers were allowed to separate and the yellow colored solvent layer was passed through Whatman filter paper (number 41, 9 cm diameter) and collected into 10 mL measuring flask. The absorbance of the yellow complex was measured at 435 nm against similarly treated reagent blank. The standard calibration curve was prepared by applying the procedure to a solution containing tin up to 13 μg per 10 mL of the aqueous volume. The tin contents were computed from this calibration curve.

Modifications of the method for V, Fe, Nb, Zr, W, Mo, Bi, and Ti: in the sample when Ti(IV), Zr(IV), and W(VI) were masked with sodium phosphate, Fe(III) and V(V) with ascorbic acid, Bi(III) with potassium iodide, Mo(VI) with sodium dithionite, and Nb(V) with sodium oxalate added prior to the addition of reagent and solvent. The respective amount of the masking agents used was mentioned in the effect of diverse ions.

3. Results and Discussion

Tin(II) reacted with 6-chloro-3-hydroxy-2-(2'-thienyl)-4-oxo-4H-1-benzopyran(CHTB) in an acid medium to form

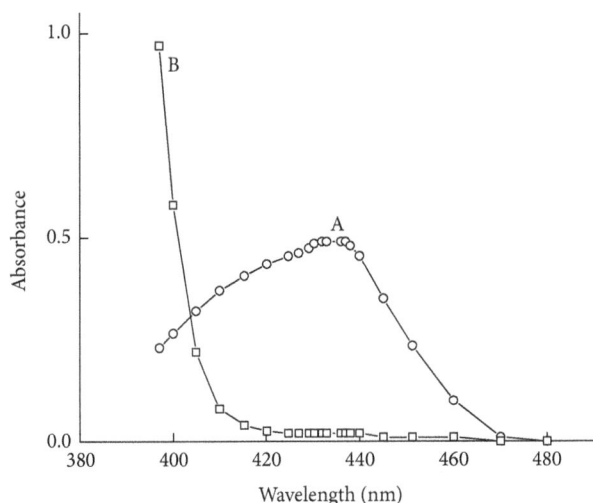

FIGURE 1: Absorption spectrum of Sn(II)-CHTB complex in dichloromethane. Curve A: 1 μg Sn/mL measured against reagent blank; Curve B: reagent blank measured against dichloromethane.

a yellow colored species, which was quantitatively extracted into dichloromethane. The absorption spectrum of the colored Sn(II)-CHTB complex in dichloromethane indicated the maximum absorbance at 432–437 nm in the visible region, where the reagent blank had hardly any absorbance (Figure 1). The effect of various parameters on the formation and absorbance of the complex are listed in Table 2.

The absorbance of the complex was found maximum in HCl medium, where as it was observed to be low in H_2SO_4, CH_3COOH, and $HClO_4$. Since the Sn(II)-CHTB complex showed maximum absorbance in 0.046–0.05 mol L^{-1} HCl, so 0.05 mol L^{-1} HCl was chosen to provide suitable acidity. Portions 0.5–2.2 mL of 0.1% CHTB solution in acetone resulting in maximum absorbance to the complex under all the conditions were stated in Table 1 and thus 1 mL was considered to be sufficient for the system. Further, the complex shows maximum absorbance when an equilibration time of up to 5 min is kept; therefore, in order to save time, 20 s is considered to be the desired contact time for the extraction of the complex from the aqueous solution.

Out of the number of the solvents studied for extraction of the Sn(II)-CHTB complex, dichloromethane was found to be most suitable because it provides a high absorbance value and stability of the complex. The absorbance showed a downward trend in the case of dichloromethane, 1,2-dichloroethane, benzene, toluene, ethyl acetate, carbon tetrachloride, isoamyl acetate, isobutyl methyl ketone, chloroform, cyclohexane, and isoamyl alcohol. So the dichloromethane was selected for the extraction of the Sn(II)-CHTB complex from the aqueous phase.

From a study of the above variables, the optimum conditions for the system have been laid down, as already stated in the procedure. The metal complex obeys Beer's law in the range 0–1.3 μg Sn(II) mL^{-1}. However, according to Ringbom plot [31], the optimum range for accurate determination of tin is 0.28–1.25 μg mL^{-1}. The molar absorptivity, specific absorptivity, and Sandell's sensitivity of the complex at 435 nm are

TABLE 2: Effect of various parameters on the absorbance of Sn(II)-CHTB complex.

HCl[a] (M)	0.043	0.044	0.045	0.046–0.060	0.065
Absorbance	0.370	0.420	0.460	0.490	0.450
CHTB[b] (mL)	0.2	0.3	0.4	0.5–2.2	2.5
Absorbance	0.310	0.375	0.440	0.490	0.430
Equilibration time[c] (sec)	0.0	2	4	5–300	—
Absorbance	0.100	0.320	0.480	0.490	—

Conditions: (a) Sn(II) = 10 μg; HCl = variable; CHTB (0.1% (m/v) in acetone) = 1 mL; aqueous volume = solvent volume = 10 mL; solvent = dichloromethane; equilibration time = 20 s; $\lambda_{max.}$ = 435 nm, (b) HCl = 0.046–0.060 mol L^{-1}; other conditions being the same as in (a) except for the variation in CHTB concentration; also b = 6-chloro-3-hydroxy-2-($2'$-thienyl)-4-oxo-4H-1-benzopyran (CHTB), and (c) 0.1% CHTB in acetone = 1 mL; other conditions being the same as in (b) except for the variation in equilibration time.

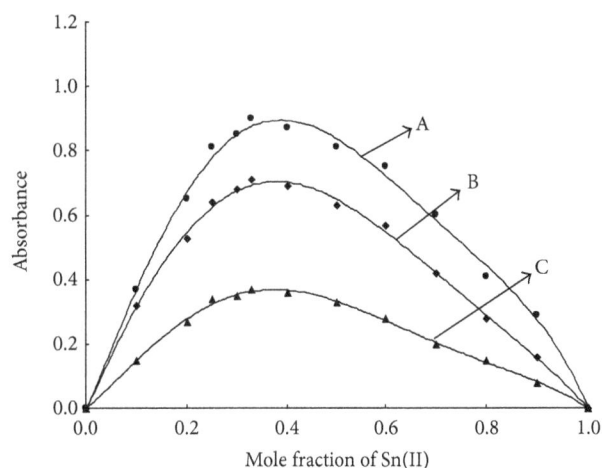

FIGURE 2: Job's method of continuous variations of Sn(II) and CHTB. Curve A:435 nm, Curve B: 410 nm, and Curve C: 450 nm.

5.81×10^4 L mol^{-1} cm^{-1}, 0.489 mL g^{-1} cm^{-1}, and 0.0020 μg Sn(II) cm^{-2}, respectively. The ratio of Sn(II) : CHTB in the extracted species is determined using their equimolar solution (8.425×10^{-4} M) at three different wavelengths, 410, 435, 450 nm, by Job's method (Figure 2) of continuous variations as modified by Vosburgh and Cooper for a two-phase system [32, 33]. The sharp break in the curves indicates a metal-to-ligand ratio of 1 : 2 stoichiometry in the extracted species. This is further supported by the mole ratio method (Figure 3) [34] by taking the concentration of Sn(II) as 4.218×10^{-4} M and measuring the absorbance again at three wavelengths, 410, 435, 450 nm. The most probable structure of the Sn-CHTB complex is given as shown in Figure 5.

3.1. Effect of Diverse Ions. Under optimum conditions of the procedure, the effect of different anions and cations has been studied on the absorbance of the Sn(II)-CHTB complex. The amount of diverse ions which caused a ≤1% error in the absorbance was taken as the tolerance limit. The tolerance limit of foreign ions tested is given in Table 3. The

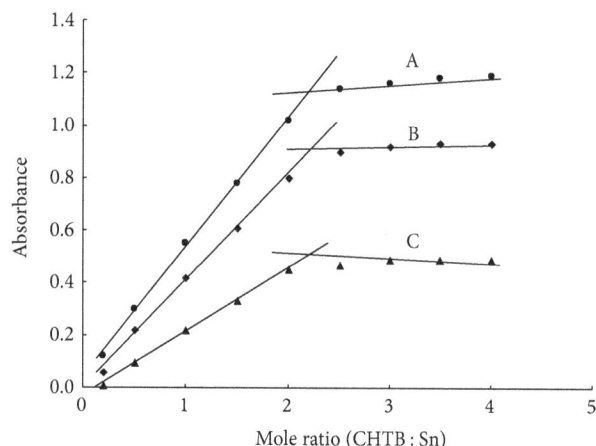

FIGURE 3: Mole ratio method of CHTB and Sn(II). Curve A:435 nm, Curve B: 410 nm, and Curve C:450 nm.

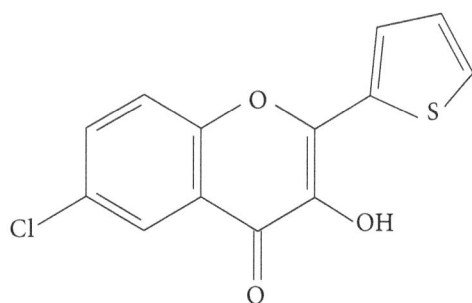

FIGURE 5: Chemical Structure of Sn(II)-CHTB complex.

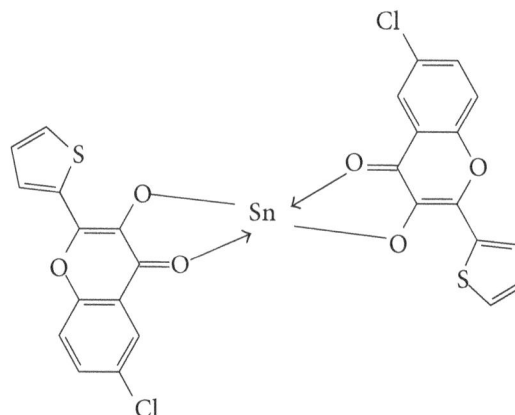

FIGURE 4: Chemical Structure of CHTB.

reported anions and cations did not influence the absorbance of the Sn(II)-CHTB complex. However, fluoride interfered seriously even in traces. The amount of sodium or potassium salts of the various anions were taken in mg while glycerol and H_2O_2 (30%, m/v) were taken in mL.

Among the study of cations it was found that cations like Fe(III), Zr(IV), Nb(V), V(V), Mo(VI), W(VI), and Ti(IV) did influence the absorbance of the Sn(II)-CHTB complex. However the interference of these metals could be prevented by making use of suitable masking agents, that is, for 1 mg of Fe(III), 100 mg ascorbic acid; for 1 mg of Nb(V), 4 mg sodium oxalate; for 1 mg of V(V), 100 mg ascorbic acid; for 0.1 mg of Mo(VI), 5 mg of sodium dithionite; for 1 mg of W(VI), 7 mg sodium phosphate; for 1.5 mg Zr(IV), 7 mg sodium phosphate; for 0.3 mg of Ti(IV), 7 mg sodium phosphate; and for 5 mg of Bi(III), 100 mg iodide added prior to the addition of CHTB in 10 mL aqueous volume under optimum condition of the procedure.

4. Conclusion

For the determination of microamounts of tin, the proposed method is simple, rapid, sensitive, and selective and free from the interference of a large number of metal ions. The wide applicability of the method is tested by the analysis of several synthetic samples, tin can, and gun metal sample with

TABLE 3: Tolerance limit of different ions in the determination of 10 μg of Sn(II).

Ions	Tolerance limit (concentration mg/10 mL)
Thiourea, sulphite, ascorbic acid, and iodide	100.0
Sulphate, nitrate	80.0
Bromide, sulfosalicylic acid	75.0
Chloride, tartrate	50.0
Acetate	40.0
Carbonate, citrate	20.0
Thiocyanate	10.0
Phosphate	7.0
Dithionite	5.0
Oxalate	4.0
EDTA "disodium salt"	2.0
Glycerol	1.0[a]
H_2O_2 (30%, m/v)	0.5[a]
Zn(II), Pb(II), and Se(II)	10.0
Cd(II)	8.0
Ba(II), Ni(II), Co(II), and Hg(II)	5.0
Be(II), Ce(IV), As(II), Mg(II) Mn(II), Sr(II), Al(III), and Os(VIII)	3.0
Ag(I), Cu(II)	2.0
U(VI)	1.0
Ta(V)	0.7
Cr(VI)	0.5
Th(IV)	0.3
Ru(III), Ir(III)	0.1

[a]Value given in mL.

satisfactory results. The high reproducibility of the method is tested by performing several sets of experiments while keeping the same amount of tin metal ions in each set; the relative standard deviation of the method is 0.98%.

A Rapid Extractive Spectrophotometric Method for the Determination of Tin with...

127

Conflict of Interests

The authors declare that there is no conflict of interests regarding the publication of this paper.

Acknowledgments

The authors' sincere thanks are due to Kurukshetra University, Kurukshetra, and Panjab University, Chandigarh, for providing the necessary facilities.

References

[1] R. Kataria, H. K. Sharma, N. Agnihotri, and J. R. Mehta, "2-(2′-Furyl)-3-hydroxy-4-oxo-4H-1-benzopyran as a highly selective and sensitive reagent for spectrophotometric determination of tin(II)," *Proceedings of the National Academy of Sciences of India*, vol. 78, pp. 31–35, 2008.

[2] T. Madrakian, A. Afkhami, R. Moein, and M. Bahram, "Simultaneous spectrophotometric determination of Sn(II) and Sn(IV) by mean centering of ratio kinetic profiles and partial least squares methods," *Talanta*, vol. 72, no. 5, pp. 1847–1852, 2007.

[3] C. Prior and G. S. Walker, "The use of the bismuth film electrode for the anodic stripping voltammetric determination of tin," *Electroanalysis*, vol. 18, no. 8, pp. 823–829, 2006.

[4] Y. Mino, "Determination of tin in canned foods by X-ray fluorescence spectrometry," *Journal of Health Science*, vol. 52, no. 1, pp. 67–72, 2006.

[5] J.-B. Liu and Y.-Z. Wu, "Rapid determination of tin in ore by atomic emission spectrometry," *Yejin Fenxi*, vol. 33, no. 3, pp. 65–68, 2013.

[6] Y. Lin, "Determination of tin in canned food with hydride-atomic fluorescence spectrometry," *Fenxi Ceshi Jishu Yu Yigi*, vol. 19, pp. 149–152, 2013.

[7] Y. Yu, Z.-Y. He, Z.-C. Mao et al., "Determination of tin in by spectral lines with different sensitivity of alternating current arc emission spectroscopy," *Yankuang Ceshi*, vol. 32, pp. 44–47, 2013.

[8] L. Pruša, J. Dědina, and J. Kratzer, "Ultratrace determination of tin by hydride generation in-atomizer trapping atomic absorption spectrometry," *Analytica Chimica Acta*, vol. 804, pp. 50–58, 2013.

[9] S. V. de Azevedo, F. R. Moreira, and R. C. Campos, "Direct determination of tin in whole blood and urine by GF AAS," *Clinical Biochemistry*, vol. 46, no. 1-2, pp. 123–127, 2013.

[10] I. Trandafir, V. Nour, and M. E. Ionica, "Determination of tin in canned foods by inductively coupled plasma-mass spectrometry," *Polish Journal of Environmental Studies*, vol. 21, no. 3, pp. 749–754, 2012.

[11] X.-L. Wang, P. Zhang, and Y. Chen, "Spectrophotometric determination of stannum in copper alloy based on fading reaction of methyl orange," *Yejin Fenxi*, vol. 32, no. 12, pp. 73–75, 2012.

[12] R. Kataria and H. K. Sharma, "3-Hydroxy-2-[1′-phenyl-3′-(4″-methoxyphenyl)-4′-pyrazoyl]-4-oxo-4H-1-benzopyran as a spectrophotometric reagent for the micro-determination of tin," *Journal of the Indian Chemical Society*, vol. 89, no. 1, pp. 121–126, 2012.

[13] R. Kataria and H. K. Sharma, "An extractive spectrophotometric determination of tin as Sn(II)-6-chloro-3-hydroxy-7-methyl-2-(4′-methoxyphenyl)-4-oxo-4H-1-benzopyran complex into

dichloromethane," *Eurasian Journal of Analytical Chemistry*, vol. 6, no. 3, pp. 140–149, 2011.

[14] B. Chen, Q. Zhang, H. Minami, M. Uto, and S. Inoue, "Spectrophotometric determination of tin in steels with 2-(5-nitro-2-pyridylazo)-5-[N-n-propyl-N-(3-sulfopropyl)amino] phenol," *Analytical Letters*, vol. 33, no. 14, pp. 2951–2961, 2000.

[15] J.-H. Tang, L.-H. Cheng, and X.-M. Wu, "Pyrocatechol violet-CPB spctrophotometric Determination of tin in copper alloys," *Guangpu Shiyanshi*, vol. 30, pp. 1925–1928, 2013.

[16] M. Abbasi-Tarighat, "Kinetic-spctrophotometric Determination of tin species using feed-forward neural network and radial basis function network in water and juices of canned fruits," *Analytical Chemistry*, vol. 12, pp. 256–263, 2013.

[17] T. Madrakian and F. Ghazizadeh, "Micelle-mediated extraction and determination of tin in soft drink and water samples," *Journal of the Brazilian Chemical Society*, vol. 20, no. 8, pp. 1535–1540, 2009.

[18] A. C. S. Costa, L. S. G. Teixeira, and S. L. C. Ferreira, "Spectrophotometric determination of tin in copper-based alloys using pyrocatechol violet," *Talanta*, vol. 42, no. 12, pp. 1973–1978, 1995.

[19] P. Huang, "Spectrophotometric determination of tin in antimony materials by using phenylflurone," *Hunan Youse Jinshu*, vol. 29, pp. 68–79, 2013.

[20] D.-X. Wang, F. Chen, and Z.-F. Liu, "Spectrophotometric determination of tin in flot glass," *The American Ceramic Society Bulletin*, vol. 84, no. 12, pp. 9401–9404, 2005.

[21] H. Yan, "Spectrophotometric determination of tin in steel with dibromo-hydroxyphenylfluorone," *Yejin Fenxi*, vol. 23, no. 6, pp. 45–46, 2003.

[22] C. Cai, Z. Zhou, S. Chen, and Y. Fang, "Research progress of tannery wastewater treatment," *Applied Mechanics and Materials*, vol. 361–363, pp. 666–669, 2013.

[23] S. P. Arya, S. C. Bhatia, A. Bansal, and M. Mahajan, "Isoamyl xanthate as a sensitive reagent for the spectrophotometric determination of tin," *Journal of the Indian Chemical Society*, vol. 79, no. 4, pp. 359–360, 2002.

[24] A. Varghese and A. M. A. Khadar, "Highly selective derivative spectrophotometric determination of tin (II) in alloy samples in the presence of cetylpyridinium chloride," *Acta Chimica Slovenica*, vol. 53, no. 3, pp. 374–380, 2006.

[25] X. Huang, W. Zhang, S. Han, and X. Wang, "Determination of tin in canned foods by UV/visible spectrophotometric technique using mixed surfactants," *Talanta*, vol. 44, no. 5, pp. 817–822, 1997.

[26] S. P. Arya and A. Bansal, "Rapid and selective method for the spectrophotometric determination of tin using potassium ethylxanthate," *Mikrochimica Acta*, vol. 116, no. 1–3, pp. 63–71, 1994.

[27] S. P. Arya, S. C. Bhatia, and A. Bansal, "Extractive-spectrophotometric determination of tin as Sn(II)-ferron complex," *Fresenius' Journal of Analytical Chemistry*, vol. 345, no. 11, pp. 679–682, 1993.

[28] A. M. Gutierrez, M. V. Laorden, A. Sanz-Medel, and J. L. Nieto, "Spectrophotometric determination of tin(IV) by extraction of the ternary tin/iodide/5,7-dichloro-8-quinolinol complex," *Analytica Chimica Acta*, vol. 184, no. C, pp. 317–322, 1986.

[29] G. H. Jeffery, J. Bassett, J. Mendham, and R. C. Denny, *Vogels Textbook of Quantitative Chemical Analysis?* Addison Wesley Longman, Singapore, 5th edition, 1989.

[30] S. C. Gupta, N. S. Yadev, and S. N. Dhawan, "Synthesis of 2, 3 diaryl 8 methyl 2, 3, 4, 10 tetrahydropyrano 3, 2 b, 1 benzopyran 10 ones photoisomerization of styrylchromones," *Indian Journal Of Chemistry Section B: Organic Chemistry Including Medicinal Chemistry*, vol. 30, no. 2, pp. 790–792, 1991.

[31] A. Ringbom, "Über die Genauigkeit der colorimetrischen Analysenmethoden I," *Fresenius Journal of Analytical Chemistry*, vol. 115, no. 9–10, pp. 332–343, 1938.

[32] P. Job, "Formation and stability of inorganic complexes in solution," *Annali di Chimica*, vol. 9, pp. 113–203, 1928.

[33] W. C. Vosburgh and G. R. Cooper, "Complex ions. I. The identification of complex ions in solution by spectrophotometric measurements," *The Journal of the American Chemical Society*, vol. 63, no. 2, pp. 437–442, 1941.

[34] J. H. Yoe and A. L. Jones, "Colorimetric determination of iron with disodium-1,2-dihydroxybenzene-3,5-disulfonate," *Industrial & Engineering Chemistry Analytical Edition*, vol. 16, no. 2, pp. 111–115, 1944.

14

Space Group Approximation of a Molecular Crystal by Classifying Molecules for Their Electric Potentials and Roughness on Their Inertial Ellipsoid Surface

Jose Fayos

Rocasolano Institute, CSIC, Rodriguez Ayuso 6, 28022 Madrid, Spain

Correspondence should be addressed to Jose Fayos; jose_fayos@yahoo.es

Academic Editor: Sailaja Krishnamurty

In order to predict the most probable space group where a molecule crystallizes, it is assumed that molecular shape and electric potential distribution on the molecular surface are the main factors or predictors. However, to compare and classify molecules by these two factors seems to be very difficult for in general such different objects. Thus, in order to compare molecules, they are reduced to their inertial ellipsoid in which surface 26 equally spaced points were chosen where a roughness factor and an electric potential due to all atomic charges of the whole molecule are calculated. By this procedure, different molecules encoded by these two predictor vectors can be compared and classified, showing that molecules that crystallize in the same space group have more similar predictor vectors. This result opens the possibility to predict the more probable spatial group associated with a molecule.

1. Introduction

The first hypothesis considered for crystal packing prediction CPP of organic molecules is that the isolated molecule contains the information of its future crystal [1]. So it is for the so-called "blind tests," the last published in 2011 [2], where several laboratories compete to find the crystal structure of a molecule by diverse calculations, where it is assumed that 95% of the molecules present no polymorph and prefer a cell with a given space group SG. Some approaches were previously done to get molecular crystal structure information by data mining [3, 4]. In the present study it is further assumed that the molecular crystal space group is mainly predetermined by the molecular form or roughness and by the electric potential distribution on the molecular surface, both factors being the best predictors for crystal packing, including the formation of hydrogen bonds. In fact electrostatic forces determine molecular reactions as can be observed by X-ray in the electron density distributions of crystalline molecules [5, 6], were interactions by electrostatic forces (including H-bonds) between equal molecules in a crystal, would determine its crystal packing. The purpose of this work is to classify the molecules into groups by similarity of the above predictors and to check if these assumed types of packing are correlated with their space group SG, or conversely, to verify that molecules crystallizing in the same SG have similar aggregation predictors. However, comparing these two predictors between usually dissimilar molecules does not seem so simple, unless the molecules could first be reduced to more comparable objects. In a previous work [7], some molecular crystal descriptors like cell axes and the presence of some symmetry elements, but not the SG, were predicted by reducing each molecule to its inertial ellipsoid (adding to each axis the hydrogen VDW radius of 1.17 A). The same molecular reduction to its inertial ellipsoid is taken here, defined by its three axes: Large, Medium, and Small (L, M, S), where 26 points are added to its surface: 2×3 on the ends of the ellipsoid axes, 3×4 in the edges centers, and 8 in the face centers: all points approximately equidistant from each other. Figure 1(a) shows the numbered 26 points with their sequence of coordinates on the ellipsoid surface, where for clarity the ellipsoid has been deformed to a cube in this figure.

The classical charges q_j for the atoms in every molecule were calculated by Chem3D Pro [8]; then the electrostatic potential V_i on those 26 points of every ellipsoid surface, due

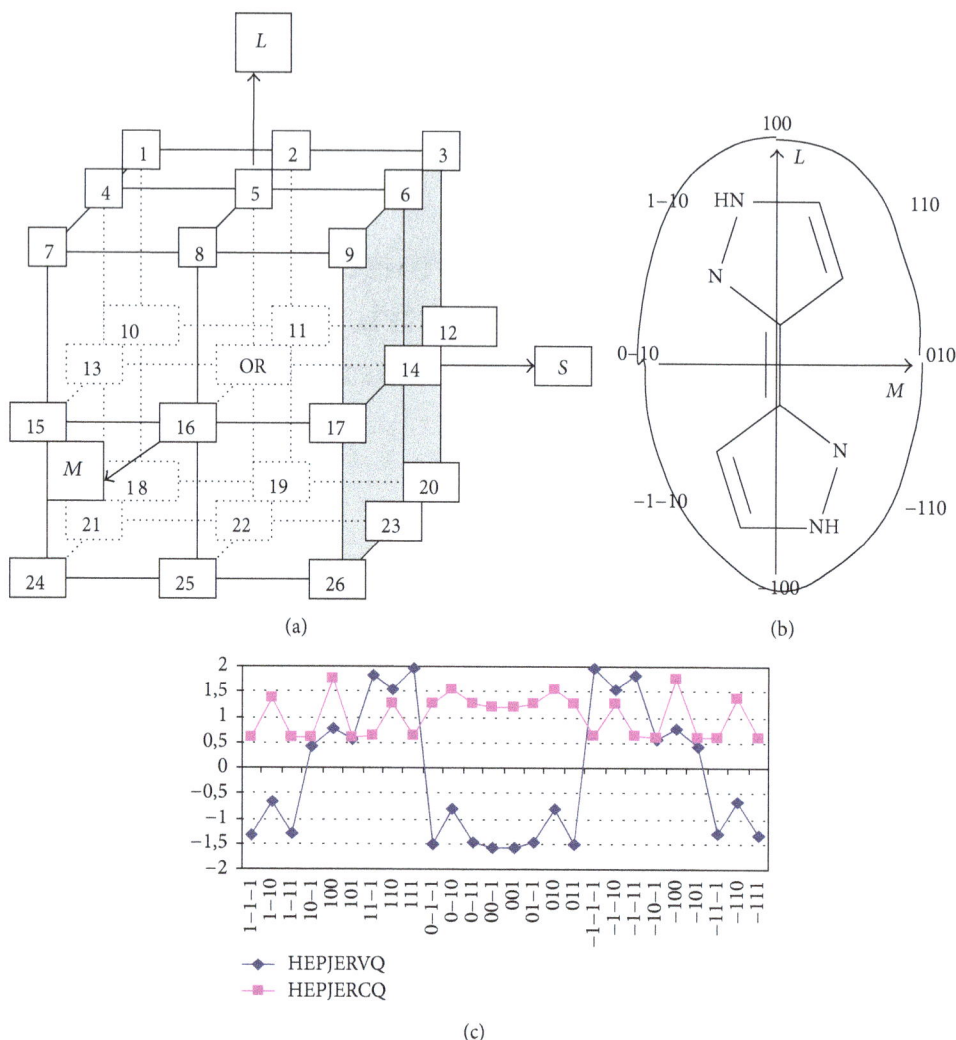

FIGURE 1: (a) Inertial ellipsoid IE reduced to a cube, only for clarity in this figure, with the 26 points on its surface. The 26 (L, M, S) coordinates: (1–1–1 1–10 1–11 10–1 100 101 11–1 110 111 0–1–1 0–10 0–11 00–1 001 01–1 010 011 −1–1–1 −1–10 −1–11 −10–1 −100 −101 −11–1 −110 −111). (b) HEPJER molecule into its inertial ellipsoid with the 8 points for S = 0. (c) Distribution of 26 potentials V and 26 concavities C of HEPJER on his inertial ellipsoid, from (1–1–1) to (−111).

to the charges q_j of all j atoms in the molecule, was calculated by $V_i = \Sigma j\,(q_j/rij)$, where rij are the distances from point i to the j atoms.

Besides, a roughness factor in those 26 points of every molecular ellipsoid surface, from now concavity factor, was calculated as the average of the distances from each point to their four closest atoms in the molecule. In total two vectors per molecule: 26V for potentials and 26C for concavities, each of 26 components, are both the space group predictors in this work. In fact, the main differences with the previous work [7] are the addition of this concavity vector, not scaling the molecular potential vectors and a more simple space group classification procedure.

Table 1 shows molecular global form parameters such as the average molecular concavity ⟨conc⟩ and its standard deviation ⟨desv⟩, the relationship between the principal inertial axes (M/L, S/L, S/M), and other parameters of internal

symmetry of the molecule, some of them used in the previous work [7].

Table 2 shows the averages of those parameters per space group SG. Although all these form parameters are somehow conditioning the SG of the crystal, it is not easy to find in these tables any similarities between those global form parameters for molecules of the same SG, which justifies extending that global form information among 26P over the inertial ellipsoid.

2. Molecular Space Group Predictors

Due to the relation between the present and previous work [7], the same 31 molecules are also chosen here. Table 3 shows those molecules all having azol group with three different substitutes: a group of 17 molecules 17P21/c with space group

TABLE 1: Global form descriptors for each molecule.

	$\langle$conc$\rangle$	desv	M/L	S/L	S/M	PL	mS	mM	mL
FAQROE P21/c	2.076	0.32	0.574	0.371	0.647	2	2	0	1
FAQSAR P21/n	2.152	0.29	0.542	0.363	0.669	2	2	0	1
FAQSEV P21/c	2.041	0.19	0.688	0.372	0.541	2	2	0	1
FAQSOF P21/n	2.334	0.42	0.987	0.516	0.523	2	1	0	0
FAQSUL P21/c	2.175	0.32	0.664	0.380	0.572	2	2	0	1
FAQTAS P21/n	2.111	0.35	0.589	0.370	0.627	1	1	0	0
HDMPYZ P21/c	2.316	0.42	0.795	0.619	0.778	0	0	0	0
HEPHUF P21/n	2.372	0.28	0.592	0.579	0.978	0	0	0	0
HEPJER P21/n	1.774	0.29	0.636	0.245	0.385	2	2	1	1
KOSFUT P21/n	2.164	0.41	0.761	0.697	0.915	0	0	0	0
LEVVAJ P21/n	1.996	0.24	0.544	0.284	0.521	2	2	0	1
RIVBAZ P21/c	2.083	0.53	0.649	0.581	0.895	0	2	0	1
RUPSEA P21/n	1.866	0.21	0.511	0.230	0.449	2	2	1	0
TEHQAY P21/c	2.181	0.46	0.883	0.353	0.399	2	2	1	0
VAXLAH P21/a	2.370	0.43	0.763	0.408	0.535	1	1	0	0
WILBAU P21/c	2.031	0.51	0.644	0.576	0.894	0	1	0	1
YAXZOM P21/n	2.073	0.23	0.439	0.282	0.643	2	2	0	1
HEPJAN Pbca	2.341	0.47	0.916	0.595	0.649	0	0	0	0
MBCPAZ Pbca	2.243	0.22	0.729	0.416	0.570	2	2	0	1
POYXUW Pbca	2.044	0.38	0.456	0.210	0.460	2	2	0	1
BIWWEJ P212121	2.453	0.47	0.361	0.311	0.861	1	1	0	1
PAZDPY P212121	1.913	0.36	0.755	0.271	0.359	2	2	1	0
RUPRID P212121	1.905	0.25	0.584	0.255	0.437	2	2	1	0
BEWLEU Pna21	2.461	0.35	0.651	0.367	0.564	2	1	0	0
HIWJIG01 Pna21	1.944	0.25	0.870	0.401	0.461	1	1	0	0
HIWJIG Pca21	2.141	0.22	0.909	0.521	0.573	1	1	0	0
TAXLOT P-1	2.721	0.58	0.683	0.496	0.727	2	2	0	1
VEHCOA P-1	2.277	0.40	0.776	0.764	0.984	1	2	0	0
BENSES P2/c	2.851	0.50	0.789	0.609	0.772	2	2	1	0
GISZIR C2/c	2.245	0.32	0.581	0.326	0.561	2	2	0	1
PYRZAL10 P21	2.176	0.22	0.606	0.529	0.873	0	0	0	0

$\langle$conc$\rangle$ = average $\langle$distance(P-4nearest_atoms)$\rangle$ in the 26P, desv = stand_deviation of $\langle$conc$\rangle$.
M/L, S/L, S/M for the inertial ellipsoid.
PL = 2, planar molec H's no considered; PL = 1, pseudo-PL; PL = 0, no PL.
mS, mM, mL = 2: molecular symmetry plane m perpendicular to S, M, L.
mS, mM, mL = 1: pseudo m perpendicular to S, M, L.
mS, mM, mL = 0: no m.

TABLE 2: Mean form factors $\langle$M/L$\rangle$, $\langle$S/L$\rangle$ of the inertial ellipsoid; mean planarity of the molecules $\langle$PL$\rangle$ (between max = 2, 1 and min = 0); and mean molecular planes of symmetry $\langle$mS$\rangle$, $\langle$mM$\rangle$, or $\langle$mL$\rangle$ (between max = 2, 1 and min = 0), with their standard deviations, for each SG.

	$\langle$M/L$\rangle$	des	$\langle$S/L$\rangle$	des	$\langle$PL$\rangle$	des	$\langle$mS$\rangle$	des	$\langle$mM$\rangle$	des	$\langle$mL$\rangle$	des
14NoP21/c	0.69	0.16	0.43	0.16	1.43	0.76	1.43	0.76	0.21	0.43	0.36	0.50
31az	0.68	0.15	0.43	0.15	1.35	0.84	1.42	0.76	0.19	0.40	0.45	0.51
17P21/c	0.66	0.14	0.43	0.14	1.29	0.92	1.41	0.80	0.18	0.39	0.53	0.51
6FAQP21c	0.67	0.16	0.40	0.06	1.83	0.41	1.67	0.52	0	0	0.67	0.52
3Pbca	0.70	0.23	0.41	0.19	1.33	1.15	1.33	1.15	0	0	0.67	0.58
3P212121	0.57	0.20	0.28	0.03	1.67	0.58	1.67	0.58	0.67	0.58	0.33	0.58
3Pna21	0.81	0.14	0.43	0.08	1.33	0.58	1	0	0	0	0	0
2P-1	0.73	0.07	0.63	0.19	1.5	0.71	2	0	0	0	0.50	0.71

TABLE 3: The 31 molecules selected for the present work, with the CSD codes.

Compound	R3	R4	R5
(1) BENSES	(a)	CN	H
(2) BEWLEU	(b)	COOMe	COOMe
(3) BIWWEJ	Ph	H	diMe-NH-CH_2-Ph
(4) FAQROE	COOEt	H	H
(5) FAQSAR	COOEt	H	Me
(6) FAQSEV	COOEt	Me	H
(7) FAQSOF	COOEt	Ph	H
(8) FAQSUL	COOEt	Br	H
(9) FAQTAS	COOEt	Br	Me
(10) GISZIR	Me	CN	NH-Ph-4-CF_3
(11) HDMPYZ	H	(c)	Me
(12) HEPHUF	H	CH_2-4-pz	H
(13) HEPJAN	Me	CH_2-3,5-diMe-4-pz	Me
(14) HEPJER	3-pz	H	H
(15) HIWJIG	H	NH_2	H
(16) HIWJIG01	H	NH_2	H
(17) KOSFUT	H	O-Si(2tBuOH)	C(tBuO)
(18) LEVVAJ	(d)	H	H
(19) MBCPAZ	Me	Br	C(O)(NH_2)
(20) PAZDPY	N=N^+=N^-	Ph	H
(21) POYXUW	NH-COPh	H	Br
(22) PYRZAL10	CH_2-C(COO^-)(NH_3^+)	H	H
(23) RIVBAZ	tBu	N=O	tBu
(24) RUPRID	NH_2	H	Ph
(25) RUPSEA	Ph-4-NO_2	H	NH_2
(26) TAXLOT	NH-CH-C(e)(COOEt)	H	H
(27) TEHQAY	H	H	3,5-dimethoxyph
(28) VAXLAH	(f)	COOMe	COOMe
(29) VEHCOA	H	NO_2	$SiMe_3$
(30) WILBAU	tBu	NO_2	tBu
(31) YAXZOM	(g)	Me	(g)

(a)   (b)   (c)

(d)   (e)   (f)   (g)

TABLE 4: Total averages $\langle 26\langle NV\rangle\rangle$ and $\langle 26\langle NC\rangle\rangle$ for potentials V and concavities C on the inertial ellipsoid IE, and the average of the 26 standard deviations of $26\langle NV\rangle$ and $26\langle NC\rangle$ for each space group SG. $\langle 26\langle NV\rangle\rangle$ is the average of the $26V$ averaged (vertical sense in Supplementary Table) between the N molecules of the SG. $\langle 26(\text{des}\langle NV\rangle)\rangle$ is the average of the 26 standard deviations for the $26\langle NV\rangle$. $\langle 26\langle NC\rangle\rangle$ is the average of the $26C$ averaged (vertical sense in Supplementary Table) between the N molecules of the SG. $\langle 26(\text{des}\langle NC\rangle)\rangle$ is the average of the 26 standard deviations for the $26\langle NC\rangle$. des$\langle 26\text{des}\rangle$ is the standard deviation of the last average $\langle 26(\text{des}\langle NC\rangle)\rangle$.

N molecules	$\langle 26\langle NV\rangle\rangle$	$\langle 26(\text{des}\langle NV\rangle)\rangle$	$\langle 26\langle NC\rangle\rangle$	$\langle 26(\text{des}\langle NC\rangle)\rangle$	des$\langle 26\text{des}\rangle$
14NoP21/c	0.09	0.10	2.26	0.47	0.12
31azols	0.005	0.09	2.18	0.42	0.11
17P21/c	0.01	0.07	2.11	0.34	0.10
6FAQP21/c	−0.05	0.05	2.15	0.25	0.08
2P-1	−0.003	0.05	2.50	0.52	0.13
3P212121	0.071	0.13	2.02	0.38	0.11
3Pna21	−0.003	0.05	2.18	0.35	0.07
3Pbca	−0.01	0.06	2.21	0.34	0.18

$\langle 26(\text{des}\langle NV\rangle)\rangle$ shows the V similarity for the N molecules in the same point of IE.
$\langle 26(\text{des}\langle NC\rangle)\rangle$ shows the C similarity for the N molecules in the same point of IE.
des$\langle 26\text{des}\rangle$ shows the similarity between the $26(\text{des}\langle NC\rangle)$.

SG P21/c (also P21/n or P21/a), six of them 6FAQP21/c sharing also a COOEt group in the same substituent, eleven molecules distributed among several SG: 3 in Pbca, 3 in P212121, 3 in Pna21, 2 in P-1, and other three molecules crystallizing in different SG.

Finally, in order to compare properly both molecular vectors, $26V$ for potentials and $26C$ for concavities, between different molecules, each molecule was reoriented within its inertial ellipsoid IE, by using the symmetry planes L = 0, M = 0, or S = 0, in order to have the molecular largest potential V_weighted octant, among the eight octants, in the IE octant (+L+M+S), were V_weighted is calculated with the potentials of the seven points of the octant in the proportion: [3V (111) +2V (110 +101 +011) + V (100 +010 +001)]. The Supplementary Table (available online at http://dx.doi.org/10.1155/2014/737480) shows the final comparable values of $26V$ and $26C$ for the reoriented 31 molecules. It is important to see in Tables 1 and 3 that the presence of two molecules in the same space group SG occurs not only for little changes in their structure like between FAQROE and FAQSAR but also for big changes like for FAQROE and KOSFUT, suggesting that the 26 potentials and 26 concavities sequences around the inertial ellipsoid will determine better the space group.

It is interesting to consider here that all the molecules could also be reoriented to have the lowest, instead of the largest, potential V_weighted octant in (+L+M+S). Although the relative orientation between the largest and lowest V_weighted octants is not the same for all molecules, however the average distribution on the inertial ellipsoid IE surface of the $26V$ and $-26V$, respectively, is similar for both orientations of the 31 molecules, which reinforces this analysis. Finally the largest V_weighted octant was taken for molecular reorientation. Figure 2(a) shows first the distribution along the IE of the $26V$ averaged for the 31 azole molecules $26\langle 31V\rangle$, together with the 26 standard deviations of these averages 26std$\langle 31V\rangle$, showing two saw-tooth peaks, around (110)_(111) and (010)_(011), as expected (see Figure 1(a)) for molecules

oriented with the largest potential V_weighted octant in (+L+M+S).

The centrosymmetric molecule HEPJER almost planar in (L, M, 0) is shown in Figure 1(b) into a schematic inertial ellipsoid IE, because it is more suitable for its simplicity to understand the sequence of concavities and potentials along the 26 points on its IE shown in Figure 1(c). The concavities C are maximal in the 8 molecular LM0 contour points, the remaining concavities on either side of plane S = 0 being lower. The symmetrical potentials V have lower values on 0MS points away from the molecule and higher peaks especially on LM0 points.

3. Analysis of Averaged Potentials and Concavities by Space Group

Figure 2(a) shows the distribution from (1–1–1) to (−111) along the inertial ellipsoid of the 26 potentials averaged for the N molecules $26\langle NV\rangle$ belonging to different groups: the total of 31az, the 14NotP21/c, the 17P21/c, and the 6FAQP21/c. It also shows the distribution of the 26 standard deviations of those averages des $(26\langle NV\rangle)$ per group, which indicate the similarity between the potentials of the N molecules in each one of the 26 points of their inertial ellipsoid. In general des$(26\langle NV\rangle)$ are maximum in singular points with maximum or minimum values of $\langle NV\rangle$ and in particular are clearly superior for the first two groups of mixed space groups compared to the last two P21/c groups, as also shown in Table 4 with the total averages under the column $\langle \text{des}(26\langle NV\rangle)\rangle$. Figure 2(a) also shows that the distribution of the $26\langle NV\rangle$ for 31az and 14azNotP21/c is similar with two peaks in saw-tooth form, although being less pronounced the second peak for 31az. The distributions of the $26\langle NV\rangle$ for the equal space groups 17P21/c and 6FAQP21/c differ from the previous with the loss of the second saw-tooth peak and an increased negative potential V on (0−10).

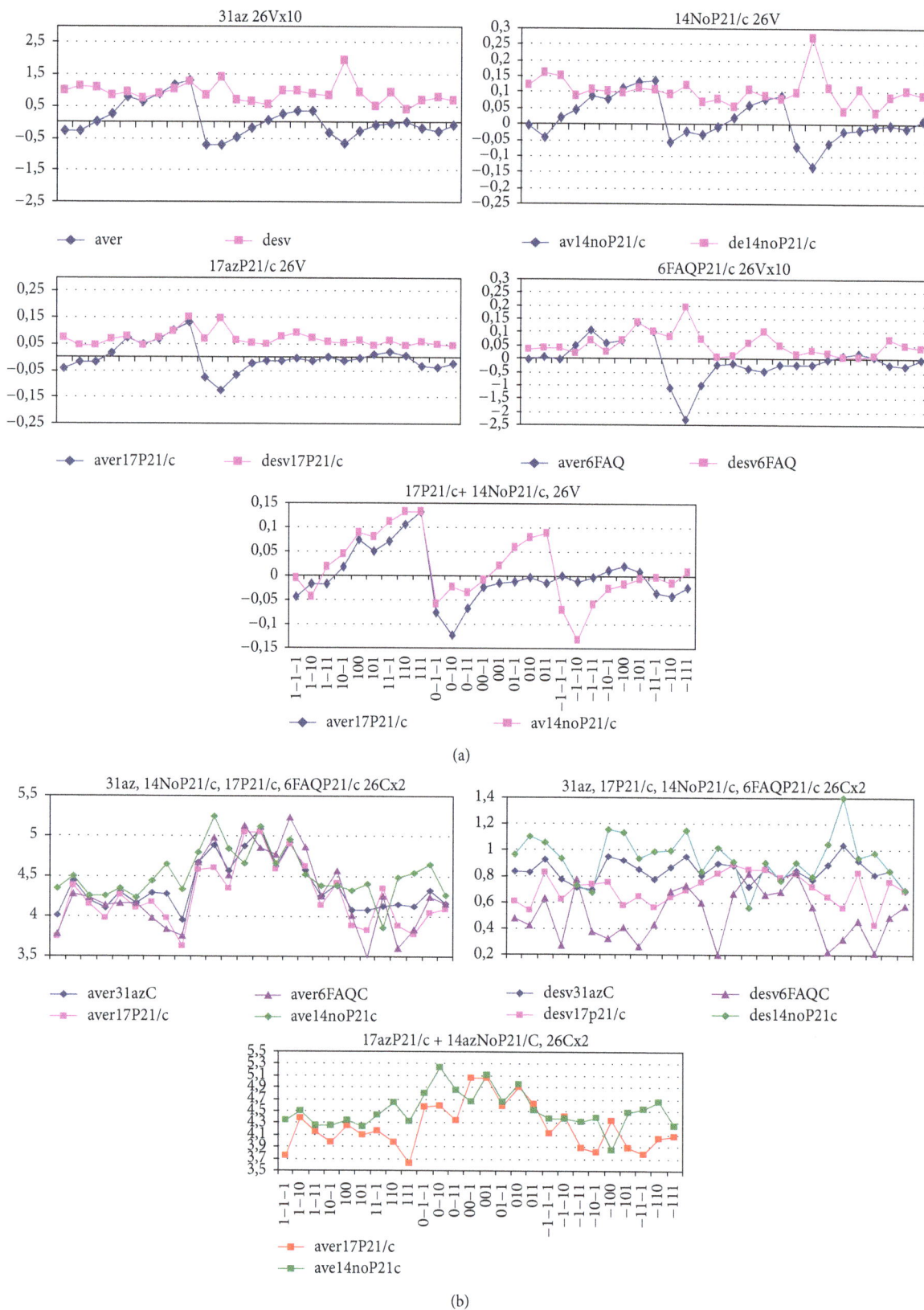

FIGURE 2: (a) Distribution of $26\langle NV \rangle$ for N molecules: 26 components of averages $\langle NV \rangle$ with the standard deviations for groups 31az, 14NotP21/c, 17P21/c, and 6FAQP21/c. Comparison between the $26\langle NV \rangle$ for 17P21/c and 14NotP21/c. (b) Distribution of the $26\langle NC \rangle$ concavities and the corresponding standard deviations: 26 components of $\langle NC \rangle$ averages for the groups 31az, 14NotP21/c, 17P21/c, and 6FAQP21/c.

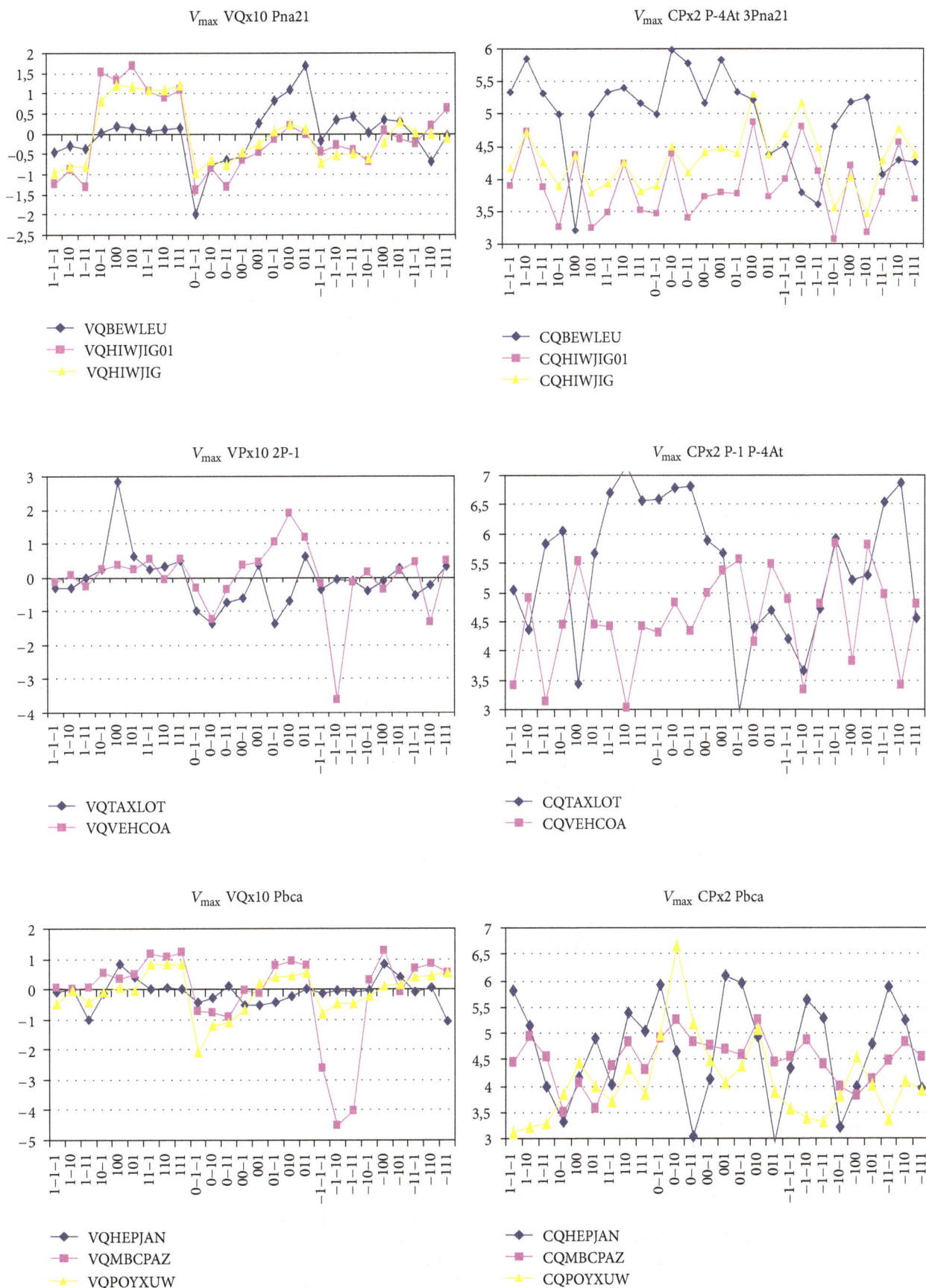

$V_{\max}$ VQx10 Pna21

$V_{\max}$ CPx2 P-4At 3Pna21

VQBEWLEU
VQHIWJIG01
VQHIWJIG

CQBEWLEU
CQHIWJIG01
CQHIWJIG

$V_{\max}$ VPx10 2P-1

$V_{\max}$ CPx2 P-1 P-4At

VQTAXLOT
VQVEHCOA

CQTAXLOT
CQVEHCOA

$V_{\max}$ VQx10 Pbca

$V_{\max}$ CPx2 Pbca

VQHEPJAN
VQMBCPAZ
VQPOYXUW

CQHEPJAN
CQMBCPAZ
CQPOYXUW

(a)

FIGURE 3: Continued.

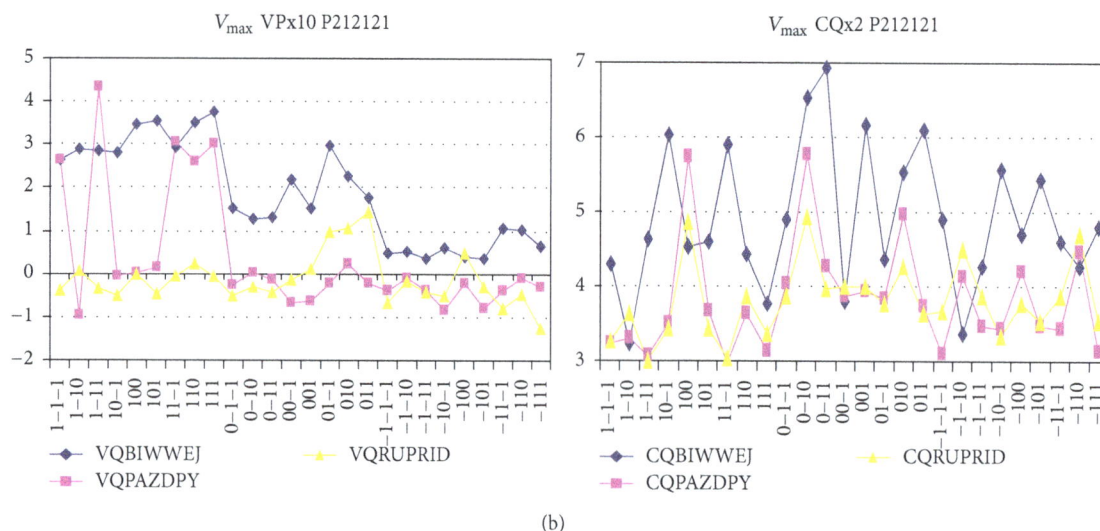

(b)

FIGURE 3: Distribution of the $26V$ potentials and $26C$ concavities for molecules of the groups 3Pbca, 3P212121, 3Pna21, and 2P-1.

Therefore, although the overall molecular shape factors of Tables 1 and 2 and the contents of the molecules in Table 3 do not seem to indicate similarity between molecules within each space group SG, however the distributions of the $26\langle NV\rangle$ potentials along the 26 points of the inertial ellipsoid appear to have some similarity between the molecules of the same SG. In fact, the analysis of Figure 2(a) and the values of $\langle 26(\text{des}\langle NV\rangle)\rangle$ in Table 4 suggest that there exists an association of the potential $26V$ predictor vector of a molecule with its molecular crystal or space group SG.

Figure 2(b) shows the average distribution of the 26 concavities $26\langle NC\rangle$ and of the corresponding standard deviations $\text{des}(26\langle NC\rangle)$ for the four groups of N molecules: 31az, 14NotP21/c,17P21/c, and 6FAQP21/c. While for 14azNotP21/c, the distribution of $26\langle NC\rangle$ is pseudosymmetrical around the center (00 ± 1), that symmetry tends to break for 17azP21/c and more for 6FAQP21/c molecules, especially in the area where their $26\langle NV\rangle$ potentials lose the second saw-tooth peak. The distribution of $26\langle NC\rangle$ for 31az involves molecules of equal and different space group being intermediate as might be expected. Table 4 also shows the quantitative differences between the average distributions of $26\langle NC\rangle$ concavities for different groups of molecules. The average values under $\langle 26(\text{des}\langle NC\rangle)\rangle$ showing more similar distributions for the 26 concavities between the molecules of the groups 6FAQP21/c and 17P21/c than between the molecules of the group 14NotP21/c, while an intermediate similarity for the total of 31az. This shows the association of the predictor vector $26C$ of molecular concavities with the space group SG of the molecular crystal, parallel to the previous association observed of the predictor vector $26V$ of potentials with the space group SG.

Figure 3 shows the molecular distributions of $26V$ potentials and $26C$ concavities for the four minority space groups: Pbca, P212121, Pna21, and P-1. Although this analysis is less significant with only 3, 3, 3, and 2 molecules each, it also notes some similarity between the distributions of its $26V$ and of

its $26C$ for molecules with the same space group SG, except for P-1 and BIWWEJ (unique molecule in Table 2 with its $26V$ "calculated" positive). Furthermore, Table 4 shows that the average standard deviation values $\langle 26(\text{des}\langle NV\rangle)\rangle$ and $\langle 26(\text{des}\langle NC\rangle)\rangle$ for these four space groups do not deviate too much from those of groups 6FAQP21/c and 17P21/c, with the exceptions described above.

Finally, Figure 2(b) also shows the differences between the average potentials $\langle V\rangle$ and average concavities $\langle C\rangle$ distributions along the inertial ellipsoid surface for the 17P21/c and 14NonP21/c molecular groups. While for L = 1 there is quit similitude between the V's and between the C's of both groups, for L = 0 the second saw-tooth disappearance for the V's is accompanied with some C's distribution variations between both groups, and for L = −1 besides the notable V's distribution differences there are drastic differences between the C's distributions for both molecular groups.

4. Conclusion

Assuming that the molecular form and the potential distribution on its surface were the major predictors for crystal packing, it is not easy to compare these properties between different molecules, in order to classify them by their space group. To enable this comparison, molecules are reduced to their inertial ellipsoids IE with 26 singular points equally spaced on the surface, in which 26 potentials and 26 concavity factors are calculated. These two molecular vectors $26V$ and $26C$ are taken as molecular packing or space group predictors for the molecular crystal (assuming no polymorphism). Comparing both predictors between 31 molecules, there is more similarity between them for molecules crystallizing in the same space group SG than between molecules with different SG. This suggests that each space group would have its own mean distribution of their $26V$ potentials and $26C$ concavities on a virtual inertial ellipsoid, which would enable predicting the probable space group of a molecular crystal by calculating the $26V$ and $26C$ distributions on its molecular inertial ellipsoid. Foreknowledge of the probable space group associated with a molecule would facilitate the total crystal prediction CPP to perform other crystal engineering calculation. For example, if the above predicted space group SG (associated with a crystalline form) were not convenient for pharmaceutical processes [1], a molecular modification simulation could be tried to change the SG avoiding that molecular aggregation.

Summary of Symbols

IE:	Inertial ellipsoid of the molecule
SG:	Crystal space group
L, M, S:	Large, medium, small axes of IE
P:	One of 26 points on the IE surface
V:	Electric potential at one P
C:	Concavity at one P
N:	Number of molecules in a group
des:	Standard deviation of an average
$26V$:	Vector with the $26V$ on IE surface
$26C$:	Vector with the $26C$ on IE surface
$\langle NV \rangle$:	NV average on a P of a molecular group
$\langle NC \rangle$:	NC average on a P of a molecular group
$\text{des}\langle NV \rangle$:	Standard deviation of average $\langle NV \rangle$
$\text{des}\langle NC \rangle$:	Standard deviation of average $\langle NC \rangle$
$\langle 26\langle NV \rangle \rangle$:	Average of the total $26\langle NV \rangle$ of a group
$\langle 26\langle NC \rangle \rangle$:	Average of the total $26\langle NC \rangle$ of a group
$\langle 26(\text{des}\langle NV \rangle) \rangle$:	Average of the $26(\text{des}\langle NV \rangle)$
$\langle 26(\text{des}\langle NC \rangle) \rangle$:	Average of the $26(\text{des}\langle NC \rangle)$.

Conflict of Interests

The author declares that there is no conflict of interests regarding the publication of this paper.

References

[1] S. L. Price, "Predicting crystal structures of organic compounds," *Chemical Society Reviews*, vol. 43, pp. 2098–2111, 2014.

[2] D. A. Bardwell, C. S. Adjiman, Y. A. Arnautova et al., "Towards crystal structure prediction of complex organic compounds—a report on the fifth blind test," *Acta Crystallographica Section B: Structural Science*, vol. 67, part 6, pp. 535–551, 2011.

[3] J. Fayos and F. H. Cano, "Crystal-packing prediction by neural networks," *Crystal Growth and Design*, vol. 2, no. 6, pp. 591–599, 2002.

[4] J. Fayos, L. Infantes, and F. H. Cano, "Neural network prediction of secondary structure in crystals: hydrogen-bond systems in pyrazole derivatives," *Crystal Growth and Design*, vol. 5, no. 1, pp. 191–200, 2005.

[5] H. Nakatsuji, S. Kanayama, S. Harada, and T. Yonezawa, "Electrostatic force theory for a molecule and interacting molecules. 7. Ab initio verification of the force concepts based on the flotating wave functions of ammonia, methyl(1+) ion, and ammonia(1+) ion," *Journal of the American Chemical Society*, vol. 100, no. 24, pp. 7528–7534, 1978.

[6] Y. Honda and H. Nakatsuji, "Force concept for predicting the geometries of molecules in an external electric field," *Chemical Physics Letters*, vol. 293, no. 3-4, pp. 230–238, 1998.

[7] J. Fayos, "Molecular crystal prediction approach by molecular similarity: data mining on molecular aggregation predictors and crystal descriptors," *Crystal Growth and Design*, vol. 9, no. 7, pp. 3142–3153, 2009.

[8] F. H. Allen, "The Cambridge structural database: a quarter of a million crystal structures and rising," *Acta Crystallographica B*, vol. 58, no. 1, part 3, pp. 380–388, 2002.

15

Synthesis and Characterization of Novel Processable and Flexible Polyimides Containing 3,6-Di(4-carboxyphenyl)pyromellitic Dianhydride

Muhammad Kaleem Khosa,[1] Muhammad Asghar Jamal,[1] Rubbia Iqbal,[1] and Mazhar Hamid[2]

[1] Department of Chemistry, Government College University Faisalabad, Faisalabad 38000, Pakistan
[2] National Engineering and Scientific Commission, P.O. Box 2801, Islamabad, Pakistan

Correspondence should be addressed to Muhammad Kaleem Khosa; mkhosapk@yahoo.com

Academic Editor: Alexandra Muñoz-Bonilla

A series of six novel polyimides containing 3,6-di(4-carboxyphenyl)pyromellitic dianhydride were synthesized via two steps condensation method. Aromatic diamines monomers, 4-(4-aminophenoxy)-N-(4-(4-aminophenoxy)benzylidene)-3-chloroaniline (DA1), 4-(4-amino-3-methylphenoxy)-N-(4-(4-amino-3-methylphenoxy)benzylidene)-3-chloroaniline (DA2), 4-(4-amino-2-methylphenoxy)-N-(4-(4-amino-2-methylphenoxy)benzylidene)-3-chloroaniline (DA3) 4-(4-aminophenoxy)-N-(4-(4-aminophenoxy)benzylidene)-2-methylaniline (DA4), 4-(4-amino-3-methylphenoxy)-N-(4-(4-amino-3-methylphenoxy)benzylidene)-2-methylaniline (DA5), and 4-(4-amino-2-methylphenoxy)-N-(4-(4-amino-2-methylphenoxy)benzylidene)-2-methylaniline (DA6) were prepared and used to synthesize new polyimides by reaction with resynthesized 3,6-di(4-carboxyphenyl)pyromellitic dianhydride by using two-step condensation method. The inherent viscosities of polyimides range from 0.68–1.04 dL gm^{-1} and were soluble in polar solvents. Polyimides have excellent thermal stability by showing 10% weight loss temperature was above 450°C. Their glass transition temperatures lie in the range of 250–335°C. Wide-angle X-ray diffractometer investigations revealed the amorphous nature of polyimides. Therefore, these polymers can be a potential candidate as processable high performance polymeric materials.

1. Introduction

Polyimides are very interesting group of amazingly strong and marvellously heat resistant polymers [1]. Aromatic polyimides have received much attention for half a century, due to their excellent combination of properties and potential applications in aerospace, microelectronics (flexible printed boards for electronic devices), photoelectronic industry (as photoresists), and separation industry [2–6]. Other applications include adhesives and matrix resins for composites. However some of their properties like limited solubility, rigid chain characteristics, strong chain-chain interaction, high glass transition, and melting temperatures create difficulties in their processing. That is why applications of these rigid polyimides are restricted in technological and industrial applications. Extensive research has been carried out to improve their solubility by synthesizing soluble polyimides without disturbing their excellent properties [7–15]. In present days a number of ways exist to alter chemical structure of synthesizing polymeric materials while maintaining the excellent level of their thermal and mechanical properties. Several modifications have been made in their chemical structure by the introduction of bulky alkyl side substitution, noncoplanar, alicyclic structures, flexible aryl, or alkyl ether linkages and asymmetric biphenyl moieties in the back bone of rigid polyimides [16–20]. By the introduction of flexible linkages progress in solubility and significant processability have been achieved by altering crystallinity and intermolecular interactions. The incorporation of aliphatic segments and noncoplanar structures helps to improve solubility of polyimides but is deleterious for thermal and mechanical properties. To achieve better quality polyimides there is need to design new monomers, that is, diamines and dianhydride with structural modifications [21–24]. Researchers are doing efforts to design and synthesize new diamines and dianhydride monomers, thus, producing a wide range of polyimides with promising processability and solubility for various technological and industrial applications. These efforts make it possible to synthesize many polyimides which are soluble and easily processable without disturbing their excellent properties. Among these polyimides fluorinated monomers have also gained attention. When bulky fluorinated groups are incorporated, soluble polyimides with excellent thermal properties can be achieved [25–28]. Asymmetric introduction of bulky substituent or linkages in the backbone of polyimides is another effective approach to improve the solubility of polyimides. Another approach is the introduction of ether linkages using a nucleophilic aromatic substitution reaction. Due to the presence of flexible moieties on the polyimide backbone; there will be a decrease in rigidity of polymer chain to improve the solubility of polyimides. To improve the solubility of polyimides carboxy groups were also introduced in the backbone of polyimides and were found very helpful in improving solubility of polyimides. The carboxy functionality is desirable in developing new applications such as fabrication of nanostructure by molecular self-assembly, ion exchange membranes polymer electrolytes, or ionomers. Thus, components into the polyimide main chain are one of the most successful approaches in attaining solubility without changing their excellent properties. These structural modifications for monomers have also led to new polyimides with several improved properties. Consequently, it was of interest to investigate the thermal stability of aromatic polyimides with various linking groups in their backbone. The azomethine linkage is of distinctive significance due to its remarkable properties such as liquid crystalline property, semiconductivity, ability to form metal chelates, fibre-forming ability, fine thermal stability, and nonlinear optical activity [26–29]. A variety of polymers with a Schiff-base structure have been synthesized, characterized, and investigated with respect to their properties [30]. Keeping in view the useful properties of polyimides, thus, a diamine monomer containing ether and azomethine linkage has been designed, synthesized, and exploited to prepare novel poly azomethine imides with good thermal stability and processability. The aromatic diamines containing azomethine moiety along with ether linkages were incorporated in polymer backbone to examine their structure-property relationship in terms of inherent viscosity, solubility in various solvents, thermal stability, glass transition temperature, and so forth. Introduction of azomethine moiety in the polymer backbone will incorporate features like semiconductivity [31], biomedical activity [32], as thermal stabilizers [33], and corrosion inhibition [34–36] in the newly synthesized polymers. The monomers and polyimides, PI (1–6), were characterized by means of elemental analyses, FTIR, ^{1}H-NMR spectroscopy.

2. Experimental

2.1. Materials. 4-amino-3-methylphenol, 4-amino-2-chlorophenol, 4-fluoronitrobenzene, 5-fluoro-2-nitrotoluene,2-fluoro-5-nitrotoluene,4-hydroxybenzaldehyde, potassium permanganate, tetrakis(triphenylphosphine) palladium, *p*-tolylboronic acid, sodium carbonate, dibromodurene, hydrazine monohydrate, and pyridine were purchased from E. Merck and Aldrich and used without further purification. All organic solvents dimethyl sulfoxide (DMSO), dimethylformamide (DMF), N,N-dimethylacetamide (DMAc), N-dimethylpyrolidone (NMP), *m*-cresol, methanol, ethanol, and toluene were purchased from E. Merck, Germany and were dried before used according to the standard methods [37].

2.2. Measurements. Melting points were determined by using capillary tube on an electrochemical melting point apparatus, model MP-D Mitamura Riken Kogyo, Japan and are uncorrected. Infrared absorption spectra were recorded as KBr disc on Bio-Rad Excalibur FT-IR Model FTS 3000 MX. Elemental analysis was performed using a Perkine Elmer 2400 CHN elemental analyzer. The ^{1}H spectra were recorded on a BRUKER Spectrometer operating at 300 MHz. Solvent used for analysis was deutrated dimethyl sulfoxide (DMSO-d6). Thermal and DSC analysis were carried out using Perkin Elmer TGA-7 and DSC 404C Netzsch under nitrogen atmosphere. Wide-angle diffractograms were obtained using 3040/60 X'Pert PRO diffractometer. Viscosities were obtained by using Gilmount falling ball viscometer. Solubility was determined in different solvents. Moisture absorption values were determined by changes in weight of the dried film before and after immersion of water at 25°C.

2.3. Synthesis of Monomers. Six new diamine monomers, i.e., 4-(4-aminophenoxy)-N-(4-(4-aminophenoxy) benzylidene)-3-chloroaniline (DA1), 4-(4-amino-3-methylphenoxy)-N-(4-(4-amino-3-methylphenoxy)benzylidene)-3-chloroaniline (DA2), 4-(4-amino-2-methylphenoxy)-N-(4-(4-amino-2-methylphenoxy)benzylidene)-3-chloroaniline (DA3), 4-(4-aminophenoxy)-N-(4-(4-aminophenoxy)benzylidene)-2-methylaniline (DA4), 4-(4-amino-3-methylphenoxy)-N-(4-(4-amino-3-methylphenoxy)benzylidene)-2-methylaniline (DA5), and 4-(4-amino-2-methylphenoxy)-N-(4-(4-amino-2-methylphenoxy)benzylidene)-2-methylaniline (DA6), were synthesized in three steps as shown in the Scheme 1. In first step Schiff bases were prepared by condensation of 0.025 mol of substituted *p*-hydroxyamines and 0.025 mol of 4-hydroxyaldehyde in dry ethanol (15–20 mL) in the presence of acetic acid which act as catalyst. After stirring of about 1h the reaction mixture was refluxed 4-5h. Reaction was monitored by thin layer chromatography (TLC). The reaction mixture was filtered and precipitates were collected. The crude product was washed with ethanol and recrystallized with methanol. In the second step 1g of dihydroxy Schiff base compound was treated with 0.01 mol of substituted 4-fluoronitrobenzene and 0.01 mol of potassium

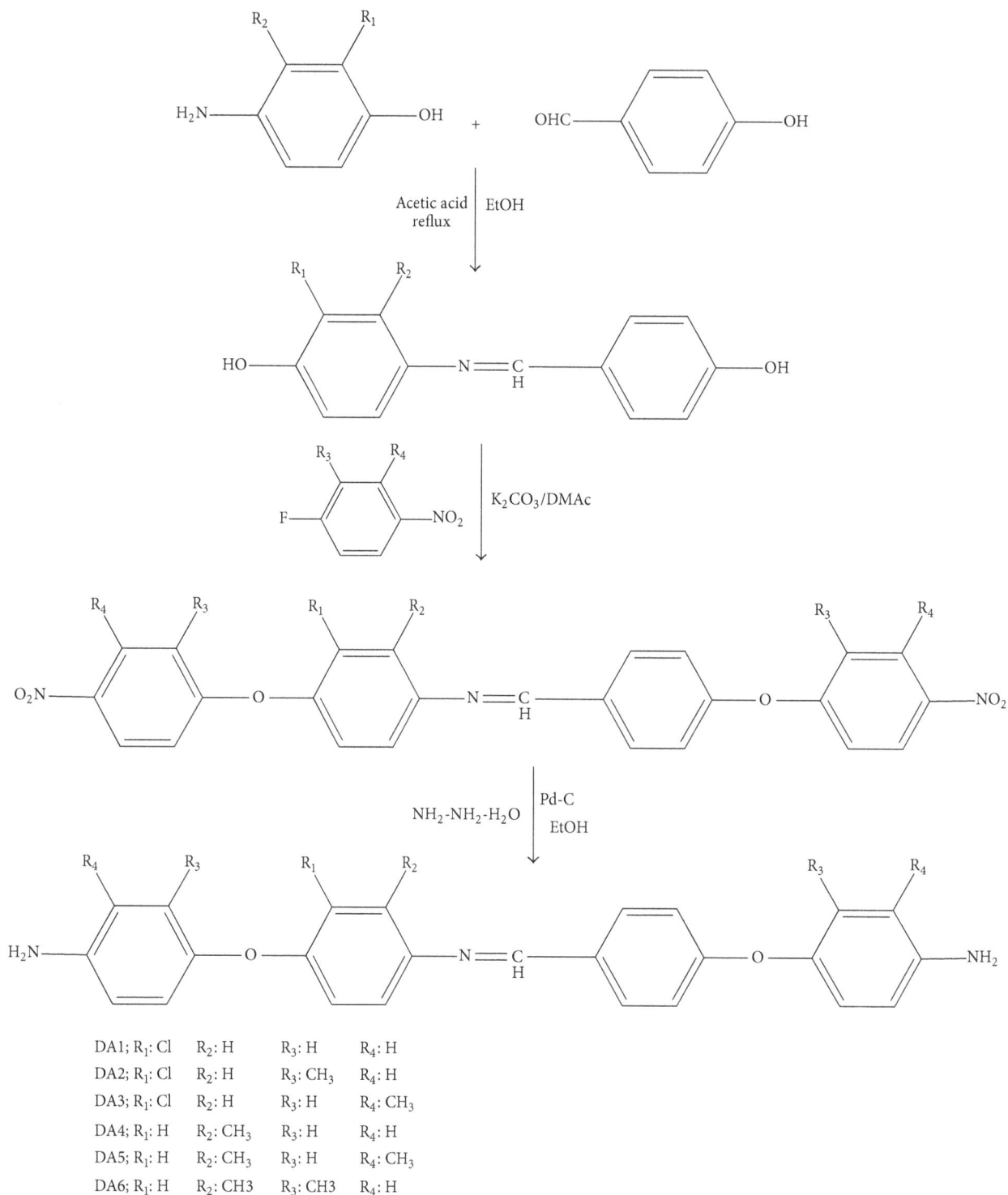

DA1; R_1: Cl	R_2: H	R_3: H	R_4: H
DA2; R_1: Cl	R_2: H	R_3: CH$_3$	R_4: H
DA3; R_1: Cl	R_2: H	R_3: H	R_4: CH$_3$
DA4; R_1: H	R_2: CH$_3$	R_3: H	R_4: H
DA5; R_1: H	R_2: CH$_3$	R_3: H	R_4: CH$_3$
DA6; R_1: H	R_2: CH3	R_3: CH3	R_4: H

SCHEME 1

carbonate in DMAc (60 mL) in two necks round bottom flask under inert conditions. The reaction mixture was heated at 100°C for 18–20 h. The reaction mixture was cooled to room temperature and poured into water. Solid obtained was washed thoroughly with water and filtered. Product obtained was recrystallized from ethanol. In third step for reduction of nitro compounds, 1 g of above synthesized dinitro compound was suspended in 250 mL of two-necked flask with 10 mL of hydrazine monohydrate and 0.06 g of 5% palladium on carbon (Pd-C) and 60 mL of dry ethanol. The reaction mixture was heated at refluxed for 16 h and then filtered to remove Pd-C. Solvent was evaporated through rotary evaporator and the solid product obtained was recrystallized from ethanol [24].

Data of DA1. Molecular formula; $C_{25}H_{20}N_3O_2Cl$, melting point; 172 ± 1, yield; 75, elemental analysis; calculated (found), %C; 69.77(69.75), %H; 4.65(4.63), %N; 9.76(9.7), IR(KBr)/cm^{-1} 3400, 3325(–NH$_2$) 1624(–C=N) 1355 (–C–N), ^{1}H NMR (DMSO-d$_6$, δ, ppm), 5.11 (–NH$_2$); 6.68–7.80 (m, 15H, aromatic); 8.65 (s, 1H, azomethine).

Data of DA2. Molecular formula; $C_{27}H_{24}N_3O_2Cl$, melting point; 156 ± 1, yield; 78, elemental analysis; calculated (found), %C; 70.74(70.73), %H; 4.94(4.92), %N; 8.65(8.5), IR(KBr)/cm^{-1} 3415, 3330(–NH$_2$), 1650(–C=N), 1345 (–C–N), ^{1}H NMR (DMSO-d$_6$, δ, ppm) 2.20 (6H, CH$_3$); 5.45 (–NH$_2$); 6.48–7.68 (m, 13H, aromatic); 8.64(s, 1H, azomethine).

Data of DA3. Molecular formula; $C_{27}H_{24}N_3O_2Cl$, melting point; 162 ± 1, yield; 82, elemental analysis; calculated (found), %C; 70.74(70.73), %H; 4.94(4.96), %N; 8.65(8.62), IR(KBr)/cm^{-1} 3405, 3340(–NH$_2$), 1618(–C=N), 1365 (C–N), ^{1}H NMR (DMSO-d$_6$, δ, ppm) 2.14 (s, 6H, CH$_3$); 6.01 (–NH$_2$); 6.61–7.60 (m, 13H, aromatic); 8.52 (s, 1H, azomethine).

Data of DA4. Molecular formula; $C_{26}H_{23}N_3O_2$, melting point; 185 ± 1, yield; 85%, elemental analysis; calculated (found), %C; 76.28(76.19), %H; 5.62(5.60), %N; 10.27(10.22), IR(KBr)/cm^{-1} 3443, 3365(–NH$_2$), 1698(–C=N)1350 (–C–N), ^{1}H NMR (DMSO-d$_6$, δ, ppm) 2.12 (s, 3H, CH$_3$); 5.74 (–NH$_2$); 6.45–7.88 (m, 15H, aromatic); 8.59 (s, 1H, azomethine).

Data of DA5. Molecular formula; $C_{28}H_{27}N_3O_2$, melting point; 160 ± 1, yield; 80, elemental analysis; calculated (found), %C; 76.88(76.85), %H; 6.17(6.16), %N; 9.61 (9.60), IR(KBr)/cm^{-1} 3405, 3355(–NH$_2$), 1645(–C=N), 1366(–C–N), ^{1}H NMR (DMSO-d$_6$, δ, ppm)2.06 (s, 9H, CH$_3$); 6.12 (–NH$_2$); 6.75–7.82 (m, 13H, aromatic); 8.66 (s, 1H, azomethine).

Data of DA6. Molecular formula; $C_{28}H_{27}N_3O_2$, melting point; 160±1, yield; 80, elemental analysis; calculated(found), %C; 76.88(76.84), %H; 6.17(6.12), %N; 9.61 (9.59), IR(KBr)/cm^{-1} 3412, 3362(–NH$_2$), 1645(–C=N), 1365(–C–N), ^{1}H NMR (DMSO-d$_6$, δ, ppm) 2.10 (s, 9H, CH$_3$); 5.52 (–NH$_2$); 6.65–7.96 (m, 13H, aromatic); 8.56 (s, 1H, azomethine).

2.4. Synthesis of 4,2,3,5,6,4-Hexamethyl-p-terphenyl. Dianhydride used for the synthesis of polyimides was resynthesized by following the reported method [10]. A 250 mL flask was charged with 4.67 g (16 mmol) dibromodurene, 100 mL toluene, 1.1 g (0.96 mmol) Pd(PPh$_3$)$_4$, and 100 mL 2 M Na$_2$CO$_3$ solution. After addition of 4.7 g (35 mmol) p-tolylboronic acid in 30 mL ethanol, the mixture was refluxed for 24 hrs. After cooled to room temperature, 12.5 mL of H$_2$O$_2$ (30%) was added with great care under vigorous stirring and the mixture was again stirred for hour. After stirring, two layers were formed in the reaction mixture. Separate the organic layer and dried with MgSO$_4$. The solvent was evaporated to obtain crude product, which was washed by cold ethanol to remove byproducts and recrystallized from

ethyl acetate. Yield: 68%; m.p.: 282–285°C; IR (KBr, cm^{-1}); 528 (para-subst. oop), 1524(–C=C), 1565, 1645, 1789, 1900 (para-subst.), 2915–3045(–CH); ^{1}HNMR (DMSO-d$_6$, ppm): 1.85(s, 6H, CH$_3$), 2.48(s, 12H, CH$_3$), 7.15–7.30(m, 8H, Ar).

2.5. Synthesis of 3,6-Di(4-carboxyphenyl)pyromellitic Acid. A 250 mL three-necked flask equipped with magnetic stirrer and condenser was charged with 3.14 g (0.01 mol) of 1,4-ditolyldurene, 120 mL of pyridine, and 10 mL of water. The mixture was heated up to 120°C then added 28.4 g (0.18 mol) of KMnO$_4$ portion wise. After complete addition of KMnO$_4$ the reaction mixture was further refluxed 12 h. This hot reaction mixture was vacuum filtered over a glass funnel to remove MnO$_2$. After evaporating the filtrate, a solid residue was left behind. That solid was put into a 250 mL three-necked flask with stirrer and condenser. It was treated with 100 mL of 4% aqueous NaOH solution under agitation. After that flask contents were heated to 100°C, 9.48 g (0.06 mol) of KMnO$_4$ was added and refluxed it for 12 hrs. Excess KMnO$_4$ was removed using ethanol until the KMnO$_4$ colour disappeared. Then filtered the reaction mixture to remove MnO$_2$ and the filtrate was acidified with HCl to get product, which was further purified by recrystallization in water. Yield: 79%; IR (KBr, cm^{-1}); 1610(C=C, Ar), 1678(–C=O), 3550–2565(–OH, acid); ^{1}H NMR (DMSO-d$_6$, ppm) 7.24–7.82(m, 8H) 13.1(s, 6H, COOH) [10].

2.6. Synthesis of 3,6-Di(4-carboxyphenyl)pyromellitic Dianhydride. A 100 mL two-necked flask was charged with white crystal of 3,6-di(4-carboxy)phenylpyromellitic acid (4.58 g, 10 mmol) and 50 mL of acetic anhydride in inert conditions. After the reflux of 5 h, acetic anhydride was evaporated out to get yellow solids which were further recrystallized from dried DMAc. Yield: 87%; elemental analysis ($C_{24}H_{10}O_{10}$), (458.33) calculated (found) %C 62.83(62.90), %H 2.18(2.23), O 34.90; 34.67 IR (KBr, cm^{-1}): 1675(C=O, acid), 1868 and 1775(C=O, anhydride), 3545–2550(O–H, acid); ^{1}H NMR (DMSO-d6, ppm): 7.65–8.26(m, 8H, Ar), 13.01(s, 2H, acid) [10].

2.7. Synthesis of Polyimide (PI). The synthesized monomers were subjected to polyimide synthesis. polyimides were synthesized by polycondensation of diamine monomers DA1, DA2, DA3, DA4, DA5, and DA6 with resynthesized dianhydride monomer 3,6-di(4-carboxyphenyl)pyromellitic dianhydride [10]. Three-necked flask equipped with nitrogen inlet and mechanical stirrer was charged with 1.2 mol of diamine in 5–10 mL of NMP. Then 1.2 moL of 3,6-di(4-carboxyphenyl)pyromellitic dianhydride was added slowly in the dissolved diamine in NMP. The reaction mixture was stirred for 24 h at room temperature to produce polyamic acid in inert conditions. The resulting polyamic acid solution was converted into polyimides through chemical imidization and thermal method. For chemical imidization, in polyamic acid solution equimolar pyridine and acetic anhydride was added. The reaction mixture was stirred at room temperature for one hour and then heated at 100°C for 3 h. The resulting solution was poured into methanol and fibrous precipitates of polyimides were collected and dried as shown in Scheme 2.

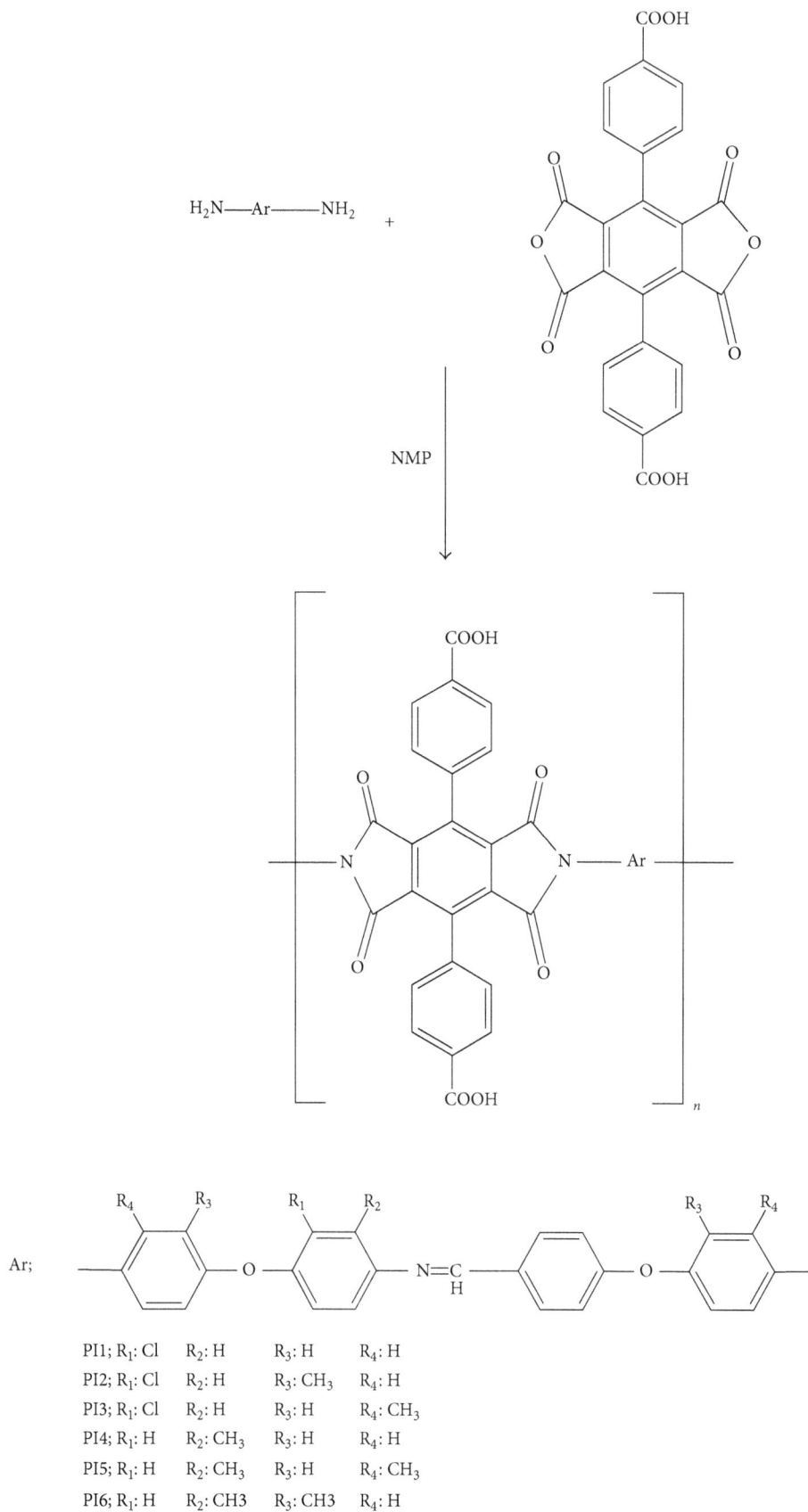

Ar;

PI1; R_1: Cl R_2: H R_3: H R_4: H

PI2; R_1: Cl R_2: H R_3: CH$_3$ R_4: H

PI3; R_1: Cl R_2: H R_3: H R_4: CH$_3$

PI4; R_1: H R_2: CH$_3$ R_3: H R_4: H

PI5; R_1: H R_2: CH$_3$ R_3: H R_4: CH$_3$

PI6; R_1: H R_2: CH3 R_3: CH3 R_4: H

SCHEME 2

TABLE 1: FTIR, ^{1}H NMR of polyimides (PI-1–PI-6).

Compound	IR (KBr)/cm^{-1}	^{1}H NMR (DMSO-d$_6$, δ, ppm) (s; singlet, m; multiplet)
PI-1	1725, (CO)$_{sym}$, 1788 (CO)$_{asym}$ 1665 (C=N)	13.01 (s, 2H, COOH) 6.78–7.90 (m, aromatic protons); 8.65 (s, 1H, azomethine)
PI-2	1718, (CO)$_{sym}$, 1772 (CO)$_{asym}$ 1650 (C=N)	13.21 (s, 2H, COOH); 6.65–7.89 (m, aromatic protons); 8.15 (s, 1H, azomethine)
PI-3	1725, (CO)$_{sym}$, 1778 (CO)$_{asym}$ 1664 (C=N)	13.12 (s, 2H, COOH) 6.41–7.90 (m, aromatic protons); 8.52 (s, 1H, azomethine)
PI-4	1715 (CO)$_{sym}$, 1782 (CO)$_{asym}$ 1657 (C=N)	12.98 (s, 2H, COOH) 6.75–7.97 (m, aromatic protons); 8.15 (s, 1H, azomethine)
PI-5	1725 (CO)$_{sym}$, 1775 (CO)$_{asym}$ 1665 (C=N)	12.99 (s, 2H, COOH) 6.45–7.78 (m, aromatic protons); 8.62 (s, 1H, azomethine);
PI-6	1717 (CO)$_{sym}$, 1780 (CO)$_{asym}$ 1638 (C=N)	13.08 (s, 2H, COOH) 6.35–7.88 (m, aromatic protons); 8.25 (s, 1H, azomethine)

In DMSO-d$_6$ at 295 K.

For thermal treatment polyamic acid solution was poured in glass petri dishes and placed in oven at 90°C overnight for film preparation. Further the film was treated by heating at 150°C for 30 min, 200°C for 30 min, and 280°C for 1 h [37].

3. Results and Discussion

Six new different polyimides were synthesized by polycondensation method. All the synthesized diamines were characterized by IR, NMR, and elemental analysis. FT-IR and NMR spectroscopic techniques confirmed the structures of diamines monomers and polyimides.

3.1. IR Spectroscopy. FT-IR analysis and spectral data confirmed the chemical structure of monomers diamines (DA1–DA6) and their respective polyimides. The data is presented in Table 1. The presence of imide ring was confirmed by the characteristic bands at 1788–1715 cm^{-1} for (CO)$_{asym}$ and (CO)$_{sym}$ stretching, respectively. The dehydration cyclization of polyamic acid (PAA) to form an imide ring was confirmed by the disappearance of the band at 1690 cm^{-1} (related to C=O of amic acid). Further the formation of polyimides was confirmed by the absence of N–H vibration at 3500 and 3200 cm^{-1}. Characteristic band at 1627–1664 cm^{-1} corresponding to the azomethine group (–CH=N–) indicated the presence of azomethine moiety in polyimides as shown in Figure 1. The disappearance of the amide and carbonyl bands indicated virtually a complete conversion of the imide ring in the resulting polyimide [38].

3.2. ^{1}H NMR Spectroscopy. Formation of diamines and their respective polyimides were confirmed by ^{1}H NMR. ^{1}H NMR confirmed the reduction of nitro compounds into

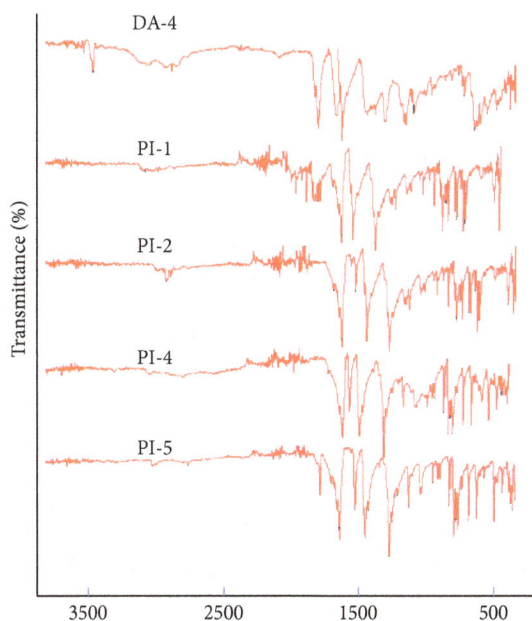

FIGURE 1: Spectra of DA-4, PI-1, PI-2, PI-4, and PI-5.

amines by high field shift of aromatic protons. Signals due to NH$_2$ protons were found near 5.11–6.12. Characteristic signals of C=N– at 8.15–8.66 in spectra of diamines and polymer confirmed the presence of azomethine moiety. The peaks between 2.06 and 2.20 ppm confirmed the presence of aliphatic proton structure in the polymer, that is, methyl. The aromatic proton peaks can be observed between 6.30 and 8.00 ppm regions. The characteristic signal of amino group disappears in the ^{1}H NMR of polyimides confirm the synthesis of polyimides and complete imidization [39].

TABLE 2: Solubility and inherent viscosity of polyimides (PI-1–PI-6).

Polyimide	DMSO	DMAc	DMF	THF	m-cresol	H_2SO_4	η_{inn} (dL/g)
PI-1	++	+	+	−	+	+++	0.68
PI-2	++	+	−	−	++	+++	0.98
PI-3	+++	−	++	+	++	+++	1.01
PI-4	++	+	+	−	++	+++	0.72
PI-5	++	+	+	+	++	+++	0.94
PI-6	+++	+	−	+	++	+++	1.04

+++ = soluble at room temperature, ++ = soluble on heating, + = slightly soluble on heating, − = insoluble.

TABLE 3: Thermal stability of the polyimides (PI-1–PI-6).

Polymer code	T_g (°C)	Thermal stability			
		T_5 (°C)[a]	T_{10} (°C)[b]	T_{max} (°C)[c]	Char yield (%)[d]
PI-1	250	455	480	500	58
PI-2	325	470	498	498	60
PI-3	270	498	525	495	65
PI-4	280	460	485	525	59
PI-5	335	490	515	520	63
PI-6	310	498	520	496	64

[a]T_5%, [b]T_{10}%: temperatures at 5, 10% weight loss, respectively.
[c]Temperature of maximum decomposition rate.
[d]Residual weight when heated to 600°C in nitrogen.

3.3. Viscosity Measurement. Viscosities of all the synthesised polyimides were obtained by using Gilmount falling ball viscometer in H_2SO_4 at 30°C. The inherent viscosities of the synthesized polyimides were calculated between the ranges of 0.68 and 1.04 dL/g indicated that polymers have moderate to higher molecular weights. Most of the polyimides showed higher value of inherent viscosity than other reported polyimides having azomethine linkages [40, 41].

3.4. Organosolubility of Polymers. Solubility of the synthesized polyimides was investigated in different solvents like DMSO, DMAc, DMF, m-cresol, and THF and summarized in Table 2. Polyimides were found to be soluble in most of the polar protic solvents. The improvement in solubility is attributed to flexibility induced by incorporated moieties along with carboxyl group in polyimide structure which is related with intermolecular interaction and packing of polyimide films. Some polyimides were soluble at room temperature and some were soluble on heating. The ether linkages along with other aliphatic substitutions like methyl and chloro also helpful in increasing solubility of polyimides [10, 24].

3.5. Moisture Absorption. The polyimides were also subjected to analyse the moisture absorption capacity through change in weight of dried polyimide before and after immersion in distilled water at 25°C for 24 hrs. The moisture absorption of the polyimides was in the range from 1.02 to 1.28%. That increase in water uptake of the polymers is might be due to the incorporation of carboxyl group [10].

3.6. Thermal Behavior. The thermal properties of the synthesized polyimides were evaluated by TGA and DSC at heating rate of 10°C min^{-1}. The obtained data is summarized in Table 3. Thermal properties of the compounds are strongly depended on their chemical structure. Polyimides showed excellent thermal properties (Figure 2). They were found thermally stable up to 480–525°C from the 10% weight loss, T_{max} of ca. 495–525°C and char yields of 58–65 at 600°C. The loss of weight in the first degradation step may be due to the −COOH group. So it may be predicted that first step was decarboxylation step and showed the decomposition of polyimide backbone. Dehydration behaviour of polymers was also observed by loss of 2-3% of weight. This might be due to water uptake ability of polyimides. Taking into account the results from TGA analysis, it was found that the presented polyimides possess good thermal stability without significant weight losses up to 380°C. This implies that no thermal decomposition occurs below this temperature and that the onset decomposition temperature was as high as 400°C for the all polyimides. The glass transition temperature (T_g) of the PIs samples, which is one of the key parameters of polymers when considering the high-temperature devices fabrication and the long-term heat releasing environment was taken as the midpoint of the change in slope of the baseline in DSC curves. The glass transition temperature T_g as a second order endothermic transition could be considered as the temperature at which a polymer undergoes extensive cooperative segmental motion along the backbone. The flexible linkages decreased the energy of internal rotation and lowered the T_g [42, 43]. Different intra- and intermolecular interactions including hydrogen bonding, electrostatic and ionic forces,

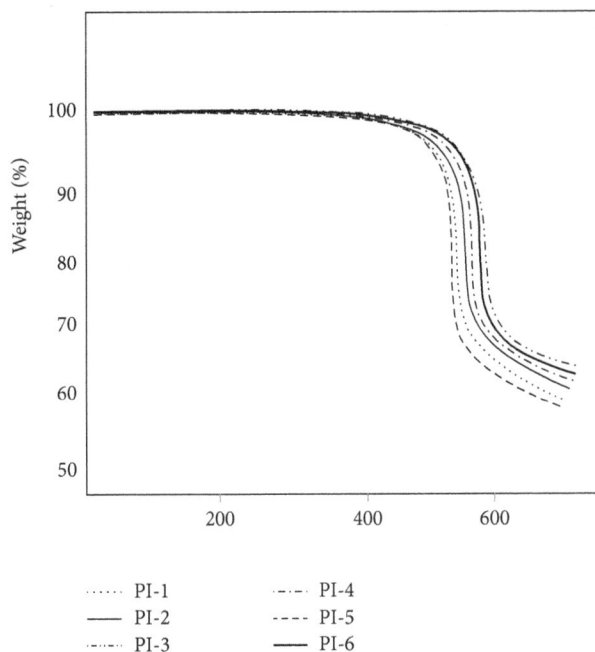

FIGURE 2: TGA curves of polyimides (PI-1 to PI-6).

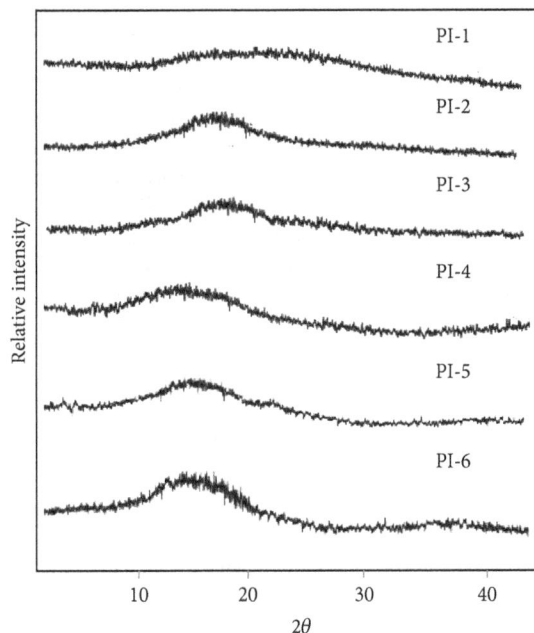

FIGURE 3: X-ray diffraction pattern of polyimides (PI-1 to PI-6).

chain packing efficiency, and chain stiffness also affected the T_g. The T_g of the polyimides (PI-1–PI-6) were observed in the range of 250–335°C, depending upon the structure of diamine component along with dianhydride and decreased with the decreasing of rigidity of the polymer backbone. (PI-5) exhibited the highest T_g (335°C) owing to the highly rigid pyromellitimide unit along with azomethine backbone with methyl substituents.

3.7. X-Ray Diffraction. X-ray diffraction analysis was also carried out for the characterization of polyimides (Figure 3). All the polyimides showed broad peaks in wide-angle X-ray diffraction pattern indicated the amorphous nature of polyimides. It might be due to strong chain interaction between the acid groups, which could not help in building regular structure and disturb chain packing of the polyimides. The strong interaction among carboxyl group hindered the chain mobility. The bulky phenyl groups inside the polymer chain with ether linkages might have introduced the amorphous characteristic in these polyimides. The amorphous nature of polyimides is responsible for the improvement in solubility of polyimides [10, 24].

4. Conclusion

A new series of polyimides were synthesized by condensation method of newly synthesized diamines having ether and azomethine linkages with resynthesized dianhydride. All the polyimides were found soluble in most of the polar aprotic solvents and have inherent viscosities in the range of 0.68–1.04 dL/g indicated the moderate to higher molecular weights of polyimides. All the polyimides were found thermally stable upto 498–523°C. All the polyimides have amorphous nature

indicated by the wide-angle X-ray diffraction analysis. On the basis of such results it is concluded that these polyimides could be considered for new processable high performance polymeric materials.

Conflict of Interests

The authors declare that there is no conflict of interests regarding the publication of this paper.

Acknowledgments

The authors are grateful to the Higher Education Commission of Pakistan for financial support under indigenous scholarship program for Ph.D. [Pin no.106-2119-PS6-029].

References

[1] M. K. Ghosh and K. L. Mittal, *Polyimides: Fundamentals and Applications*, Marcel Dekker, New York, NY, USA, 1996.

[2] C. E. Sroog, "Polyimides," *Journal of Polymer Science Macromolecular Reviews*, vol. 11, no. 1, pp. 161–208, 1976.

[3] F. Li, S. Fang, J. J. Ge et al., "Diamine architecture effects on glass transitions, relaxation processes and other material properties in organo-soluble aromatic polyimide films," *Polymer*, vol. 40, no. 16, pp. 4571–4583, 1999.

[4] R. Rubner, "Photoreactive polymers for electronics," *Advanced Materials*, vol. 2, pp. 452–457, 1990.

[5] T. Fukushima, Y. Kawakami, T. Oyama, and M. Tomoi, "Photosensitive polyetherimide (Ultem) based on reaction development patterning (RDP)," *Journal of Photopolymer Science and Technology*, vol. 15, no. 2, article 191, 2002.

[6] T. Fukushima, T. Oyama, T. Iijima, M. Tomoi, and H. Itatani, "New concept of positive photosensitive polyimide: Reaction

development patterning (RDP)," *Journal of Polymer Science A: Polymer Chemistry*, vol. 39, pp. 3451–3463, 2001.

[7] H. S. Li, J. G. Liu, J. M. Rui, L. Fan, and S. Y. Yang, "Synthesis and characterization of novel fluorinated aromatic polyimides derived from 1,1-bis(4-amino-3,5-dimethylphenyl)-1-(3,5-ditrifluoromethylphenyl)-2,2,2-trifluoroethane and various aromatic dianhydrides," *Journal of Polymer Science A: Polymer Chemistry*, vol. 44, no. 8, pp. 2665–2674, 2006.

[8] C. P. Yang, Y. Y. Su, and F. Z. Hsiao, "Synthesis and properties of organosoluble polyimides based on 1,1-bis[4-(4-amino-2-trifluoromethylphenoxy)phenyl]cyclohexane," *Polymer*, vol. 45, no. 22, pp. 7529–7538, 2004.

[9] K. H. Choi, K. H. Lee, and J. G. C. Jung, "Synthesis of new poly(amide imide)s with (n-alkyloxy)phenyloxy side branches," *Journal of Polymer Science Part A: Polymer Chemistry*, vol. 39, no. 21, pp. 3818–3825, 2001.

[10] J. C. Choia, H. S. Kima, B. H. Sohna, W. C. Zina, and M. Reea, "Synthesis and characterization of new alkali-soluble polyimides and preparation of alternating multilayer nanofilms therefrom," *Polymer*, vol. 45, no. 5, pp. 1517–1524, 2004.

[11] T. Fukushima, K. Hosokawa, T. Oyama, T. Iijima, M. Tomoi, and H. Itatani, "Synthesis and positive-imaging photosensitivity of soluble polyimides having pendant carboxyl groups," *Journal of Polymer Science A: Polymer Chemistry*, vol. 39, no. 6, pp. 934–946, 2001.

[12] M. Ueda and T. A. Nakayama, "A new negative-type photosensitive polyimide based on poly(hydroxyimide), a cross-linker, and a photoacid generator," *Macromolecules*, vol. 29, no. 20, pp. 6427–6431, 1996.

[13] D. J. Liaw, F. C. Chang, M. K. Leung, M. Y. Chou, and M. Klaus, "High thermal stability and rigid rod of novel organosoluble polyimides and polyamides based on bulky and noncoplanar naphthalene-biphenyldiamine," *Macromolecules*, vol. 38, pp. 4024–4029, 2005.

[14] B. Liu, W. Hu, T. Matsumoto, Z. Jiang, and S. J. Ando, "Synthesis and characterization of organosoluble ditrifluoromethylated aromatic polyimides," *Journal of Polymer Science Part A: Polymer Chemistry*, vol. 43, no. 14, pp. 3018–3029, 2005.

[15] S. Tamai, A. Yamaguchi, and M. Ohta, "Melt processible polyimides and their chemical structures," *Polymer*, vol. 37, no. 16, pp. 3683–3692, 1996.

[16] C.-P. Yang, S.-H. Hsiao, and M.-F. Hsu, "Organosoluble and light-colored fluorinated polyimides from 4,4'-bis(4-amino-2-trifluoromethylphenoxy)biphenyl and aromatic dianhydrides," *Journal of Polymer Science A: Polymer Chemistry*, vol. 40, no. 4, pp. 524–534, 2002.

[17] C. P. Yang, R. S. Chen, and K. H. Chen, "Effects of diamines and their fluorinated groups on the color lightness and preparation of organosoluble aromatic polyimides from 2,2-bis[4-(4-amino-2-trifluoromethylphenoxy)phenyl]-hexafluoropro-pane," *Journal of Polymer Science A: Polymer Chemistry*, vol. 41, no. 7, pp. 922–938, 2005.

[18] T. M. Moy, C. D. Deporter, and J. E. McGrath, "Synthesis of soluble polyimides and functionalized imide oligomers via solution imidization of aromatic diester-diacids and aromatic diamines," *Polymer*, vol. 34, no. 4, pp. 819–824, 1993.

[19] J. Xu, C. He, and T. S. Chung, "Synthesis and characterization of soluble polyimides derived from [1,1';4',1'']terphenyl-2',5'-diol and biphenyl-2,5-diol," *Journal of Polymer Science Part A: Polymer Chemistry*, vol. 39, no. 17, pp. 2998–3007, 2001.

[20] H. B. Zhang and Z. Y. Wang, "Polyimides derived from novel unsymmetric dianhydride," *Macromolecules*, vol. 33, no. 12, pp. 4310–4312, 2000.

[21] D. M. Hergenrother, K. A. Watson, J. G. Smith, J. W. Connel, and R. Yokota, "Polyimides from 2,3,3',4'-biphenyltetracarboxylic dianhydride and aromatic diamines," *Polymer*, vol. 43, no. 19, pp. 5077–5093, 2002.

[22] X. Z. Fang, Q. X. Li, Z. Wang, Z. H. Yang, L. X. Gao, and M. X. J. Ding, "Synthesis and properties of novel polyimides derived from 2,2,3,3-benzophenonetetracarboxylic dianhydride," *Journal of Polymer Science A: Polymer Chemistry*, vol. 42, pp. 2130–2144, 2004.

[23] S. H. Hsiao and K. H. J. Lin, "Polyimides derived from novel asymmetric ether diamine," *Journal of Polymer Science A: Polymer Chemistry*, vol. 43, pp. 331–341, 2005.

[24] Y. Shao, Y. F. Li, X. Zhao, X. L. Wang, T. Ma, and F. C. Yang, "Synthesis and properties of fluorinated polyimides from a new unsymmetrical diamine: 1,4-(2'-Trifluoromethyl-4',4'-diaminodiphenoxy)benzene," *Journal of Polymer Science Part A: Polymer Chemistry*, vol. 44, no. 23, pp. 6836–6846, 2006.

[25] D. S. Reddy, C. H. Chou, C. F. Shu, and G. H. Lee, "Synthesis and characterization of soluble poly(ether imide)s based on 2,2'-bis(4-aminophenoxy)-9,9'-spirobifluorene," *Polymer*, vol. 44, no. 3, pp. 557–563, 2003.

[26] G. I. Rusu, A. Airinei, M. Rusu et al., "On the electronic transport mechanism in thin films of some new poly(azomethine sulfone)s," *Acta Materials*, vol. 55, pp. 433–442, 2007.

[27] A. Zabulica, E. Perju, M. Bruma, and L. Marin, "Novel luminescent liquid crystalline polyazomethines. Synthesis and study of thermotropic and photoluminescent propertie," *Liquid Crystal*, vol. 41, no. 2, pp. 252–262, 2014.

[28] L. Marin, V. Cozan, and M. Bruma, "Comparative study of new thermotropic polyazomethines," *Polymers for Advanced Technologies*, vol. 17, pp. 664–672, 2006.

[29] L. Marin, M. Dana, and D. Damaceanu, "New thermotropic liquid crystalline polyazomethines containing luminescent mesogens," *Soft Materials*, vol. 7, pp. 1–20, 2009.

[30] C. H. Li and T. C. Chang, "Thermotropic liquid crystalline polymer. III. Synthesis and properties of poly(am ide-azomethineester)," *Journal of Applied Polymer Science A: Polymer Chemistry*, vol. 29, pp. 361–367, 1991.

[31] A. Atta, "Alternating current conductivity and dielectric properties of newly prepared poly(bis thiourea sulphoxide)," *International Journal of Polymeric Materials*, vol. 52, no. 5, pp. 361–372, 2003.

[32] J. D. D'Cruz, T. K. Venkatachalam, and F. M. Uckun, "Novel thiourea compounds as dual-function microbicides," *Biology of Reproduction*, vol. 63, no. 1, pp. 196–205, 2000.

[33] M. W. Sabaa, R. R. Mohamed, and A. A. Yassin, "Organic thermal stabilizers for rigid poly(vinyl chloride) VIII. Phenylurea and phenylthiourea derivatives," *Polymer Degradation and Stability*, vol. 81, no. 1, pp. 37–45, 2003.

[34] I. Dehri and M. Ozcan, "The effect of temperature on the corrosion of mild steel in acidic media in the presence of some sulphur-containing organic compounds," *Materials Chemistry and Physics*, vol. 98, pp. 316–323, 2006.

[35] M. Özcan, I. Dehri, and M. Erbil, "Organic sulphur-containing compounds as corrosion inhibitors for mild steel in acidic media: correlation between inhibition efficiency and chemical structure," *Applied Surface Science*, vol. 236, pp. 155–164, 2004.

[36] T. K. Venkatachalam, E. Sudbeck, and F. M. Uckun, "Structural influence on the solid state intermolecular hydrogen bonding of substituted thioureas," *Journal of Molecular Structure*, vol. 751, no. 1–3, pp. 41–54, 2005.

[37] D. D. Perrin, W. L. F. Armarego, and D. R. Perrin, *Drying of Solvents and Laboratory Chemicals. Purification of Laboratory Chemicals*, Pergamon, 2nd edition, 1980.

[38] S. H. Hsio, C. P. Yang, and S. H. Chen, "Synthesis and properties of ortho-linked aromatic polyimides based on 1,2-bis(4-aminophenoxy)-4-tert-butylbenzene," *Journal of Polymer Science A: Polymer Chemistry*, vol. 38, pp. 1551–1559, 2000.

[39] S. J. Zhang, Y. F. Li, X. L. Wang, D. X. Yin, Y. Shao, and X. Zhao, "Synthesis and characterization of novel polyimides based on pyridine-containing diamine," *Chinese Chemical Letters*, vol. 16, no. 9, pp. 1165–1168, 2005.

[40] D. L. Pavia, G. M. Lampman, and G. S. Kriz, *Introduction to Spectroscopy*, Harcourt Brace College, San Diego, Calif, USA, 1996.

[41] H. S. Hsiao, C. P. Yang, and C. L. Chung, "Synthesis and characterization of novel fluorinated polyimides based on 2,7-bis(4-amino-2-trifluoromethylphenoxy)naphthalene," *Journal of Polymer Science A: Polymer Chemistry*, vol. 41, p. 2001, 2003.

[42] G. R. Srinivasa, S. N. Narendra Babu, C. Lakshmi, and D. C. Gowda, "Conventional and microwave assisted hydrogenolysis using zinc and ammonium formate," *Synthetic Communications*, vol. 3, pp. 1831–1837, 2004.

[43] H. Behniafar and H. Ghorbani, "New heat stable and processable poly(amide–ether–imide)s derived from 5-(4-trimellitimidophenoxy)-1-trimellitimido naphthalene and various diamines," *Polymer Degradation and Stability*, vol. 93, no. 3, pp. 608–617, 2008.

16

Structural Conformational Study of Eugenol Derivatives Using Semiempirical Methods

Radia Mahboub

Department of Chemistry, Faculty of Sciences, University of Tlemcen, BP 119, 13000 Tlemcen, Algeria

Correspondence should be addressed to Radia Mahboub; radiamahboub@yahoo.com

Academic Editor: Maria Roca

We investigated the conformational structure of eugenol and eugenyl acetate under torsional angle effect by performing semiempirical calculations using AM1 and PM3 methods. From these calculations, we have evaluated the strain energy of conformational interconversion. To provide a better estimate of stable conformations, we have plotted the strain energy versus dihedral angle. So, we have determined five geometries of eugenol (three energy minima and two transition states) and three geometries of eugenyl acetate (two energy minima and one transition state). From the molecular orbital calculations, we deduce that the optimized *trans* form by AM1 method is more reactive than under PM3 method. We can conclude that both methods are efficient. The AM1 method allows us to determine the reactivity and PM3 method to verify the stability.

1. Introduction

Eugenol (4-allyl-2-methoxyphenol) is a phenylpropene, an allyl chain-substituted guaiacol. It is the main phenolic compound extracted from certain essential oils especially from clove oil, nutmeg, cinnamon, basil, and bay leaf [1–14]. Eugenol is a phenol derivative used in many areas such as perfumes, flavorings agent, and dental materials. It is used as an antiseptic, analgesic, fungicide, bactericide, insecticide, anticarcinogenic, antiallergic, antioxidant, anti-inflammatory, and so forth [15–17]. As derivative, the eugenyl acetate was characterized and its structural properties have investigated by Dos Santos et al. [18, 19].

To our knowledge, a study of the conformational structure as a function of the dihedral angle was not reported. In the present paper, we investigated the conformational structure of eugenol and eugenyl acetate under torsional angle effect by performing semiempirical calculations using AM1 and PM3 methods. From these calculations, we have evaluated the strain energy of conformational interconversion to provide a better estimate of stable conformations. These results can be used to make future applications possible (Figure 1).

2. Methodology

Molecular modeling of the optimized eugenol and eugenyl acetate was carried out with the use of an efficient program for molecular mechanics (MM). Calculations are performed for all optimized geometries using AM1 and PM3 methods. The main molecular properties to characterize the geometry structures and the molecular orbital of the eugenyl acetate were calculated and compared. For each method, the geometry of the compound was optimized by using the Polak-Ribiere conjugate gradient algorithm with a gradient of 0.01 Kcal/mol (RMS). The following quantum chemical results are considered: heat of formation (ΔH_f), total energy (E_t), minimum energy of conformation ($E_{min,conf}$), strain energy of conformational interconversion (E_s), energy of highest occupied molecular orbital (HOMO), energy of lowest unoccupied molecular orbital (LUMO), and HOMO-LUMO energy gaps (EG).

3. Results and Discussion

Molecular geometries of eugenol and eugenyl acetate were optimized by semiempirical molecular orbital method (AM1

TABLE 1: Main calculated properties of eugenol and eugenyl acetate with semiempirical methods.

Entry	Properties	Eugenol							
		AM1				PM3			
		Trans	Eclipsed	Gauche	Cis	Trans	Eclipsed	Gauche	Cis
1	ΔH_f	−43.562	−45.560	−45.540	−38.360	−44.474	−45.230	−45.214	−40.236
2	E_t	−48090.984	−48092.980	−48092.960	−48085.781	−45111.215	−45111.973	−45111.957	−45106.976
3	$E_{min,conf}$	−2496.805	−2498.802	−2498.782	−2491.602	−2497.716	−2498.472	−2498.456	−2493.478
4	E_s	5.202	7.200	7.180	0.000	4.238	4.994	4.978	0.000
5	HOMO	−8.599	−8.614	−8.606	−8.592	−8.670	−8.703	−8.701	−8.666
6	LUMO	0.327	0.332	0.342	0.338	0.254	0.250	0.260	0.263
7	EG	8.926	8.946	8.948	8.930	8.924	8.953	8.961	8.929
Entry	Properties	Eugenyl acetate							
		AM1				PM3			
		Trans	Eclipsed	Gauche	Cis	Trans	Eclipsed	Gauche	Cis
8	ΔH_f	−68.176	−77.344	−77.356	−77.356	−77.728	−83.323	−83.330	−83.328
9	E_t	−62008.605	−62017.772	−62017.785	−62017.785	−58055.633	−58061.226	−58061.234	−58061.230
10	$E_{min,conf}$	−3026.961	−3036.129	−3036.140	−3036.140	−3036.514	−3042.108	−3042.115	−3042.113
11	E_s	9.180	0.012	0.000	0.000	5.600	0.005	−0.002	0.000
12	HOMO	−8.852	−9.215	−9.205	−9.205	−8.925	−9.302	−9.297	−9.299
13	LUMO	0.001	−0.134	−0.130	−0.130	−0.044	−0.191	−0.188	−0.189
14	EG	8.852	9.081	9.075	9.075	8.881	9.101	9.109	9.110

The strain energy (E_s) for each geometry of a molecule is defined as the difference between the minimum energy of conformation for that geometry and the most stable conformation of the molecule.

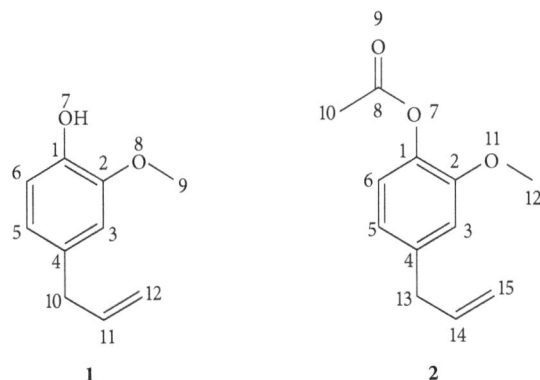

FIGURE 1: Conformation structures of *cis*-eugenol and *cis*-eugenyl acetate.

and PM3). The semiempirical simulations results for structure optimization of eugenol and eugenyl acetate are given in Table 1. The conformational interconversion energy-minimum of eugenyl acetate and eugenol was investigated in detail by changing different torsional angles. From these studies, we have determined five geometries of eugenol (three energy minima and two transition states) and three geometries of eugenyl acetate (two energy minima and one transition state). These geometries are important in the description of the conformational properties of our systems (Figure 2).

We have obtained the curves plotted in Figure 2 from the simulated data using the nonlinear fitting process based on the Levenberg-Marquardt algorithm implemented in the Origin v. 6.0. Software [20].

The dihedral angle for rotation about C_4–C_{10} bond in eugenol has several stationary points. A/A′, C, C′, and E/E′ are minima and B, B′ and D, D′ are maxima. Only the structures at the minima represent stable species and of these, the *syn* conformation is more stable than the *anti*. The *gauche* and the *eclipsed* represent the transition states. In eugenyl acetate, the stationary points A/A′, B/B′, C/C′, E/E′, F/F′, and G/G′ are minima and D, D′ are maxima. The *anti* conformation represents the transition state while the *syn*, *gauche*, and *eclipsed* conformations are stable species. So, the deformation around C_4–C_{13} in eugenyl acetate remains unchanged and is not influenced by torsional angle effect.

To provide a better estimate of conformations, we should search the conformational space in reasonable computing time. So, we run the simulations; then we run a geometry optimization on each structure. Thus, we have grouped the resulting structures in Figure 3. First, we observe that all geometries from *cis* conformations obtained after optimization present deformation mainly on branching allyl. The torsional angle value varies around 136°. This situation is due to methylene group (sp³ hybridization) which gives a non-coplanar final geometry. Then, the π-bond of branching allyl is situated in the same side that acetate group. Second, we note that the geometry in *anti* conformations stays unchanged after optimization. All the substituents of aromatic ring, acetate and methoxy groups then branching allyl, are situated in the same plane (Φtrans: 180°).

From our molecular orbital calculations, we want to deduce the structure-reactivity relationship depending on different conformations. First, AM1 and PM3 calculations show that the *cis* forms are favored (Table 1, entry 7, EG: 9.075, 9.110 ev). On the other hand, the same calculations show that

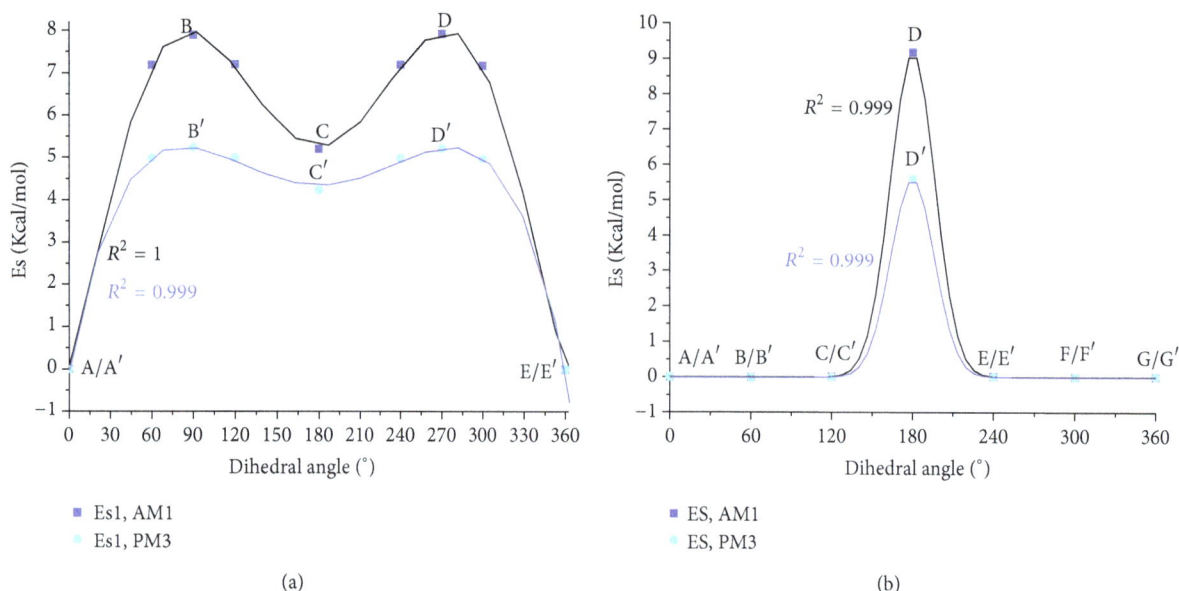

FIGURE 2: Calculated strain energy for conformational interconversion with semiempirical AM1 and PM3 methods. (a) Eugenol and (b) eugenyl acetate.

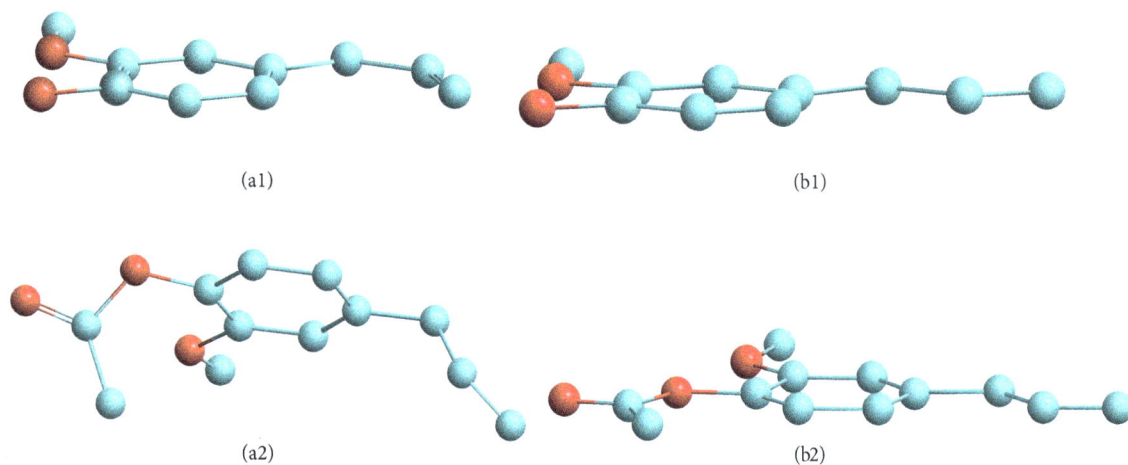

FIGURE 3: Estimated conformation structures of eugenol and eugenyl acetate; (a1) and (a2) *cis* forms, (b1) and (b2) *trans* forms. Molecule orientation was chosen around *x*-axis. Hydrogen atoms are omitted for clarity.

the most active site of the nucleophilic reaction is located on the oxygens of the acetate group and methoxy, and the most active site of the electrophilic reaction is C8 position of eugenyl acetate.

The AM1 and PM3 calculations show that the *trans* form is most active (Table 1, entry 7, EG: 8.852; 8.881 ev). Thus, these results reveal that the instability is caused by the high activity of methylene group in the strand allyl which provides to the aromatic ring another nucleophilic reaction site. This one is stabilized by resonance with the double bonds of the aromatic ring and the allylic radical (Figure 4). So, we note that the value of the strain energy obtained with AM1 method (Table 1, entry 4, Es: 9.180 ev) is higher than that obtained by PM3 calculation (Table 1, entry 4, Es: 5.600 ev). Consequently, we deduce that the optimized *trans* form by AM1 method is more

reactive than under PM3 method. We can conclude that the efficient method for the eugenyl acetate is the semiempirical method AM1.

These observations remain the same for eugenol, except for the energy gaps. From Figure 2(a), we note that the *trans* form (C or C′) is near the transition states B and D, respectively, B′ and D′. This situation is clearer when the eugenol is optimized by PM3 (Table 1, entries 5 and 6, PM3: HOMO −8.599, LUMO 0.327 ev; AM1: HOMO −8.670, LUMO 0.254 ev). So, this *trans* form (C or C′) corresponds to the reaction intermediate present in eugenol and absent in eugenyl acetate (AM1: HOMO −8.606, LUMO 0.343; PM3: HOMO −8.702, LUMO 0.260 ev).

Furthermore, we also see that the HOMO is located at the oxygen sites whereas for the molecule the Homo is

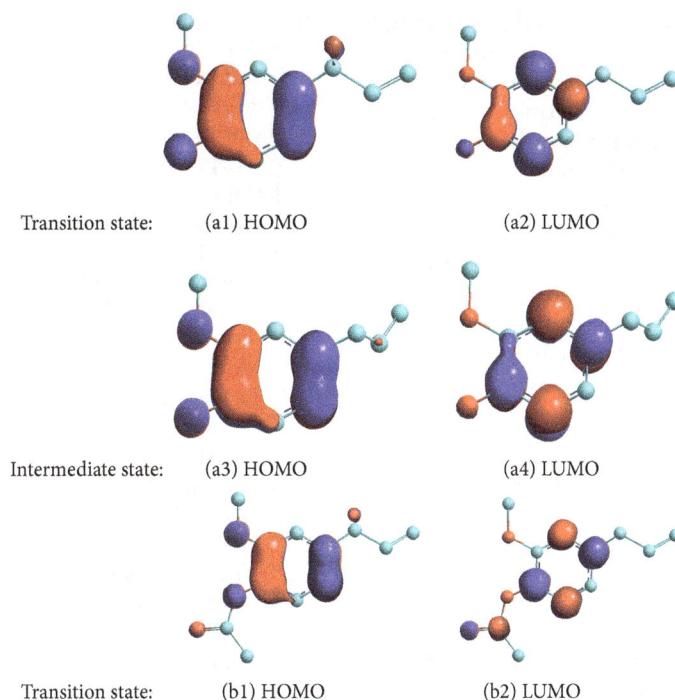

FIGURE 4: Molecular orbital calculated for eugenol and eugenyl acetate by semiempirical methods (AM1 and PM3). Contour values: 0.05 Å^{-3}. Blue lines represent positive contours. Red lines represent negative contours. Hydrogen atoms are omitted for clarity.

TABLE 2: Mulliken charges of the optimized structures of eugenol and eugenyl acetate.

Compound	State	Method	Type charge
Eugenol	Transition	AM1	C_1 0.065, C_2 0.058, C_3 −0.148, C_4 −0.078, C_5 −0.103, C_6 −0.165, O_7 −0.208, O_8 −0.166, C_9 0.047, C_{10} −0.019, C_{11} −0.134, C_{12} −0.168.
		PM3	C_1 0.063, C_2 0.063, C_3 −0.164, C_4 −0.072, C_5 −0.095, C_6 −0.169, O_7 −0.208, O_8 −0.166, C_9 0.047, C_{10} −0.017, C_{11} −0.143, C_{12} −0.170.
	Intermediate	AM1	C_1 0.054, C_2 0.058, C_3 −0.170, C_4 −0.067, C_5 −0.124, C_6 −0.185, O_7 −0.230, O_8 −0.187, C_9 0.079, C_{10} −0.096, C_{11} −0.156, C_{12} −0.224.
		PM3	C_1 0.065, C_2 0.058, C_3 −0.147, C_4 −0.078, C_5 −0.095, C_6 −0.169, O_7 −0.208, O_8 −0.166, C_9 0.047, C_{10} −0.019, C_{11} −0.134, C_{12} −0.168.
Eugenyl acetate	Transition	AM1	C_1 0.057, C_2 0.079, C_3 −0.197, C_4 −0.043, C_5 −0.136, C_6 −0.168, O_7 −0.190, C_8 0.306, O_9 −0.291, C_{10} −0.253, O_{11} −0.178, C_{12} −0.081, C_{13} −0.098, C_{14} −0.157, C_{15} −0.226.
		PM3	C_1 0.042, C_2 0.081, C_3 −0.174, C_4 −0.054, C_5 −0.113, C_6 −0.161, O_7 −0.156, C_8 0.351, O_9 −0.327, C_{10} −0.139, O_{11} −0.160, C_{12} 0.046, C_{13} −0.020, C_{14} −0.145, C_{15} −0.166.

distributed along the aromatic cycle site (Figure 4, Table 2). This clearly shows the high reactivity of eugenol compared to its corresponding acetate. This reactivity is due to mobility of hydrogen and the nucleophilicity on the aromatic ring.

From calculating wave functions, we observe that the charge distributions are mainly located on electrowithdrawing oxygen atoms in each molecule. They also are situated on aromatic ring and the strand allyl. The charge density is much higher under AM1 than under PM3. So, these results are in accordance with their energy properties (see Table 1) and the electronic properties of each substituent. The acetate group is an electron withdrawing type which reduces the aromatic cycle charge while hydroxyl group is an electron donor type that can provide its charge to the aromatic ring and thus increases its nucleophilic effect (Table 2).

4. Conclusion

In the present work, we have studied the conformational structure of eugenol and eugenyl acetate under torsional angle effect by performing semiempirical calculations using AM1 and PM3 methods. From quantum calculations, we have evaluated the strain energy of conformational interconversion. To provide a better estimate of stable conformations, we have plotted the strain energy versus dihedral angle. So, we have determined five geometries of eugenol (three energy minima and two transition states) and three geometries of eugenyl acetate (two energy minima and one transition state). We have verified the presence of the intermediate form of eugenol which corresponds to the *trans* form (C or C′).

From the molecular orbital calculations, we deduce that the optimized *trans* form by AM1 method is more reactive than under PM3 method. We note that the charge distributions are mainly located on the aromatic ring and the strand allyl in each molecule. We can conclude that both methods are efficient. The AM1 method allows us to determine the reactivity and PM3 method to verify the stability.

Conflict of Interests

The author declares that there is no conflict of interest regarding the publication of this paper.

References

[1] E. Reverchon, "Supercritical fluid extraction and fractionation of essential oils and related products," *The Journal of Supercritical Fluids*, vol. 10, no. 1, pp. 1–37, 1997.

[2] W. Guan, S. Li, R. Yan, S. Tang, and C. Quan, "Comparison of essential oils of clove buds extracted with supercritical carbon dioxide and other three traditional extraction methods," *Food Chemistry*, vol. 101, no. 4, pp. 1558–1564, 2007.

[3] M. N. I. Bhuiyan, J. Begum, N. C. Nandi, and F. Akter, "Constituents of the essential oil from leaves and buds of clove (*Syzygium caryophyllatum* (L.) Alston)," *African Journal of Plant Science*, vol. 4, pp. 451–454, 2010.

[4] S. M. Palacios, A. Bertoni, Y. Rossi, R. Santander, and A. Urzúa, "Efficacy of essential oils from edible plants as insecticides against the house fly, *Musca domestica* L.," *Molecules*, vol. 14, no. 5, pp. 1938–1947, 2009.

[5] M. H. Alma, M. Ertaş, S. Nitz, and H. Kollmannsberger, "Chemical composition and content of essential oil from the bud of cultivated Turkish clove (*Syzygium aromaticum* L.)," *BioResources*, vol. 2, no. 2, pp. 265–269, 2007.

[6] A. K. Srivastava, S. K. Srivastava, and K. V. Syamsundar, "Bud and leaf essential oil composition of *Syzygium aromaticum* from India and Madagascar," *Flavour and Fragrance Journal*, vol. 20, no. 1, pp. 51–53, 2005.

[7] K.-G. Lee and T. Shibamoto, "Antioxidant property of aroma extract isolated from clove buds [*Syzygium aromaticum* (L.) Merr. et Perry]," *Food Chemistry*, vol. 74, no. 4, pp. 443–448, 2001.

[8] A. A. Clifford, A. Basile, and S. H. R. Al-Saidi, "A comparison of the extraction of clove buds with supercritical carbon dioxide and superheated water," *Fresenius' Journal of Analytical Chemistry*, vol. 364, no. 7, pp. 635–637, 1999.

[9] G. Della Porta, R. Taddeo, E. D'Urso, and E. Reverchon, "Isolation of clove bud and star anise essential oil by supercritical CO_2 extraction," *LWT—Food Science and Technology*, vol. 31, no. 5, pp. 454–460, 1998.

[10] F. N. Lugemwa, "Extraction of betulin, trimyristin, eugenol and carnosic acid using water-organic solvent mixtures," *Molecules*, vol. 17, no. 8, pp. 9274–9282, 2012.

[11] B. Jayawardena and R. M. Smith, "Superheated water extraction of essential oils from *Cinnamomum zeylanicum* (L.)," *Phytochemical Analysis*, vol. 21, no. 5, pp. 470–472, 2010.

[12] S. Ghosh, D. Chatterjee, S. Das, and P. Bhattacharjee, "Supercritical carbon dioxide extraction of eugenol-rich fraction from Ocimum sanctum Linn and a comparative evaluation with other extraction techniques: process optimization and phytochemical characterization," *Industrial Crops and Products*, vol. 47, pp. 78–85, 2013.

[13] S. Ghosh, D. Roy, D. Chatterjee, P. Bhattacharjee, and S. Das, "SFE as a superior technique for extraction of eugenol-rich fraction from *Cinnamomum tamala* Nees (Bay Leaf)-process analysis and phytochemical characterization," *International Journal of Biological, Life Science and Engineering*, vol. 8, no. 1, pp. 9–17, 2014.

[14] F. Memmou and R. Mahboub, "Composition of essential oil from fresh flower of clove," *Journal of Scientific Research in Pharmacy*, vol. 1, pp. 33–35, 2012.

[15] M. He, M. Du, M. Fan, and Z. Bian, "In vitro activity of eugenol against Candida albicans biofilms," *Mycopathologia*, vol. 163, no. 3, pp. 137–143, 2007.

[16] S. A. Guenette, F. Beaudry, J. F. Marier, and P. Vachon, "Pharmacokinetics and anesthetic activity of eugenol in male Sprague-Dawley rats," *Journal of Veterinary Pharmacology and Therapeutics*, vol. 29, no. 4, pp. 265–270, 2006.

[17] G. Blank, A. A. Adejumo, and J. Zawistowski, "Eugenol induced changes in the fatty acid content two *Lactobacillus* species," *Lebensmittel-Wissenschaft. Technology*, vol. 24, pp. 231–235, 1991.

[18] A. L. Dos Santos, G. O. Chierice, K. Alexander, and A. Riga, "Crystal structure determination for eugenyl acetate," *Journal of Chemical Crystallography*, vol. 39, no. 9, pp. 655–661, 2009.

[19] A. L. Dos Santos, G. O. Chierice, A. T. Riga, K. Alexander, and E. Matthews, "Thermal behavior and structural properites of plant-derived eugenyl acetate," *Journal of Thermal Analysis and Calorimetry*, vol. 97, no. 1, pp. 329–332, 2009.

[20] D. W. Marquardt, "An algorithm for least-squares estimation of nonlinear parameters," vol. 11, pp. 431–441, 1963.

17

Baphia nitida Leaves Extract as a Green Corrosion Inhibitor for the Corrosion of Mild Steel in Acidic Media

V. O. Njoku,[1] E. E. Oguzie,[2] C. Obi,[3] and A. A. Ayuk[2]

[1] *Department of Chemistry, Faculty of Science, Imo State University, PMB 2000, Owerri, Nigeria*
[2] *Department of Chemistry, Federal University of Technology Owerri, PMB 1526, Owerri, Nigeria*
[3] *Department of Pure and Industrial Chemistry, University of Port Harcourt, PMB 5323, Port Harcourt, Nigeria*

Correspondence should be addressed to V. O. Njoku; viconjoku@yahoo.com

Academic Editor: Ana Mornar

The inhibiting effect of *Baphia nitida* (BN) leaves extract on the corrosion of mild steel in $1\,M\,H_2SO_4$ and $2\,M\,HCl$ was studied at different temperatures using gasometric and weight loss techniques. The results showed that the leaves extract is a good inhibitor for mild steel corrosion in both acid media and better performances were obtained in $2\,M\,HCl$ solutions. Inhibition efficiency was found to increase with increasing inhibitor concentration and decreasing temperature. The addition of halides to the extract enhanced the inhibition efficiency due to synergistic effect which improved adsorption of cationic species present in the extract and was in the order $KCl < KBr < KI$ suggesting possible role of radii of the halide ions. Thermodynamic parameters determined showed that the adsorption of BN on the metal surface is an exothermic and spontaneous process and that the adsorption was via a physisorption mechanism.

1. Introduction

One of the most practical methods of preventing electrochemical corrosion is to isolate the metal surface from corrosive agents [1]. Of the many methods available, the use of corrosion inhibitors is usually the most appropriate method to achieve this objective [2–9]. These inhibitors could be in the form of organic, inorganic, precipitating, passivating, or volatile species. Generally, corrosion inhibitors may be divided into three broad classes, namely, oxidizing, precipitating, and adsorption inhibitors [10]. Adsorption inhibitors are usually organic substances containing heteroatoms with high electron density such as nitrogen, sulfur, and oxygen [11, 12] and the presence of unsaturated bonds or aromatic rings in the molecular structure of the inhibitor favors adsorption on corroding metal surface [13]. The adsorption is influenced by the nature and the surface charge of the metal, the type of corrosion media, and the molecular structure of the inhibitor [4]. Some corrosion inhibitors used in different media and for different metals and alloys decrease considerably the oxidation states of the corroding metals. In acid corrosion, inhibitor adsorption may lead to structural changes in the double layer, which could reduce the rates of either the anodic metal dissolution and the cathodic hydrogen ion reduction or both.

It is known that some corrosion inhibitors and their derivatives are toxic and pollute the environment [14]. There is therefore the need to explore new nontoxic, environmental friendly, ecologically acceptable and inexpensive corrosion inhibitor substitutes. Among the alternative corrosion inhibitors, natural products of plant origin have been shown to be quite efficient as corrosion inhibitors [15–19].

In this work, the inhibitory properties of leaf extracts of *Baphia nitida* on mild steel corrosion in $1\,M\,H_2SO_4$ and $2\,M\,HCl$ have been studied using gasometric and weight loss techniques. The plant popularly called camwood and also known as African sandalwood belongs to the family of *Leguminosae* and its wood is commonly used to make a red dye. Phytochemical analysis of the leaves detected tannins, flavonoids, and saponin glycosides [20]. The actions of the leaf extract as inhibitor in both acid media over a range of inhibitor concentration and solution temperature, as well as synergistic effects of halides, have been studied.

2. Experimental

2.1. Materials Preparation

2.1.1. Metal Specimen. Mild steel strips of compositions 0.05% C, 0.6% Mn, 0.36% P, and 0.03% Si remainder iron and dimensions 3 cm × 1.5 cm × 0.14 cm were used for gasometric and weight loss studies. The specimens were washed with distilled water, degreased by soaking in absolute ethanol, dried in acetone, and stored in moisture-free desiccators prior to use.

2.1.2. Reagents. Analytical grade reagents were utilized to prepare 1 M H_2SO_4 and 2 M HCl using distilled water.

2.1.3. Plant Extracts. The *Baphia nitida* leaves used were obtained locally and were dried to a constant weight in an oven at a temperature of 110°C, and then ground to fine powder. The *Baphia nitida* (BN) extract was prepared by adding 10 g of the powder into 250 mL of 1 M H_2SO_4 in a round bottom flask. The same was repeated for 2 M HCl. The resulting solutions were heated under reflux for 2 h and left to cool overnight, and then filtration was carried out using filter paper. From the respective stock solutions, inhibitor test solutions were prepared in the concentration range 5–100 mg/L.

2.2. Methods

2.2.1. Gasometric Experiments. The gasometric setup is essentially an apparatus that measures the volume of gas evolved from a reaction system as described by Onuchukwu [21]. The reaction vessel was connected to a burette via a delivery tube, which was in turn connected to a reservoir of paraffin oil. Fifty milliliters of the test solutions were introduced, respectively, into the reaction vessel for blank determinations and the initial volumes of air in the burette, taken against that of the paraffin oil, were recorded. Thereafter, two mild steel coupons were introduced into the reaction vessel and the flask quickly closed. The volume of hydrogen gas evolved by the corrosion reaction was monitored by the drop in the volume of the paraffin oil level in the gasometric gauge. The progress of the corrosion reaction was monitored by careful volumetric measurement of the evolved hydrogen gas at fixed time intervals. The temperature of the experiment was controlled at 30 ± 1 and 60 ± 1°C. The experiments were performed separately employing 20 and 100 mg/L inhibitor concentrations in 1 M H_2SO_4 and 2 M HCl. The effects of halide ions on the inhibitive action of the BN extract were studied by adding KCl, KBr, and KI separately to 1 M H_2SO_4 and 2 M HCl with and without 100 mg/L BN extract solutions to yield 0.5 mM concentrations of the halides in each case.

2.2.2. Weight Loss Experiments. The prepared and weighed mild steel coupons were immersed in beakers containing 200 mL of the test solutions with and without the addition of BN extract of concentrations ranging from 5 to 100 mg/L inhibitor concentrations in 1 M H_2SO_4 and 2 M HCl at 30, 40, 50, and 60°C. The metal strips were suspended in the

beakers using glass rods and hooks. After 3 h, the specimens were removed from the solutions, washed appropriately, dried, and reweighed. The weight loss was taken to be the difference between the weight of the coupons after the 3 h period of immersion in the solutions and the initial weight of the coupons. Gravimetric experiments were performed in triplicate and the results showed good reproducibility. The average values were taken and used in subsequent calculations.

3. Results and Discussion

3.1. Gasometric Measurements

3.1.1. Effect of Immersion Time on Corrosion Rate. The spontaneous dissolution of mild steel in acidic media is accompanied by the cathodic reduction of hydrogen ions as shown in (1)

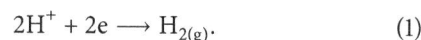

$$2H^+ + 2e \longrightarrow H_{2(g)}. \qquad (1)$$

The corrosion of iron and steel in acidic solutions is controlled by the hydrogen evolution reaction [22]. Thus, the corrosion rates of the test coupons in absence and presence of inhibitor were assessed using hydrogen evolution measurements. Previous workers have demonstrated the effectiveness of the gas-volumetric technique in monitoring any modifications in the double layer resulting from the action of an adsorbed inhibitor in a metal/corrodent system [23–25]. Results obtained by this technique are corroborated by other well established methods including weight loss and thermometry, potentiostatic polarization, and impedance spectroscopy [24, 26–28].

Gasometric measurements of mild steel subjected to the effect of acidic media in the absence and presence of BN extract were made at various time intervals. Figures 1(a) and 1(b) present plots of evolved hydrogen gas as a function of time for mild steel corrosion in 1 M H_2SO_4, in absence and presence of 20 and 100 mg/L BN extract concentrations at 30 and 60°C, respectively. Similar plots are shown in Figures 2(a) and 2(b) for 2 M HCl at 30 and 60°C. The plots in Figures 1 and 2 show a remarkable decrease in hydrogen evolution with the introduction of the inhibitor indicating that BN extract inhibits corrosion of mild steel in acidic environments. The rates of hydrogen evolution were observed to decrease with increasing inhibitor concentration, suggesting that the inhibiting effectiveness of the BN extract depends on the inhibitor concentration. This dependence was almost linear throughout the time interval studied in the absence and presence of BN extract indicating that the inhibitor acts rapidly and does not lose its inhibitory properties with time. However, the kinetic parameters indicate satisfactory inhibitor efficiencies even at low concentration of BN extract.

In all the cases, the dissolution of steel was characterized by a linear increase in the evolution of hydrogen with time. The reaction rate was characterized by differentiating the volume of hydrogen evolved with time and was obtained from the slope of the linear portions of Figures 1(a) and 1(b). Table 1 shows the values of corrosion rates obtained for the different

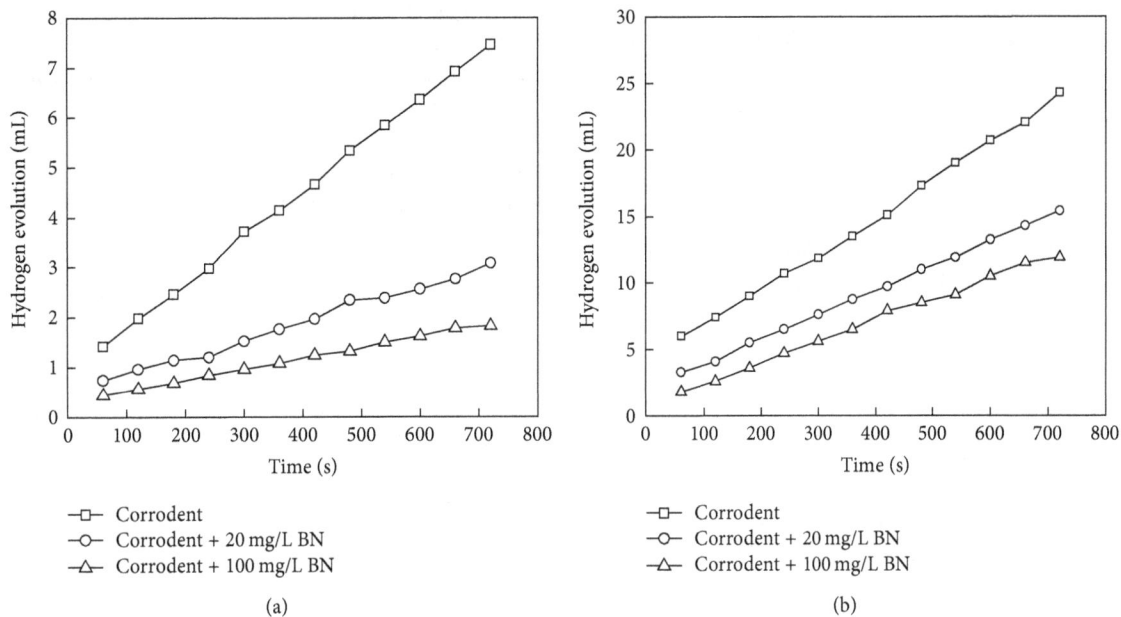

FIGURE 1: Hydrogen evolution during mild steel corrosion in 1 M H_2SO_4 in absence and presence of *Baphia nitida* extract at (a) 30°C and (b) 60°C.

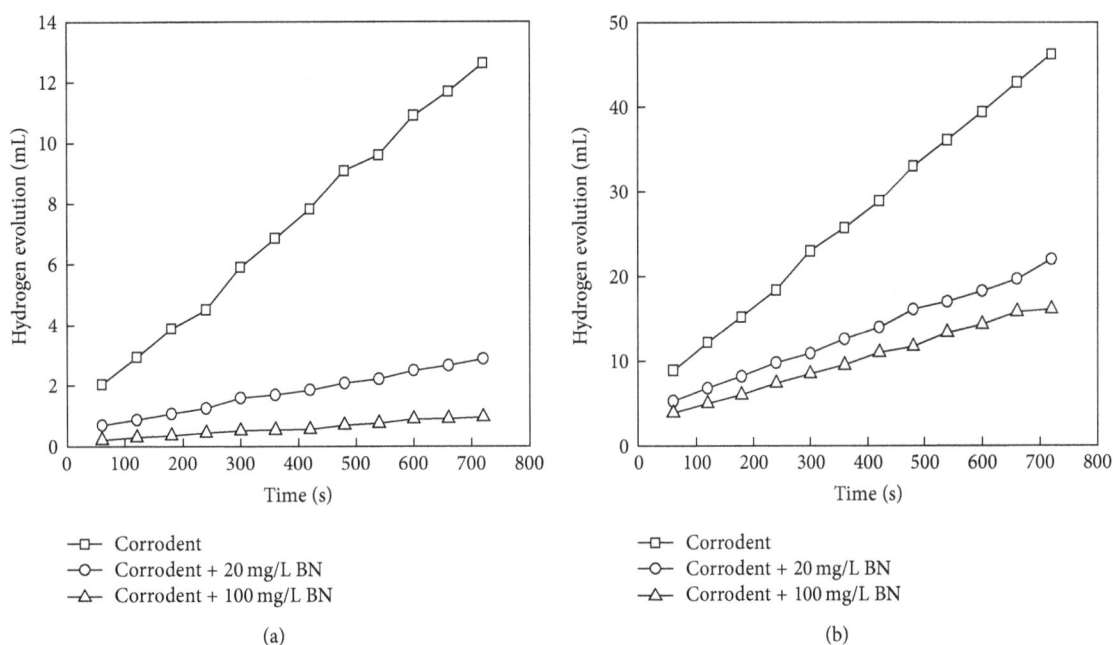

FIGURE 2: Hydrogen evolution during mild steel corrosion in 2 M HCl in absence and presence of BN extract at (a) 30°C and (b) 60°C.

test solutions. The results show that corrosion rate reduced in the presence of the inhibitor and was found to decrease with increasing BN extract concentration in both corrodents. Also, mild steel was observed to exhibit higher corrosion susceptibility in 2 M HCl than in 1 M H_2SO_4.

3.1.2. Inhibition Efficiency. For the gasometric experiments, the degree of surface coverage (θ) and the inhibition

efficiency (IE, %) of BN extract on mild steel in acidic media was evaluated from (2). Consider

$$\theta = 1 - \frac{v_{inh}}{v_{free}},$$

$$IE = 100 \times \left(1 - \frac{v_{inh}}{v_{free}}\right),$$

(2)

TABLE 1: Corrosion rates of mild steel in $1\,M\,H_2SO_4$ and $2\,M\,HCl$ in absence and presence of BN extract obtained from gasometric method.

System	Corrosion rate ($\times10^{-3}\,H_2$ gas vol. $mL\,s^{-1}$)	
	30°C	60°C
$1\,M\,H_2SO_4$		
Corrodent alone	9.20	27.65
Corrodent + 20 mg/L inhibitor	3.52	18.49
Corrodent + 100 mg/L inhibitor	2.17	15.83
$2\,M\,HCl$		
Corrodent alone	16.34	56.96
Corrodent + 20 mg/L inhibitor	3.28	24.68
Corrodent + 100 mg/L inhibitor	1.14	19.25

TABLE 2: Degree of surface coverage and inhibition efficiencies of different concentrations of BN extract during mild steel corrosion in $1\,M\,H_2SO_4$ and $2\,M\,HCl$ obtained from gasometric method.

Concentration (mg/L)	Surface coverage, θ		Inhibition efficiency, IE (%)	
	30°C	60°C	30°C	60°C
$1\,M\,H_2SO_4$				
Blank	—	—	—	—
20	0.62	0.33	61.7	33.1
100	0.76	0.43	76.4	42.8
$2\,M\,HCl$				
Blank	—	—	—	—
20	0.80	0.57	79.9	56.7
100	0.93	0.66	93.0	66.2

where v_{free} and v_{inh} are the corrosion rates in the absence and presence, respectively, of a given concentration of BN extract. Table 2 shows values of the degree of surface coverage (θ) and the inhibition efficiency (IE, %) obtained for different concentrations of BN extract in both corrodents using the gasometric technique. The result shows that the extract retarded acid corrosion of the mild steel and surface coverage and hence inhibition efficiencies increased with increasing inhibitor concentration at both temperatures.

Phytochemical analysis of leaves of BN detected tannins, flavonoids, and saponin glycosides and these are known to possess corrosion inhibitory properties [20]. The inhibitive effect could be attributed to the net adsorption of the organic matter on the steel/acid solution interface thereby reducing the surface area available for corrosion reaction and the degree of protection increases with increasing inhibitor concentration. It can be clearly noticed that the inhibition efficiency of BN extract was higher in $2\,M\,HCl$ than in $1\,M\,H_2SO_4$ over the concentration range studied, suggesting that the nature of the acid anion influences metal-inhibitor interactions. In the presence of strong acids, some inhibitor species become protonated. The surface charge on iron in acidic solution is positive at the corrosion potential and specific adsorption of chloride ions of HCl renders the metal surface more negative and susceptible to adsorption

of protonated inhibitor species compared to H_2SO_4 [29, 30]. Thus, the adsorption of the protonated inhibitor species on the metal surface will be enhanced in HCl, leading to higher inhibition efficiencies.

3.2. Weight Loss Measurements. The inhibition efficiency (IE, %) for the weight loss measurements was calculated using (3) as follows:

$$IE = \frac{(W_2 - W_1)}{W_1} \times 100, \qquad (3)$$

where W_1 (mg) is the weight loss of steel in uninhibited solutions and W_2 (mg) the weight loss of steel in inhibited solutions. IE (%) and θ for mild steel exposed to $1\,M\,H_2SO_4$ and $2\,M\,HCl$ at various temperatures as a function of BN concentration are shown in Tables 3 and 4. It is observed that at all temperatures inhibition efficiency increased on increasing BN concentration. This observation is in support of results obtained with the gasometric experiments.

The effect of temperature on the corrosion behavior of mild steel in the absence and presence of BN extract was investigated by performing weight loss experiments at 30, 40, 50, and 60°C. The results as shown in Tables 3 and 4 demonstrate that the weight loss increased and the inhibition efficiency decreased with increase in temperature. The decrease in inhibition efficiency with increasing temperature suggests weak adsorption interaction between the metal and the extract organic matter. Such behavior corresponds to physical adsorption, such that at higher temperatures, there is a possible shift of the adsorption-desorption equilibrium towards desorption of adsorbed inhibitor [31]. The increase in solution agitation resulting from higher rates of H_2 gas evolution as well as the agitation of the interface and roughening of the metal surface as a result of enhanced corrosion all contribute to the reduced stability of the adsorbed inhibitor at higher temperature [15, 32].

3.3. Adsorption Isotherm Behaviour. Experimental and theoretical studies have shown that the protective action of organic substances during metal corrosion is based on the adsorption ability of their molecules, where the resulting adsorption film isolates the metal surface from the corrosive medium [4–16]. Therefore, Langmuir adsorption isotherm expression (4), which relates the surface coverage θ defined by IE/100 and the inhibitor concentration (C), could be applied to determine adsorption equilibrium constant, K, at the different temperatures. Consider

$$\frac{C}{\theta} = \frac{1}{K} + C. \qquad (4)$$

The plots of C/θ versus C for $1\,M\,H_2SO_4$ and $2\,M\,HCl$ are shown in Figures 3(a) and 3(b), respectively, and the values of K subsequently calculated from the intercept are shown in Table 5. The adjusted correlation coefficient (R^2) values which were all above 0.990 shows a good fit of the experimental data and suggests that the adsorption of BN extract on metal surface followed the Langmuir adsorption

TABLE 3: Corrosion parameters obtained from weight loss of mild steel in 1 M H_2SO_4 containing various concentrations of BN extract at different temperatures.

Concentration of BN (mg/L)	Temperature											
	30°C			40°C			50°C			60°C		
	W (mg/cm²)	IE (%)	θ	W (mg/cm²)	IE (%)	θ	W (mg/cm²)	IE (%)	θ	W (mg/cm²)	IE (%)	θ
0	5.82	—	—	7.47	—	—	14.57	—	—	23.87	—	—
5	2.56	56.01	0.5601	3.96	46.99	0.4699	9.19	36.93	0.3693	16.81	29.58	0.2958
10	2.39	58.93	0.5893	3.86	48.33	0.4833	8.96	38.5	0.3850	16.48	30.96	0.3096
20	2.21	62.03	0.6203	3.7	50.47	0.5047	8.58	41.11	0.4111	16.02	32.89	0.3289
40	1.93	66.84	0.6684	3.45	53.82	0.5382	8.21	43.65	0.4365	15.35	35.69	0.3569
60	1.69	70.96	0.7096	3.19	57.3	0.5730	7.69	47.22	0.4722	14.98	37.24	0.3724
80	1.54	73.54	0.7354	2.95	60.51	0.6051	7.19	50.65	0.5065	14.24	40.34	0.4034
100	1.41	75.77	0.7577	2.75	63.19	0.6319	6.98	52.09	0.5209	13.55	43.23	0.4323

TABLE 4: Corrosion parameters obtained from weight loss of mild steel in 2 M HCl containing various concentrations of BN extract at different temperatures.

Concentration of BN (mg/L)	Temperature											
	30°C			40°C			50°C			60°C		
	W (mg/cm^2)	IE (%)	θ	W (mg/cm^2)	IE (%)	θ	W (mg/cm^2)	IE (%)	θ	W (mg/cm^2)	IE (%)	θ
0	4.44	—	—	5.43	—	—	9.38	—	—	13.66	—	—
5	1.18	73.42	0.7342	1.92	64.64	0.6464	4.00	57.36	0.5736	6.60	51.68	0.5168
10	1.07	75.9	0.7590	1.83	66.3	0.6630	3.82	59.28	0.5928	6.27	54.1	0.5410
20	0.88	80.18	0.8018	1.68	69.06	0.6906	3.58	61.83	0.6183	5.86	57.1	0.5710
40	0.65	85.36	0.8536	1.48	72.74	0.7274	3.24	65.46	0.6546	5.43	60.25	0.6025
60	0.52	88.29	0.8829	1.31	75.87	0.7587	2.90	69.08	0.6908	5.01	63.32	0.6332
80	0.41	90.77	0.9077	1.13	79.19	0.7919	2.70	71.26	0.7126	4.77	65.08	0.6508
100	0.29	93.47	0.9347	0.97	82.14	0.8214	2.53	73.03	0.7303	4.59	66.4	0.6640

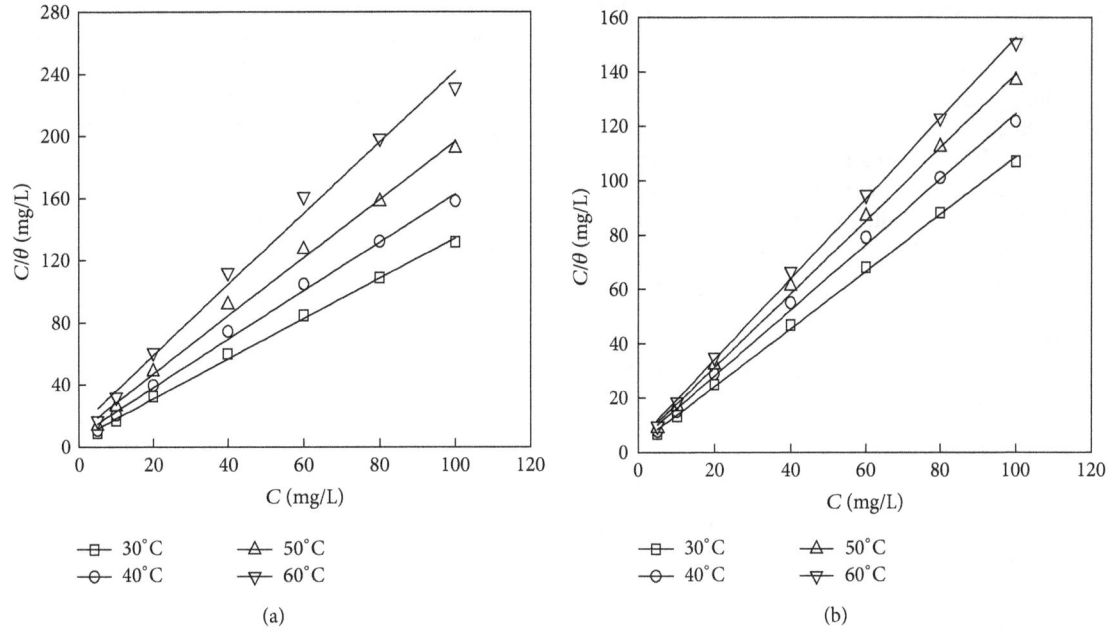

FIGURE 3: Langmuir adsorption plots for mild steel corrosion with BN extract as inhibitor at different temperatures in (a) 1 M H_2SO_4 and (b) in 2 M HCl.

TABLE 5: Thermodynamic parameters for the adsorption of BN extract in 1 M H_2SO_4 and 2 M HCl on the mild steel at different temperatures.

Temperature	K (L/mg)	R^2	ΔG (kJ/mol)	ΔH (kJ/mol)	ΔS (J/mol K)
1 M H_2SO_4					
30	0.194	0.997	−1.028	−26.21	−83.11
40	0.141	0.995	−1.270	−26.21	−79.68
50	0.103	0.994	−1.517	−26.21	−76.45
60	0.076	0.990	−1.772	−26.21	−73.39
2 M HCl					
30	0.316	0.999	−0.723	−9.61	−29.33
40	0.241	0.997	−0.921	−9.61	−27.76
50	0.228	0.998	−0.988	−9.61	−26.69
60	0.221	0.999	−1.041	−9.61	−25.73

isotherm. The results show that the adsorption equilibrium constant (K) decreased with increasing temperature, indicating better adsorption of BN extract onto the steel surface at lower temperatures. However, at higher temperatures, the equilibrium tends towards desorption.

3.4. Thermodynamic Parameters. The thermodynamic parameters, the standard free energy of adsorption ($\Delta G°$), the standard heat of adsorption ($\Delta H°$), and the standard entropy of adsorption ($\Delta S°$) give an insight into the mechanism of the corrosion inhibition process. From the van't Hoff equation (5), $\Delta H°$ was determined by a linear regression between $\ln K$ and $1/T$ [33]. Consider

$$\ln K = \frac{-\Delta H}{RT} + C. \tag{5}$$

The values of $\Delta G°$ were evaluated from (6) [34]

$$K = \frac{1}{55.5} \exp\left(-\frac{\Delta G°}{RT}\right). \tag{6}$$

The $\Delta S°$ values were then obtained from the basic thermodynamic equation

$$\Delta G° = \Delta H° - T\Delta S°. \tag{7}$$

The thermodynamic parameters obtained are listed in Table 5. The negative values of $\Delta H°$ show that the adsorption of BN on the metal surface is an exothermic process, supporting the previous observation that IE decreases with increase in temperature. Exothermic process signifies either physisorption or chemisorption while endothermic process is indicative solely of chemisorptions [35]. For an exothermic process, the absolute value of $\Delta H°$ of the process is used to distinguish physisorption from chemisorptions. If $\Delta H°$ is lower than 41.86 kJ/mol, physisorption is involved, while if $\Delta H°$ approaches 100 kJ/mol, it is a chemisorption process [36]. The absolute value of $\Delta H°$ in the present study is lower than 41.86 kJ/mol, confirming the physisorption mechanism proposed involving electrostatic interactions between charged BN molecules and charged metal. The negative values of $\Delta G°$ indicate the stability of the adsorbed layer on the steel surface and that the adsorption of BN molecules onto the steel surface is spontaneous. The negative values of $\Delta S°$ show that the adsorption process is accompanied by a decrease in entropy. The explanation is that the adsorption of BN molecules onto the steel surface reduces the level of chaos on the steel surface leading to a decrease in entropy.

3.5. Apparent Activation Energy (E_a).

It has been reported that for the acid corrosion of mild steel, the natural logarithm of the corrosion rate is a linear function of $1/T$. Therefore, the apparent activation energy of the corrosion inhibition process could be obtained by the application of the Arrhenius-type equation (8) [1]. Consider

$$\ln \nu = \ln A - \frac{E_a}{RT},\qquad(8)$$

where ν (mg/cm^2 h) is the corrosion rate, E_a (J/mol) is the apparent activation energy, R (8.314 J/mol K) is the universal gas constant, T(K) is the absolute temperature, and A (mg/cm^2 h) is the pre-exponential factor. The corrosion rate, ν, was obtained from the relation

$$\nu = \frac{W}{t},\qquad(9)$$

where W (mg/cm^2) is the weight loss per area and t is the immersion (corrosion) time (3 h). From the slope and intercept of the linear regression of $\ln \nu$ versus $1/T$ (Figure not shown), the values of E_a and A for mild steel corrosion in different concentrations of BN in both 1 M H$_2$SO$_4$ and 2 M HCl were calculated, respectively, as shown in Table 6. The results show that the value of E_a increased with the addition of BN extract. Increase in E_a with increasing concentration of inhibitor indicates physisorption mechanism [21]. Similar results have been reported in some previous studies [9, 37].

Analysis of the temperature dependence of inhibition efficiency as well as comparison of corrosion activation energies in absence and presence of inhibitor gives some insight into the possible mechanism of inhibitor adsorption. A decrease in inhibition efficiency with rise in temperature, with analogous increase in corrosion activation energy in the presence of inhibitor compared to its absence, is frequently interpreted as being suggestive of formation of an adsorption film of physical (electrostatic) nature. The reverse effect, corresponding to an increase in inhibition efficiency with rise in temperature and lower activation energy in the presence of inhibitor, suggests a chemisorption mechanism [22]. The results for both acid media show higher activation energy values in the presence of the extract compared to the blank acids, suggesting physical adsorption [30–32, 38].

TABLE 6: Calculated values of apparent activation energy (E_a) and preexponential factor (A) for mild steel corrosion in 1 M H$_2$SO$_4$ and 2 M HCl with BN extract as inhibitor.

C (mg/L)	E_a (kJ/mol)	A (mg/cm^2 h)	R^2
1 M H$_2$SO$_4$			
0	40.97	2.02×10^7	0.950
5	54.30	1.79×10^9	0.974
10	55.55	2.78×10^9	0.980
20	56.81	4.26×10^9	0.984
40	59.40	1.05×10^{10}	0.989
60	62.24	2.87×10^{10}	0.992
80	63.39	4.13×10^{10}	0.992
100	64.72	6.44×10^{10}	0.992
2 M HCl			
0	32.75	6.06×10^5	0.951
5	49.44	1.26×10^8	0.988
10	50.66	1.87×10^8	0.992
20	54.11	6.24×10^8	0.994
40	60.12	5.22×10^9	0.992
60	63.84	1.87×10^{10}	0.989
80	69.26	1.28×10^{11}	0.987
100	77.82	2.82×10^{12}	0.979

TABLE 7: Effect of halide ions on the mild steel corrosion in 1 M H$_2$SO$_4$ and 2 M HCl in the absence and presence of BN extract.

System	Inhibition efficiency, E (%)			
	1 M H$_2$SO$_4$		2 M HCl	
	30°C	60°C	30°C	60°C
100 mg/L BN	76.4	42.8	93.0	66.2
0.5 mM KCl	32.7	26.2	26.4	20.9
0.5 mM KBr	52.9	45.7	52.4	40.3
0.5 mM KI	69.6	60.0	65.0	54.8
100 mg/L BN + 0.5 mM KCl	79.2	44.8	94.4	67.3
100 mg/L BN + 0.5 mM KBr	81.7	47.0	95.8	68.4
100 mg/L BN + 0.5 mM KI	89.5	54.5	97.8	69.8

3.6. Synergism Considerations.

The influence of halide ions on the inhibitive action of BN extract was assessed. Table 7 illustrates the effects of 0.5 mM KCl, KBr, and KI without or with 100 mg/L BN extract on the corrosion of mild steel in 1 M H$_2$SO$_4$ and 2 M HCl. The inhibition efficiency of BN extract was significantly improved in the presence of halide ions in both acid media and at both temperatures studied. This suggests that adsorption of protonated species in the BN extract is enhanced through ion pair interactions, with the halide ions forming an intermediate bridge between the positively charged metal surface and the inhibitor [22, 39]. From the inhibition efficiencies, corrosion inhibition efficiencies of the halide alone as well as in combination with BN extract increased in the order KCl < KBr < KI. This is in accordance with the findings of other researchers [15, 39–42]. This observation could be explained on the basis of the

TABLE 8: Synergism parameter (S_I) for the various halides at 30 and 60°C.

Halides	Synergism parameter (S_I)			
	1 M H_2SO_4		2 M HCl	
	30°C	60°C	30°C	60°C
KCl	1.38	1.55	1.27	1.29
KBr	1.59	1.82	1.52	1.57
KI	1.64	1.90	1.62	1.74

halide ion radii, which increases in the order Cl^- (0.09 nm) < Br^- (0.114 nm) < I^- (0.135 nm), with the highest ionic radius being more predisposed to adsorption.

The data in Table 7 also shows that inhibition efficiencies in the presence of halides were better improved in 1 M H_2SO_4 than in 2 M HCl which implies that more halide ions are adsorbed on the metal surface in 1 M H_2SO_4. This could be attributed to comparatively more positive charge on the steel surface in 1 M H_2SO_4 [43].

At 60°C, all the halides in combination with BN extract exhibited reduced inhibition efficiencies, indicating that the synergistic effect of BN extract and halide ions is diminished at higher temperatures. The decrease in inhibition efficiency of the BN extract and halide complex with rise in temperature as shown in Table 7 still supports the physisorption mechanism for BN extract on the mild steel surface which is in line with the observation in the absence of halides.

The synergism parameters, S_I, were calculated using the relationship (10) given by Aramaki and Hackermann [44]:

$$S_I = \frac{1 - I_{1+2}}{1 - I'_{1+2}}, \tag{10}$$

where $I_{1+2} = (I_1 + I_2)$; I_1 is the inhibition efficiency of the halide; I_2 is the inhibition efficiency of BN extract; I'_{1+2} is the inhibition efficiency of BN extract in combination with halide. The calculated values are presented in Table 8 for the different halides at 30 and 60°C. From Table 8, it could be seen that all values of S_I are greater than unity, clearly showing that the corrosion inhibition brought about by the complex of BN extract and halide is due mainly to synergistic effect [41, 45].

4. Conclusion

B. nitida leaf extract inhibited mild steel corrosion in 1 M H_2SO_4 and 2 M HCl at the temperatures studied. Inhibition efficiency increased with increase in BN extract concentration and synergistically increased in the presence of halide ions. Temperature studies revealed a decrease in inhibition efficiency with rise in temperature and corrosion activation energies being higher in the presence of the plant extract. Comparative analyses of the results from both acid solutions suggest that protonated species in the extract play a predominant role in inhibitive behavior observed, with the predominant effect being the physical adsorption of protonated species.

Conflict of Interests

The authors declare that there is no conflict of interests regarding the publication of this paper.

References

[1] E. S. Ferreira, C. Giacomelli, F. C. Giacomelli, and A. Spinelli, "Evaluation of the inhibitor effect of L-ascorbic acid on the corrosion of mild steel," *Materials Chemistry and Physics*, vol. 83, no. 1, pp. 129–134, 2004.

[2] D. A. Jones, *Principles and Prevention of Corrosion*, Prentice Hall, Upper Saddle River, NJ, USA, 2nd edition, 1996.

[3] M. G. Fontana, *Corrosion Engineering*, McGraw-Hill, Singapore, 3rd edition, 1986.

[4] A. Popova, M. Christov, and T. Deligeorgiev, "Influence of the molecular structure on the inhibitor properties of benzimidazole derivatives on mild steel corrosion in 1M hydrochloric acid," *Corrosion*, vol. 59, no. 9, pp. 756–764, 2003.

[5] E. E. Oguzie, C. K. Enenebeaku, C. O. Akalezi, S. C. Okoro, A. A. Ayuk, and E. N. Ejike, "Adsorption and corrosion-inhibiting effect of *Dacryodis edulis* extract on low-carbon-steel corrosion in acidic media," *Journal of Colloid and Interface Science*, vol. 349, no. 1, pp. 283–292, 2010.

[6] A. R. Hosein Zadeh, I. Danaee, and M. H. Maddahy, "Thermodynamic and adsorption behaviour of medicinal nitramine as a corrosion inhibitor for AISI steel alloy in HCl solution," *Journal of Materials Science and Technology*, vol. 29, no. 9, pp. 884–892, 2013.

[7] I. Lukovits, E. Kalman, and F. Zuchi, "Corrosion inhibitors-correlation between electronic structure and efficiency," *Corrosion*, vol. 57, no. 1, pp. 3–8, 2001.

[8] P. Mohan and G. P. Kalaignan, "1, 4-Bis (2-nitrobenzylidene) thiosemicarbazide as effective corrosion inhibitor for mild steel," *Journal of Materials Science & Technology*, vol. 29, no. 11, pp. 1096–1100, 2013.

[9] E. E. Oguzie, V. O. Njoku, C. K. Enenebeaku, C. O. Akalezi, and C. Obi, "Effect of hexamethylpararosaniline chloride (crystal violet) on mild steel corrosion in acidic media," *Corrosion Science*, vol. 50, no. 12, pp. 3480–3486, 2008.

[10] E. E. Oguzie, "Inhibition of acid corrosion of mild steel by Telfaria occidentalis," *Pigment and Resin Technology*, vol. 34, no. 6, pp. 321–326, 2005.

[11] N. O. Eddy, P. A. Ekwumemgbo, and P. A. P. Mamza, "Ethanol extract of *Terminalia catappa* as a green inhibitor for the corrosion of mild steel in H_2SO_4," *Green Chemistry Letters and Reviews*, vol. 2, no. 4, pp. 223–231, 2009.

[12] E. S. H. El Ashry, A. El Nemr, S. A. Esawy, and S. Ragab, "Corrosion inhibitors. Part II: quantum chemical studies on the corrosion inhibitions of steel in acidic medium by some triazole, oxadiazole and thiadiazole derivatives," *Electrochimica Acta*, vol. 51, no. 19, pp. 3957–3968, 2006.

[13] M. Lebrini, F. Bentiss, H. Vezin, and M. Lagrenée, "The inhibition of mild steel corrosion in acidic solutions by 2,5-bis(4-pyridyl)-1,3,4-thiadiazole: structure-activity correlation," *Corrosion Science*, vol. 48, no. 5, pp. 1279–1291, 2006.

[14] S. E. Manahan, *Environmental Chemistry*, CRC Press, Boca Raton, Fla, USA, 1999.

[15] E. E. Oguzie, "Studies on the inhibitive effect of *Occimum viridis* extract on the acid corrosion of mild steel," *Materials Chemistry and Physics*, vol. 99, no. 2-3, pp. 441–446, 2006.

[16] O. K. Abiola, J. O. E. Otaigbe, and O. J. Kio, "*Gossipium hirsutum* L. extracts as green corrosion inhibitor for aluminum in NaOH solution," *Corrosion Science*, vol. 51, no. 8, pp. 1879–1881, 2009.

[17] A. K. Satapathy, G. Gunasekaran, S. C. Sahoo, K. Amit, and P. V. Rodrigues, "Corrosion inhibition by *Justicia gendarussa* plant extract in hydrochloric acid solution," *Corrosion Science*, vol. 51, no. 12, pp. 2848–2856, 2009.

[18] S. K. Sharma, A. Mudhoo, G. Jain, and J. Sharma, "Corrosion inhibition and adsorption properties of *Azadirachta indica* mature leaves extract as green inhibitor for mild steel in HNO_3," *Green Chemistry Letters and Reviews*, vol. 3, p. 7, 2010.

[19] E. I. Ating, S. A. Umoren, I. I. Udousoro, E. E. Ebenso, and A. P. Udoh, "Leaves extract of ananas sativum as green corrosion inhibitor for aluminium in hydrochloric acid solutions," *Green Chemistry Letters and Reviews*, vol. 3, no. 2, pp. 61–68, 2010.

[20] N. D. Onwukaeme, "Anti-inflammatory activities of flavonoids of *Baphia nitida* Lodd. (Leguminosae) on mice and rats," *Journal of Ethnopharmacology*, vol. 46, no. 2, pp. 121–124, 1995.

[21] A. I. Onuchukwu, "Corrosion inhibition of aluminum in alkaline medium. I: influence of hard bases," *Materials Chemistry and Physics*, vol. 20, no. 4-5, pp. 323–332, 1988.

[22] A. Popova, E. Sokolova, S. Raicheva, and M. Christov, "AC and DC study of the temperature effect on mild steel corrosion in acid media in the presence of benzimidazole derivatives," *Corrosion Science*, vol. 45, no. 1, pp. 33–58, 2003.

[23] B. Muller, "Corrosion inhibition of aluminium and zinc pigments by saccharides," *Corrosion Science*, vol. 44, pp. 1583–1591, 2002.

[24] A. Aytac, U. Ozmen, and M. Kabasakaloglu, "Investigation of some Schiff bases as acidic corrosion of alloy AA3102," *Materials Chemistry and Physics*, vol. 89, no. 1, pp. 176–181, 2005.

[25] E. E. Ebenso and E. E. Oguzie, "Corrosion inhibition of mild steel in acidic media by some organic dyes," *Materials Letters*, vol. 59, no. 17, pp. 2163–2165, 2005.

[26] M. N. Moussa, A. S. Fouda, A. I. Taha, and A. Elnenaa, "Some Thiosemicarbazide derivatives as corrosion inhibitors for aluminium in sodium hydroxide solution," *Bulletin of the Korean Chemical Society*, vol. 9, no. 4, pp. 191–195, 1988.

[27] A. Y. El-Etre, "Inhibition of aluminum corrosion using Opuntia extract," *Corrosion Science*, vol. 45, no. 11, pp. 2485–2495, 2003.

[28] M. Abdallah, "Antibacterial drugs as corrosion inhibitors for corrosion of aluminium in hydrochloric solution," *Corrosion Science*, vol. 46, no. 8, pp. 1981–1996, 1981.

[29] E. E. Oguzie, Y. Li, and F. H. Wang, "Effect of surface nanocrystallization on corrosion and corrosion inhibition of low carbon steel: synergistic effect of methionine and iodide ion," *Electrochimica Acta*, vol. 52, no. 24, pp. 6988–6996, 2007.

[30] M. S. S. Morad, A. E. A. Hermas, and M. S. Abdel Aal, "Effect of amino acids containing sulfur on the corrosion of mild steel in phosphoric acid solutions polluted with Cl^-, F^- and Fe^{3+} ions–behaviour near and at the corrosion potential," *Journal of Chemical Technology and Biotechnology*, vol. 77, pp. 486–494, 2002.

[31] K. Orubite-Okorosaye and N. C. Oforka, "Corrosion inhibition of zinc on HCl using *Nypa fruticans Wurmb* extract and 1,5 diphenyl carbazone," *Journal of Applied Sciences & Environmental Management*, vol. 8, pp. 56–61, 2004.

[32] S. Martinez and M. Matikos-Hukovic, "A nonlinear kinetic model introduced for the corrosion inhibitive properties of some organic inhibitors," *Journal of Applied Electrochemistry*, vol. 33, pp. 1137–1142, 2003.

[33] T. Zhao and G. Mu, "The adsorption and corrosion inhibition of anion surfactants on aluminium surface in hydrochloric acid," *Corrosion Science*, vol. 41, no. 10, pp. 1937–1944, 1999.

[34] M. Bouklah, B. Hammouti, M. Lagrenée, and F. Bentiss, "Thermodynamic properties of 2,5-bis(4-methoxyphenyl)-1,3,4-oxadiazole as a corrosion inhibitor for mild steel in normal sulfuric acid medium," *Corrosion Science*, vol. 48, no. 9, pp. 2831–2842, 2006.

[35] W. Durnie, R. de Marco, A. Jefferson, and B. Kinsella, "Development of a structure-activity relationship for oil field corrosion inhibitors," *Journal of the Electrochemical Society*, vol. 146, no. 5, pp. 1751–1756, 1999.

[36] S. Martinez and I. Stern, "Thermodynamic characterization of metal dissolution and inhibitor adsorption processes in the low carbon steel/mimosa tannin/sulfuric acid system," *Applied Surface Science*, vol. 199, no. 1–4, pp. 83–89, 2002.

[37] G. Mu and X. Li, "Inhibition of cold rolled steel corrosion by Tween-20 in sulfuric acid: Weight loss, electrochemical and AFM approaches," *Journal of Colloid and Interface Science*, vol. 289, pp. 184–192, 2005.

[38] O. Olivares, N. V. Likhanova, B. Gómez et al., "Electrochemical and XPS studies of decylamides of α-amino acids adsorption on carbon steel in acidic environment," *Applied Surface Science*, vol. 252, no. 8, pp. 2894–2909, 2006.

[39] A. I. Onuchukwu and S. P. Trasatti, "Hydrogen permeation into aluminium AA1060 as a result of corrosion in an alkaline medium. Influence of anions in solution and of temperature," *Corrosion Science*, vol. 36, no. 11, pp. 1815–1817, 1994.

[40] E. E. Oguzie, G. N. Onuoha, and A. I. Onuchukwu, "Inhibitory mechanism of mild steel corrosion in 2 M sulphuric acid solution by methylene blue dye," *Materials Chemistry and Physics*, vol. 89, no. 2-3, pp. 305–311, 2005.

[41] G. K. Gomma, "Corrosion of low-carbon steel in sulphuric acid solution in presence of pyrazole—halides mixture," *Materials Chemistry and Physics*, vol. 55, no. 3, pp. 241–246, 1998.

[42] E. E. Ebenso, "Synergistic effect of halide ions on the corrosion inhibition of aluminium in H_2SO_4 using 2-acetylphenothiazine," *Materials Chemistry and Physics*, vol. 79, no. 1, pp. 58–70, 2003.

[43] E. E. Oguzie, "Evaluation of the inhibitive effect of some plant extracts on the acid corrosion of mild steel," *Corrosion Science*, vol. 50, no. 11, pp. 2993–2998, 2008.

[44] K. Aramaki and N. Hackermann, "Inhibition mechanism of medium-sized polymethyleneimine," *Journal of the Electrochemical Society*, vol. 116, no. 5, pp. 568–574, 1969.

[45] L. Tang, X. Li, G. Mu et al., "The synergistic inhibition between hexadecyl trimethyl ammonium bromide (HTAB) and NaBr for the corrosion of cold rolled steel in 0.5 M sulfuric acid," *Journal of Materials Science*, vol. 41, pp. 3063–3069, 2006.

18

GC Analyses of *Salvia* Seeds as Valuable Essential Oil Source

Mouna Ben Taârit,[1] **Kamel Msaada,**[1] **Karim Hosni,**[2] **and Brahim Marzouk**[1]

[1] *Laboratoire des Substances Bioactives, Centre de Biotechnologie, Technopôle de Borj-Cédria, BP 901, 2050 Hammam-Lif, Tunisia*
[2] *Laboratoire des Substances Naturelles, Institut National de Recherche et d'Analyse Physico-Chimique (INRAP), Sidi Thabet, 2020 Ariana, Tunisia*

Correspondence should be addressed to Mouna Ben Taârit; taaritmouna@yahoo.fr

Academic Editor: Constantinos Pistos

The essential oils of seeds of *Salvia verbenaca*, *Salvia officinalis*, and *Salvia sclarea* were obtained by hydrodistillation and analyzed by gas chromatography (GC) and GC-mass spectrometry. The oil yields (w/w) were 0.050, 0.047, and 0.045% in *S. verbenaca*, *S. sclarea*, and *S. officinalis*, respectively. Seventy-five compounds were identified. The essential oil composition of *S. verbenaca* seeds showed that over 57% of the detected compounds were oxygenated monoterpenes followed by sesquiterpenes (24.04%) and labdane type diterpenes (5.61%). The main essential oil constituents were camphor (38.94%), caryophyllene oxide (7.28%), and 13-*epi*-manool (5.61%), while those of essential oil of *S. officinalis* were α-thujone (14.77%), camphor (13.08%), and 1,8-cineole (6.66%). In samples of *S. sclarea*, essential oil consists mainly of linalool (24.25%), α-thujene (7.48%), linalyl acetate (6.90%), germacrene-D (5.88%), bicyclogermacrene (4.29%), and α-copaene (4.08%). This variability leads to a large range of naturally occurring volatile compounds with valuable industrial and pharmaceutical outlets.

1. Introduction

The genus *Salvia* (Lamiaceae) comprises nearly 900 species widely spread throughout the world, which display marked morphological and genetic variations according to their geographical origin [1]. Several *Salvia* species, namely, *Salvia officinalis*, *Salvia sclarea*, and *Salvia verbenaca*, are widely used in folk medicine [2]. Potential therapeutic activities of these *Salvia* species are due to their essential oils [3], since these species are known to possess antioxidant, antimicrobial, antifungal, and aromatic properties [4]. Chemical composition of essential oils reveals differences among these *Salvia* species [5–7]. Numerous investigations on *Salvia officinalis* show that 1,8-cineole, α-thujone, β-thujone, and camphor are the main compounds of the essential oil [8–10]. Linalool, linalyl acetate, and germacrene-D characterize *S. sclarea* plants [11]. *Salvia* species also display great intraspecific essential oil variations according to geographical origin, since sabinene, cadinene, terpinen-4-ol, and pinene are shown to be typical compounds of *S. verbenaca* essential oil originated from Saudi Arabia [4], while β-phellandrene and (*E*)-caryophyllene prevail in essential oil from Greece [7].

In Tunisia, *S. verbenaca* essential oil shows variations of composition according to the region origin [12, 13] and in respect to the studied plant part [14].

These numerous studies are focused on aerial parts of these species, while works interested in seeds are scanty in spite of their interest. In fact, *Salvia* seeds provide dietary and healthy oil rich in essential fatty acids (linolenic and linoleic acids) [12, 15, 16] that promote decrease in coronary heart diseases [17]. Besides, the seeds of *Salvia* species often produce mucilage on wetting [18]. This mucilage is used for lacquerware [19]. In eastern countries, the mucilage is used for the treatment of eye diseases [20].

In our continuing research on essential oil with pharmacological potential and food industry applications, we report in this paper chemical analysis of *S. verbenaca*, *S. sclarea*, and *S. officinalis* seeds as a new valuable essential oil source.

2. Materials and Methods

2.1. Plant Material. The seeds of *Salvia verbenaca* (accession PI 420430) were originated from Spain; *Salvia officinalis* (accession W6 20659) and *Salvia sclarea* (accession W6

20660) which are from Italy were kindly supplied by the US National Plant Germplasm System (NPGS).

2.2. Essential Oil Isolation. The seeds of each species were ground in a mortar before essential oil isolation [12] and were subjected to conventional hydrodistillation for 90 min followed by a liquid-liquid extraction using diethyl ether and *n*-pentane mixture (v/v) as solvent. The concentration step was carried out at 35°C using a Vigreux column and the essential oils obtained were dried over anhydrous sodium sulphate and stored in amber vials at −18°C until they were analyzed.

3. Chromatographic Analysis

3.1. Gas Chromatography (GC-FID). The essential oils were analysed by gas chromatography using a Hewlett-Packard 6890 gas chromatograph (Palo Alto, CA, USA) equipped with a flame ionization detector (FID) and an electronic pressure control (EPC) injector. A polar HP Innowax (PEG) column and an apolar HP-5 column (30 m × 0.25 mm, 0.25 μm film thicknesses) were used. The carrier gas was N_2 with a flow rate of 1.6 mL/min; split ratio was 60 : 1. The analysis was performed using the following temperature program: oven temps isotherm at 35°C for 10 min, from 35 to 205°C, at the rate of 3°C/min, and isotherm at 205°C during 10 min. Injector and detector temperatures were held, respectively, at 250 and 300°C. The volume injected was 1 μL.

3.2. Gas Chromatography-Mass Spectrometry (GC-MS). GC-MS analysis was performed on a gas chromatograph HP 5890 (II) interfaced with a HP 5972 mass spectrometer with electron impact ionization (70 eV). A HP-5MS capillary column (30 m × 0.25 mm, 0.25 μm film thickness) was used. The column temperature was programmed from 50°C to rise to 240°C at a rate of 5°C/min. The carrier gas was helium with a flow rate of 1.2 mL/min; split ratio was 60 : 1. Scan time and mass range were 1 s and 40–300 m/z, respectively.

3.3. Compounds Identification. The identification of the essential oil constituents was based on the comparison of their retention indexes relative to (C_8–C_{22}) *n*-alkanes with those of literature or with those of authentic compounds available in our laboratory. Further identification was made by matching their recorded mass spectra with those stored in the Wiley/NBS mass spectral library of the GC-MS data system and other published mass spectra [21].

3.4. Statistical Analyses. Data were subjected to statistical analysis using "Statistica" statistical program package [22]. The percentages of volatile compounds are means of three experiments; the one-way analysis of variance (ANOVA) followed by Duncan multiple range test was employed and the differences between individual means were deemed to be significant at $P < 0.05$.

4. Results

Hydrodistillation of full ripened seeds of *S. verbenaca*, *S. sclarea*, and *S. officinalis* offered essential oils with average yields of 0.050, 0.047, and 0.045% (w/w on the dry weight basis), respectively.

Essential oil constituents of *Salvia* seeds were presented in Table 1. The results of analysis of essential oil of *S. verbenaca* seeds by GC and GC-MS techniques revealed the occurrence of thirty-two compounds. The essential oil composition showed that over 57% of the detected compounds were oxygenated monoterpenes followed by sesquiterpenes (24.04%) and labdane type diterpenes (5.61%). Apart from camphor, the main essential oil constituents of this sample were caryophyllene oxide (7.28%), 13-*epi*-manool (5.61%), δ-elemene (3.97%), and β-eudesmol (3.76%).

The essential oil of *S. officinalis* seeds showed a higher percentage of monoterpenes (56.59%) than sesquiterpenes (17.32%). Oxygenated derivatives were major among the monoterpenes (50.14%), while they represented only 5.93% of sesquiterpenes. Seeds essential oils were characterised by the predominance of α-thujone (14.77%), camphor (13.08%), and 1,8-cineole (6.66%). Viridiflorol (2.66%) and α-humulene (3.71%) were also detected in seeds essential oil of sage. Furthermore, other minor compounds were found, especially the labdane type diterpene 13-*epi*-manool.

Essential oil compounds of *S. sclarea* seeds were representing 80.79% of total essential oil components. The essential oil is predominated by monoterpenes accounting for 47.98%; their oxygenated derivatives (38.38%) prevailed on hydrocarbon ones (9.60%). The sesquiterpenes pool is less numerous (29.39%). In addition, one diterpene compound, 13-*epi*-manool, is detected at a level of 0.59% and some phenols (2.03%) such as thymol and carvacrol were produced in small amounts (0.1% and 1.93%, resp.). The oxygenated monoterpenes were characterised by linalool (24.25%), geraniol (2.79%), and their ester derivates (linalyl acetate (6.90%) and geranyl acetate (1.94%)). The sesquiterpenes characteristics of seeds were germacrene-D (5.88%), bicyclogermacrene (4.29%), and α-copaene (4.08%).

5. Discussion

As for *S. verbenaca*, the oil yield was higher than that offered by seeds originated from Tunisia [12]. Compared with leaves of *S. verbenaca* from Tunisia [14], seeds herein studied appeared as moderately rich in volatile oil. Recovered essential oil from *S. sclarea* seeds appeared to be near to literature data which showed that inflorescences oil yield of a cultivated strain developed in India [23]. In *S. officinalis*, the fruits of sage cultivated in Tunisia are distinguished by oil yields of 0.39% [24]. Thereby seeds appeared as oil-moderate organs contrary to the different other parts of the species owing to the genus *Salvia* its aromatic reputation amongst Lamiaceae family.

As regards essential oil composition of *S. verbenaca*, it is worthy to note that tricyclene and camphor are also common to the seeds sample from Tunisia [12] so we may suggest that these two compounds are significant markers of

TABLE 1: Essential oil composition (%w/w) of *Salvia verbenaca* (Sv), *Salvia officinalis* (So), and *Salvia sclarea* (Ss) seeds.

Number	Compounds*	RI[a]	RI[b]	Sv	So	Ss	Identification
1	Hexanal	800	1093	0.42	—	—	GC-MS, Co-GC
2	(*E*)-2-Hexenal	852	1232	—	—	0.03	GC-MS, Co-GC
3	1-Hexanol	878	1360	—	1.29	—	GC-MS, Co-GC
4	Tricyclene	927	1014	0.96[a]	0.23[b]	0.08[c]	GC-MS
5	α-Thujene	930	1035	—	3.08[b]	7.48[a]	GC-MS
6	α-Pinene	939	1032	0.44[b]	1.26[a]	0.27[c]	GC-MS
7	δ-3-Carene	1012	1159	0.50[a]	0.03[b]	—	GC-MS
8	p-Cymene	1022	1280	0.37[b]	1.52[a]	0.12[c]	GC-MS
9	Limonene	1030	1203	—	—	0.17	GC-MS, Co-GC
10	β-Phellandrene	1030	1218	—	—	0.03	GC-MS
11	1,8-Cineole	1033	1213	—	6.66[a]	0.12[b]	GC-MS, Co-GC
12	*cis*-Sabinene hydrate	1068	1556	—	0.19	—	GC-MS
13	*cis*-Linalool oxide	1072	1478	—	0.13	—	GC-MS
14	α-Fenchone	1087	1406	0.26	—	—	GC-MS
15	*trans*-Linalool oxide	1087	1450	1.05	—	—	GC-MS
16	Linalool	1098	1553	0.84[b]	0.68[b]	24.25[a]	GC-MS, Co-GC
17	*n*-Undecane	1100	1100	2.65[a]	0.48[c]	0.75[b]	GC-MS, Co-GC
18	α-Thujone	1102	1429	0.52[b]	14.77[a]	—	GC-MS
19	*cis*-*allo*-Ocimene	1113	1382	1.55[a]	0.14[b]	1.46[b]	GC-MS
20	β-Thujone	1114	1451	—	4.30	—	GC-MS
21	Geijerene	1140	1338	—	1.01	—	GC-MS
22	Camphor	1143	1532	38.94[a]	13.08[b]	—	GC-MS
23	Borneol	1165	1719	—	3.54	—	GC-MS
24	Terpinen-4-ol	1177	1611	0.88[a]	0.09[c]	0.37[b]	GC-MS
25	p-Cymen-8-ol	1185	1864	—	0.17	—	GC-MS
26	δ-Terpineol	1187	1682	0.49[b]	2.42[a]	0.20[c]	GC-MS
27	α-Terpineol	1189	1706	2.03[a]	0.91[b]	0.20[c]	GC-MS
28	Myrtenal	1190	1648	—	0.55	—	GC-MS
29	Myrtenol	1194	1804	—	0.28	—	GC-MS
30	*cis*-Sabinol	1210	1800	—	0.18	—	GC-MS
31	Nerol	1228	1797	—	—	0.98	GC-MS, Co-GC
32	Linalyl acetate	1239	1565	2.53[b]	0.08[c]	6.90[a]	GC-MS
33	Geraniol	1254	1857	—	0.33[b]	2.79[a]	GC-MS, Co-GC
34	Bornyl acetate	1285	1597	0.79[a]	0.16[b]	—	GC-MS
35	Thymol	1293	2198	—	0.37[a]	0.10[b]	GC-MS
36	Carvacrol	1296	nd	—	0.83[b]	1.93[a]	GC-MS
37	δ-Elemene	1337	1479	3.97[a]	0.07[c]	0.85[b]	GC-MS
38	α-Cubebene	1348	1456	—	—	2.86	GC-MS
39	α-Terpinyl acetate	1353	1709	4.77[a]	1.81[b]	0.29[c]	GC-MS
40	Eugenol	1357	2192	—	0.83	—	GC-MS, Co-GC
41	Neryl acetate	1366	1733	2.40[a]	—	0.36[b]	GC-MS, Co-GC
42	α-Ylangene	1372	1493	—	0.04[b]	0.24[a]	GC-MS
43	α-Copaene	1376	1497	—	0.01[b]	4.08[a]	GC-MS
44	β-Damascenone	1381	1838	—	0.06	—	GC-MS
45	Geranyl acetate	1383	1765	—	—	1.94	GC-MS, Co-GC
46	β-Bourbonene	1384	1533	1.73[a]	—	0.35[b]	GC-MS
47	β-Elemene	1389	1600	—	0.16	—	GC-MS
48	Methyl eugenol	1402	2030	—	0.18	—	GC-MS, Co-GC
49	β-Caryophyllene	1413	1612	0.27[a]	0.19[b]	0.24[a]	GC-MS
50	β-Cubebene	1419	1549	1.50[a]	—	0.72[b]	GC-MS
51	Aromadendrene	1439	1628	0.66[a]	0.18[b]	—	GC-MS

TABLE 1: Continued.

Number	Compounds*	RI[a]	RI[b]	Sv	So	Ss	Identification
52	(Z)-β-Farnesene	1441	1668	1.76	—	—	GC-MS
53	α-Humulene	1454	1687	0.67[b]	3.71[a]	0.08[c]	GC-MS, Co-GC
54	allo-Aromadendrene	1460	1661	1.29[b]	1.43[a]	0.06[c]	GC-MS
55	α-Amorphene	1474	1680	—	0.47	—	GC-MS
56	γ-Muurolene	1477	1704	0.05	—	—	GC-MS
57	Germacrene-D	1479	1726	—	1.18[b]	5.88[a]	GC-MS
58	epi-Cubebol	1491	1900	—	0.34[a]	0.24[b]	GC-MS
59	Bicyclogermacrene	1494	1755	—	1.29[b]	4.29[a]	GC-MS
60	(E,E)-α-Farnesene	1506	1758	—	0.24	—	GC-MS
61	β-Bisabolene	1508	1741	1.10[a]	0.72[b]	—	GC-MS
62	γ-Cadinene	1513	1776	—	0.08	—	GC-MS
63	δ-Cadinene	1517	1773	—	0.53[a]	0.24[b]	GC-MS
64	α-Calacorene	1542	1942	—	—	0.10	GC-MS
65	Germacrene-B	1558	1854	—	0.08[b]	1.29[a]	GC-MS
66	(E)-Nerolidol	1563	2050	—	1.41	—	GC-MS
67	Spathulenol	1575	2144	—	0.08[a]	0.03[b]	GC-MS
68	Caryophyllene oxide	1579	2008	7.28[a]	0.16[c]	3.18[b]	GC-MS
69	Viridiflorol	1592	2104	—	2.66	—	GC-MS
70	Humulene epoxide I	1596	2045	—	—	1.55	GC-MS
71	Humulene epoxide II	1606	2071	—	0.25	—	GC-MS
72	T-Cadinol	1642	2187	—	0.15[b]	0.52[a]	GC-MS
73	α-Cadinol	1643	nd	—	0.41[b]	0.88[a]	GC-MS
74	β-Eudesmol	1649	nd	3.76[a]	0.47[c]	1.72[b]	GC-MS
75	13-epi-Manool	Nd	nd	5.61[a]	2.22[b]	0.59[c]	GC-MS
Compound classes							
	Alcohols			2.46[a]	0.42[b]	0.03[c]	
Aliphatic hydrocarbons	Alkanes			—	1.29	—	
	Aldehydes			0.84[a]	0.48[c]	0.75[b]	
Monoterpene hydrocarbons				3.82[c]	6.45[b]	9.60[a]	
Oxygenated monoterpenes				57.30[a]	50.14[b]	38.38[c]	
Sesquiterpene hydrocarbons				13.00[b]	11.39[b]	21.27[a]	
Norisoprenoids with 13 carbons				—	0.06	—	
Phenylpropanes				—	1.01	—	
Phenols				—	1.10[b]	2.03[a]	
Oxygenated sesquiterpenes				11.04[a]	5.93[c]	8.12[b]	
Labdane type diterpenes				5.61[a]	2.22[b]	0.59[c]	
Total				**92.03[a]**	**80.17[c]**	**80.79[b]**	

*Components are listed according to their elution on apolar column (HP-5). RI: retention indices relative to C_8–C_{22} n-alkanes on the [a]HP-5 and [b]HP-Innowax columns; nd: not detected; GC/MS: identification based on comparison of mass spectra; Co-GC: identification based on retention time comparison to authentic compounds. Values (means of three replicates) in the same lines with different letters (a–c) are significantly different at $P < 0.05$.

essential oil compounds of *S. verbenaca* seeds whatever the sample origin is. Interestingly, the essential oil of the studied seeds could be employed as antimicrobial agent since their high percentage of camphor is associated with an efficient antimicrobial activity according to Magiatis et al. [25] and Bougatsos et al. [26].

In *S. officinalis*, the predominance of α-thujone, camphor, and 1,8-cineole endowed to the essential oil an antimicrobial activity [27]. Viridiflorol and α-humulene identified in several *S. officinalis* essential oils showed antiacetyl-cholinesterase activity used in the treatment of Alzheimer's

disease [28], antifungal property [29], and cytotoxic activity against some tumor cell lines [30]. Furthermore, the labdane type diterpene 13-*epi*-manool displays *in vitro* a cytotoxic activity against human leukemic cell lines [31]. We noted that seeds had a similar qualitative composition to aerial parts with the predominance of α-thujone, β-thujone, camphor, and 1,8-cineole. These monoterpenes are taken as significant parameter to differentiate *S. officinalis* from other species [32]. Their amount is lower than that in leaves but matches the ranges of the standard ISO 9909 for official sage oil except for the toxic ketone α-thujone which had a lower amount (14.77%). These findings promote the use of the seeds essential oil in food industry.

Similar to the studied *S. sclarea* essential oil, the sesquiterpenes were mainly composed of germacrene-D, α-copaene, and bicyclogermacrene in plant inflorescence according to Carrubba et al. [11] and Lorenzo et al. [33]. In our sample, the essential oil of *S. sclarea* flowering shoots raised in experimental plots in India was characterised by linalool (36.6–41.9%) and linalyl acetate (13.2–19.2%) as main compounds [23]. The wild-growing *S. sclarea* collected at flowering stage from central Greece [7] and Spain [6] showed a close similar composition regarding linalool and linalyl acetate (30.43%, 32.97% and 19.75%, 16.85%, resp.). Generally, *S. sclarea* essential oil is extracted from flowering shoots which are found to be rich in linalool and linalyl acetate. According to Carrubba et al. [11] high amounts of linalool and linalyl acetate are typical of good quality oil suitable for flavouring purposes. Furthermore, linalool plays a major role as anti-inflammatory suggesting that linalool-producing species are potentially anti-inflammatory agents [34]. Moreover, linalool has an ecological role since it constitutes one of the common components of floral scent that can attract a large variety of insects that convey pollen [35]. The present study showed that essential oil derived from seeds had a similar composition to the flowering parts. Thus, seeds seemed to display the same enzymatic patterns of the essential oil biosynthesis as the flowers, while the essential oil composition of *S. sclarea* vegetative organs displayed different qualitative trends from reproductive parts.

6. Conclusions

Overall, it emerges that tricyclene and camphor were biochemical markers of the essential oil of *S. verbenaca* seeds. Being rich in camphor, seeds could be used as antimicrobial agent. Another point that should be highlighted is that *S. officinalis* seeds had the same α-thujone chemotype as leaves, whereas these two organs showed some quantitative differences leading to the safe use of seeds essential oil in food industry. From a qualitative standpoint, seeds of *S. sclarea* seemed to have the same enzymatic trend as flowers characterized by the prevalence of linalool. It is noteworthy to mention that linalool-producing seeds as *S. sclarea* were suitable for flavouring purposes and constitute potential anti-inflammatory agents.

Conflict of Interests

The authors declare that there is no conflict of interests regarding the publication of this paper.

References

[1] I. C. Hedge, *A Global Survey of the Biogeography of Labiatae*, Royal Botanical Gardens, Kew, UK, 1992.

[2] G. Penso, *Index Plantarum Medicinalium Totius Mundi Eorumque Synonymorum*, OEMF, Milano, Italy, 1983.

[3] A. Y. Leung and S. Foster, *Encyclopedia of Common Natural Ingredients Used in Food, Drugs and Cosmetics*, John Wiley & Sons, New York, NY, USA, 2nd edition, 1996.

[4] T. A. Al-Howiriny, "Chemical composition and antimicrobial activity of essential oil of *Salvia verbenaca*," *Biotechnology*, vol. 1, pp. 45–48, 2002.

[5] F. Chialva and F. Monguzzi, "Composition of the essential oils of five *Salvia* species," *Journal of Essential Oil Research*, vol. 4, pp. 447–455, 1992.

[6] M. E. Torres, A. Velasco-Negueruela, M. J. Pérez-Alonso, and M. G. Pinilla, "Volatile constituents of two Salvia species grown wild in Spain," *Journal of Essential Oil Research*, vol. 9, no. 1, pp. 27–33, 1997.

[7] D. Pitarokili, O. Tzakou, and A. Loukis, "Essential oil composition of *Salvia verticillata*, *S. verbenaca*, *S. glutinosa* and *S. candidissima* growing wild in Greece," *Flavour and Fragrance Journal*, vol. 21, no. 4, pp. 670–673, 2006.

[8] V. Radulescu, S. Chiliment, and E. Oprea, "Capillary gas chromatography-mass spectrometry of volatile and semi-volatile compounds of Salvia officinalis," *Journal of Chromatography A*, vol. 1027, no. 1-2, pp. 121–126, 2004.

[9] P. Avato, I. M. Fortunato, C. Ruta, and R. D'Elia, "Glandular hairs and essential oils in micropropagated plants of *Salvia officinalis* L.," *Plant Science*, vol. 169, no. 1, pp. 29–36, 2005.

[10] S. Marie, M. Maksimovic, and M. Milos, "The impact of the locality altitudes and stages of development on the volatile constituents of *Salvia officinalis* L. from Bosnia and Herzegovina," *Journal of Essential Oil Research*, vol. 18, no. 2, pp. 178–180, 2006.

[11] A. Carrubba, R. La Torre, R. Piccaglia, and M. Marotti, "Characterization of an Italian biotype of clary sage (*Salvia sclarea* L.) grown in a semi-arid Mediterranean environment," *Flavour and Fragrance Journal*, vol. 17, no. 3, pp. 191–194, 2002.

[12] M. B. Taarit, K. Msaada, and B. Marzouk, "Chemical composition of fatty acids and essential oils of *Salvia verbenaca* L. Seeds from Tunisia," *Agrochimica*, vol. 54, no. 3, pp. 129–141, 2010.

[13] M. B. Taarit, K. Msaada, K. Hosni, T. Chahed, and B. Marzouk, "Essential oil composition of *Salvia verbenaca* L. growing wild in Tunisia," *Journal of Food Biochemistry*, vol. 34, no. 1, pp. 142–151, 2010.

[14] M. Ben Taarit, K. Msaada, K. Hosni, N. Ben Amor, B. Marzouk, and M. E. Kchouk, "Chemical composition of the essential oils obtained from the leaves, fruits and stems of *Salvia verbenaca* L. from the northeast region of Tunisia," *Journal of Essential Oil Research*, vol. 22, no. 5, pp. 449–453, 2010.

[15] R. Ayerza, W. Coates, and M. Lauria, "Chia seed (*Salvia hispanica* L.) as an ω-3 fatty acid source for broilers: influence on fatty acid composition, cholesterol and fat content of white and dark meats, growth performance, and sensory characteristics," *Poultry Science*, vol. 81, no. 6, pp. 826–837, 2002.

[16] B. Heuer, Z. Yaniv, and I. Ravina, "Effect of late salinization of chia (*Salvia hispanica*), stock (*Matthiola tricuspidata*) and evening primrose (*Oenothera biennis*) on their oil content and quality," *Industrial Crops and Products*, vol. 15, no. 2, pp. 163–167, 2002.

[17] J. E. Kinsella, K. S. Broughton, and J. W. Whelan, "Dietary unsaturated fatty acids: interactions and possible needs in relation to eicosanoid synthesis," *Journal of Nutritional Biochemistry*, vol. 1, no. 3, pp. 123–141, 1990.

[18] I. C. Hedge and L. Salvia, *Flora of Turkey and the East Aegean Islands*, Edinburgh University Press, Edinburgh, UK, 1982.

[19] A. Estilai, A. Hashemi, and K. Truman, "Chromosome number and meiotic behavior of cultivated chia, *Salvia hispanica* (Lamiaceae)," *Hortscience*, vol. 25, pp. 1646–1647, 1990.

[20] T. Baytop, *Türkiye'de bitkilerle tedavi (geçmiste ve bugün)*, Baskı, Nobel Tıp Kitapevleri, Çapa-Ýstanbul, Konak-Ýzmir, Sıhhıye-Ankara, Turkey, 1999.

[21] R. P. Adams, *Identification of Essential Oil Components by Gas Chromatography/Quadrupole Mass Spectroscopy*, Allured, Carol Stream, Ill, USA, 2001.

[22] Statsoft, *STATISTICA for Windows (Computer Program Electronic Manual)*, StatSoft, Tulsa, Okla, USA, 1998.

[23] S. K. Lattoo, R. S. Dhar, A. K. Dhar, P. R. Sharma, and S. G. Agarwal, "Dynamics of essential oil biosynthesis in relation to inflorescence and glandular ontogeny in *Salvia sclarea*," *Flavour and Fragrance Journal*, vol. 21, no. 5, pp. 817–821, 2006.

[24] M. Ben Taarit, K. Msaada, K. Hosni, M. Hammami, M. E. Kchouk, and B. Marzouk, "Plant growth, essential oil yield and composition of sage (*Salvia officinalis* L.) fruits cultivated under salt stress conditions," *Industrial Crops and Products*, vol. 30, no. 3, pp. 333–337, 2009.

[25] P. Magiatis, A. L. Skaltsounis, I. Chinou, and S. Haroutounian, "Chemical composition and *in vitro* antimicrobial activity of the essential oils of three Greek *Achillea* species," *Zeitschrift für Naturforschung*, vol. 57, pp. 287–290, 2002.

[26] C. Bougatsos, O. Ngassapa, D. K. B. Runyoro, and I. B. Chinou, "Chemical composition and *in vitro* antimicrobial activity of the essential oils of two *Helichrysum species* from Tanzania," *Zeitschrift fur Naturforschung*, vol. 59, no. 5-6, pp. 368–372, 2004.

[27] V. Jalsenjak, S. Peljnjak, and D. Kuštrak, "Microcapsules of sage oil: essential oils content and antimicrobial activity," *Pharmazie*, vol. 42, no. 6, pp. 419–420, 1987.

[28] M. Miyazawa, H. Watanabe, K. Umemoto, and H. Kameoka, "Inhibition of acetylcholinesterase activity by essential oils of *Mentha* species," *Journal of Agricultural and Food Chemistry*, vol. 46, no. 9, pp. 3431–3434, 1998.

[29] H. J. M. Gijsen, J. B. P. A. Wijnberg, G. A. Stork, A. de Groot, M. A. De Waard, and J. G. M. Van Nistelrooy, "The synthesis of mono- and dihydroxy aromadendrane sesquiterpenes, starting from natural (+)-aromadendrene-III," *Tetrahedron*, vol. 48, no. 12, pp. 2465–2476, 1992.

[30] S. L. da Silva, P. M. Figueiredo, and T. Yano, "Cytotoxic evaluation of essential oil from *Zanthoxylum rhoifolium* Lam. leaves," *Acta Amazonica*, vol. 37, no. 2, pp. 281–286, 2007.

[31] K. Dimas, C. Demetzos, M. Marsellos, R. Sotiriadou, M. Malamas, and D. Kokkinopoulos, "Cytotoxic activity of labdane type diterpenes against human leukemic cell lines *in vitro*," *Planta Medica*, vol. 64, no. 3, pp. 208–211, 1998.

[32] J. Bruneton, *Pharmacognosy, Phytochemistry, Medicinal Plants*, Tech. & Doc. Lavoisier, Paris, France, 1999.

[33] D. Lorenzo, D. Paz, P. Davies et al., "Characterization and enantiomeric distribution of some terpenes in the essential oil of a Uruguayan biotype of *Salvia sclarea* L.," *Flavour and Fragrance Journal*, vol. 19, no. 4, pp. 303–307, 2004.

[34] A. T. Peana, M. D. L. Moretti, and C. Juliano, "Chemical composition and antimicrobial action of the essential oils of *Salvia desoleana* and *S. sclarea*," *Planta Medica*, vol. 65, no. 8, pp. 752–754, 1999.

[35] E. Pichersky, J. P. Noel, and N. Dudareva, "Biosynthesis of plant volatiles: nature's diversity and ingenuity," *Science*, vol. 311, no. 5762, pp. 808–811, 2006.

19

Synthesis of Disodium Salt of Sulfosuccinate Monoester from the Seed Oil of *Terminalia catappa* and Its Inhibitive Effect on the Corrosion of Aluminum Sheet in 1 M HCl

Adewale Adewuyi,[1] **Adewale Dare Adesina,**[2] **and Rotimi A. Oderinde**[2]

[1] *Department of Chemical Sciences, Faculty of Natural Sciences, Redeemer's University, Mowe, Ogun State, Nigeria*
[2] *Industrial Unit, Department of Chemistry, University of Ibadan, Ibadan, Oyo State, Nigeria*

Correspondence should be addressed to Adewale Adewuyi; walexy62@yahoo.com

Academic Editor: Claudio Cameselle

Oil was extracted from the seed of *Terminalia catappa* and used to synthesize disodium salt of sulfosuccinate monoester using simple reaction mechanism. The disodium salt of sulfosuccinate monoester was applied as corrosion inhibitor of aluminum sheet in 1 M HCl via weight loss method. The adsorption was found to obey Langmuir isotherm. The results presented disodium salt of sulfosuccinate monoester as an efficient inhibitor of aluminum sheet corrosion in 1 M HCl.

1. Introduction

Corrosion is most commonly referred to as the degradation of a material due to its reaction with its environment. Such degradation may mean deterioration of the physical properties of the material which may be in form of weakening of the material due to loss of cross-sectional area, shattering due to hydrogen embrittlement, or cracking due to sunlight exposure. Corrosion is usually found in several materials but most especially in metals; these materials have both domestic and industrial uses but the existence of corrosion which can take place under acidic or alkaline medium has resulted in limitation to their use. Importance of protection against corrosion in acidic or alkaline solutions is known to be increased by the fact that metals are more susceptible to be attacked in aggressive media, most of which are the commonly exposed metals (such as mild steel) in industrial environments [1]. The corrosion process is usually slowed down in various ways one of which is the use of corrosion inhibitors which when added in small amounts to a corroding environment decreases the rate of attack by such environment on material [2–4].

Being the third most abundant element and the most abundant metal, aluminum has found several industrial applications which may be due to its economical considerations and the fact that its corrosion falls into general attack [5]. Thermodynamically, aluminum is expected to have a low corrosion resistance. The high corrosion resistance is due to the presence of a thin, compact film of adherent aluminum oxide on the surface which is formed on exposure to either air or water. This aluminum oxide dissolves in some chemicals, notably strong acids and alkaline solutions. When the oxide film is removed, the metal corrodes rapidly by uniform dissolution. So study of aluminum sheet corrosion phenomena has become important particularly in acidic media because of the increased industrial applications of acid solutions [6–9].

In the past time, use of inhibitors has been one of the most common different protective means used to control corrosion. Most inhibitors reported are synthetic organic compounds containing heteroatoms, such as O, N, and S, and multiple bonds [10, 11]; these heteroatoms have been established to have high electron density that contributes to the inhibitory capacity of such organic compounds. Although these synthetic organic compounds are widely used, their use as corrosion inhibitor has limitations such as being expensive, being nonrenewable, and being toxic to both plant and animal in the environment [12, 13]. Some efforts have been made to develop cheap and nontoxic corrosion inhibitors but quite a number of them have reduced inhibitory activity at low concentration or are toxic at high concentration when they get into the environment. Due to superb environmental stability, ease of sustainability, and low level of toxicity, several plant extracts have been considered and reported as corrosion inhibitors [14–16] but it has been established that their efficiency may be improved upon by simple modification in terms of chemical functionality. This has also shown the need for green corrosion inhibitors and their importance over synthetic chemical products. The use of plant extract with little modification is of much importance and economically viable because, aside from being ecofriendly, they are renewable, easy to modify, and inexpensive [17–20]. Thus, they can be used as feed stock for oleochemicals which can serve as green corrosion inhibitors. Terminalia catappa seed oil is an example of plant extract that can be utilized to achieve such purpose.

Terminalia catappa is a large tropical tree in the leadwood family, Combretaceae. It is commonly called almond, a small deciduous tree, growing 4–10 m (13–33 feet) in height, with trunk of up to 30 cm (12 inches) in diameter. The leaves are 3–5 inches long with a serrated margin and a 2.5 cm (1 inch) petiole. The flowers are white or pale pink, 3–5 cm (1-2 inches) diameter with five petals, produced singly or in pairs before the leaves in early spring [21]. The antioxidant property of the solvent extract of the leaves has been reported [22]. The almond fruit is about 3.5–6 cm long. The seed has been reported to contain 6% water, 31% lipid, 29% protein, 25% carbohydrate, 3% mineral, 2% vitamins, and 4% sugars; saturated fat (palmitic acid) was 6%, monosaturated fat (oleic acid) was 64%, and polyunsaturated fat (linoleic acid) was 26% while major minerals were Ca-14%, Mg-16%, P-27%, and K-42% [23]. The oil has been reported to contain high levels of unsaturated fatty acids, especially oleic and linoleic; thus Terminalia catappa oil can be classified in the oleic-linoleic acid group [24].

Apart from the domestic use of plant products, they have also found wide application as sources of oleochemicals [25]. Oleochemicals are completely biodegradable and so could replace a number of petrochemicals. Sulfosuccinate is an example of an oleochemical that is produced from a renewable source, biodegradable and environmentally friendly. Sulfosuccinates have been reported to exhibit compatibility with chromium and have no adverse effect on the textile strength of processed fibre with wide range of applications which includes household formulations, textiles, polymers, paints and coating, agriculture, and production of shampoos [26]. Sulfosuccinates are known with excellent wetting properties which suggest them as possible corrosion inhibitors. At present, there is no report on their use as corrosion inhibitors. Since they are biodegradable, ecofriendly, and relatively cheap with the presence of heteroatoms in their structure, it will be important to determine their anticorrosion capacity.

In continuation of our search for cheap oleochemicals that can be used as corrosion inhibitors, the present study synthesized disodium salt of sulfosuccinate monoester from Terminalia catappa seed oil and investigated its inhibiting effect on aluminum sheet corrosion in strong acidic solution using weight loss method.

2. Materials and Methods

2.1. Materials. Seeds of Terminalia catappa were collected from the Botanical Garden, University of Ibadan. They were manually cracked, air-dried, and milled in a blender. The powdered seeds were finally extracted with hexane in a soxhlet extractor as described by Adewuyi and Oderinde [27].

2.2. Synthesis of Fatty Ethanolamide from the Seed Oil of Terminalia catappa. This was achieved as previously described by Adewuyi et al. [28] with little modification. Briefly, the oil of Terminalia catappa was reacted with diethanolamine in ratio 1 (oil) : 3 (diethanolamine) in a 250 mL round bottom flask equipped with a magnetic stirrer, a thermometer, and a condenser. The flask was placed in an oil bath while the reaction temperature was gradually increased and maintained at 140°C. The reaction mixture was continuously stirred for 10 h to form fatty ethanolamide. At the end of the reaction, the mixture was concentrated on a rotary evaporator after which the product formed was dissolved in a mixture of methanol and chloroform [50/50 (v/v)]. The solvent was later removed in a rotary evaporator. Then, acetonitrile was added to the resultant solid and the solution was cooled in an ice bath. The amide precipitated out and was subsequently recovered by filtration using Whatman filter paper [29]. This is shown in Scheme 1.

2.3. Synthesis of Disodium Salt of Sulfosuccinate Monoester. The fatty ethanolamide synthesized was transferred into a round bottom flask and heated to 110°C while maleic acid anhydride (10 g, 0.1 mole) was gently added and stirred and the temperature was kept constant at 110°C. The reaction mixture was continuously stirred for 3 h while an aqueous solution of 30% sodium bisulphite (15.71 g, 0.1 mol) was added to the reaction mixture. The reaction temperature was gradually raised to 130°C with continuous stirring at this temperature for 1 h while the pH of the reaction mixture was adjusted using aq. NaOH. The product obtained was disodium salt of sulfosuccinate monoester as illustrated in Scheme 2. The synthesized disodium salt of sulfosuccinate monoester was purified by washing with petroleum ether for about 2 to 3 times. This removes any unwanted impurities and unreacted materials. The obtained product was filtered, dried, and analyzed using FTIR.

2.4. Corrosion Study. The corrosion inhibition study of disodium salt of sulfosuccinate monoester on aluminum sheet was carried out in 1 M HCl solution using weight loss measurement method. In this case, HCl was prepared to initiate the corrosion while disodium salt of sulfosuccinate monoester was used as the corrosion inhibitor. A cold rolled aluminum sheet of dimensions 5.0 cm by 5.0 cm with an area of 25.0 cm^2 was washed, dried, and accurately weighed. After weighing accurately, the aluminum sheets were immersed in a beaker which contained 1 M HCl with and without addition of

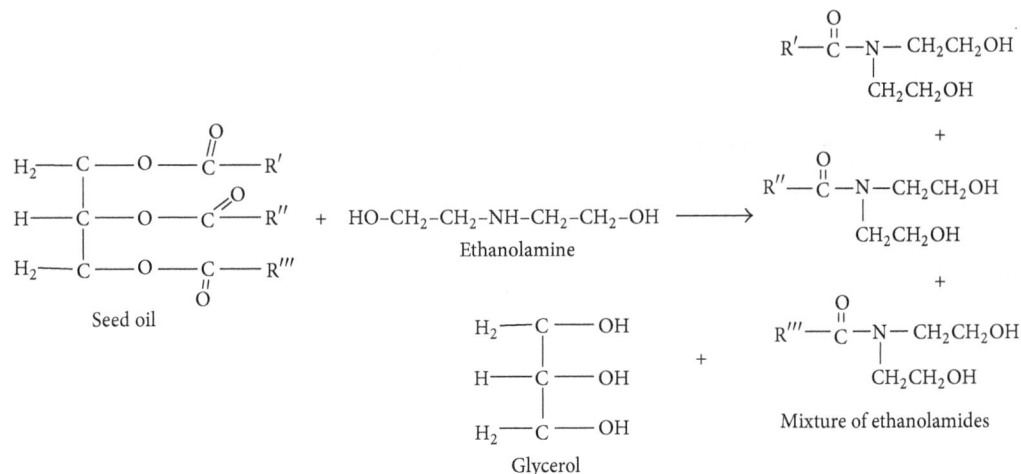

SCHEME 1: Synthesis of ethanolamide.

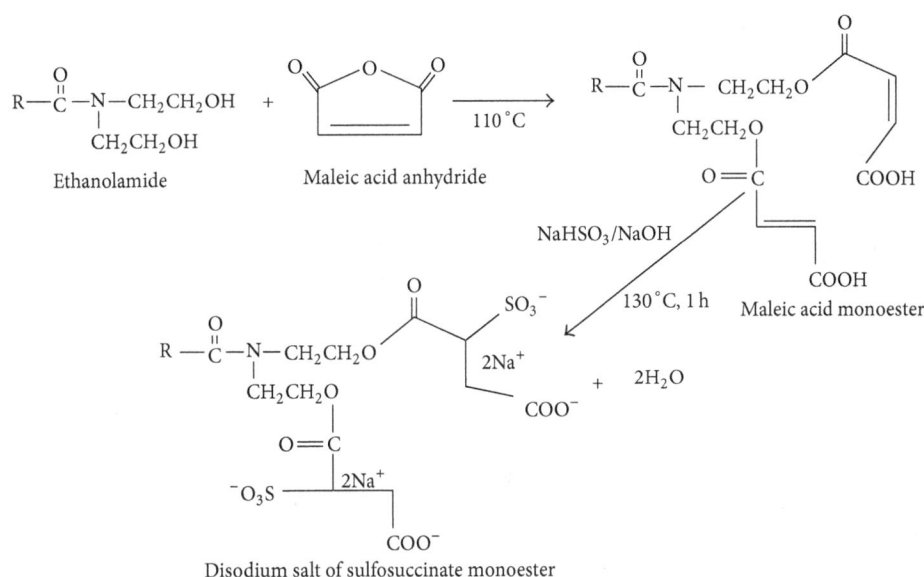

SCHEME 2: Synthesis of disodium salt of sulfosuccinate monoester.

disodium salt of sulfosuccinate monoester. The solution of acid without the disodium salt of sulfosuccinate monoester was used as the control in this study while the concentration of disodium salt of sulfosuccinate monoester in the other solution varied from $0.50\,g/L$ to $3.00\,g/L$. All the aggressive acid solutions were opened to air for a period of 6 h and at an interval of 1 h; the aluminum sheets were taken out of solution, washed, dried, and reweighed accurately. The experiments were carried out in duplicate, and the average weight loss of the cold rolled aluminum sheets was obtained and recorded.

3. Results and Discussion

3.1. Synthesis of Disodium Salt of Sulfosuccinate Monoester.
Figure 1 shows the peaks for the FTIR analysis carried out on the oil of Terminalia catappa (a), fatty ethanolamide

(b), and disodium salt of sulfosuccinate monoester (c) using Shimadzu FTIR-400S. It was observed that the Terminalia catappa oil, ethanolamide, and disodium salt of sulfosuccinate monoester showed characteristic absorption bands at $2924\,cm^{-1}$ and $2852\,cm^{1}$ corresponding to the C–H stretching of methyl (–CH_3) and methylene (–CH_2) functional groups, respectively. The absorption band present at $721\,cm^{-1}$ in the oil, ethanolamide, and disodium salt of sulfosuccinate monoester spectra can be attributed to the rocking motion associated with –CH_2 groups in an open chain while $1465\,cm^{-1}$ suggests –CH_2 of bending vibrations of alkanes. The absorption bands representing the C=O stretching of ester occurred at $1745\,cm^{1}$ in Terminalia catappa oil. This C=O stretching band of ester disappeared in the ethanolamide and disodium salt of sulfosuccinate monoester with the appearance of a new peak at $1634\,cm^{1}$ corresponding to the C=O stretching of amide functional group. The –OH

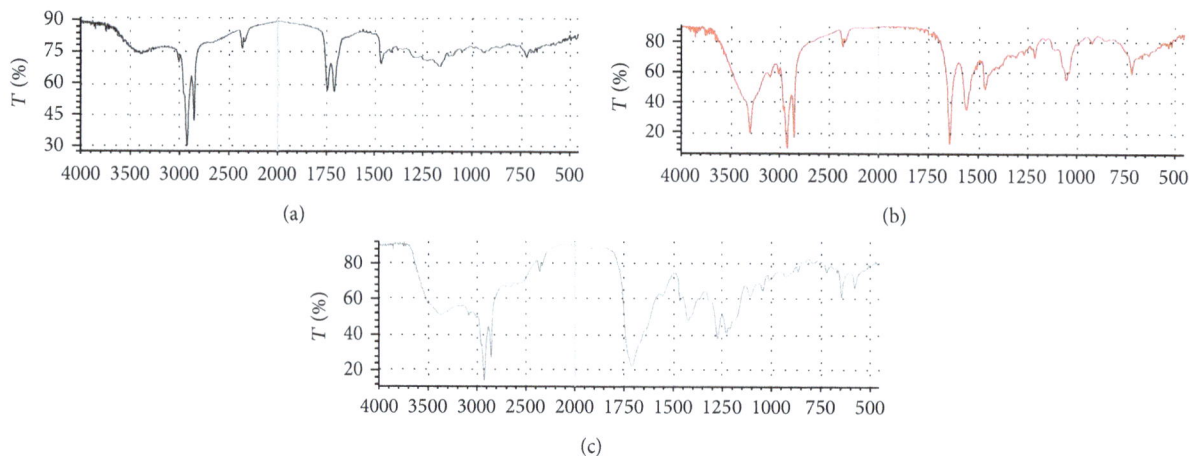

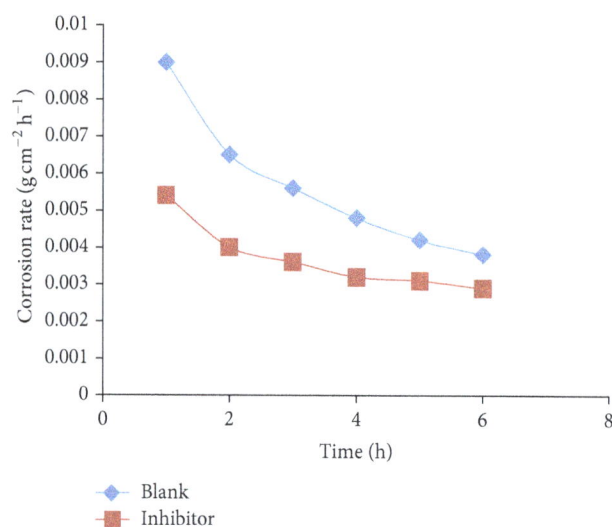

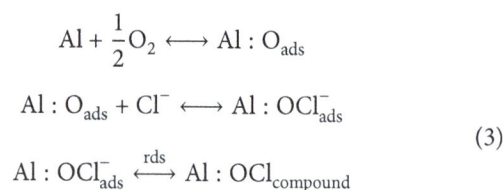

FIGURE 1: FTIR result of the oil (a), fatty ethanolamide (b), and disodium salt of sulfosuccinate monoester (c).

functional group vibrational frequency in the ethanolamide was found at 3296 cm^{-1}. The N–H stretching vibration was also observed at 3371 cm^{-1}. The C–H bending vibration of alkane was observed at 1454 cm^{-1} while the O–H stretching vibration was absent in the sulfosuccinate monoester indicating the formation of the product.

3.2. Corrosion Study. The corrosion rate was determined using the following expression [30]:

$$R = \frac{\Delta W}{At}, \tag{1}$$

where R (g cm^{-2} h^{-1}) is the corrosion rate, ΔW is the average weight loss after immersion, A is the surface area of the aluminum sheet, and t is the total time (6 h) of immersion. The inhibition efficiency (% E_w) was also calculated using the following equation [31]:

$$\%E_w = \left(R_o - \frac{R_t}{R_o}\right) \times 100, \tag{2}$$

where R_t and R_o are corrosion rates of aluminum sheet with and without inhibitor, respectively.

The corrosion rate of the aluminum sheet immersed into the blank (solution without the inhibitor) was faster than that of the solution with the inhibitor; this is shown in Figure 2.

At the initial stage, the weight loss of aluminum sheet in the blank was almost double that of the solution with the inhibitor but as time went on the weight loss of aluminum reduced in the blank but was still higher than what was observed in the case of the inhibitor. This observation may be due to the fact that corrosion started immediately on exposure of the aluminum sheet to the aggressive HCl solution but with time there may have been the formation of a protective covering on the surface of the aluminum sheet which reduced the rate but later lost its protective capacity

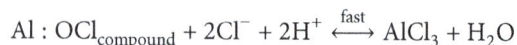

FIGURE 2: Corrosion rate versus time for both blank and inhibitor.

with time as shown in equation below with the formation of AlCl$_3$ [32]:

$$Al + \frac{1}{2}O_2 \longleftrightarrow Al : O_{ads}$$

$$Al : O_{ads} + Cl^- \longleftrightarrow Al : OCl^-_{ads}$$

$$Al : OCl^-_{ads} \overset{rds}{\longleftrightarrow} Al : OCl_{compound} \tag{3}$$

$$Al : OCl_{compound} + 2Cl^- + 2H^+ \overset{fast}{\longleftrightarrow} AlCl_3 + H_2O$$

Corrosion inhibitors or a mixture of corrosion inhibitors have been reported to form a protective film as a result of the reaction of the aggressive solution with the corroding surface of which inhibitors may impede the anodic, the cathodic, or both electrochemical reactions [4]. Disodium salt of sulfosuccinate monoester used as inhibitor may have formed a protective covering on the surface of the aluminum sheet. As shown in Scheme 2, disodium salt of sulfosuccinate

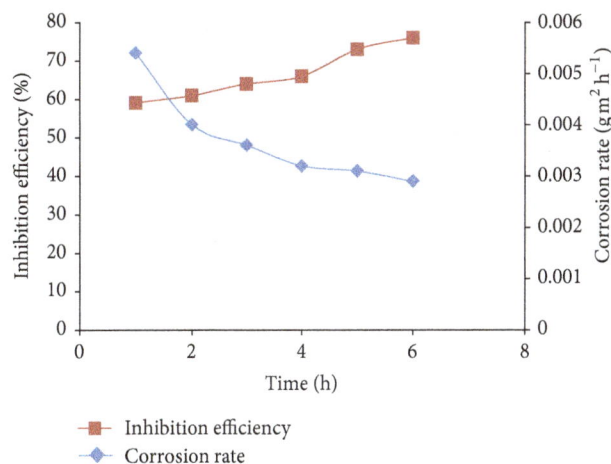

Inhibition efficiency
Corrosion rate

FIGURE 3: Comparison of the inhibition efficiency with corrosion rate over a period of time.

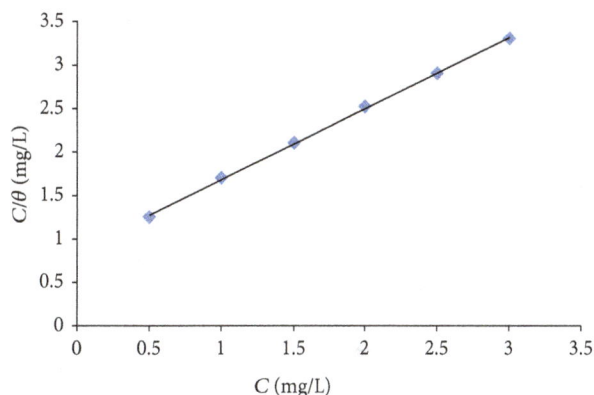

FIGURE 4: Langmuir adsorption plot for aluminum sheet in 1 M HCl containing disodium salt of sulfosuccinate monoester.

TABLE 1: Values of corrosion rate, inhibition efficiency, and surface covering.

Inhibitor (g/dm^3)	Surface coverage (θ)	IE (%)	Corrosion rate (gm^2 h^{-1})
0.5	0.400	40.0	0.0054
1.0	0.588	58.8	0.0037
1.5	0.714	71.4	0.0026
2.0	0.794	79.4	0.0019
2.5	0.862	86.2	0.0012
3.0	0.909	90.9	0.0008

monoester has heteroatoms such as oxygen and nitrogen and also the presence of π (pie) electron systems which have been reported in the past to play active role in adsorption [33]. These heteroatoms and the π electron systems are rich in electrons and may have interacted with the surface of the aluminum via this electron density to form the protective covering at the surface of the aluminum sheet.

Figure 3 presents the correlations between the inhibition efficiencies and the corrosion inhibition rates over a period of 6 h for the inhibitor. The inhibition efficiency of disodium salt of sulfosuccinate monoester was found to increase with time while the corrosion rate reduced with time in the presence of disodium salt of sulfosuccinate monoester. This observation must have been due to the fact that the inhibitor adsorbed on the surface of aluminum and was able to reduce the interaction between the aluminum surface and the aggressive acid solution [34].

The different values of corrosion rate, inhibition efficiency, and surface covering at various concentrations of disodium salt of sulfosuccinate monoester are presented in Table 1.

It was observed that the inhibition efficiency increased just as the concentration of disodium salt of sulfosuccinate monoester increased. This was also noticed in the surface covering of the aluminum sheet which increased as the concentration of disodium salt of sulfosuccinate monoester

in solution increased; this trend has also been reported by Wang et al. [35]. This may, apparently, be accounted for as the inhibitor interacting with the surface of the metal, thus blocking the active sites to form a barrier against infiltration of the aggressive electrolyte solution since the process of corrosion is considered to be electrolytic in nature [36, 37].

Attempt was made to fit the values of the surface coverage (θ) into different adsorption isotherms but the best fit was obtained with Langmuir adsorption isotherm using the following equation as proposed by Langmuir [38]:

$$\frac{C}{\theta} = \frac{1}{K_{ads}} + C, \tag{4}$$

where C is the concentration of the inhibitor and K_{ads} is the adsorptive equilibrium constant.

As shown in Figure 4, a plot of C/θ versus C gave straight lines with R^2 value of 0.999 which suggested a monolayer adsorption of the inhibitor at the surface of aluminum [34].

The essential characteristic of this isotherm can be expressed with the following equation:

$$K_R = \frac{1}{1 + K_{ads}C}, \tag{5}$$

where K_R is the equilibrium parameter, K_{ads} is the Langmuir constant, and C is the inhibitor concentration. K_R describes the type of the isotherm accordingly. If $K_R > 1$ the process is unfavourable, if $K_R = 1$, the process is linear, if $0 < K_R < 1$, the process is favourable, and when $K_R = 0$ the process is irreversible [39, 40]. In the present study, K_R value was found to be less than 1 and greater than 0 indicating that the adsorption process was favourable and reversible.

4. Conclusion

Disodium salt of sulfosuccinate monoester was synthesized from the seed oil of *Terminalia catappa* which was found to contain oleic acid as the most dominant fatty acid. The disodium salt of sulfosuccinate monoester had good inhibitive capacity against the corrosion of aluminum sheet in 1 M HCl with inhibition efficiency increasing as the concentration of

174

Textbook of Chemistry

disodium salt of sulfosuccinate monoester increased while the corrosion rate decreased.

Conflict of Interests

The authors declare that there is no conflict of interests regarding the publication of this paper.

Acknowledgments

The authors are most grateful to the Department of Chemistry, University of Ibadan, Ibadan, Oyo State, and also the Department of Chemical Sciences, Redeemer's University, Mowe, Ogun State, for supplying chemicals, equipment, and laboratory space for this research work.

References

[1] M. A. Amin, K. F. Khaled, Q. Mohsen, and H. A. Arida, "A study of the inhibition of iron corrosion in HCl solutions by some amino acids," *Corrosion Science*, vol. 52, no. 5, pp. 1684–1695, 2010.

[2] D. A. Jones, "Coatings and inhibitors," in *Principles and Prevention of Corrosion*, A. D. Jones, Ed., pp. 477–512, Prentice Hall, Upper Saddle River, NJ, USA, 2nd edition, 1996.

[3] M. A. Ash, *Handbook of Corrosion Inhibitors*, NACE International, Houston, Tex, USA, 2001.

[4] H. A. Videla and L. K. Herrera, "Understanding microbial inhibition of corrosion. A comprehensive overview," *International Biodeterioration and Biodegradation*, vol. 63, no. 7, pp. 896–900, 2009.

[5] M. Heydari and M. Javidi, "Corrosion inhibition and adsorption behaviour of an amido-imidazoline derivative on API 5L X52 steel in CO_2-saturated solution and synergistic effect of iodide ions," *Corrosion Science*, vol. 61, pp. 148–155, 2012.

[6] J. L. Mora-Mendoza and S. Turgoose, "Fe_3C influence on the corrosion rate of mild steel in aqueous CO_2 systems under turbulent flow conditions," *Corrosion Science*, vol. 44, no. 6, pp. 1223–1246, 2002.

[7] D. S. Carvalho, C. J. B. Joia, and O. R. Mattos, "Corrosion rate of iron and iron-chromium alloys in CO_2 medium," *Corrosion Science*, vol. 47, no. 12, pp. 2974–2986, 2005.

[8] A. Ostovari, S. M. Hoseinieh, M. Peikari, S. R. Shadizadeh, and S. J. Hashemi, "Corrosion inhibition of mild steel in 1 M HCl solution by henna extract: a comparative study of the inhibition by henna and its constituents (Lawsone, Gallic acid, α-d-Glucose and Tannic acid)," *Corrosion Science*, vol. 51, no. 9, pp. 1935–1949, 2009.

[9] G. A. Zhang and Y. F. Cheng, "Corrosion of X65 steel in CO_2-saturated oilfield formation water in the absence and presence of acetic acid," *Corrosion Science*, vol. 51, no. 8, pp. 1589–1595, 2009.

[10] M. Elayyachy, A. El Idrissi, and B. Hammouti, "New thio-compounds as corrosion inhibitor for steel in 1 M HCl," *Corrosion Science*, vol. 48, no. 9, pp. 2470–2479, 2006.

[11] S. H. S. Dananjaya, M. Edussuriya, and A. S. Dissanayake, "Inhibition action of lawsone on the corrosion of mild steel in acidic media," *The Online Journal of Science and Technology*, vol. 2, pp. 32–36, 2012.

[12] A. Y. El-Etre, "Khillah extract as inhibitor for acid corrosion of SX 316 steel," *Applied Surface Science*, vol. 252, no. 24, pp. 8521–8525, 2006.

[13] Y. Ren, Y. Luo, K. Zhang, G. Zhu, and X. Tan, "Lignin terpolymer for corrosion inhibition of mild steel in 10% hydrochloric acid medium," *Corrosion Science*, vol. 50, no. 11, pp. 3147–3153, 2008.

[14] S. H. Khalid and P. Sisodia, "Paniala (F *lacourtia Jangomas*) plant extract as eco friendly inhibitor on the corrosion of mild steel in acidic media," *Rasayan Journal of Chemistry*, vol. 4, no. 3, pp. 548–553, 2011.

[15] M. R. Singh and G. Singh, "Hibiscus cannabinus extract as a potential green inhibitor for corrosion of mild steel in 0.5M H_2SO_4 solution," *Journal of Materials and Environmental Science*, vol. 3, no. 4, pp. 698–705, 2012.

[16] A. Khadraoui, A. Khelifa, H. Hamitouche, and R. Mehdaoui, "Inhibitive effect by extract of Mentha rotundifolia leaves on the corrosion of steel in 1 M HCl solution," *Research on Chemical Intermediates*, vol. 40, pp. 961–972, 2014.

[17] O. K. Abiola and A. O. James, "The effects of Aloe vera extract on corrosion and kinetics of corrosion process of zinc in HCl solution," *Corrosion Science*, vol. 52, no. 2, pp. 661–664, 2010.

[18] J. C. da Rocha, J. A. da Cunha Ponciano Gomes, and E. D'Elia, "Corrosion inhibition of carbon steel in hydrochloric acid solution by fruit peel aqueous extracts," *Corrosion Science*, vol. 52, no. 7, pp. 2341–2348, 2010.

[19] P. Kalaiselvi, S. Chellammal, S. Palanichamy, and G. Subramanian, "Artemisia pallens as corrosion inhibitor for mild steel in HCl medium," *Materials Chemistry and Physics*, vol. 120, no. 2-3, pp. 643–648, 2010.

[20] D. Ben Hmamou, R. Salghi, L. Bazzi et al., "Prickly pear seed oil extract: a novel green inhibitor for mild steel corrosion in 1 M HCl Solution," *International Journal of Electrochemical Science*, vol. 7, no. 2, pp. 1303–1318, 2012.

[21] M. D. Griffiths and J. H. Anthony, *The New Royal Horticultural Society Dictionary of Gardening*, Macmillan Press, London, UK, 1992.

[22] C.-C. Chyau, S.-Y. Tsai, P.-T. Ko, and J.-L. Mau, "Antioxidant properties of solvent extracts from *Terminalia catappa* leaves," *Food Chemistry*, vol. 78, no. 4, pp. 483–488, 2002.

[23] U. D. Akpabio, "Evaluation of proximate composition, mineral element and anti-nutrient in almond (Terminalia catappa) seeds," *Research Journal of Applied Sciences*, vol. 3, no. 4, pp. 2247–2252, 2012.

[24] L. Matos, J. M. Nzikou, A. Kimbonguila et al., "Composition and nutritional properties of seeds and oil from *Terminalia catappa* L," *Advance Journal of Food Science and Technology*, vol. 1, no. 1, pp. 72–77, 2009.

[25] W. H. Morrison, R. J. Hamilton, and C. Kalu, "Sunflower seed oil," in *Developments in Oils and Fats*, R. J. Hamilton, Ed., pp. 132–152, Chapman and Hall, London, UK, 1995.

[26] A. Domsch and B. Irrgang, "Sulfosusscinates," in *Anionic Surfactants: Organic Chemistry*, H. W. Stache, Ed., vol. 56 of *Surfactant Science Series*, pp. 501–547, Marcel Dekker, New York, NY, USA, 1996.

[27] A. Adewuyi and R. A. Oderinde, "Analysis of the lipids and molecular speciation of the triacylglycerol of the oils of *Luffa cylindrical* and *Adenopus breviflorus*," *CYTA—Journal of Food*, vol. 10, no. 4, pp. 313–320, 2012.

[28] A. Adewuyi, R. A. Oderinde, B. V. S. K. Rao, and R. B. N. Prasad, "Synthesis of alkanolamide: a nonionic surfactant from the oil

of gliricidia sepium," *Journal of Surfactants and Detergents*, vol. 15, no. 1, pp. 89–96, 2012.

[29] D. Myers, *Surfactant Science and Technology*, John Wiley & Sons, New York, NY, USA, 3rd edition, 2006.

[30] A. K. Maayta, M. B. Bitar, and M. M. Al-Abdallah, "Inhibition effect of some surface active agents on dissolution of copper in nitric acid," *British Corrosion Journal*, vol. 36, no. 2, pp. 133–135, 2001.

[31] L. Tang, G. Mu, and G. Liu, "The effect of neutral red on the corrosion inhibition of cold rolled steel in 1.0 M hydrochloric acid," *Corrosion Science*, vol. 45, no. 10, pp. 2251–2262, 2003.

[32] A. M. Abdel-Gaber, B. A. Abd-El-Nabey, I. M. Sidahmed, A. M. El-Zayady, and M. Saadawy, "Kinetics and thermodynamics of aluminium dissolution in 1.0 M sulphuric acid containing chloride ions," *Materials Chemistry and Physics*, vol. 98, no. 2-3, pp. 291–297, 2006.

[33] T. H. Ibrahim and M. A. Zour, "Corrosion inhibition of mild steel using fig leaves extract in hydrochloric acid solution," *International Journal of Electrochemical Science*, vol. 6, no. 12, pp. 6442–6455, 2011.

[34] A. Adewuyi, A. Göpfert, and T. Wolff, "Succinyl amide gemini surfactant from *Adenopus breviflorus* seed oil: a potential corrosion inhibitor of mild steel in acidic medium," *Industrial Crops and Production*, vol. 52, pp. 439–449, 2014.

[35] X. Wang, Y. Wan, Q. Wang, F. Shi, Z. Fan, and Y. Chen, "Synergistic inhibition between bisbenzimidazole derivative and chloride ion on mild steel in 0.25 M H_2SO_4 solution," *International Journal of Electrochemical Science*, vol. 8, no. 2, pp. 2182–2195, 2013.

[36] A. Zarrouk, I. Warad, B. Hammouti, A. Dafali, S. S. Al-Deyab, and N. Benchat, "The effect of temperature on the corrosion of Cu/HNO3 in the presence of organic inhibitor: part-2," *International Journal of Electrochemical Science*, vol. 5, no. 10, pp. 1516–1526, 2010.

[37] F. El-Hajjaji, R. A. Belkhmima, B. Zerga et al., "Time and temperature elucidation on steel corrosion inhibition by 3-methyl-1-prop-2 ynylquinoxalin-2(1H)-one in molar hydrochloric acid: part 2," *Journal of Material Environmental Science*, vol. 5, pp. 263–270, 2014.

[38] H. Keleş, M. Keleş, and I. Dehri, "Adsorption and inhibitive properties of aminobiphenyl and its Schiff base on mild steel corrosion in 0.5 M HCl medium," *Colloids and Surfaces A: Physicochemical and Engineering Aspects*, vol. 320, no. 1–3, pp. 138–145, 2008.

[39] E. A. Noor and A. H. Al-Moubaraki, "Thermodynamic study of metal corrosion and inhibitor adsorption processes in mild steel/1-methyl-4[4′(-X)-styryl pyridinium iodides/hydrochloric acid systems," *Materials Chemistry and Physics*, vol. 110, no. 1, pp. 145–154, 2008.

[40] A. S. Patel, V. A. Panchal, G. V. Mudaliar, and N. K. Shah, "Impedance spectroscopic study of corrosion inhibition of Al-Pure by organic Schiff base in hydrochloric acid," *Journal of Saudi Chemical Society*, vol. 17, no. 1, pp. 53–59, 2013.

20

The Inhibition Effect of Potassium Iodide on the Corrosion of Pure Iron in Sulphuric Acid

Tarik Attar,[1,2] Lahcène Larabi,[1] and Yahia Harek[1]

[1] *Laboratory of Analytical Chemistry and Electrochemistry, Department of Chemistry, Faculty of Sciences, P.O. Box 119, University Abou-Bekr Belkaïd, 13000 Tlemcen, Algeria*
[2] *University Center of Naâma, BP 66, 45000 Naâma, Algeria*

Correspondence should be addressed to Tarik Attar; t_attar@mail.univ-tlemcen.dz

Academic Editor: Armando Zarrelli

The use of inorganic inhibitors as an alternative to organic compounds is based on the possibility of degradation of organic compounds with time and temperature. The inhibition effect of potassium iodide on the corrosion of pure iron in 0.5 M H_2SO_4 has been studied by weight loss. It has been observed from the results that the inhibition efficiency (IE%) of KI increases from 82.17% to 97.51% with the increase in inhibitor concentration from $1{\cdot}10^{-4}$ to $2{\cdot}10^{-3}$ M. The apparent activation energy (E_a) and the equilibrium constant of adsorption (K_{ads}) were calculated. The adsorption of the inhibitor on the pure iron surface is in agreement with Langmuir adsorption isotherm.

1. Introduction

Corrosion is the deterioration of materials by chemical interaction with their environment [1]. Corrosion can cause disastrous damage to metal and alloy structures causing economic consequences in terms of repair, replacement, product losses, safety, and environmental pollution [2]. Several protective measures are taken to control and prevent corrosion. One of these is the use of corrosion inhibitors, which are usually chemical substances; when added in a small concentration to a corrosive medium, they reduce effectively the corrosion of the metal and/or alloy [3, 4]. The use of corrosion inhibitors constitutes one of the most economical ways to mitigate the corrosion rate [5]. The corrosion and corrosion protection of iron in corrosive environments have attracted the attention of many investigators [6–8]. Iron plays a central role as one of the most widely used materials in our daily life because of its so many applications [9]. The inhibition efficiency depends on the parameters of the corrosive system [pH, temperature, duration, metal composition, etc.] and on the nature of the inhibitor [10]. Sulfuric acid is one of the most aggressive acids for iron and its alloys and is often used during cleaning, pickling, descaling, acidizing, and so forth [11, 12]. The inhibitor molecules get bonded to the metal surface by chemisorption, physisorption, or complexation with the polar groups acting as the reactive centers in the molecules [13].

2. Experimental

2.1. Materials. The weight loss experiments were conducted in a 150 mL beaker; the electrolyte volume is 100 mL. Julabo thermostat brand keeps the electrolyte at the desired temperature (±0.1°C).

The test pieces were mechanically polished with emery paper (a coarse paper was used initially and then progressively finer grades were employed, 400 to 1200 grade). The specimens were weighed by electronic digital analytical balance with five decimal accuracies before and after exposure.

2.2. Electrolyte. The corrosive solution, 0.5 M H_2SO_4, was obtained by dilution of analytical grade 98% sulphuric acid with bidistilled water. The concentration range of inhibitor employed was $1{\cdot}10^{-4}$ to $2{\cdot}10^{-3}$ M in the sulphuric acid.

2.3. Weight Loss Method. The test pieces were washed with bidistilled water, degreased with acetone, washed again with bidistilled water, dried between two filter papers, and weighed. After specified periods of time, 3 test pieces were taken out of the test solution, rinsed with bidistilled water, dried as before, and weighed again. The average weight loss at a certain time for each set of three samples was taken. The weight loss experiments were performed after an exposure of 2 h. The inhibition efficiency of potassium iodide was expressed in terms of percentage inhibition, calculated using

$$\text{IE} \, (\%) = \left(\frac{w_{\text{corr}} - w_{\text{inh}}}{w_{\text{corr}}} \right) \times 100, \tag{1}$$

where W_{corr} is the corrosion rate of blank sulfuric acid and W_{inh} is the corrosion rate after adding inhibitor.

The corrosion rate (W) was calculated from the following equation:

$$w = \left(\frac{m_1 - m_2}{S \cdot t} \right), \tag{2}$$

where m_1 is the mass of the specimen before corrosion, m_2 is the mass of the specimen after corrosion, S is the total area of the specimen, t is the corrosion time, and W is the corrosion rate.

The degree of surface coverage (Θ) was calculated using the following equation:

$$\theta = 1 - \frac{w_{\text{inh}}}{w_{\text{corr}}}. \tag{3}$$

3. Results and Discussion

3.1. Effect of Inhibitor Concentration. The values of percentage inhibition efficiency (IE%) and corrosion rate (w) at different concentrations of KI at 303 K are summarized in Table 1. Figure 1 shows the results obtained from weight loss measurements for pure iron in 0.5 M H_2SO_4 solutions in the absence and presence of different concentrations of KI. It has been observed from the results that the IE% of KI increases from 82.17% to 97.51% with the increase in inhibitor concentration from $1 \cdot 10^{-4}$ to $2 \cdot 10^{-3}$ M. The optimum concentration of this effect is $2 \cdot 10^{-3}$ M.

Figure 2 shows that the corrosion rate decreases with increasing concentration of inhibitor, which explains the effect of protection against the corrosion by the type of inhibitor selected.

3.2. Effect of Immersion Time. The weight loss measurements were performed in 0.5 M H_2SO_4 in absence and presence of KI at $2 \cdot 10^{-3}$ M concentration for 30 min to 6 h immersion time at temperature of 303 K. Inhibition efficiencies were plotted against immersion time as seen in Figure 3. This figure shows that inhibition efficiency of the potassium iodide was increased with increasing immersion time. The increase in inhibition efficiency up to 2 h reflects the inorganic inhibitor adsorption of constituents on the pure iron surface. This result indicates a stabilization of the inhibition rate from 2 hours of immersion. According to this study, it was found that KI is a very effective inhibitor for pure iron in H_2SO_4 0.5 M because after an immersion time of half an hour the power of protection already achieved 94.91%.

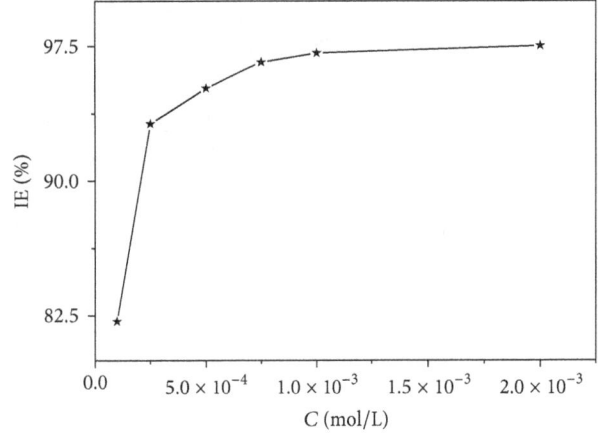

FIGURE 1: Variation of inhibition efficiency with KI concentration.

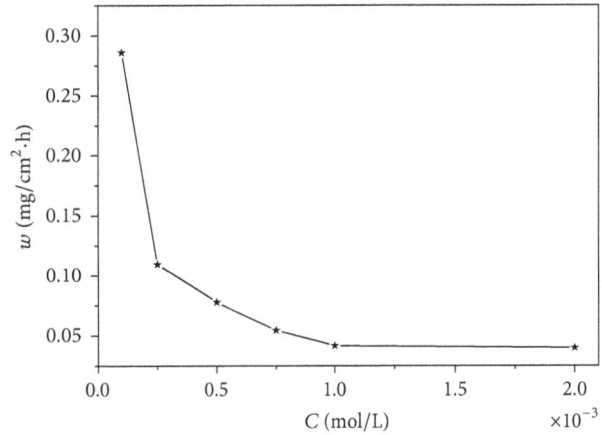

FIGURE 2: Plot of corrosion rate versus inhibitor concentration for pure iron test specimens after 2 hrs of exposure.

3.3. Effect of Temperature. In order to study the effect of temperature on the inhibition efficiencies of potassium iodide, weight loss measurements were carried out in the temperature range 293–323 K in absence and presence of inhibitor at optimum concentration during 2 hours of immersion. Table 2 shows the effect of temperature on the corrosion rate of pure iron in absence and presence of inhibitor. It is evident from this table that inhibition efficiency increases with increasing temperature.

The apparent activation energy E_a for pure iron corrosion in 0.5 M H_2SO_4 in the absence and presence of inhibitors was evaluated from Arrhenius equation [14]:

$$\ln(w) = -\frac{E_a}{RT} + A, \tag{4}$$

where w is the corrosion rate determined from gravimetric measurements, A is the Arrhenius frequency factor, R is the molar gas constant, and T is the absolute temperature.

The plots of $\ln(w)$ against $1/T$ were linear, as shown in (Figure 4); E_a values were obtained from the slope and are as presented in Table 3.

TABLE 1: Corrosion parameters for pure iron in aqueous solution of $0.5\,M\,H_2SO_4$ in presence and absence of different concentrations of KI at 303 K for 2 h.

	C (mol/L)	w (mg·cm^{-2}·h^{-1})	IE (%)
H_2SO_4	0.5	1.60383	—
KI	$1\cdot10^{-4}$	0.28593	82.17
	$2.5\cdot10^{-4}$	0.10956	93.16
	$5\cdot10^{-4}$	0.07787	95.15
	$7.5\cdot10^{-4}$	0.05445	96.60
	$1\cdot10^{-3}$	0.04169	97.11
	$2\cdot10^{-3}$	0.03986	97.51

TABLE 2: Effect of temperature on pure iron in the presence and absence of KI, at 2 h.

T (K)	w_0 (mg/cm^2·h^1)	w_{inh} (mg/cm^2·h^1)	IE (%)
293	0.15107	0.00622	95.88
303	0.34152	0.01329	96.10
313	1.32256	0.05035	96.19
323	3.43109	0.11280	96.71

TABLE 3: The values of activation parameters for pure iron in $0.5\,M\,H_2SO_4$ in the absence and the presence of inhibitor of $2\cdot10^{-3}$ M concentration at 2 h.

Concentration of inhibitor (M)	E_a (kJ/mol)	ΔH_a (kJ/mol)	$E_a - \Delta H_a$ (kJ/mol)
$0.5\,H_2SO_4$	84.32	81.65	2.67
$2\cdot10^{-3}$ KI	78.89	76.22	2.67

A plot of $\ln(w/T)$ versus $1/T$ gave a straight line (Figure 5) with a slope of $-(\Delta H_a/R)$ from which the value of ΔH_a was calculated and listed in Table 3.

The E_a values in the presence of inhibitor are lower than in the absence of inhibitor indicating that the inhibition efficiency increases with increases in temperature. The positive sign of enthalpy of activation reflects the endothermic nature of the steel dissolution process.

3.4. Adsorption Isotherm. Adsorption isotherms provide information about the interaction of the adsorbed molecules with the electrode surface [15]. The adsorption of the inhibitors can be described by two main types of interaction: physical adsorption and chemisorptions [16, 17]. These are influenced by the chemical structure of the inhibitor, the type of the electrolyte, pH, the charge and nature of the metal, and temperature [18]. The phenomenon of interaction between the metal surface and the inhibitor can be better understood in terms of adsorption isotherm. The plots of C_{inh}/θ against C (Figure 6) yield a straight line with approximately unit slope, indicating that the inhibitor under study obeys Langmuir adsorption isotherm. According to this isotherm, θ is related to C_{inh} by [19]:

$$\frac{C_{inh}}{\theta} = \frac{1}{K_{ads}} + C_{inh}. \tag{5}$$

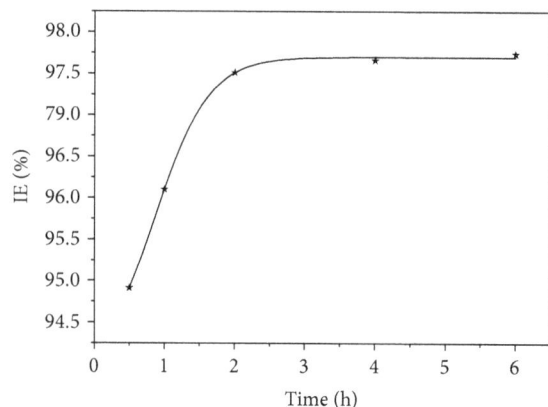

FIGURE 3: Variation of inhibition efficiency of KI with immersion time.

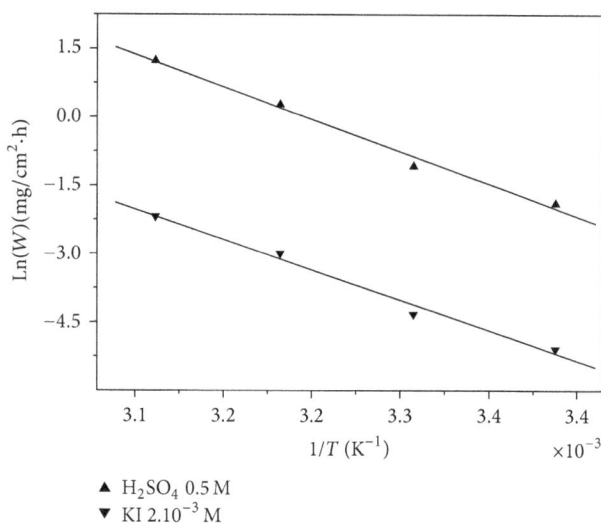

▲ H_2SO_4 0.5 M
▼ KI 2.10^{-3} M

FIGURE 4: Adsorption isotherm plots for ln w versus $1/T$.

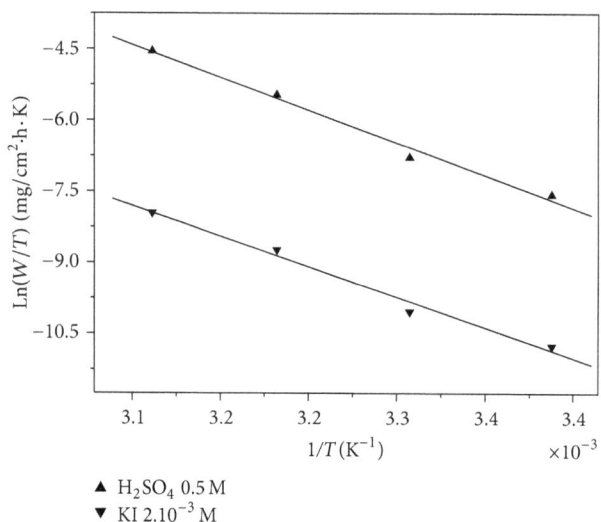

▲ H_2SO_4 0.5 M
▼ KI 2.10^{-3} M

FIGURE 5: Adsorption isotherm plots for $\ln(w/T)$ versus $1/T$.

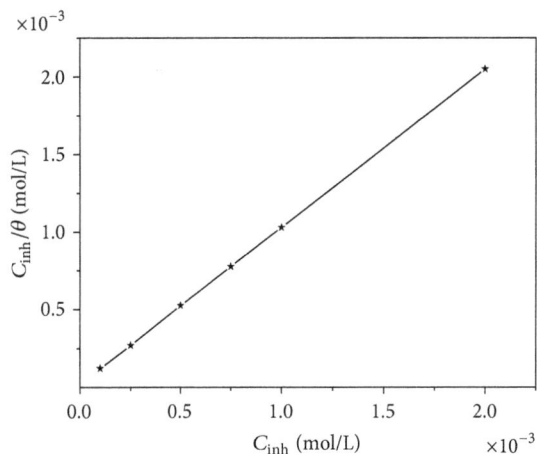

FIGURE 6: Relationship between (C_{inh}/θ) and inhibitor concentration C_{inh}.

The K_{ads} values can be calculated from the intercept lines on the C_{inh}/θ-axis. This is related to the standard free energy of adsorption (ΔG_{ads}) with the following equation [20]:

$$K_{ads} = \left(\frac{1}{55.5}\right) \exp\left(-\frac{\Delta G_{ads}}{RT}\right), \qquad (6)$$

where

$$\Delta G_{ads} = -RT \ln\left(55.5 K_{ads}\right), \qquad (7)$$

where R is the gas constant and T is the absolute temperature. The constant value of 55.5 is the concentration of water in solution in mol/dm3 [21].

The intercept permits the calculation of the equilibrium constant K_{ads} which is 6.17×10^4 L/mol, respectively. The value of K_{ads} which indicates the binding power of the inhibitor to the pure iron surface leads to calculation of adsorption energy. Value of ΔG_{ads} is -37.55 kJ/mol, respectively. The negative values of ΔG_{ads} showed that the adsorption of inhibitor molecules on the metal surface is spontaneous [22].

Generally, the standard free energy values of -20 kJ/mol or less negative are associated with an electrostatic interaction between charged molecules and charged metal surface (physical adsorption); those of -40 kJ/mol or more negative involves charge sharing or transfer from the inhibitor molecules to the metal surface to form a co-ordinate covalent bond (chemical adsorption).

Note that our measures were carried out without inert atmosphere. In 1 M H^+, the iodide ion becomes hydrogen iodide which reacts with oxygen to form molecule iodide I_2. The formed molecule could be adsorbed onto iron and occupying the surface, it strikes the adsorption of iodide. At the end, a lower efficiency of adsorption of iodide was obtained but the inhibition efficiency of the studied inhibitor KI increases due to adsorption of I_2.

Otherwise the value of ΔG_{ads} (-37.55 kJ/mol) near to -40 kJ/mol indicates that, in our case, the adsorption is neither typical chemisorption nor typical physisorption but

it is a complex mixed type. That is, the adsorption inhibitor molecule on the iron surface in the present study involves both chemisorption (of I_2) and physisorption (of I^-) but chemisorption is the predominant mode of adsorption. This assumption is supported by the data obtained from temperature dependence of inhibition process, reported in Tables 2 and 3, which show that the inhibition efficiency of the studied compound as inhibitor increases with increase in temperature and that the value of E_a in absence of the inhibitor is lower than that in its presence [23]. On the other hand it is known that the iron surface acquires positive charge in H_2SO_4 0.5 M [24], while iodide ion is negatively charged, as a result the physisorption (electrostatic attraction) of the iodide ion occurs onto iron surface.

4. Conclusion

On the basis of the experimental results obtained in the present study, the following conclusions can be drawn.

(1) Potassium iodide is a good inhibitor for pure iron corrosion in 0.5 M H_2SO_4 solution. The inhibition efficiency increases with increased KI concentration to attain a maximum value of 97.51% at $2 \cdot 10^{-3}$ M.

(2) The adsorption of KI on pure iron obeyed Langmuir adsorption isotherm.

(3) At higher experimental temperature, inhibitor molecules are adsorbed into the metal surface.

(4) The negative value of ΔG_{ads} is a sign of spontaneous adsorption on the metal surface.

Conflict of Interests

The authors declare that there is no conflict of interests regarding the publication of this paper.

References

[1] B. Eker and E. Yuksel, "Solutions to corrosion caused by agricultural chemicals," *Trakia Journal of Sciences*, vol. 3, no. 7, pp. 1–6, 2005.

[2] N. Patni, S. Agarwal, and P. Shah, "Greener Approach towards Corrosion Inhibition," *Chinese Journal of Engineering*, vol. 2013, Article ID 784186, 10 pages, 2013.

[3] T. H. Ibrahim and M. A. Zour, "Corrosion inhibition of mild steel using fig leaves extract in hydrochloric acid solution," *International Journal of Electrochemical Science*, vol. 6, no. 12, pp. 6442–6455, 2011.

[4] E.-S. M. Sherif, "Electrochemical and gravimetric study on the corrosion and corrosion inhibition of pure copper in sodium chloride solutions by two azole derivatives," *International Journal of Electrochemical Science*, vol. 7, no. 2, pp. 1482–1495, 2012.

[5] P. Rajeev, A. O. Surendranathan, and C. S. N. Murthy, "Corrosion mitigation of the oil well steels using organic inhibitors—a review," *Journal of Materials and Environmental Science*, vol. 3, no. 5, pp. 856–869, 2012.

[6] E.-S. M. Sherif, "Corrosion and corrosion inhibition of pure iron in neutral chloride solutions by 1,1′-thiocarbonyldiimidazole," *International Journal of Electrochemical Science*, vol. 6, no. 8, pp. 3077–3092, 2011.

[7] H. Amar, A. Tounsi, A. Makayssi, A. Derja, J. Benzakour, and A. Outzourhit, "Corrosion inhibition of Armco iron by 2-mercaptobenzimidazole in sodium chloride 3% media," *Corrosion Science*, vol. 49, no. 7, pp. 2936–2945, 2007.

[8] E.-S. M. Sherif, R. M. Erasmus, and J. D. Comins, "In situ Raman spectroscopy and electrochemical techniques for studying corrosion and corrosion inhibition of iron in sodium chloride solutions," *Electrochimica Acta*, vol. 55, no. 11, pp. 3657–3663, 2010.

[9] E. S. M. Sherif, "Corrosion and corrosion inhibition of pure iron in neutral chloride solutions by 1,1′-thiocarbonyldiimidazole," *International Journal of Electrochemical Science*, vol. 6, pp. 3077–3092, 2011.

[10] R. T. Loto, C. A. Loto, and A. P. I. Popoola, "Corrosion inhibition of thiourea and thiadiazole derivatives: a review," *Journal of Materials and Environmental Science*, vol. 3, no. 5, pp. 885–894, 2012.

[11] T. Poornima, N. Jagannatha, and A. Nityananda Shetty, "Studies on corrosion of annealed and aged 18 Ni 250 grade maraging steel in sulphuric acid medium," *Portugaliae Electrochimica Acta*, vol. 28, no. 3, pp. 173–188, 2010.

[12] P. Kumar and A. N. Shetty, "Electrochemical investigation on the corrosion of 18%Ni M250 grade maraging steel under welded condition in sulfuric acid medium," *Surface Engineering and Applied Electrochemistry*, vol. 49, no. 3, pp. 253–260, 2013.

[13] I. Lukovits, E. Kálmán, and F. Zucchi, "Corrosion inhibitors—correlation between electronic structure and efficiency," *Corrosion*, vol. 57, no. 1, pp. 3–8, 2001.

[14] M. A. Quraishi and S. Khan, "Thiadiazoles-A potential class of heterocyclic inhibitors for prevention of mild steel corrosion in hydrochloric acid solution," *Indian Journal of Chemical Technology*, vol. 12, no. 5, pp. 576–581, 2005.

[15] E. A. Noor and A. H. Al-Moubaraki, "Thermodynamic study of metal corrosion and inhibitor adsorption processes in mild steel/1-methyl-4[4′(-X)-styryl pyridinium iodides/hydrochloric acid systems," *Materials Chemistry and Physics*, vol. 110, no. 1, pp. 145–154, 2008.

[16] L. Larabi, O. Benali, and Y. Harek, "Corrosion inhibition of copper in 1 M HNO_3 solution by *N*-phenyl oxalic dihydrazide and oxalic *N*-phenylhydrazide *N′*-phenylthiosemicarbazide," *Portugaliae Electrochimica Acta*, vol. 24, no. 3, pp. 337–346, 2006.

[17] L. Larabi, Y. Harek, M. Traisnel, and A. Mansri, "Synergistic influence of poly(4-vinylpyridine) and potassium iodide on inhibition of corrosion of mild steel in 1M HCl," *Journal of Applied Electrochemistry*, vol. 34, no. 8, pp. 833–839, 2004.

[18] M. A. Ameer and A. M. Fekry, "Inhibition effect of newly synthesized heterocyclic organic molecules on corrosion of steel in alkaline medium containing chloride," *International Journal of Hydrogen Energy*, vol. 35, no. 20, pp. 11387–11396, 2010.

[19] M. Benabdellah, A. Aouniti, A. Dafali et al., "Investigation of the inhibitive effect of triphenyltin 2-thiophene carboxylate on corrosion of steel in 2 M H_3PO_4 solutions," *Applied Surface Science*, vol. 252, no. 23, pp. 8341–8347, 2006.

[20] A. M. Fekry and M. A. Ameer, "Corrosion inhibition of mild steel in acidic media using newly synthesized heterocyclic organic molecules," *International Journal of Hydrogen Energy*, vol. 35, no. 14, pp. 7641–7651, 2010.

[21] O. Olivares, N. V. Likhanova, B. Gómez et al., "Electrochemical and XPS studies of decylamides of α-amino acids adsorption on carbon steel in acidic environment," *Applied Surface Science*, vol. 252, no. 8, pp. 2894–2909, 2006.

[22] A. Dabrowski, "Adsorption-from theory to practice," *Advances in Colloid and Interface Science*, vol. 93, no. 1–3, pp. 135–224, 2001.

[23] T. Szauer and A. Brandt, "On the role of fatty acid in adsorption and corrosion inhibition of iron by amine-fatty acid salts in acidic solution," *Electrochimica Acta*, vol. 26, no. 9, pp. 1257–1260, 1981.

[24] B. S. Prathibha, P. Kottees waram, and R. V. Bheema, "Study on the inhibition of mild steel corrosion by quaterwarey Ammonium compound in H_2SO_4 medium," *Research Journal of Recent Sciences*, vol. 2, no. 4, pp. 1–10, 2013.

21

XPS Study of the Chemical Structure of Plasma Biocopolymers of Pyrrole and Ethylene Glycol

Maribel González-Torres,[1,2] **Ma. Guadalupe Olayo,**[1] **Guillermo J. Cruz,**[1] **Lidia Ma. Gómez,**[1,2]
Víctor Sánchez-Mendieta,[2] **and Francisco González-Salgado**[1,3]

[1] *Departamento de Física, Instituto Nacional de Investigaciones Nucleares, km 36.5 Carretera México-Toluca,*
  52750 Ocoyoacac, MEX, Mexico
[2] *Posgrado en Ciencia de Materiales, Facultad de Química, Universidad Autónoma del Estado de México,*
  Paseos Tollocan y Colón, 52000 Toluca, MEX, Mexico
[3] *Departamento de Posgrado, Instituto Tecnológico de Toluca, Avenida Tecnológico s/n, 52760 Metepec, MEX, Mexico*

Correspondence should be addressed to Guillermo J. Cruz; guillermoj.cruz@hotmail.com

Academic Editor: Zhengcheng Zhang

An XPS study about the structure of plasma biocopolymers synthesized with resistive radio frequency glow discharges and random combinations of ethylene glycol, pyrrole, and iodine, as a dopant, is presented in this work. The collisions of molecules produced structures with a great variety of chemical states based in the monomers, their combinations, crosslinking, doping, fragmentation, and oxidation at different levels in the plasma environment. Iodine appears bonded in the copolymers only at high power of synthesis, mainly as C–I and N–I chemical bonds. Multiple bonds as C≡C, C≡N, C=O, and C=N were found in the copolymers, without belonging to the initial reagents, and were generated by dehydrogenation of intermediate compounds during the polymerization. The main chemical states on PEG/PPy/I indicate that all atoms in pyrrole rings participate in the polymerization resulting in crosslinked, partially fragmented, and highly oxidized structures. This kind of analysis can be used to modify the synthesis of polymers to increase the participation of the most important chemical states in their biofunctions.

1. Introduction

Polymers formed with oxygenated and/or nitrogenated chemical groups, such as polyethylene glycol (PEG) and polypyrrole (PPy), are studied as biomaterials to be implanted in the central nervous system to reduce possible side effects in the spinal cord after a severe injury. PEG is an oxygenated polymer with the potential to influence or repair the membrane permeability caused by injuries or diseases [1, 2] and PPy is one of the most studied nitrogenated biocompatible polymers used as a biosensor, cell growth supporter for nerve cells, and substrate for junction between neurons and microelectrodes [3, 4]. The PPy potential for transferring electric charges is related to the alternated multiple-single chemical bonds in the rings of its structure.

The chemical structure for random plasma combinations of ethylene glycol (EG) and pyrrole (Py) copolymers (PEG/PPy) is studied in this work to produce polymers capable of interacting with neuronal cells. Other components can be added to the mix, for example, iodine with the aim of increasing the conductive properties [5]. In the spinal cord, many polymers have caused rejection due to their noncompatible physicochemistry or to the residues of solvents, catalysts, or other reagents used in the synthesis which irritate the delicate tissues causing adverse reactions that destroy healthy nerve cells. The polymers of this work reduce this problem because the plasma synthesis only uses the monomers and dopants involved in the process, without any other foreign material, producing clean and sterile polymers.

The structure of similar random plasma copolymers of EG and allylamine has been studied before using IR spectroscopy finding OH, NH and multiple bonds originated during the plasma polymerization. These copolymers also showed signals of electrical charge transference important in the ionic processes within the human body [6]. In the field of semiconductor materials, plasma random copolymers of Py and aniline have been studied finding that the electrical properties of the copolymers are between the levels of both homopolymers [7].

Morphological, hydrophilical, and electrical studies of random plasma Py and EG copolymers have also been studied obtaining porous and rough layers [8]. It has been also reported that these plasma combinations of Py and EG can be used as implants in the spinal cord of rats after a severe injury to prevent secondary destruction in the spinal cord tissues and to partially recover the lost motor functions [9, 10]. In view of this important use, in this work, the main atomic chemical states of plasma PEG/PPy/I copolymers are studied by XPS considering the energetic distribution of C1s, O1s, N1s, and I3d atomic orbitals with the purpose to identify and quantify the structure in the copolymers. This work goes beyond the study of individual chemical bonds because it includes the whole atomic bonding.

2. Experimental

The copolymers were synthesized in a vacuum glass tubular reactor with 9 cm diameter and 26 cm length. The central tube has stainless steel flanges at the ends with three access ports each. On the central ports two stainless steel electrodes were inserted with a diameter of 7 cm and a separation of 6 cm between them. The electrodes were connected to a RFX-600 Advanced Energy power supply. In another port, a pressure gauge, an Alcatel Pascal 2010C1 vacuum pump, and an Alcatel LNT 25S condenser for residual vapors were collocated.

The monomers used in the polymerization were EG (Tecsiquim, 99.5%) and Py (Aldrich, 98%) in separate containers connected to the reactor through individual entrances. During the synthesis, EG was maintained at $50°C$ to produce vapors that entered the reactor, while Py and iodine were supplied at room temperature. All vapors mixed freely inside the reactor to produce random copolymers. The synthesis conditions were 0.1 mbar, 13.5 MHz, and 80 W during 180 min.

The resulting PEG/PPy/I copolymers were obtained as thin films attached to the reactor walls and electrodes. To detach the material it was necessary to apply distilled water, ethanol or acetone on the walls and dry the films and if they were released from the surface, remove them carefully with a spatula. If not, another cycle of wetting and drying was applied until the film was released from the surface. The films had an average thickness of 20 μm.

The structural chemical analyses of the bulk copolymers were performed on an infrared (IR) spectrophotometer Thermo Scientific Nicolet iS5 on ATR mode using 64 scans. Superficial X-ray photoelectron analyses (XPS) were done on a Thermo K-Alpha photoelectron spectrometer equipped with a monochromatic Al X-ray source (1486.6 eV). The diameter of the analysis area was 400 μm. The base pressure of the analysis chamber was 10^{-9} mbar; however, a beam of Ar ions was applied to the samples to reduce the electrostatic charges that increased the pressure up to 10^{-7} mbar in which the analyses were performed. The copolymers were located

in Al tapes on stainless steel sample holders, which remained in a preanalysis chamber for approximately 1 hr at 10^{-3} mbar before entering the analysis chamber. The step energy of the survey mode was 1 eV, but in the C1s, N1s, O1s, and I3d orbital scans, the energy step was adjusted to obtain approximately 200 points per each distribution, unimodal 200, bimodal 400, and so on. This means energy steps between 0.05 and 0.02 eV, depending on the element and the energetic region. Once the total energetic distribution of the orbitals in study is obtained, the specific energetic atomic states were evaluated adjusting the main distribution with internal Gaussian curves.

3. Result and Discussion

3.1. Chemical Structure of PEG/PPy/I Copolymers. Figure 1 shows the IR spectra of PEG/PPy/I synthesized between 40 and 100 W. The data were taken in ATR mode directly from the copolymer films, which show basically the same absorption, indicating that at this level of power the chemical structures do not have great differences. The widest band is located between 3750 and 3000 cm^{-1} centered approximately at 3269 cm^{-1}, which includes O–H, N–H and =C–H bonds [7, 11]. These groups are part of both monomeric structures used in the copolymers. The peak at 2944 cm^{-1} corresponds to –C–H aliphatic groups of EG and to some saturated fragments of Py rings. The wide absorption centered in 733 cm^{-1} indicates also the presence of –C–H and =C–H groups in different combinations.

In 2212 cm^{-1}, a signal of multiple bonds is found, which may be combinations of C≡C, C≡N, C=C, C=O, and C=N [12]. Triple bonds are not part of the monomeric structures and may originate from strong dehydrogenation and fragmentation caused by the high kinetic energy of particles in the plasma [13]. This absorption has a medium intensity and appears in many plasma polymers.

The most intense absorption is centered in 1613 cm^{-1} in all copolymers belonging to C=C double bonds of Py molecules; however, as this absorption becomes wider, more complex interactions occur in the structure. In this way, although the center belongs to the C=C bonds of the Py structure, at higher wavenumbers in the same curve, C=O and C=N chemical groups appear. C=O may form with the dehydrogenation of two neighboring atoms of EG molecules; however, C=N needs two consecutive effects, the breaking of Py molecules and the dehydrogenation of such fragments in similar process of C=O groups. Oxygen in the copolymers can also be observed at 1421 and 1029 cm^{-1} with the absorption of C–O groups.

All the discussed data indicate that the copolymers have hydrogenated groups, such as C–H, O–H, and N–H that survived the energy of the discharges, and multiple bonds such as C≡C, C≡N, C=O, and C=N, created as a consequence of the dehydrogenation produced by the plasma.

3.2. Superficial Elemental Analysis. The bulk and superficial structures of any material are different, because on the surface there is a complex balance of forces due to the end of the solid phase and the starting of another, usually the atmospheric gas. As a consequence, the contact with the environment

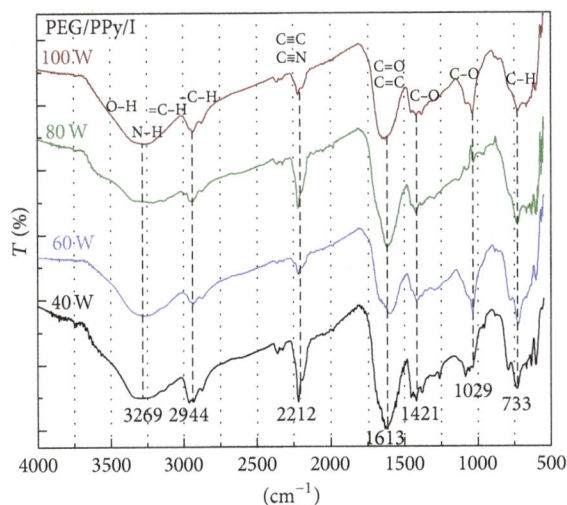

FIGURE 1: IR spectra of plasma PEG/PPy/I synthesized between 40 and 100 W.

produces characteristics that depend not only on the structure of the material, but also on the surrounding fluids. In biomaterials, the surface is especially important because it is the face exposed to cells. In the following sections, the structure of PEG/PPy/I surface is studied with XPS techniques.

The survey spectra of PEG/PPy/I synthesized at different power are presented in Figure 3 in which C1s, N1s, O1s, I3d, and Si2p orbitals can be observed representing the content of their respective elements. The atomic percentage is included in the graph. As Si is not part of the copolymers, it was considered superficial contamination.

Extending the atomic ratio in the monomers to the respective homopolymers, PPy would have N/C = 0.25 with O/C = 0, because it does not have oxygen in its structure. For similar reasons PEG would have N/C = 0 with O/C = 1; see Figure 2. Thus, in plasma PEG/PPy/I, N/C should be between 0 and 0.25 and O/C should be between 0 and 1; and in a random copolymer with 1/1 combination of both monomers, both ratios would be N/C = 0.17 and O/C = 0.33.

The experimental N/C in the copolymers synthesized at different power is between 0.1 and 0.2 suggesting that both monomers participate in approximately the same proportion. However, O/C has values between 0.03 and 0.2 which indicates that Py has much more participation than EG in the copolymers. Considering the analyses of both ratios, Py participates with at least half of the content in the copolymers. On its part, I/C has very low values, from 0 to 0.0005, because it has participation of a dopant, however, at low power, this ratio has so small value that could not be measured. In some way, I/C indicates that the dopant needs power of synthesis higher than 80 W to survive in the copolymers.

3.3. Superficial Chemical Structure.
To obtain the chemical composition of a solid surface, the electronic binding energy (BE) of each element on the surface was used. This is the electronic energy of the orbital in study before a perturbation, for example, the energy in 1s orbitals for C, N, and O elements.

BE can be calculated if the kinetic energy (KE) of some released electrons is known, as a consequence of an incident X-ray with carefully measured energy (hv). The relationship among these variables is an energy balance in the released electrons; see (1), where W is a work function that depends on the level of energy reached. XPS spectrometers calculate these variables in different ways giving whole distributions of BE of each element on the surface as the most valuable information. With these data, the chemical atomic environment on the surface can be studied [14]

$$BE = hv - KE - W. \tag{1}$$

Figures 4, 6, 8, and 10 show the energetic distribution of BE in the C1s, N1s, O1s, and I3d orbitals, respectively, to study the main superficial chemical states of PEG/PPy/I copolymers, where x-axis represents BE and y-axis represents its intensity. Each orbital curve was adjusted with several interior Gaussian curves that represent at least one atomic chemical state per curve. The maximum point of each curve is used to identify energetic state. Figures 5, 7, and 9 show the maximum position of each curve and the associated % area, which can be identified as the percentage of that state.

The curve fitting was done considering the full width at half maximum (FWHM) parameter based on the Crist work for advanced fitting of monochromatic XPS spectra [15]. The baseline of each BE distribution was set manually with the trend of data. Each energetic state was associated with a specific atomic chemical state involving all bonding orbitals shared in the atoms. For example, C shares 4 suborbitals, N shares 3, O shares 2, and I shares only 1 which can be associated with their respective valencies, but although the valence orbitals are in the exterior electronic shell and the orbitals analyzed in XPS are further inside the atomic structure (1s in C, N, and O and 3d in I), any modification in the valence orbitals modifies the energetic equilibrium of the entire atom, modifying the interior orbitals studied in the XPS analysis.

In many atoms, the energetic changes in the valence orbitals are less than 1 eV when the atom bonds with another one. However, this small modification is reflected in different magnitude in the interior orbitals which are hundreds of eV higher. Neighboring atoms also exert influence on the energetic distribution of the surrounding atoms, although they are not directly bonded. Thus, in analyzing the energetic distribution of orbitals, the whole chemical environment has to be considered and each case is different.

The atomic chemical states in this work were constructed including most of the possible bonding combinations in the copolymers. The notation indicates that the atom in bold face is bonded with all atoms in the formula. For example, the most common chemical state of C atoms in polyethylene is C–$\mathbf{CH_2}$–C, in which the central C atom is bonded with two C and two H atoms in their specific spatial distribution. Thus, the previous configuration could also be expressed as C_2–$\mathbf{C}$–H_2 with the same meaning.

3.4. Carbon Chemical States.
Figure 4 shows a detailed analysis of the energetic distribution of C1s orbitals in PEG/PPy/I synthesized at 80 W. This copolymer was used as a case of study; however, the other copolymers synthesized at different

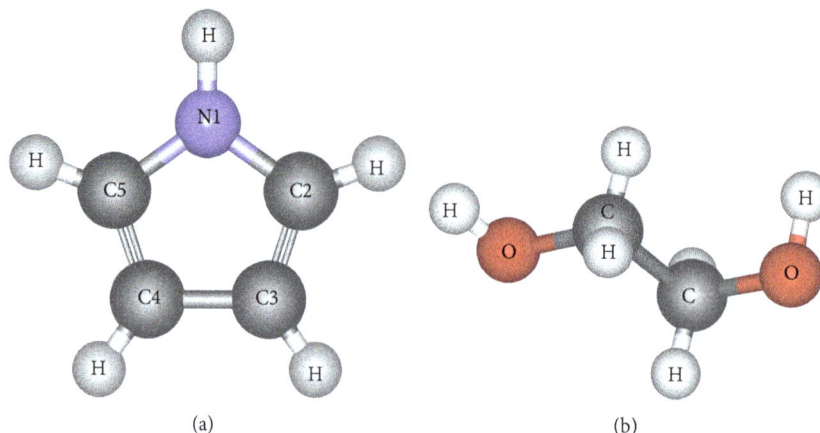

(a)   (b)

FIGURE 2: Atomic structure of pyrrole (a) with N/C = 0.25 and ethylene glycol (b) with O/C = 1. See the positions in the pyrrole ring and the respective chemical state: N1 (C–**N**H–C), C2 (C=**C**H–N), C3 (C=**C**H–C), C4 = C3, and C5 = C2. In ethylene glycol, C and O have only one chemical state, which are C–CH_2O and C–O–H, respectively.

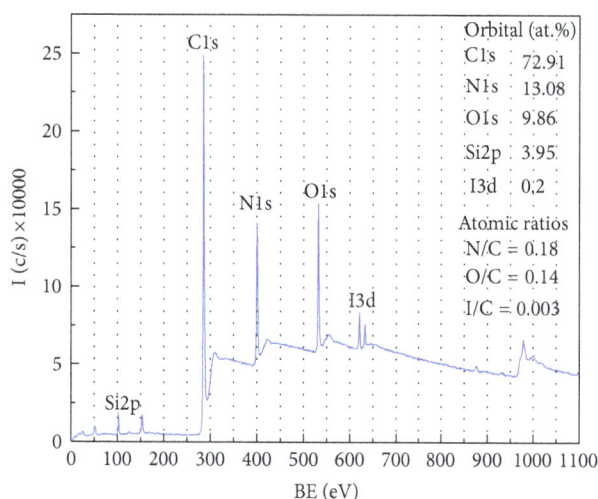

FIGURE 3: XPS survey scan of PEG/PPy/I synthesized in the 40–100 W interval with elemental content and atomic ratios.

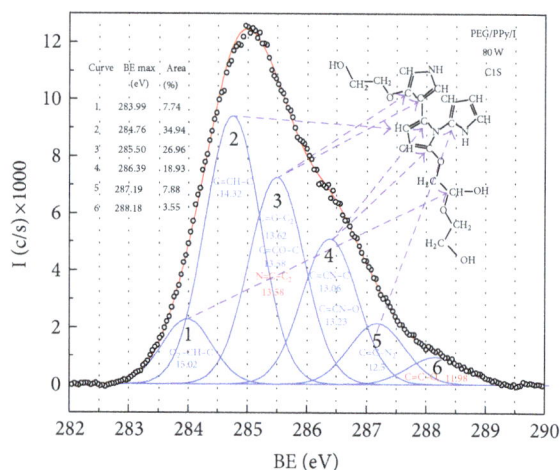

FIGURE 4: High-resolution C1s scan. The chemical formula represents the possible union of three pyrrole and three ethylene glycol molecules in PEG/PPy/I. The legends indicate the probable chemical states with their approximated formation energy in eV. The red legends indicate chemical states that do not belong to the monomers.

power were analyzed in the same way, but as the data generated is huge, their most important condensed data is presented in a comparative way in Figure 5. The legends inside the curves indicate the probable chemical state with its respective approximated formation energy in eV. The area percentage of each curve and its maximum BE is also presented in the graph. The molecular formula in Figure 4 represents a possible union of three Py and three EG molecules in PEG/PPy/I and contains most of the chemical states associated with the curves.

C1s was fitted with 6 curves with FWHM = 1.0 ± 0.1 eV, where each curve is represented by its maximum BE in the discussion. The association between the energetic curves with the most probable chemical states was done considering the formation energy of each state calculated with the sum of all atomic bonding energies [16–18] which varies according to the oxidation level. The association started with the most hydrogenated, or least oxidized, state in the copolymers

located in the lowest BE region, and as the hydrogen atoms are gradually substituted for other elements, more oxidized states appear towards higher BE zones reaching combinations of double and/or triple bonds.

The first fitted curve (7.74%) with the lowest BE and peak at 283.99 eV was identified with the main C state of PEG, O_2–CH–C, which is the union of two EG molecules; see Figure 4. The second curve has the maximum area of C1s (34.94%) with center at 284.76 eV and can be assigned to C=CH–C. This configuration is part of the Py structure in carbons in C3 and C4 positions; see Figure 3. In PPy, chains grow preferentially substituting hydrogen atoms in C2 and C5 leaving C3 and C4 almost untouched, which is the configuration associated with curve 2. Curve 3 with maximum BE at 285.5 (29.96%) can be assigned to C=C–C_2 and C=CO–C, both with approximately the same formation energy, 13.6 eV, originated from the union

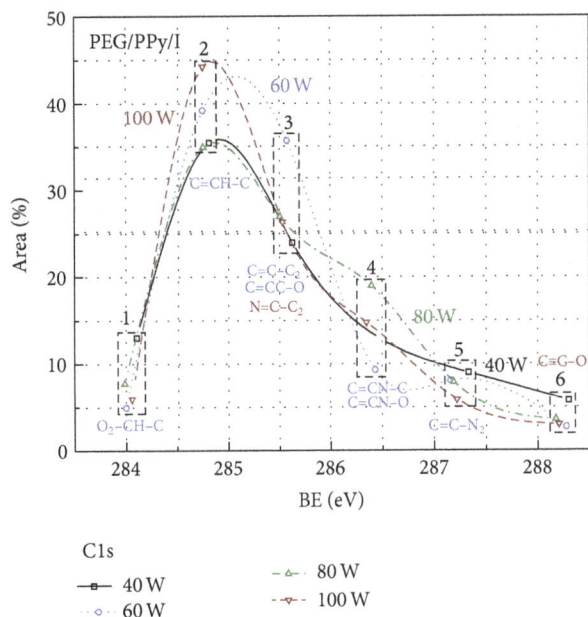

FIGURE 5: C1s orbital energy distribution comparison of copolymers synthesized at 40, 60, 80, and 100 W. The points enveloped in each box represent the same chemical state.

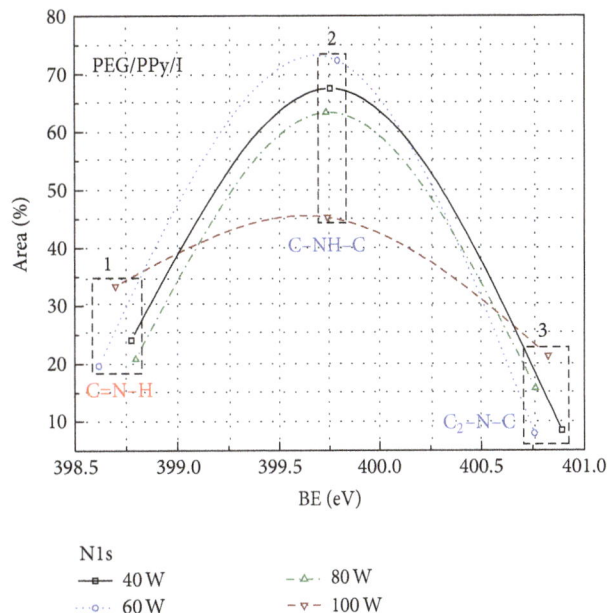

FIGURE 7: Comparison of nitrogen chemical states in copolymers synthesized at 40, 60, 80, and 100 W. The red legend shows a chemical state that does not belong to the monomers.

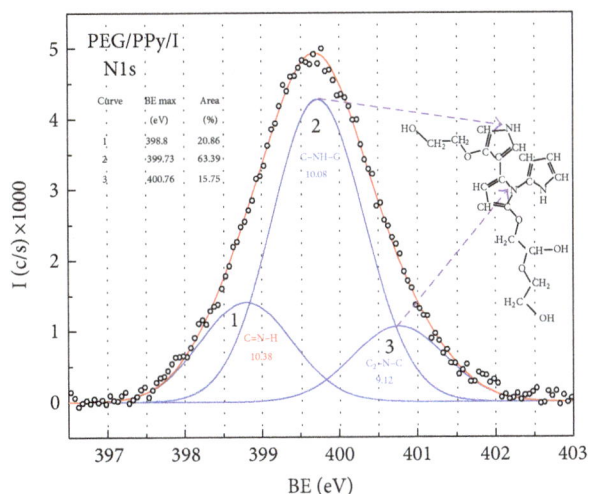

FIGURE 6: High-resolution N1s scan. The legends indicate the probable chemical state with its respective formation energy in eV. The red legend shows a chemical state that does not belong to the monomers.

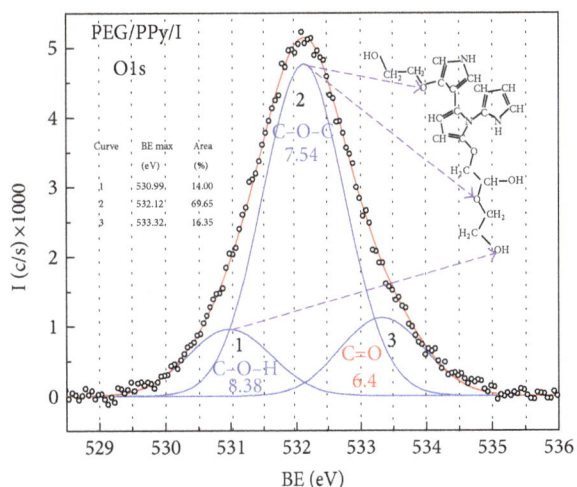

FIGURE 8: High-resolution O1s scan. The legends indicate the probable chemical states with their respective formation energy in eV. The red legend shows a chemical state that does not belong to the monomers.

between two pyrrole rings in C4 position, and with the union of Py and EG in the copolymers, respectively. There is another chemical state in red with approximately the same formation energy identified with the N=C-C2 configuration that may appear in oxidized fragments of Py. At 286.39 eV curve 4 (18.93%) has its maximum with the C=CN-C and C=CN-O possible states which are the bonds of Py-Py and Py-EG, respectively, both involving C2 position in Py. This is typical growing configuration of polypyrroles. The fifth curve with maximum at 287.19 eV (7.88%) was assigned to $C=C-N_2$ which is part of the Py-Py bond through the nitrogen atoms

and is another combination of a crosslinked polymerization of pyrrole. The last curve (3.55%) has its maximum at 288.18 eV and can be assigned to combinations of double and triple bonds with the possible chemical states: $C≡C-O$ and $N≡C-N$. Triple bonds are undoubtedly expressions of maximum oxidation in PEG/PPy/I and can be created in complex structures or in molecular fragments of the polymers in progress. As these chemical states do not belong to the structure of the monomers, they are not included in the structure of the copolymers discussed in Figures 4, 6, and 8.

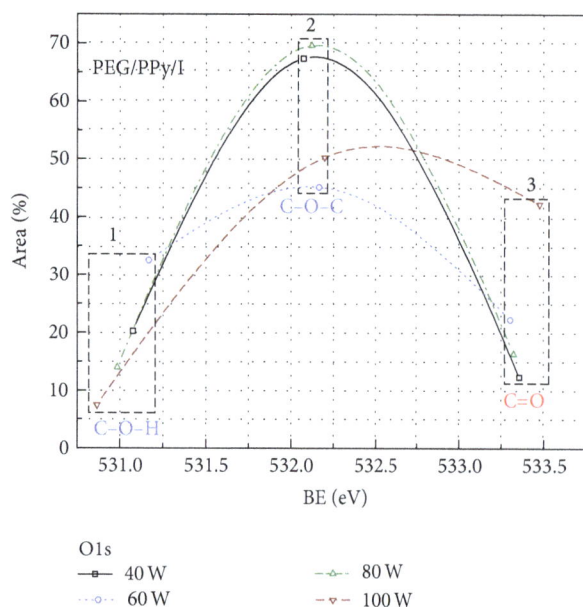

FIGURE 9: Comparison of oxygen chemical states in the copolymers synthesized at 40, 60, 80 and 100 W. The red legend shows a chemical state that does not belong to the monomers.

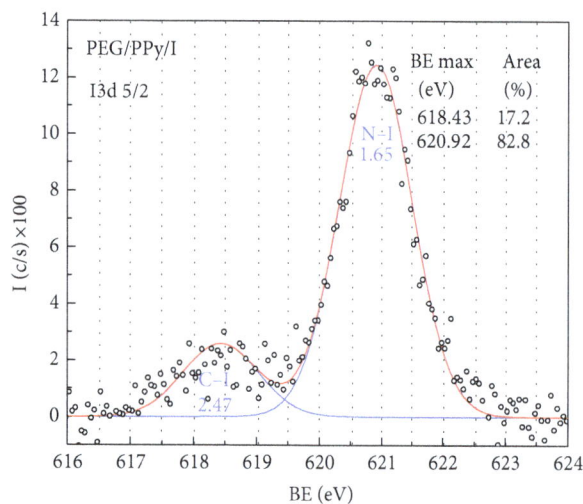

FIGURE 10: High-resolution I3d 5/2 scan. The legends indicate the probable chemical states with their respective formation energy in eV.

The sum of Py or EG percentages of their respective chemical states would give their participation in the copolymers. In the curves involving both monomers, the percentage can be divided equally between them. In this scheme, the percentages of Py and EG in the copolymers are 65.76% and 30.69%, respectively. This means that Py doubles the participation of EG. This is another indicative that Py has more participation in the copolymers than EG, as discussed in the superficial elemental analysis section. The molecular fragmentation belongs to the last curve and is 3.55% in the copolymers synthesized at 80 W.

Figure 5 condenses the C1s data in copolymers synthesized at different power to study the evolution of chemical states. The graph shows the maximum BE of each chemical state in x-axis and the area percentage of that state in y-axis. There are 4 curves, one for each power of synthesis. The points in the same box indicate that they belong to the same chemical state. Note that the differences in BE of all chemical states in the same box are small indicating that the power does not change substantially the chemical states; it only modifies the participation of them in the copolymers without a specific pattern respect to the power.

3.5. Nitrogen Chemical States. The energetic distribution of N1s orbitals is studied in Figure 6, which was adjusted with 3 Gaussian curves using FWHM = 1.3 ± 0.1 eV. The transformation of energetic states to chemical states in this orbital was done following similar dehydrogenation criteria to those in C1s. The first curve with peak at 398.8 eV (20.86%) with the lowest oxidation can be assigned to C=**N**–H (10.38 eV) originated in hydrogenated fragments of Py molecules formed during the collisions of particles during the synthesis.

The second curve, with maximum at 399.73 eV (63.39%) and the maximum area, represents the typical chemical state of nitrogen in Py, C–**N**H–C (10.08 eV). The third curve with maximum at 400.76 eV (15.75%) was assigned to C_2–**N**–C (9.12 eV) and probably with triple bonds C≡**N** (9.11 eV) of highly oxidized nitrogen in the copolymers; both have approximately the same formation energy and cannot be differentiated in this analysis. The first chemical state belongs to tertiary amines and it is a signal that the N1 positions of Py also participate in the chemical reactions to crosslink the copolymers, EG may also join Py through this position. The second state is a fragment of Py with maximum oxidation of N atoms; this kind of bonds was also found in the IR spectra. Near this curve another chemical state has been found in melamine ($C_3N_6H_6$), C=N–C (9.42 eV) [19], with energy that could be partially located between the second and third curves. According to these data, N states show that N bonds preferably with C in the copolymers, avoiding bonding with oxygen and that there are signals of fragmentation and crosslinking in the copolymers.

Figure 7 shows a comparison between N1s chemical states in copolymers obtained at different power. As in C1s, the power of synthesis does not change the chemical states, only the percentage of their participation. From 40 to 80 W the percentages of chemical states are similar, but at 100 W the percentage of C–**N**H–C state is much lower than in the other syntheses suggesting that C–**N**H–C dehydrogenates in a greater extent as the power of synthesis increases. This effect is reflected in the participation of tertiary amines, C2–**N**–C and/or triple bonds C≡N, which also increase with the power.

3.6. Oxygenated Chemical States. Figure 8 shows the high resolution scan for O1s orbitals of PEG/PPy/I. The first peak is located at 530.99 eV (14%) and was assigned to the C–**O**–H (8.38 eV) state which is part of EG structure. The greatest curve (69.65%) with peak located at 532.12 eV was assigned to C–**O**–C (7.12 eV) groups. This state may appear from at least two sources, the union of two EG molecules and from the union of EG with Py. Both are part of the copolymerization

[20]. The third curve, located at 533.32 eV (16.34%), can be assigned to C=O (6.4 eV) with possible origin in the dehydrogenation of C–O–H groups discussed in the first curve.

Figure 9 presents a comparison among the adjusted O1s orbitals for each copolymer synthesized at different power. There are 3 chemical states in each curve, although with different percentage showing that oxygenated states are more sensitive to the power of synthesis than to N and C groups. However, the lack of a specific tendency suggests that the superficial interaction influences the participation of O and N chemical states in the study due to the additional oxidation and nitridation of the copolymers. This is usually observed in the chemical states with the highest oxidation.

3.7. Iodine Chemical States. Figure 10 shows a high-resolution BE scanning of I3d 5/2 orbitals in the 616–624 eV interval. As iodine was used as a dopant, its content was very small, 0.02% in the copolymer synthesized at 80 W, and the data was collected with noise, but 2 Gaussian curves can be clearly adjusted to the data with FWHM in the 1.35–1.4 eV interval. The first curve has a maximum at 618.43 eV (17.2%) and was assigned to C–I (2.47 eV). The second curve has a maximum at 620.92 eV (82.8%) and was assigned to N–I (1.65 eV), which indicates some affinity to form N–I bonds under the energetic conditions in plasmas. This effect can be seen at 80 W or higher power of synthesis. No signals of iodine could be measured in the copolymers at lower power.

4. Conclusions

The structure of plasma random bio-copolymers of ethylene glycol and pyrrole doped with iodine is studied in this work; the aim is to reduce possible side effects in the spinal cord after a severe injury. The structure of such copolymers has not been studied at the light of the XPS energetic orbitals to identify and quantify the main chemical states. The N/C and O/C atomic ratios indicate that pyrrole has more participation than ethylene glycol in the copolymers. On its part, the content of iodine indicates that the dopant appears in the copolymers at 80 W or higher power of synthesis, mainly as N–I chemical bonds.

The chemical states of PEG/PPy/I copolymers can be associated with the structure of the traditional homopolymers of polypyrrole and polyethylene glycol, with the combination of pyrrole and ethylene glycol, and with a big variety of other additional states due to crosslinking, doping, fragmentation, and oxidation at different levels. Multiple bonds not present in the typical homopolymers of pyrrole and ethylene glycol were formed by dehydrogenation and oxidation caused by the constant collisions of accelerated particles in plasmas where the syntheses occurred. With this information, the synthesis of PEG/PPy/I copolymers can be modified to increase the participation of the most important chemical states in their biofunctions.

Conflict of Interests

The authors declare that there is no conflict of interests regarding the publication of this paper.

Acknowledgments

The authors thank CONACyT for providing the financial support to this work with the Projects 130190 and 154757.

References

[1] P. J. Photos, L. Bacakova, B. Discher, F. S. Bates, and D. E. Discher, "Polymer vesicles in vivo: correlations with PEG molecular weight," *Journal of Controlled Release*, vol. 90, no. 3, pp. 323–334, 2003.

[2] R. B. Borgens and R. Shi, "Immediate recovery from spinal cord injury through molecular repair of nerve membranes with polyethylene glycol," *The FASEB Journal*, vol. 14, no. 1, pp. 27–35, 2000.

[3] G. Ruggeri, M. Bianchi, G. Puncioni, and F. Ciardelli, "Molecular control of electric conductivity and structural properties of polymers of pyrrole derivatives," *Pure and Applied Chemistry*, vol. 69, no. 1, pp. 143–149, 1997.

[4] H. P. Wong, B. C. Dave, F. Leroux, J. Harreld, B. Dunn, and L. F. Nazar, "Synthesis and characterization of polypyrrole/vanadium pentoxide nanocomposite aerogels," *Journal of Materials Chemistry*, vol. 8, no. 4, pp. 1019–1027, 1998.

[5] J. Morales, M. G. Olayo, G. J. Cruz, and R. Olayo, "Plasma polymerization of random polyaniline-polypyrrole-iodine copolymers," *Journal of Applied Polymer Science*, vol. 85, no. 2, pp. 263–270, 2002.

[6] L. M. Gomez, P. Morales, G. J. Cruz et al., "Plasma copolymerization of ethylene glycol and allylamine," *Macromolecular Symposia*, vol. 283-284, no. 1, pp. 7–12, 2009.

[7] J. Morales, M. G. Olayo, G. J. Cruz, and R. Olayo, "Synthesis by plasma and characterization of bilayer aniline-pyrrole thin films doped with iodine," *Journal of Polymer Science B: Polymer Physics*, vol. 40, no. 17, pp. 1850–1856, 2002.

[8] E. Colín, M. G. Olayo, G. J. Cruz, L. Carapia, J. Morales, and R. Olayo, "Affinity of amine-functionalized plasma polymers with ionic solutions similar to those in the human body," *Progress in Organic Coatings*, vol. 64, no. 2-3, pp. 322–326, 2009.

[9] R. Olayo, C. Ríos, H. Salgado-Ceballos et al., "Tissue spinal cord response in rats after implants of polypyrrole and polyethylene glycol obtained by plasma," *Journal of Materials Science: Materials in Medicine*, vol. 19, no. 2, pp. 817–826, 2008.

[10] G. J. Cruz, R. Mondragón-Lozano, A. Diaz-Ruiz et al., "Plasma polypyrrole implants recover motor function in rats after spinal cord transection," *Journal of Materials Science: Materials in Medicine*, vol. 23, no. 10, pp. 2583–2592, 2012.

[11] R. M. Silverstein, F. X. Webster, and D. J. Kiemle, *Spectrometric Identification of Organic Compounds*, John Wiley & Sons, New York, NY, USA, 7th edition, 2005.

[12] D. L. Pavia, G. M. Lampman, and G. S. Kriz, *Introduction to Spectroscopy*, Stamford, Conn, USA, Brooks/Cole Thomson Learning, 3rd edition, 2001.

[13] G. J. Cruz, M. G. Olayo, O. G. López, L. M. Gómez, J. Morales, and R. Olayo, "Nanospherical particles of polypyrrole synthesized and doped by plasma," *Polymer*, vol. 51, no. 19, pp. 4314–4318, 2010.

[14] B. V. Crist, "Advanced peak-fitting of monochromatic XPS spectra," *Journal of Surface Analysis*, vol. 4, no. 3, pp. 428–434, 1998.

[15] J. F. Watts and J. Wolstenholme, *An Introduction to Surface Analysis by XPS and AES*, John Wiley & Sons, west sussex, UK, 2003.

[16] P. M. A. Sherwood, "Analysis of the X-ray photoelectron spectra of transition metal compounds using approximate molecular orbital theories," *Journal of the Chemical Society, Faraday Transactions 2: Molecular and Chemical Physics*, vol. 72, pp. 1791–1804, 1976.

[17] F. A Cotton and G. Wilkinson, *Química Inorgánica Avanzada*, Limusa, México City, Mexico, 2nd edition, 1980.

[18] J. A. Dean, *Lange's Handbook of Chemistry*, McGraw-Hill, New York, NY, USA, 15th edition, 1999.

[19] A. P. Dementjev, A. de Graaf, M. C. M. van de Sanden, K. I. Maslakov, A. V. Naumkin, and A. A. Serov, "X-ray photoelectron spectroscopy reference data for identification of the C_3N_4 phase in carbon-nitrogen films," *Diamond and Related Materials*, vol. 9, no. 11, pp. 1904–1907, 2000.

[20] M. G. Olayo, F. González-Salgado, G. J. Cruz et al., "Chemical structure of TiO organometallic particles obtained by plasmas," *Advances in Nanoparticles*, vol. 2, pp. 229–235, 2013.

22

A Review on Current Status of Stability and Knowledge on Liquid Electrolyte-Based Dye-Sensitized Solar Cells

Frédéric Sauvage[1,2]

[1] *Laboratoire de Réactivité et Chimie des Solides, Université de Picardie Jules Verne, CNRS UMR 7314, 33 rue Saint Leu, 80039 Amiens, France*
[2] *Institut de Chimie de Picardie (ICP), CNRS FR 3085, 33 rue Saint Leu, 80039 Amiens, France*

Correspondence should be addressed to Frédéric Sauvage; frederic.sauvage@u-picardie.fr

Academic Editor: Mohamed Sarakha

The purpose of this review is to gather the current background in materials development and provide the reader with an accurate image of today's knowledge regarding the stability of dye-sensitized solar cells. This contribution highlights the literature from the 1970s to the present day on nanostructured TiO_2, dye, Pt counter electrode, and liquid electrolyte for which this review is focused on.

1. Introduction

The photovoltaic effect was discovered by the French scientist Antoine Cesar Becquerel in 1839. His observations were presented two subsequent times at the Academie des Sciences in Paris by his son Alexandre Edmond [1]. In their experiences, they shared the observations of an intriguing phenomenon related to the generation of current flowing when setting two platinum plates in contact with an aqueous acidic galvanic cell under illumination. This generation of a galvanic current under light occurs when the surface of one of the two electrodes is modified by halide vapor (iodide, chloride, and bromide). Depending on the experimental procedure for this deposition, they noticed a current flowing in one or in the other way when placed under illumination, likely owing to a halide film exhibiting n or p characteristics. For a long time, this starting point of the photovoltaic effect was considered as a scientific curiosity. A mechanistic explanation to account for this effect was drawn by A. Einstein in 1905, who obtained his first Nobel Prize in Physics for his theory describing the origin of the photoelectric effect in 1921. At that time nobody envisioned that the photonic field would become a new area of strong fundamental and applied researches from which the underlying technologies would be a stake for human beings. The starting point for PV applications goes back to

the Cold War and the two superpowers' irrational objective to become the first nation to invade the Moon. Thanks to a worldwide support for billions of dollars, Chapin Pearson and Fuller from Bell Laboratories pioneered the silicon p-n technology, paving the way to a substantial enhancement in the light-to-electricity power conversion efficiency (PCE) compared to the selenium-based electrodes exhibiting at best *ca.* 1%. The principal objective of this new p-n Si technology was to supply the spatial vehicles in energy by making use of extraterrestrial sun power. With its 1.1 eV indirect bandgap affording panchromatic light absorption, they first demonstrated 4.5% power conversion efficiency in 1954 and 6% a few months later and even 8% was reported in the pioneering patent published in 1957 [2, 3]. Solely one year was required for technological transfer to Western Electric. They were the first to commercialize the product to a "large" public with small systems of 14 mW exhibiting 2% conversion efficiency for 25 $ (1500 $ per W).

Photovoltaic research intended public democratization after the two petrol crises and the ensuing financial crises in 1973 and 1979. Because of the soaring price for crude petrol and the public awareness of its limitations in terms of price stability and availability on Earth, research on alternative low-cost energy production and storage has been prompted. This has largely contributed to the development of PV but also

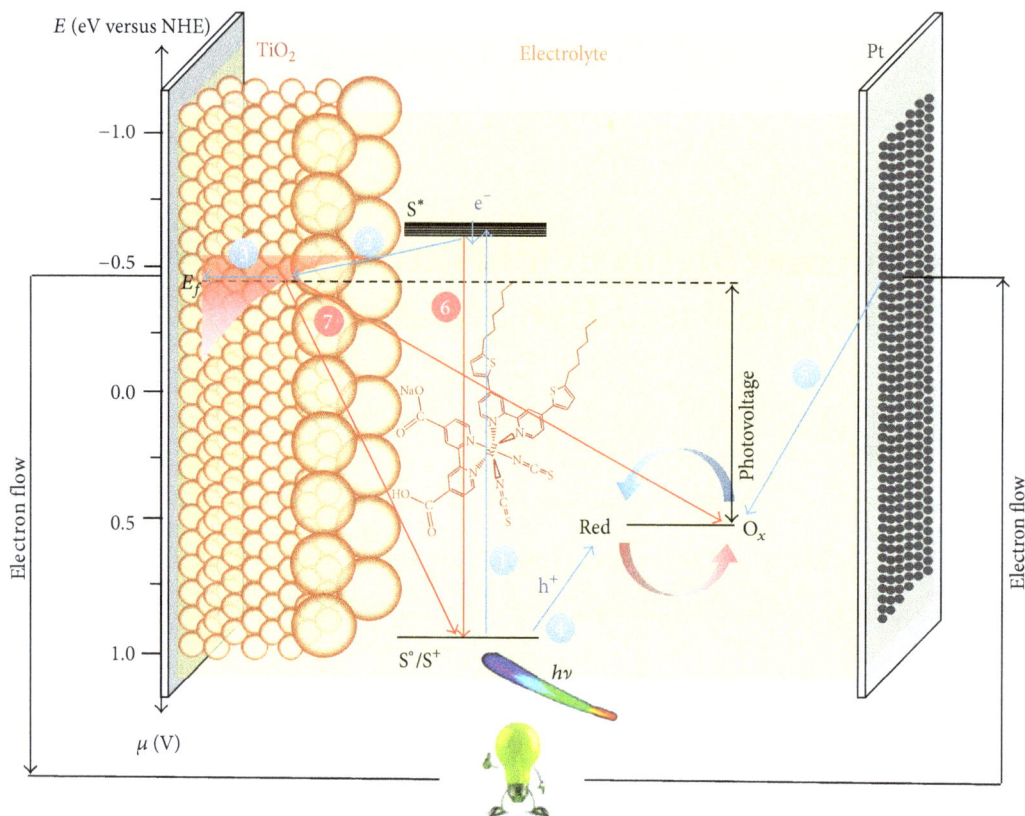

FIGURE 1: Schematic of dye-sensitized solar cell principle including the favorable charge transfer pathways in blue and the loss pathways in red.

of electrochemical batteries through the discovery of electrochemical insertion compounds which are currently used in lithium and lithium-ion technology. The most prospective low-cost PV technology is the third generation developed at the end of the 1980s. Organic PV (OPV) and hybrid organic/inorganic dye-sensitized solar cells are fulfilling not only these low-cost expectations and low environmental footprint requirements but also the easy processing, promising a fast technological transfer. What the history tends to forget is that forward-thinking Elliot Bermann envisioned already organic PV back in the late 1960s. He was supported by Exxon Company which integrated prospection for PV cutting-edge research relevant for 30 years ahead in their development projects. They rapidly cost-cut the conversion price by 5 times for the first two years' activity (from 100 to 20 $/W$_p$).

This review is dedicated to the second member of the third generation, namely, dye-sensitized solar cells (DSCs) [4]. Its working principle lies in complete disruption with the other technologies: polycrystalline materials of large bandgap are used, high purity of the materials is not necessarily required, nanocrystalline semiconductors contain large amount of point defects, the working principle relies on electrochemical processes, exciton separation and transport are detached processes, and finally no internal electric fields onset the charge separation (Figure 1). The original idea of providing light sensitiveness to larger bandgap semiconductors has more than one hundred years' history. It has been

prompted by Professor Vogel who pioneered a dye-sensitized silver halide emulsion for practical use in argentic photographic films [5]. This principle was turned away from its primary scope to energy related applications in particular by Gerischer and Tributsch in 1968 and 1972 who used ZnO single crystals sensitized by chlorophyll [6–8]. New impetus was finally given by O'Regan and Graetzel who published on the sensitization of mesoscopic film of nanocrystalline anatase TiO$_2$ particles enabling, in association with red heteroleptic ruthenium (+II) polypyridyl complexes, a remarkable enhancement in the light harvesting properties of the photoanode [4].

High photoelectrochemical stability of DSC was rapidly supported by the fact that Ru(+II) polypyridyl complexes in association with TiO$_2$ or ZnO can endure over 10^7-10^8 turnovers of electron injection from dye excited state to electron acceptor levels in the semiconductor which in turn would guarantee more than 10 years' lifetime of the panel [9]. Rapidly passing the threshold of 10% power conversion efficiency (PCE) in 1993 [10], despite endless efforts on material development, the efficiency had stagnated for a long time in the range of 11.0–11.5% [11–16] before scoring three new subsequent records, first 12.3% and 13.0% under A.M. 1.5 G conditions (100 mW/cm^2) using organic dye molecules in association with a stronger oxidant redox active cobalt polypyridyl complex [17, 18]. More recently, certified 14.1% and even certified 17.9% PCE have been achieved by

(a)

(b)

(c)

FIGURE 2: Scanning electron micrographs from (a) top view and (b) cross section of a typical mesoporous TiO_2 film used as a photoanode in dye-sensitized solar cells. (c) High resolution transmission electron micrograph of TiO_2 nanocrystal prepared by hydrothermal synthesis used in dye-sensitized solar cells.

replacing the dye with a hybrid organic/inorganic lead halide perovskite absorber and the liquid electrolyte by a solid hole transporting material (HTM) [19, 20] giving birth to a new technology called perovskite solar cells, for which progresses are remarkably fast.

Beside these excellent performances of light-to-electricity power conversion efficiency which clearly meets the standard requirement for market introduction, the race for highest PCE outshined the requirement for high stability and ageing predictability for mass industrialization in many aspects. The much lower number of publications in this area witnesses this deficiency. Matsumura et al. in 1980 already suspected issues of dye stability in devices artificially mimicking the photosynthetic process as dye-sensitized solar cells [21].

The scope of this review is to provide the reader with an overview about the current material progress which has been realized and the knowledge acquired on the stability issues of dye-sensitized solar cell technology. In this review, we will outline the recent development made in dyes and in electrolyte formulation. We will also gather the actual knowledge regarding the degradation mechanisms which are the source of efficiency decreases under accelerated ageing protocol (IEC61646). The review is then separated into three independent sections: a rapid reminder about the principle of dye-sensitized solar cells including research progress in the different cell components, an overview about the development of new electrolytes based on solvent and solvent-free

formulation, and a description of the current understanding on the degradation mechanisms in dye-sensitized solar cells.

2. Dye-Sensitized Solar Cells Principle: Role and Progress in the Different Components

A schematic presentation of the operating principles of the DSC is given in Figure 1. At the heart of the electrochemical system, a mesoscopic layer of anatase TiO_2 is sintered together to provide efficient electronic transport. The layer is typically composed of well-dispersed 20–30 nm size particles of anatase TiO_2 sheltered by a scattering layer constituted of ca. 400 nm particles whose role is to scatter unabsorbed photons (Figure 2). Attached to the surface of the nanocrystalline film is a monolayer of the light sensitive dye. Photoexcitation of the latter induces charge separation, ultrafast femto/picosecond electron injection into the conduction band of the oxide, and hole capture by the reductive redox species composing the electrolyte. This regeneration reaction enables recovering the original oxidation state of the dye. For a sufficient concentration of reductive species, this reaction is fast enough to prevent the recapture of the conduction band electron by the oxidized dye and to hamper dye degradation. The role of the dye is then pivotal for the cell operation as it will govern light absorption ability of the device. The main family of redox couples used in DSC is the two-electron system I_3^-/I^-. Subsequently for dye regeneration, tri-iodide diffuses towards the counter electrode and collects two electrons from the external circuit to form back iodide. The counter electrode is typically supported by electrocatalytic platinum nanoparticles reducing the charge transfer resistance for I–I bond breaking.

Alternatives to anatase TiO_2 nanoparticles have been investigated. The most evident approach was to explore the two other main polymorphs of TiO_2: brookite [22] and rutile, where the latter can enhance the scattering characteristics of the reflecting layer owing to its higher refractive index [23]. Alternatively, ZnO [24–43], SnO_2 [44–48], Nb_2O_5 [49–55], and In_2O_3 [56] were also proposed along three other ternary metal oxides $SrTiO_3$ [57–59] and $BaSnO_3$ perovskites [60] or the spinel $ZnSn_2O_4$ [61, 62]. So far, the anatase TiO_2 remains the leading contender photoanode as it affords the best performances of charge collection. The optimization of this charge collection efficiency has been the stake in research activities in the mesostructuration of the photoanode or by introducing additional point defects through aliovalent doping. The reduction of particles dimensionality offers faster electron transport when the photoanode is composed of vertically aligned nanowires or nanotubes [63–71]. The synthesis of TiO_2 beads leads to the most spectacular improvements resulting from the excellent particles interconnections, shortened mass transport pathways into the mesopores, and combination of light scattering ability and high surface area offering improved light confining properties of the photoanode [16, 72–74]. Remarkable efficiencies exceeding 11% (A.M. 1.5 G) were obtained on optimized beads in combination with high molar extinction coefficient C101 or C106 ruthenium dyes.

The strategy to introduce point defects relies on the high electronic sensitiveness to doping of the $3d^0$ electronic configuration adopted by the Ti^{4+}. The literature in this topic for DSC extends rapidly, even though this approach was proscribed for a long time as it could lead to a harmful increase in the density of free carriers. However, close to thermodynamic synthetic methods are employed which maintain TiO_2 a n-type semiconductor regardless of the dopant. It also allows charge compensation mechanism to maintain a low concentration carriers [75]. We can list a series of hypervalent cations which have been successfully incorporated into the anatase crystal structure: W^{6+} [76], Nb^{5+} [77–80], and Ta^{5+} [81], subvalent cations: Zn^{2+} [82], Cr^{3+} [83], Fe^{3+} [84], Sc^{3+} [85], Y^{3+} [86], and Ga^{3+} [87], or isovalent cations such as Sn^{4+} [88] and Ce^{4+} [89]. The common observation is that hypervalent doping affects the energy of acceptor trap states. A careful control of dopant concentration can therefore adjust the trap energetics and thus is favorable in some cases to accelerate the charge injection. By contrast, subvalent dopants influence neither the energy nor the distribution of traps (Figure 3). With the appropriate type of dopant and concentration, improvement of charge collection efficiency can be achieved, thus contributing to improve the power conversion efficiency, even though the efficiencies reported so far are still lying below 10%.

Efforts to replace iodine/iodide redox couple have been motivated to reduce the nearby 600 mV energy loss from improper energy alignment between the dye HOMO level and the thermodynamic redox potential of iodine/iodide. It is secondarily motivated by the strong corrosive character of iodine and its deep orange coloration penalizing the blue light conversion of the solar spectrum. A significant reduction of this loss-in potential can bring the DSC to pass the threshold of 20% power conversion efficiency [90]. Figure 4 gathers the actual redox molecules and hole transporting materials proposed so far, listed in energy scale. The closely related two electrons Br_3^-/Br^- have been suggested despite their high corrosive character [91–93]. In combination with carbazole-based sensitizers, Li et al. achieved DSC cells going beyond the 1 V photovoltage threshold. The sulfide/polysulfide redox couple is typically used for quantum-dot solar cells [94] and the alternate disulfide/thiolate redox couple showed appealing features in combination with ruthenium and organic dyes [95–97]. Another one-electron redox system was investigated, namely, the cobalt (+III/+II) polypyridyl complexes [98–103], copper [104], iron [105], or nickel [106]. Fast redox couple based on nitroxide was proposed by Zhang et al. [107].

The output voltage is limited in principle by the dye absorption bandgap value (1.5 eV < Eg < 2.5 eV). The integration of cobalt complexes has been beneficial to boost the photovoltage of champion devices. It is interesting to comment that the braking efficiency record in DSC by means of photovoltage improvement does not differ from the other single junction technology for which new records were systematically reached from a photovoltage improvement.

The central role of the sensitizing dye molecule generates demands in terms of the triptych chemical, electrochemical, and photoelectrochemical properties. The dye should of course combine chemical and photoelectrochemical stability but should also possess a high molar extinction coefficient with panchromatic light absorption and adequate HOMO/LUMO energy alignments. The pioneering work at EPFL presented a series of ruthenium polypyridyl complexes among the ubiquitous N719 and Z907 dyes which are benchmark sensitizers still today [108]. Molecular ligand engineering for ruthenium dyes was judiciously scrutinized by different groups to tune the spectral response of the sensitizers either by introducing a ligand with a low-lying π^* molecular orbital and by destabilization of the metal t_{2g} orbitals through the introduction of a strong donor unit (Figure 5). N719 and related sensitizers are typically anchored onto TiO_2 according to a bidentate mode [109]. It signifies that two of the four carboxylic acid functions remain unbounded, thus not participating efficiently in the photon-induced charge releasing into TiO_2. To address this issue, one approach lies in the replacement of one of the two dicarboxybipyridyl units by an ancillary ligand displaying a slight donating character at the excited state while including a long alkyl chain to improve the dye solubility, to provide a hydrophobic character for longer term stability [110, 111], and to reduce iodine access to the surface for hampering recombination processes. It was also found that enhancement of

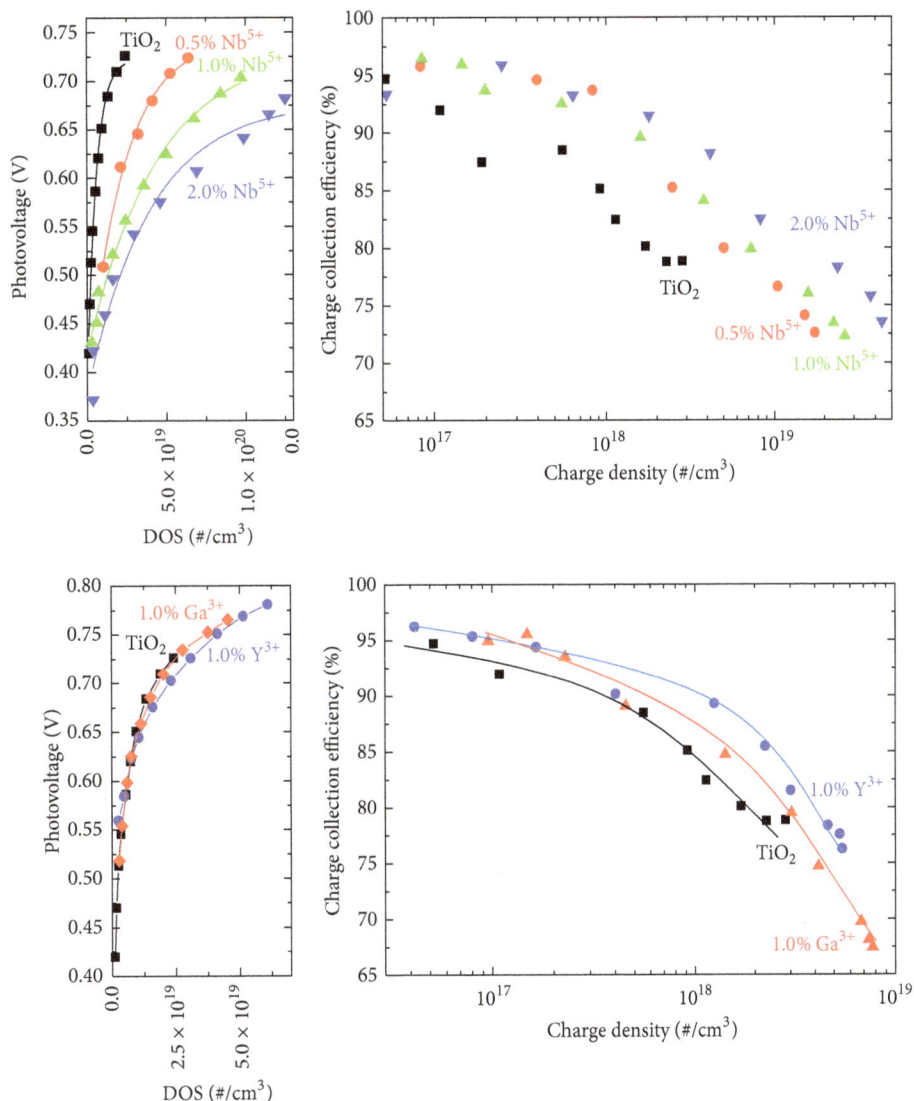

FIGURE 3: Effect of doping TiO_2 by Nb^{5+}, Ga^{3+}, and Y^{3+} on energy and distribution of trap states and on charge collection efficiency.

the π-conjugation in the ancillary ligand not only exerts an energetic stabilization of the LUMO levels, thus red-shifting the dye absorption, but also favorably affects the molar extinction coefficient of the dye reaching more than $20000\,M^{-1}\cdot cm^{-1}$ [12, 15, 112–115]. The integration of this second generation of dyes paved the way to improved performances in terms of light harvesting and power conversion efficiency. Well above 11% power conversion efficiency was achieved and excellent stability data reported at $60°C/100\,mW\cdot cm^{-2}$ conditions. The extension of conjugation in the acceptor ligand by going from bipyridine, terpyridine (e.g., N749) to quaterpyridine (e.g., N1044) contributes to a broadening in the absorption of the dye affording visible light panchromatic response [116–118], unfortunately at the expense of the molar extinction coefficient.

The second option to tune the complex absorption towards the red consists in destabilizing the t_{2g} orbitals of the ruthenium by the introduction of a stronger donor unit

than thiocyanate. Many attempts to replace the thiocyanate donor ligands have been made and motivated also by the fact that the monodentate NCS is believed to be the weakest part of the complex from a chemical point of view. For long time, these efforts had yielded limited success as the conversion efficiency remained well below few percent [119]. A paradigm arose from the YE series developed by Bessho et al. who achieved 10.1% power conversion efficiency under standard illumination conditions (A.M. 1.5 G) [120]. The lowest energy MLCT band is red-shifted by 25 nm and contains a new absorption band at 485 nm with a remarkable molar extinction coefficient. This new band is characterized by an electronic transition from HOMO, which has a sizable π-orbital contribution from the cyclometalated ligand, and the LUMO formed by a set of π^* orbitals localized on the bipyridines ligands evidencing strong electronic coupling with acceptor states in TiO_2. These features explain the high photocurrent in the range of 17 mA/cm² and the photovoltage

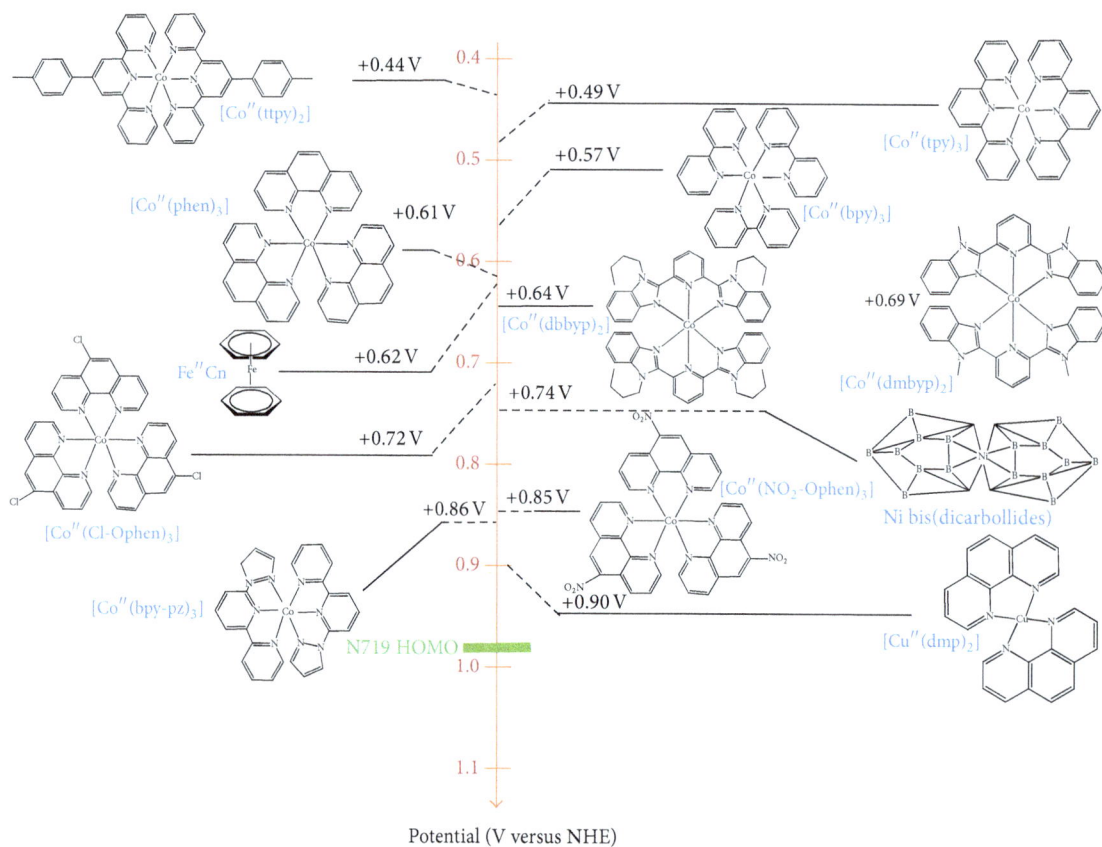

Potential (V versus NHE)

(a)

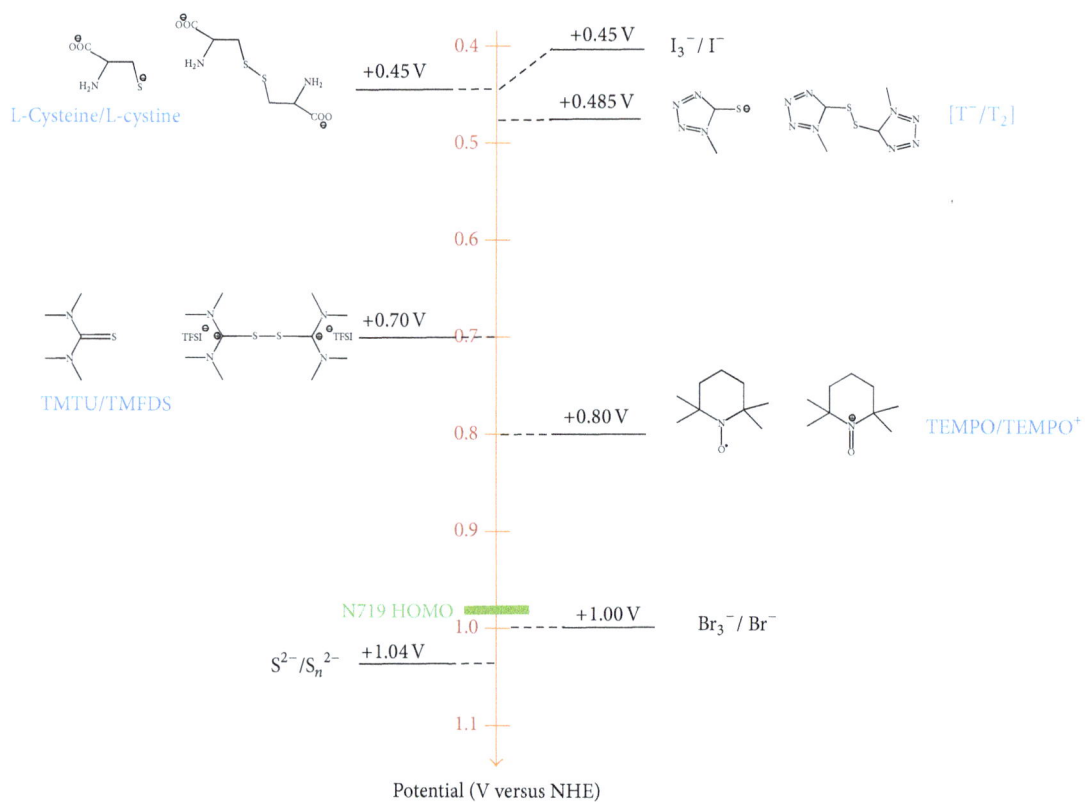

Potential (V versus NHE)

(b)

FIGURE 4: Scale of chemical potential for the main (a) redox complexes and (b) redox molecules used in dye-sensitized solar cells.

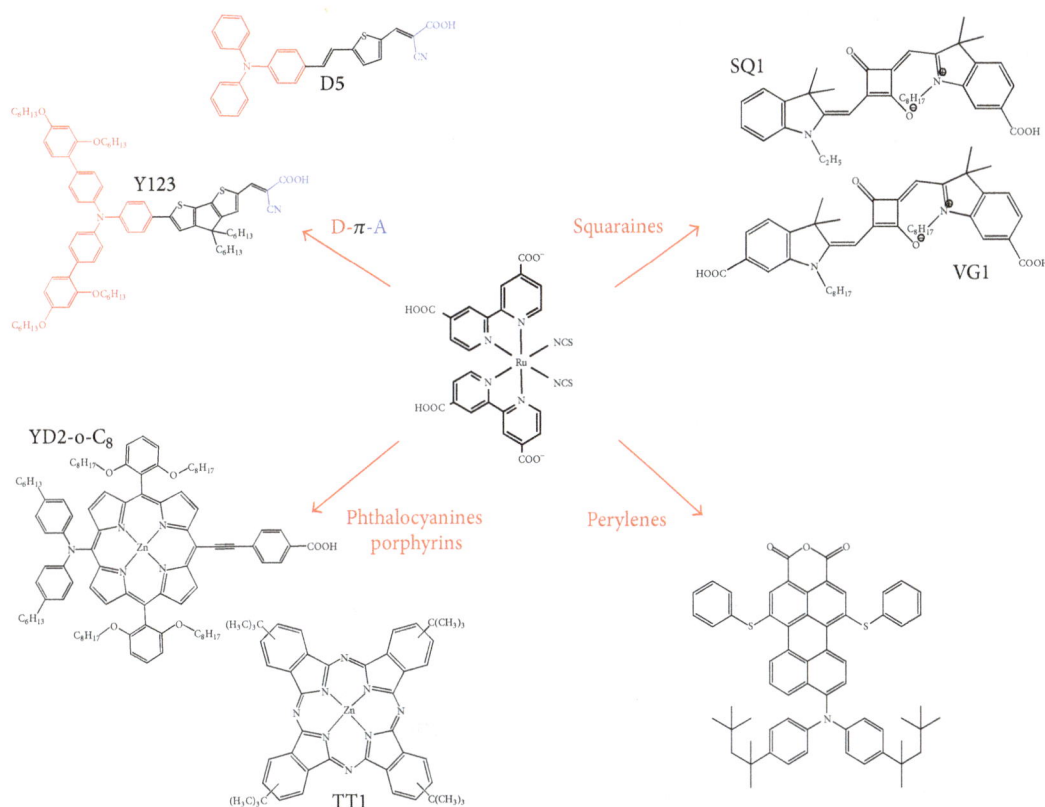

FIGURE 5: Main families of different organic dyes developed for dye-sensitized solar cells.

exceeding the threshold of 800 mV. After this work, another family of NCS-free ruthenium structures was developed by Bomben et al., who reached close to 10% efficiency [121]. The removal of NCS ligands altogether in tridentate cyclometalated ligands also provides an enhancement in the absorption ability of the complex. This was demonstrated in the series of TF-dyes compared to the thiocyanate-related N749 (so-called black dye) [122]. With TF-dyes, Chou et al. reached 10.7% power conversion efficiency with photocurrent densities attaining as high as 22 mA/cm^2. We can regret the actual lack of knowledge about the stability of all these new complexes, either because they are not offering good stability in cells or simply because they were not investigated.

In order to bypass the price for ruthenium metal which continuously beats records and to address the synthetic difficulties intrinsic to organometallic compounds often requiring lengthy and costly purification processes incompatible for real mass production, metal-free organic chromophores have been proposed. Some of them are actually meeting the ruthenium sensitizers in terms of power conversion efficiency and even stability. They exhibit significantly higher molar extinction coefficient than ruthenium dyes, lying in the range of 20000 to 400000 M^{-1}·cm^{-1}. Four main families have been reported: porphyrins and the closely related phthalocyanines (coordinated with zinc or not), donor-π-acceptor (D-π-A) assembly, squaraines, and perylenes (Figure 5). Note that other families of organic chromophores have also been reported: coumarins which approached $\eta = 7.7\%$ [123], indolines, and natural pigments among which are anthocyanin, flavonoid, carotenoid, and chlorophyll [124].

D-π-A structures actually stand out from the others as they combine high power conversion efficiencies beyond 10% [125, 126] and stability standard passing even the accelerated ageing test of 85°C/dark (Y123 dye). One key unit in the design of D-π-A originated from the introduction of cyanoacrylic function as an acceptor group which enabled extending light absorption ability of the dye compared to the related coumarin D-π–A [127]. It provides around 25% more short-circuit current density and exhibits a molar extinction coefficient close to 100000 M^{-1}·cm^{-1} at 552 nm. Sensitizers having an intense absorption band in the near-infrared region can be typically acquired through phthalocyanines which are reaching absorption maxima to 700 nm. Squaraines dyes can even get close to 800 nm [128, 129]. Despite these interesting absorption profiles, NIR dyes have typically showed unimpressive power conversion efficiencies due to either too strong aggregation and lack of directionality in the excited state for the phthalocyanines [130] or issues for fast electron injection in the case of the NIR squaraines. This is explained by a too strong energetic stabilization of the π^* orbital delocalized throughout the dye structure. For these reasons, this call for the development of new semiconductor having lower lying energy conduction band edge for endorsing fast electron transfer. A very recent in-depth review dedicated to metal-free sensitizers can be found in [131].

3. Liquid Electrolyte Development Based on Solvent and Solvent-Free Formulation for Stable Devices

Champion efficiencies reported above are systematically obtained in conjunction with an electrolyte based on a volatile solvent. Pure acetonitrile or acetonitrile/valeronitrile solvent mixtures are often preferred. Their intrinsic characteristics, mostly in terms of volatility, cannot guarantee enough stability to pass the accelerated IEC61646 protocol. The excessive vapor pressure exerted shortcomings to the cell sealing even at temperatures well below 60°C, besides additional chemical/electrochemical reactions in cell which may occur. This volatility issue was circumvented by replacing acetonitrile with a different class of lower volatile solvents. γ-Butyrolactone (GBL), propylene carbonate (PC), propionitrile (PN), sulfolane, butyronitrile (BN), or 3-methoxypropionitrile (MPN) had been proposed. Their integration led to substantial device lifetime prolonging. It seems that the last two actually show the best compromise between high efficiencies and stability. The sulfolane-based electrolyte is promising for extreme conditions even though its high viscosity impedes high photocurrent production owing to mass transport limitations [132].

The electrolyte is not solely composed of iodide and iodine. It also contains two main additives. The beneficial impact of these additives on the cell performance is undeniable. However, the exact role/action in the complete cell remains speculative in some extent. Table 1 gathers the benchmark composition of nitrile-based electrolyte which is used to obtain high efficiencies or alternatively good stability for low-volatile MPN or BN solvent. Note that solvent-free ionic liquids are composed of binary or ternary eutectics melts including different components and composition than the reported in Table 1. A review was dedicated to this topic by Zakeeruddin and Grätzel [133]. The two additives typically incorporated in electrolyte are guanidium thiocyanate and a Brönsted base, namely, terbutylpyridine for high efficiency electrolyte or replaced by the N-butyl benzimidazole or the benzimidazole in some cases for stable electrolytes. The utilization of a Brönsted base affords the deprotonation of the sensitized TiO_2 surface. It contributes to up-shift the quasi-Fermi level in TiO_2 and therefore the open circuit voltage of the cell. N-Butyl benzimidazole in stable electrolyte composition is preferred over terbutylpyridine, regardless of whether it is in MPN or BN (Figure 6). The comparison of the two molecules suggests that they have similar strength for deprotonation since the distribution of the subconduction band energetic levels is not modified. The charge collection efficiency is similar or slightly better in the case of the terbutylpyridine for which a marginally higher electron lifetime is compensated by a lower electron transport (Figure 6). The guanidinium thiocyanate has a more subtle function. The literature often refers to the guanidinium cation going to the surface of TiO_2 between the dye molecules in order to explain the improvement of charge collection efficiency. However, it is also straightforward that this additive is beneficial for the cell stability, in particular to maintain a high fill factor.

An impedance study carried out on TCO-Pt/electrolyte/Pt-TCO symmetric cells, performed at the laboratory, has highlighted that the charge transfer resistance evolution becomes strongly dependent on whether the electrolyte contains or not this guanidinium thiocyanate additive. Indeed, whereas the benchmark electrolyte composition containing 0.1 M of guanidinium thiocyanate shows a slight decrease in R_{ct} to ca. $1\,\Omega\cdot cm^2$, such evolution appears in complete contrast with an electrolyte composition free of guanidinium thiocyanate (Figure 7). Beside its aforementioned function, we concluded that the guanidinium thiocyanate molecule also plays the role of a protective agent by molecular self-assembling onto Pt to prevent its dissolution through soluble PtI_6^{2-} complexes. This is further supported by visual observation of the symmetric cell during ageing which tends to bleach at the specific area in contact with the electrolyte.

In terms of stability performances, 10% efficiency on lab cells (<1 cm^2) with PCE retention over 95% after 1000 hours under 60°C/100 mW/cm^2 light soaking experiments currently stands for the best values reported so far on lab devices. This was achieved by using either BN [134] or alternatively gelified MPN-based electrolyte [135]. Dyesol recently reported its HSS electrolyte passing even 95°C/dark accelerating test for 1000 hours in combination with N719 dye with efficiencies in the range of 5% under standard illumination conditions (A.M. 1.5 G) [132].

The standard accelerated ageing tests protocol used for evaluating the stability of dye-sensitized solar cells is by default the protocol established for terrestrial thin film PV (IEC61646):

(i) 1000 hours at 85°C ($\pm2°C$) in dark under 85% ($\pm5\%$) humidity;

(ii) 1000 hours at 60°C under light illumination (800–1000 W/m^2);

(iii) 300 thermal cycles between −40°C and +85°C.

Stable performances were reported over 2000 hours at 60°C in dark maintaining 5.5% PCE [136], 2.5 years in outdoor conditions without specification about module efficiency [137], or 2280 hours at room temperature under 80 mW/cm^2 light soaking [138], or more recently Dyesol Ltd. achieved as long as 25600 hours at 55°C–60°C under continuous light soaking with preserving 4% efficiency [139]. Since the seminal publication by Graetzel, only very few articles have reported stability at higher temperatures, requiring, as it seems, more viscous liquids to alleviate sealing issues, as, for instance, gelified electrolyte or solvent-free ionic liquids [140–143]. Apart from the very few reports announcing this achievement, still no consensual publications are actually passing the challenging threshold of 1000 hours at 85°C in dark condition with efficiencies above 5% [132, 144, 145]. Nevertheless, this assessment emphasizes on the one hand the high credibility of DSC for larger scale applications, but on the other hand, it also point out the important gap to fill in between the champion efficiencies and the stable devices for which this gap closing will require more focus on the understanding of the ageing mechanisms leading to the cell failure, and, in a next step to design specifically more

TABLE 1: Composition and structure of main liquid electrolyte components.

Name	Acetonitrile	3-Methoxypropionitrile	1,3-Dialkylimidazolium iodide	Iodine	Ter-butylpyridine	N-Butyl benzimidazole	Guanidinium thiocyanate
Structure							
High efficiency (concentration)	Solvent	—	1 M	0.03 M	0.5 M	—	0.1 M
Stable efficiency (concentration)	—	Solvent	1 M	0.15 M	—	0.5 M	0.1 M

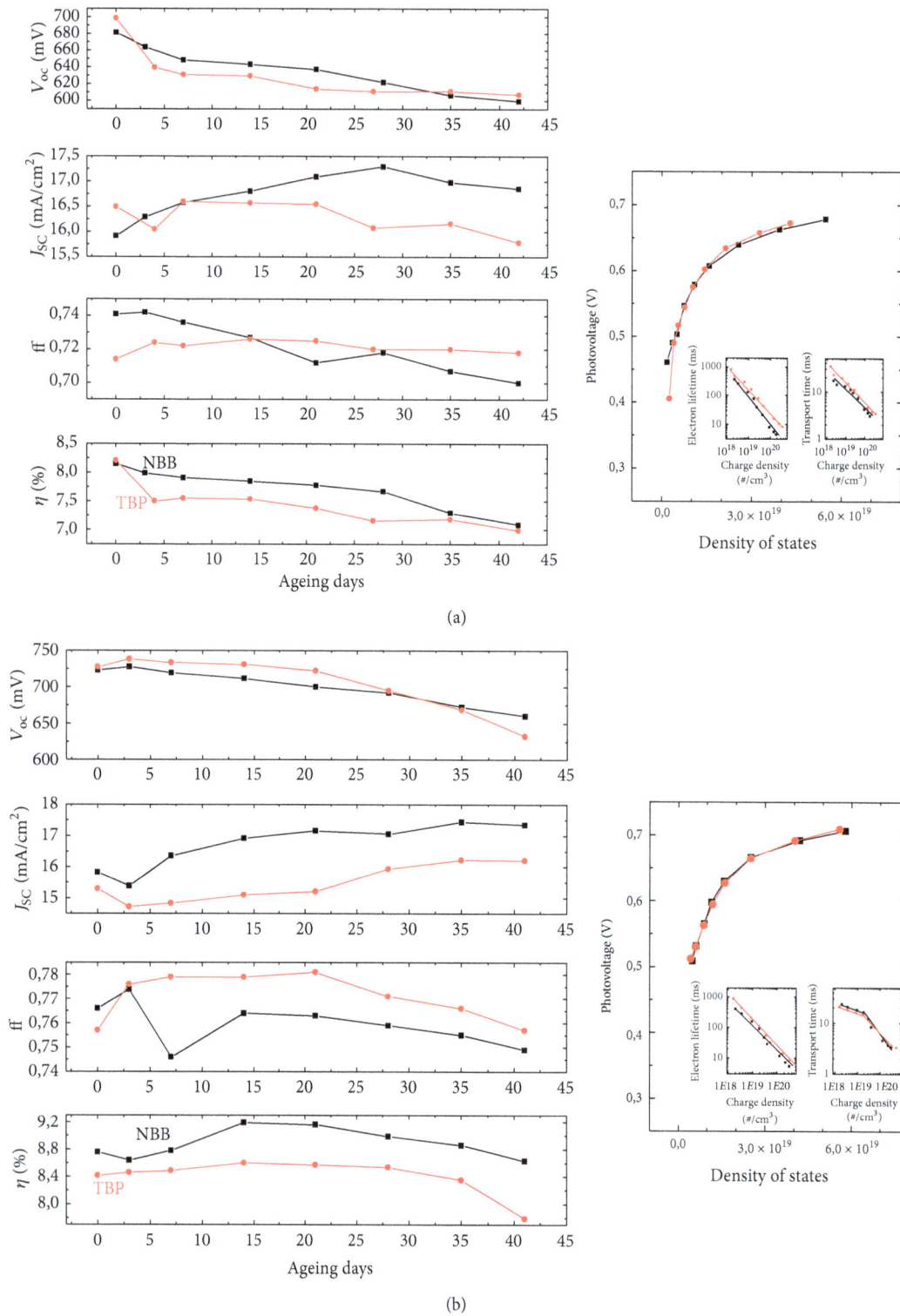

FIGURE 6: Comparison of cell characteristics during ageing at 60°C/100 mW·cm⁻² using an electrolyte constituted of TBP or NBB (composition: 1 M DMII, 0.1 M GuNCS, 0.15 M I$_2$, and 0.5 M NBB/TBP) in (a) 3-MPN solvent and in (b) BN solvent. A comparison of the trap state distribution and energy and a comparison of electron lifetime and transport time as a function of charge density are also included.

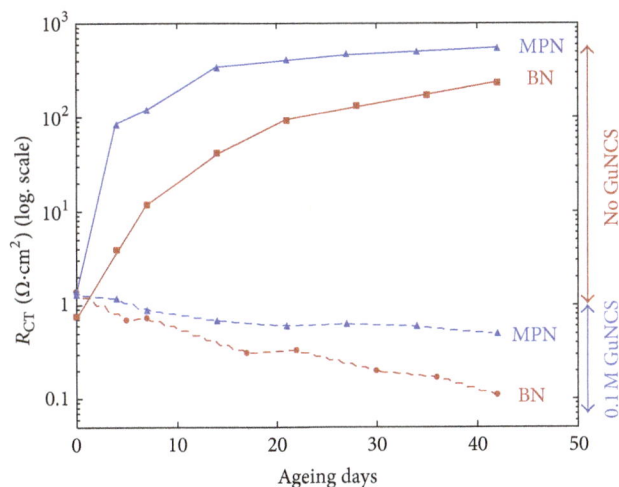

FIGURE 7: Evolution of the charge transfer resistance as a function of ageing time at $60°C/100\,mW\cdot cm^{-2}$ measured in TCO-Pt/electrolyte/Pt-TCO symmetric cell configuration comparing BN and 3-MPN-based electrolyte, with $0.1\,M$ GuNCS and without GuNCS.

robust and better performing molecules (i.e., ligand, solvent, additive(s) in electrolyte, etc.).

4. Current Understanding of Chemical/Photoelectrochemical Degradation Paths

Whereas the durability of single-crystalline and amorphous silicon modules is established to a lifetime attaining 20 years without question, the stability and ageing predictability of emerging PV technologies are still under close observation to predict at least 10 years' lifetime under working environment. Mastering the dynamics of charge transport and charge recombination that take place in dye-sensitized solar cells is a crucial issue for reaching high performances. Nevertheless the stability of each cell component taken separately and the understanding of all the chemical, electrochemical, and photochemical reactions interplay at the different material interfaces or components are a key to bring further development on DSC, not only to improve the device stability under severe ageing conditions but also to close the gap between champion and stable efficiencies. Such complete understanding of the ageing mechanisms will pave the way to the specific design of more robust components. Integrated in an advanced generation, these new materials will give impetus to DSC commercialization and broader integration into the PV panorama. Only a very few groups are actively working in this more fundamental but crucial domain. Beside the intrinsic sealing and permeability issues of the device which have been revealed to influence the device stability as will be discussed below, six distinct intrinsic features contributing to the cell degradation have been spotted in the literature (Figure 8).

(1) Likely one of the most preeminent degradation path stems from the well-known iodine consumption in

electrolyte: this reaction takes place during ageing. It translates into a well-visible bleaching in coloration of the electrolyte. It is crucial to overcome this side reaction as it is one important source of failure since iodine concentration controls the kinetic for tri-iodide mass transport. Excessive depletion will turn to short-circuit current density limitation of the device.

(2) Dye desorption at elevated temperatures (>60°C): this possible event takes its origin from the rivalry between dye solubility in the electrolyte and the binding strength of the anchoring group upon TiO_2. It can also be promoted by the ruthenium hexacoordination rupture as it will be discussed below. Water intake in the electrolyte has also been proposed to assist this dye desorption mechanism. The development of hydrophobic dyes was anticipated to go against this reaction and explained the better long term stability of Z907 versus N719.

(3) There is lack of chemical and photochemical stability of the monodentate thiocyanate ligand in heteroleptic ruthenium (+II) complexes which tends to undergo substitution reactions with different external components.

(4) There is platinum dissolution for electrolytes free of thiocyanate (cf. Figure 7).

(5) Formation of a polymeric solid electrolyte interphase (SEI) forming on TiO_2 and sensitized-TiO_2 (see Figure 10 inset).

(6) UV irradiation causes direct bandgap excitation of TiO_2 leading to conceivable dye or electrolyte component oxidation [137, 146]. Addition of MgI_2 or CaI_2 in the electrolyte was proposed to improve the electrolyte tolerance to UV irradiation [135]. It is also typical to cover the glass photoanode by an antireflecting polymer coating which also plays the role of UV filters [146].

In the following we will review in more details the two most severe issues to overcome, namely, the iodine depletion in electrolyte and the dye chemical/photoelectrochemical stability.

4.1. Iodine Consumption. Iodine consumption has been spotted by numerous groups who gave different attributions to explain this depletion. A total of eight hypotheses/observations, sometimes controversial, can be listed.

(i) Formation of IO_3^- induced by water traces [147–150]: in this case it is postulated that I_2 reacts with ingress water or residual water contained in the electrolyte. It leads to the formation of the iodate anion (IO_3^-). As commented also by some of these authors, the formation of IO_3^- has, however, never been really detected so far either by spectroscopic methods (UV-Vis, Raman, etc.) or by chromatography (LC/MS). Figure 9 reports an example of withdrawn aged electrolyte analyzed by LC/MS after 500 hours' ageing at

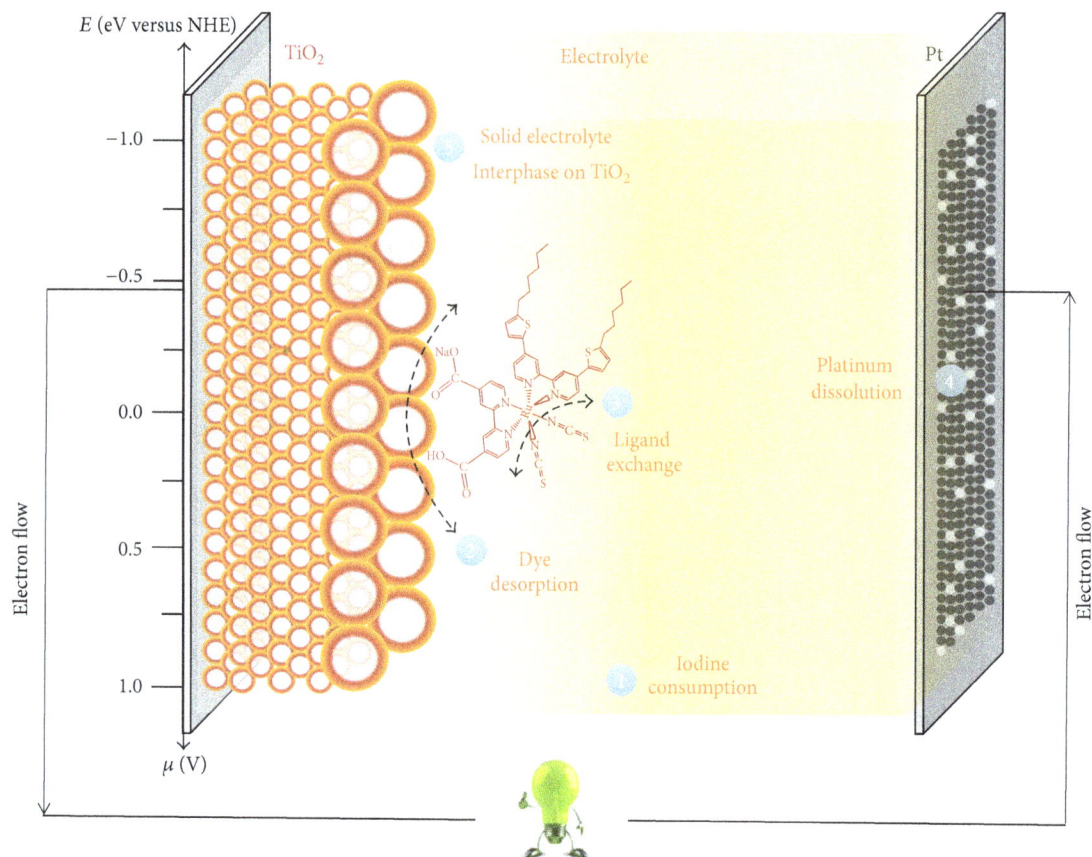

FIGURE 8: Schematic gathering the known stability issues in dye-sensitized solar cells.

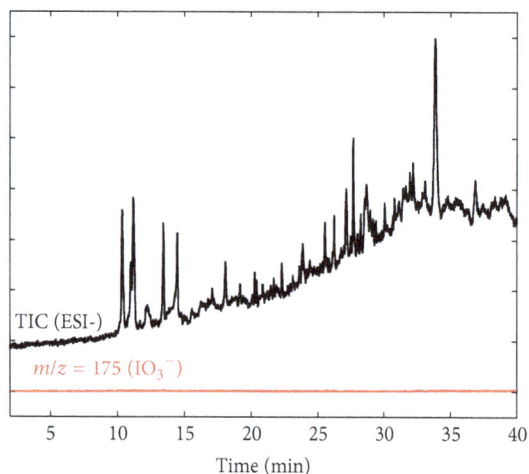

FIGURE 9: LC/MS chromatogram of aged MPN-based electrolyte in contact with TiO$_2$ during 500 hours at 85°C/dark. The chromatogram for m/z value of 175 is reported for comparison.

85°C in dark and compared to the specific signal of m/z value of IO$_3^-$ [151].

(ii) I$_2$ reacts with glass frit used in some technologies as a sealant [152]. The experimental evidences reported in the reference are conclusive. However, this depletion

is experienced regardless of the type of sealant utilized (*i.e.*, Bynel or Surlyn polymers, glass frit, etc.). This is therefore not the original path for this reaction even though it will contribute to this consumption.

(iii) Sublimation of molecular I$_2$ has been proposed by different groups [145, 153, 154]. This hypothesis has been controverted by *ex situ* and *in situ* spectroscopic methods such as Raman on aged electrolyte and GC/MS or TGA/MS performed also on aged electrolyte or on electrolyte heated at 60°C or 85°C with gas output analysis. The results were suggesting the absence of any spectroscopic band or mass which could be attributed to iodine or iodide-based compounds even in form of traces [151].

(iv) Bandgap excitation of TiO$_2$ leads to the photoelectrochemical reduction of tri-iodide [155]. This reaction would be particularly unexpected knowing that TiO$_2$ has a strong n-type semiconducting character with direct bandgap excitation. This reaction takes place faster without than with dye and therefore could not be attributed to a recombination procedure between iodine and electron in conduction band. Last but not least [152], also pointed out that iodine consumption proceeds in dark and is only thermally activated as discussed below.

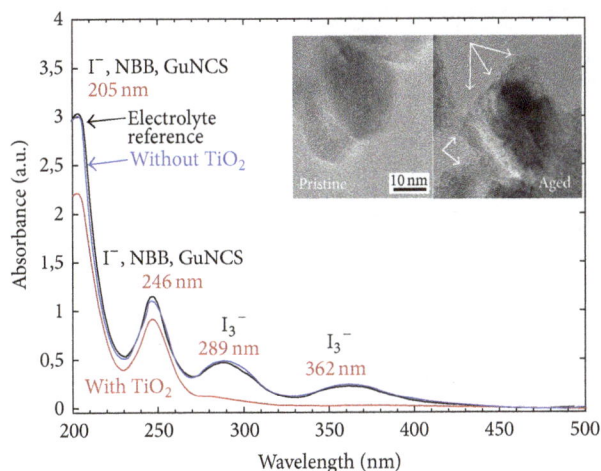

FIGURE 10: Evolution of UV-Vis absorption spectrum for reference MPN-based electrolyte (in black), electrolyte aged alone during 500 hours at 85°C/dark (in blue), and electrolyte aged in contact with TiO$_2$ (in red). In inset are reported high resolution transmission electron micrographs comparing the TiO$_2$ nanoparticles before and after ageing showing the formation of SEI layer.

(v) The electrolyte bleaching is solvent dependent. This has been clearly highlighted from the work led at Dyesol by Jiang et al. [132, 156]. The authors compared γ-butyrolactone, 3-methoxypropionitrile, and the so-called nitrile-free solvent HSS. They have concluded that iodine depletion is significantly hampered when going from γ-butyrolactone to HSS. They have also reported slower iodine consumption using tetraglyme compared to 3-MPN [156].

(vi) Iodine consumption is not an intrinsic reaction taking place in the electrolyte alone. It is triggered by the surface of TiO$_2$ which plays a catalytic role in the electrolyte degradation [151]. The authors concluded that iodine depletion is exclusively activated by temperature while light has no action on this reaction, at least for 3-MPN-based electrolyte (Figure 10).

(vii) Iodine reacts with the electrolyte additives such as the 4-*tert*-butylpyridine (TBP) leading to an iodo-pyridinate complex and with thiocyanate ligand of the dye to form I$_2$NCS$^-$ species in the particular case the electrolyte is exempt of any TBP [157, 158].

(viii) Formation of a solid electrolyte interphase (SEI) nucleating on the surface of TiO$_2$ in the dye mono-layer pinholes (Figure 10 inset) [151]: the authors highlighted that SEI formation traps or solvates iodine (and other electrolyte components) explaining its depletion from bulk electrolyte. It is supported by UV-Vis and XPS spectroscopies combined to ToF-SIMS which revealed on the one hand its very complex composition and on the other hand its high concentration of iodine/iodide-based species along other degraded components of the electrolyte. To the current understanding, this SEI layer is originating from the polymerization between acrylonitrile

radicals leading to the formation of a very cohesive polyacrylonitrile polymer. GC/MS experiments showed that in presence of TiO$_2$ the 3-MPN thermally disrupts to form two highly volatile compounds: acrylonitrile ($bp.$ = 77°C) and methanol ($bp.$ = 65°C) [159]. The free radical polymerization of acrylonitrile is well-established to take place spontaneously by mild thermal activation such as temperatures in a range of 50°C. The polymerization rate can get substantially faster in the presence of metal halide catalyst [160]. The formation of these two degradation compounds can also give a more rational explanation to the sudden electrolyte evaporation pointed out by many authors in the field, which operates at a more or less long term in the ageing, an event typically the culprit of poor cell sealing.

The literature often refers to iodine consumption in electrolyte as it is the chromatic component of the electrolyte with its characteristic strong purple color. At HOPV2014 conference our group presented that many other electrolyte components are in reality consumed. We determined by cyclic voltamperometry using a platinum microelectrode that about 50% of iodide is depleted after 500 hours' ageing at 85°C (Figure 11(a)), FT-IR carried out on aged electrolyte indicates that thiocyanate from guanidinium thiocyanate is almost completely depleted (Figure 11(b)), and last but not least N-butyl benzimidazole is also quantitatively consumed (*ca.* 50%) by LC/MS (Figure 11(c)) [159]. These experiments were realized looking at the interface between naked TiO$_2$ and electrolyte. When sensitized, these reactions are still occurring but at slower rate. This was similarly experienced in the case of iodine consumption for which Asghar et al. have made careful comparison between unsensitized TiO$_2$ and sensitized TiO$_2$ highlighting that dye monolayer only slows down this reaction [161]. MPN-based electrolyte stability is strongly affected by the surface of TiO$_2$ which tends to destabilize significantly the thermal stability of electrolyte components by about 50°C (Figure 12). The surface of TiO$_2$ also collaterally induces gas formation as aforementioned in this review [151, 159].

4.2. Chemical/Photochemical Stability of Benchmark Ruthenium Sensitizers. This part is more focused on ruthenium polypyridyl complexes. Much less is known about the core stability of organic dyes. Tanaka et al. reported that major degradation in indoline-based D-π-A dye, namely, the yellow D131 chromophore, was decarboxylation of the cyanoacrylic acid anchoring group [162]. This decarboxylation reaction is contributing to dye desorption, lowering significantly the device performance whether the ageing is at 60°C/1sun or 85°C/dark. This reaction has been attributed to harmful action of iodine or amine components to the dye. This reaction can likely be extended to some degree to the family of "push-pull" D-π-A dye, even though some of other dyes in this family have been reported to be very stable upon 60°C/light and 85°C/dark ageing conditions. Most of organic dyes are known to cause stability issues owing to their empathy to undergo fast photooxidation when they are

(a)

(b)

(c)

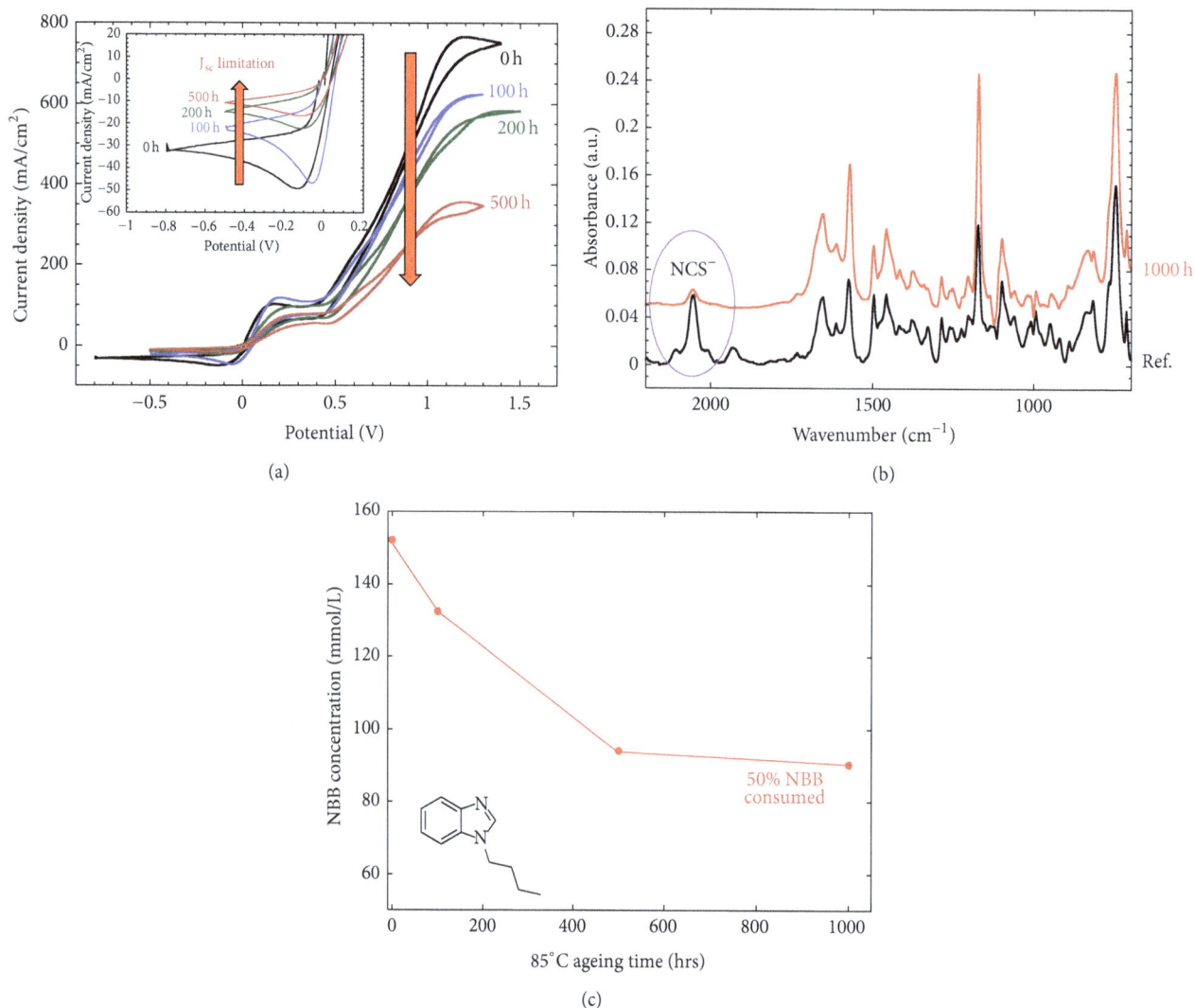

FIGURE 11: (a) Comparison of cyclic voltamperometry recorded on Pt microelectrode for MPN-based electrolyte aged in contact with TiO_2 during 100, 200, and 500 hours at 85°C/dark. (b) ATR-FT-IR spectra comparison of MPN-based reference electrolyte before and after 1000 hours ageing at 85°C/dark in contact with TiO_2. (c) Quantification by LC/MS of NBB concentration as a function of ageing time at 85°C/dark in contact with TiO_2.

anchored onto TiO_2. Following this hypothesis, increased stability will thus require a fast recovery of the reduced form to prevent this side reaction to occur.

The chemical/photo-(electro-)chemical durability of ruthenium polypyridyl-based sensitizers in contact with TiO_2 had been questioned well before the seminal paper from Anderson et al. in 1979 [163]. The underlying effect of the nanostructuration of TiO_2 on the ruthenium-complex stability was discussed [4]. The authors presented photocurrent stability over 2 months subjected to continuous visible light illumination stress with less than 10% degradation corresponding to stability over 5.10^6 turnovers. Similar arguments were reported on monomeric dye without noticeable degradation [164].

These intriguing results are in relative contradiction with other reports on cis-Ru(bpy)$_2$(SCN)$_2$ (bpy = 2,2′ bipyridine) but also on trimeric ruthenium dyes [165]. The lack of

structural robustness of N719 and Z907 had been reported in particular by Tributsch, Hagfeldt, and Lund. These authors became consensually alarmed on the vulnerability of the strong electron donor thiocyanate ligand which can easily exchange with electrolyte components as a result of the antibonding character of these orbitals. This ligand exchange reaction is triggered not only by light action (photolysis) but also can be activated by temperatures above 80°C. The monodentate thiocyanate ligand can then exchange with acetonitrile or 3-methoxypropionitrile when used as a solvent (in integral or fragmented part such as C≡N), with 4-tert-butylpyridine, water residues, or even iodide [165–173]. The different exchange reactions reported are summarized in Figure 14. This reactivity of the monodentate thiocyanate ligand was explained by Grünwald and Tributsch on the basis of an incomplete reduction of the photooxidized dye by I_3^-/I^- redox couple in combination with a certain lack of stability

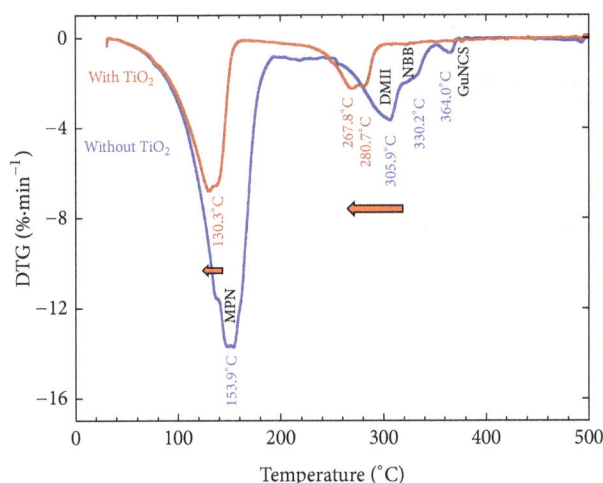

FIGURE 12: Comparison of TGA derivative curve as a function of temperature comparing the intrinsic MPN-based electrolyte stability and the same in presence of TiO_2.

of the oxidized form of the complex [165]. This explanation was further confirmed by Kohle et al. who showed in solution that dye instability can also get prompted by a too low concentration of iodide, in other words when decreasing the rate constant for the dye regeneration [174]. The results provided by Lund et al. suggest again that TiO_2 plays a significant catalytic activity towards this ligand exchange reaction which is accelerated by a factor of 2.

It is also suggested that some degradation products of N719 and Z907 dyes are in equilibrium together [171] (Figure 13). It was for instance evoked in three interrelated equilibrium reactions:

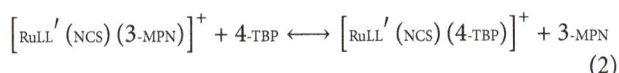

$$\left[\text{RuLL}' \, (\text{NCS})_2 \right] + 3\text{-MPN} \longleftrightarrow \left[\text{RuLL}' \, (\text{NCS}) \, (3\text{-MPN}) \right]^+ + \text{NCS}^- \quad (1)$$

$$\left[\text{RuLL}' \, (\text{NCS}) \, (3\text{-MPN}) \right]^+ + 4\text{-TBP} \longleftrightarrow \left[\text{RuLL}' \, (\text{NCS}) \, (4\text{-TBP}) \right]^+ + 3\text{-MPN} \quad (2)$$

$$\left[\text{RuLL}' \, (\text{NCS})_2 \right] + 4\text{-TBP} \longleftrightarrow \left[\text{RuLL}' \, (\text{NCS}) \, (4\text{-TBP}) \right]^+ + \text{NCS}^- \quad (3)$$

The rate for this ligand exchange reaction can be lowered by a factor of 2 when including a buffer concentration of thiocyanate anions in the electrolyte (e.g., guanidinium thiocyanate) [169, 175]. The degradation products based on 4-TBP can be prevented by replacing this latter with N-butyl benzimidazole or the closely related benzimidazole which enhances the device stability. This can explain in part or in whole the better stability achieved with NBB (Figure 6(a)). Although such ligand exchange reactions will affect the optical MLCT (metal-to-ligand charge transfer) contribution, entailing either slight bathochromic or hypsochromic shift of absorption [132], the real implication of this *in situ/in operando* dye structure modification on the practical cell characteristics remains unclear. Finally, again in the work published by Grünwald and Tributsch, the authors exposed the possibility that the ruthenium hexacoordination can get fragmented, leading to the desorption of the metal core unit

and the retaining of the anchoring bipyridine group attached upon TiO_2 [165].

Beside these intrinsic degradation mechanisms highlighted to date, more external contributions to the cell degradation have also been exposed, namely, sealing conditions, substrate corrosion, and sealing issues. Assessment in these technical aspects is relatively complex to detail since they are rarely published. An interesting work reported by Fredin et al. aims at alarming the importance to consider the temperature of sealing as it will determine the device efficiency [176]. This work was examined by comparing N719 with the organic D5 sensitizer. The authors found that N719 appears more sensitive to thermal degradation than electrodes sensitized with the organic D5. Under temperatures lying between 120°C and 250°C for 5 minutes, the cell efficiency can vary from a maximum of *ca.* 4% to almost nothing when cells were sealed with an excessive sealing temperature (>200°C). By using IMPS/IMVS spectroscopies, they also highlighted that the electron diffusion length, referring to the charge collection efficiency, can get as low as 10%.

The encapsulation of the cells should prevent the exchange of material between the inner part of the cell and the ambient environment. It is crucial for the cell stability that the encapsulation creates a barrier against water and oxygen ingress which can be assimilated by the electrolyte and the mesoporous sensitized TiO_2 layer. Conversely water has also been reported to enhance the power conversion efficiency performances of DSC based on ionic liquids [177]. The proportion of water in the cell can drastically modify the proton concentration in the electrolyte. Note that proton is a potential determining cation; in other words it tends to form specific adsorption on the surface of TiO_2 leading to downward shift in energy of the Fermi level and conduction band edge [178]. A critical threshold value of water concentration to maintain high stability and improve the cell efficiency is difficult to evaluate although it is an important factor governing the device stability. One difficulty to assess this threshold value results from the inappropriateness of DSC's electrolyte towards Karl-Fischer titration because of iodine. On the other hand, this threshold value is expected to strongly depend not only on the type of dye used but also on the type of electrolyte.

Finally, the general photocatalytic properties of TiO_2 should be mentioned since it could trigger the formation of hydrogen peroxide from oxygen and water, which itself can destructively oxidize organic compounds [179, 180]. The production of H_2O_2 is initiated by the reduction of O_2 from conduction band electrons in TiO_2 leading to superoxide anion $O_2^{-\bullet}$ which in turn will react with water. It should also be stressed that because of the instability in nature of the superoxide anion radical, it can also directly participate in internal chemical reactions to other organic cell components [181].

Related to the sealing of the cell is the conductive substrate (transparent conducting oxide, TCO) itself which in some specific conditions can sustain a number of events leading to its conduction loss properties. The indium-doped tin oxide (ITO) layer can irreversibly lose its conductivity when voltages above 1.5 V are applied, needless to remind that

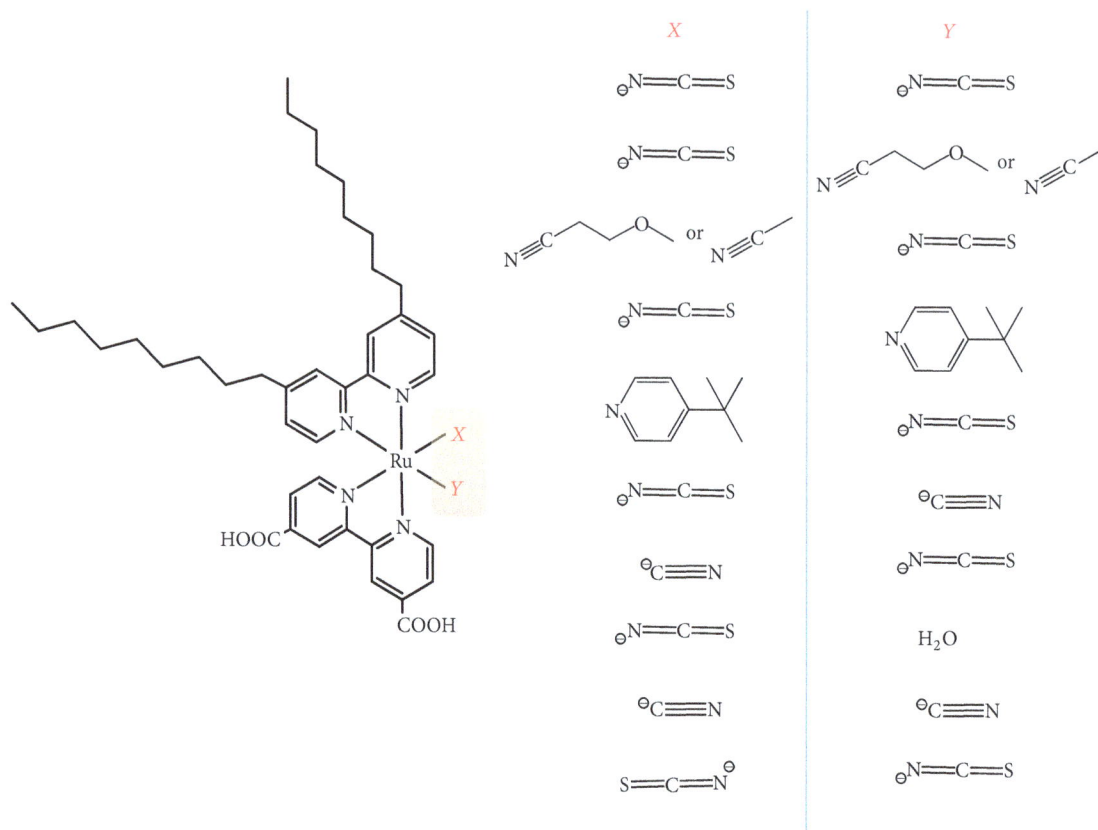

FIGURE 13: Resume of the different dye side-products characterized on either N719 or Z907Na after ageing (adapted from [172, 173]).

such voltages can be reached in series modules [182]. This may bring issues, for instance, for flexible devices for which the substrate is typically made of ITO deposited on polyethylene naphthalate (PEN) or polyethylene terephthalate (PET).

5. Conclusive Remarks

After being strongly questioned subsequently to the seminal publication of Graetzel et al. describing nanocrystalline dye-sensitized solar cells, the macroscopic stability of DSCs is now reported in hundreds of publications. To date IEC61646 accelerated protocol is passed by a few companies although the power conversion efficiency performances are still lagging well behind the champion numbers reported by the public research laboratories. The development of stable DSCs was principally relying on beliefs and excellent intuitions from the community gathered after more than 20 years of research. We can only regret that yet very little is known as a whole about the interrelated chemical reactions responsible for the device ageing and the stress factors responsible for this degradation. Which is the most critical parameter: UV, visible light, or temperature? The experiments gathered in our group on electrolyte stability pointed out the temperature to be the most critical parameter.

In practice, stability investigation papers mainly report the evolution of the cell characteristics as a function of time.

Less often, IMVS-IMPS, EIS, and transient photovoltage/ photocurrent decay were techniques used to monitor the influence of ageing on the charge transfer kinetics and distribution and on the energy of the surface trap states in the nanocrystalline TiO_2. For further improvement, not only for stable power conversion efficiency but also for prolonging significantly the lifetime of DSC to get competitive with silicon technology, there is an urgent need to establish a set of experiments and to develop techniques for careful analyses and characterization of the degradation products which are formed upon device ageing. In this area, only a very few groups are actively contributing to puzzling the complexity of dye-sensitized solar cells.

In this review, we spotted that TiO_2 plays a major role in the device stability. This goes particularly against the preconceived idea that the inorganic part of the cell is the most robust compared to organic and organometallic compounds. In fact TiO_2 displays a catalytic role not only for the dye degradation as it had been clearly demonstrated by the pioneering work of Lund's group but also for the electrolyte stability, in particular those based on alkyl and alkoxynitrile solvents which we are currently scrutinizing. Understanding the multifaceted and interplay chemistry of dye-sensitized solar cells and developing new materials based on such knowledge will indisputably open up significant stability breakthrough in this mystic artificial photosynthetic solar cell (Figure 14).

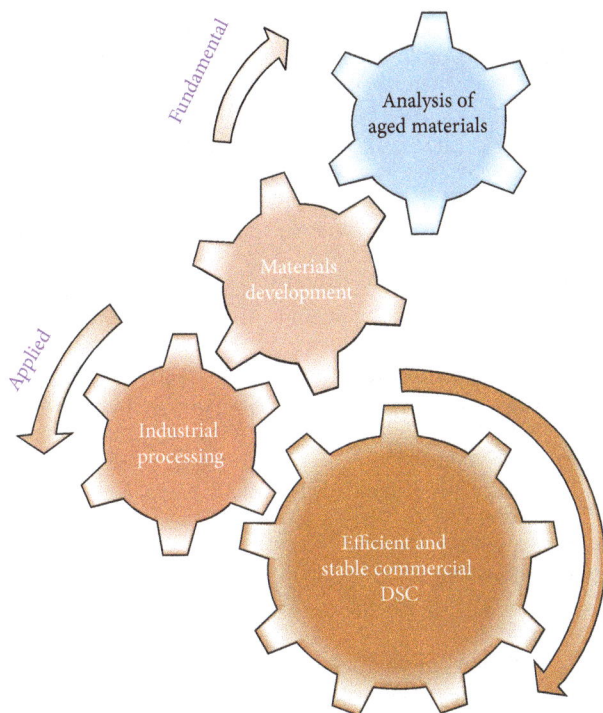

FIGURE 14

Conflict of Interests

The author declares that there is no conflict of interests regarding the publication of this paper.

Acknowledgments

The author wishes to thank the contributors involved in dye-sensitized solar cell group at LRCS, Miguel Flasque, Vittoria Novelli, Iryna Sagaidak, Guillaume Huertas, Dr. Nadia Barbero, Dr. Gregory Gachot, and Dr. Albert Nguyen Van Nhien (LG2A CNRS FRE 3517). He also expresses his gratitude to Sarine Chhor (EPFL) for her contribution to butyronitrile work, Dr. Aravind Kumar Chandiran (EPFL) for his doping work, Dr. Robin Humphry-Baker (EPFL) for FT-IR analysis of aged electrolyte, Dr. Carine Davoisne (LRCS) for TEM measurements, and Mathieu Courty (LRCS) for TGA analysis. Carine Lenfant (LRCS) is acknowledged for reading the paper. The author is also indebted to the FEDER and Région Picardie for financial support of ROBUST research program.

References

[1] A. E. Becquerel, "Mémoire sur les effets électriques produits sous l'influence des rayons solaires," *Comptes Rendus des Séances Hebdomadaires*, vol. 9, pp. 561–567, 1839.

[2] D. M. Chapin, C. S. Fuller, and G. L. Pearson, "A new silicon p-n junction photocell for converting solar radiation into electrical power," *Journal of Applied Physics*, vol. 25, no. 5, pp. 676–677, 1954.

[3] D. M. Chapin, B. Ridge, C. S. Fuller, and G. L. Pearson, Solar Energy Converting Apparatus, US Patent no. 2,780,765, 1957.

[4] B. O'Regan and M. Graetzel, "Low-cost, high-efficiency solar cell based on dye-sensitized colloidal TiO_2 films," *Nature*, vol. 353, pp. 737–740, 1991.

[5] H. W. Vogel, "On the sensitiveness of bromide of silver to the so-called chemically inactive colours," *Berichte Deutsche Chemische Gesellschaft*, vol. 6, p. 1320, 1873.

[6] H. Gerischer, M. E. Michel-Beyerle, F. Rebentrost, and H. Tributsch, "Sensitization of charge injection into semiconductors with large band gap," *Electrochimica Acta*, vol. 13, no. 6, pp. 1509–1515, 1968.

[7] H. Tributsch and M. Calvin, "Electrochemistry of excited molecules: photo-electrochemical reactions of chlorophylls," *Photochemistry and Photobiology*, vol. 14, no. 14, pp. 95–112, 1971.

[8] H. Tributsch, "Reaction of excited chorophyll molecules at electrodes and in photosynthesis," *Photochemistry and Photobiology*, vol. 16, no. 16, pp. 261–269, 1972.

[9] M. Graetzel, "Solar energy conversion by dye-sensitized photovoltaic cells," *Inorganic Chemistry*, vol. 44, pp. 6841–6851, 2005.

[10] M. K. Nazeeruddin, A. Kay, J. Rodicio et al., "Conversion of light to electricity by cis-X2bis(2,2′-bipyridyl-4,4′-dicarboxylate) ruthenium(II) charge-transfer sensitizers (X = Cl-, Br-, I-, CN-, and SCN-) on nanocrystalline titanium dioxide electrodes," *Journal of the American Chemical Society*, vol. 115, no. 14, pp. 6382–6390, 1993.

[11] Y. Chiba, A. Islam, Y. Watanabe, R. Komiya, N. Koide, and L. Han, "Dye-sensitized solar cells with conversion efficiency of 11.1%," *Japanese Journal of Applied Physics*, vol. 45, no. 25, pp. L638–L640, 2006.

[12] F. Gao, Y. Wang, D. Shi et al., "Enhance the optical absorptivity of nanocrystalline TiO_2 film with high molar extinction coefficient ruthenium sensitizers for high performance dye-sensitized solar cells," *Journal of the American Chemical Society*, vol. 130, no. 32, pp. 10720–10728, 2008.

[13] F. Gao, Y. Wang, J. Zhang et al., "A new heteroleptic ruthenium sensitizer enhances the absorptivity of mesoporous titania film for a high efficiency dye-sensitized solar cell," *Chemical Communications*, no. 23, pp. 2635–2637, 2008.

[14] Y. Cao, Y. Bai, Q. Yu et al., "Dye-sensitized solar cells with a high absorptivity ruthenium sensitizer featuring a 2-(hexylthio) thiophene conjugated bipyridine," *Journal of Physical Chemistry C*, vol. 113, no. 15, pp. 6290–6297, 2009.

[15] F. Sauvage, J.-D. Decoppet, M. Zhang et al., "Effect of sensitizer adsorption temperature on the performance of dye-sensitized solar cells," *Journal of the American Chemical Society*, vol. 133, no. 24, pp. 9304–9310, 2011.

[16] F. Sauvage, D. Chen, P. Comte et al., "Dye-sensitized solar cells employing a single film of mesoporous TiO_2 beads achieve power conversion efficiencies over 10%," *ACS Nano*, vol. 4, no. 8, pp. 4420–4425, 2010.

[17] A. Yella, H. W. Lee, H. N. Tsao et al., "Porphyrin-sensitized solar cells with cobalt (II/III)-based redox electrolyte exceed 12 percent efficiency," *Science*, vol. 334, no. 6056, pp. 629–634, 2011.

[18] S. Mathew, A. Yella, P. Gao et al., "Dye-sensitized solar cells with 13% efficiency achieved through the molecular engineering of porphyrin sensitizers," *Nature Chemistry*, vol. 6, no. 3, pp. 242–247, 2014.

[19] J. Burschka, N. Pellet, S.-J. Moon et al., "Sequential deposition as a route to high-performance perovskite-sensitized solar cells," *Nature*, vol. 499, no. 7458, pp. 316–319, 2013.

[20] S. I. Soek, "Efficieny enhancement in inorganic/organic hybrid solar cells," *Hybrid and Organic Photovoltaics*, p. I16, 2014.

[21] M. Matsumura, S. Matsudaira, H. Tsubomura, M. Takata, and H. Yanagida, "Dye sensitization and surface structures of semiconductor electrodes," *Industrial & Engineering Chemistry Product Research and Development*, vol. 19, no. 3, pp. 415–421, 1980.

[22] C. Magne, S. Cassaignon, G. Lancel, and T. Pauporté, "Brookite TiO$_2$ nanoparticle films for dye-sensitized solar cells," *ChemPhysChem*, vol. 12, no. 13, pp. 2461–2467, 2011.

[23] H.-J. Koo, J. Park, B. Yoo, K. Yoo, K. Kim, and N.-G. Park, "Size-dependent scattering efficiency in dye-sensitized solar cell," *Inorganica Chimica Acta*, vol. 361, no. 3, pp. 677–683, 2008.

[24] K. Tennakone, G. R. R. A. Kumara, I. R. M. Kottegoda, and V. P. S. Perera, "An efficient dye-sensitized photoelectrochemical solar cell made from oxides of tin and zinc," *Chemical Communications*, no. 1, pp. 15–16, 1999.

[25] M. Quintana, T. Edvinsson, A. Hagfeldt, and G. Boschloo, "Comparison of dye-sensitized ZnO and TiO$_2$ solar cells: Studies of charge transport and carrier lifetime," *Journal of Physical Chemistry C*, vol. 111, no. 2, pp. 1035–1041, 2007.

[26] J. Elias, M. Parlinska-Wojtan, R. Erni et al., "Passing the limit of electrodeposition: "gas template" H$_2$ nanobubbles for growing highly crystalline nanoporous ZnO," *Nano Energy*, vol. 1, no. 5, pp. 742–750, 2012.

[27] J. Fan, Y. Hao, A. Cabot, E. M. J. Johansson, G. Boschloo, and A. Hagfeldt, "Cobalt(II/III) redox electrolyte in ZnO nanowire-based dye-sensitized solar cells," *ACS Applied Materials and Interfaces*, vol. 5, no. 6, pp. 1902–1906, 2013.

[28] R. L. Willis, C. Olson, B. O'Regan, T. Lutz, J. Nelson, and J. R. Durrant, "Electron dynamics in nanocrystalline ZnO and TiO$_2$ films probed by potential step chronoamperometry and transient absorption spectroscopy," *The Journal of Physical Chemistry B*, vol. 106, no. 31, pp. 7605–7613, 2002.

[29] K. Keis, J. Lindgren, S.-E. Lindquist, and A. Hagfeldt, "Studies of the adsorption process of Ru complexes in nanoporous ZnO electrodes," *Langmuir*, vol. 16, no. 10, pp. 4688–4694, 2000.

[30] B. Liu and H. C. Zeng, "Hydrothermal synthesis of ZnO nanorods in the diameter regime of 50 nm," *Journal of the American Chemical Society*, vol. 125, no. 15, pp. 4430–4431, 2003.

[31] C. Bauer, G. Boschloo, E. Mukhtar, and A. Hagfeldt, "Electron injection and recombination in Ru(dcbpy)2(NCS)2 sensitized nanostructured Zno," *Journal of Physical Chemistry B*, vol. 105, no. 24, pp. 5585–5588, 2001.

[32] P. Tiwana, P. Docampo, M. B. Johnston, H. J. Snaith, and L. M. Herz, "Electron mobility and injection dynamics in mesoporous ZnO, SnO$_2$, and TiO$_2$ films used in dye-sensitized solar cells," *ACS Nano*, vol. 5, no. 6, pp. 5158–5166, 2011.

[33] C. Y. Jiang, X. W. Sun, G. Q. Lo, D. L. Kwong, and J. X. Wang, "Improved dye-sensitized solar cells with a ZnO-nanoflower photoanode," *Applied Physics Letters*, vol. 90, no. 26, Article ID 263501, 2007.

[34] Q. Zhang, C. S. Dandeneau, X. Zhou, and C. Cao, "ZnO nanostructures for dye-sensitized solar cells," *Advanced Materials*, vol. 21, no. 41, pp. 4087–4108, 2009.

[35] K. Westermark, H. Rensmo, H. Siegbahn et al., "PES studies of Ru(dcbpyH$_2$)$_2$(NCS)$_2$ adsorption on nanostructured ZnO for solar cell applications," *The Journal of Physical Chemistry B*, vol. 106, no. 39, pp. 10102–10107, 2002.

[36] R. Schoölin, M. Quintana, E. M. J. Johansson et al., "Preventing dye aggregation on ZnO by adding water in the dye-sensitization process," *Journal of Physical Chemistry C*, vol. 115, no. 39, pp. 19274–19279, 2011.

[37] Q. Zhang, T. P. Chou, B. Russo, S. A. Jenekhe, and G. Cao, "Aggregation of ZnO nanocrystallites for high conversion efficiency in dye-sensitized solar cells," *Angewandte Chemie*, vol. 120, no. 13, pp. 2436–2440, 2008.

[38] Y. Shi, C. Zhu, L. Wang et al., "Ultrarapid sonochemical synthesis of ZnO hierarchical structures: from fundamental research to high efficiencies up to 6.42% for quasi-solid dye-sensitized solar cells," *Chemistry of Materials*, vol. 25, no. 6, pp. 1000–1012, 2013.

[39] A. B. F. Martinson, J. W. Elam, J. T. Hupp, and M. J. Pellin, "ZnO nanotube based dye-sensitized solar cells," *Nano Letters*, vol. 7, no. 8, pp. 2183–2187, 2007.

[40] H. Rensmo, K. Keis, H. Lindström et al., "High light-to-energy conversion efficiencies for solar cells based on nanostructured ZnO electrodes," *Journal of Physical Chemistry B*, vol. 101, no. 14, pp. 2598–2601, 1997.

[41] M. Law, L. E. Greene, J. C. Johnson, R. Saykally, and P. Yang, "Nanowire dye-sensitized solar cells," *Nature Materials*, vol. 4, no. 6, pp. 455–459, 2005.

[42] J. B. Baxter and E. S. Aydil, "Nanowire-based dye-sensitized solar cells," *Applied Physics Letters*, vol. 86, Article ID 053114, 2005.

[43] A. K. Chandiran, M. Abdi-Jalebi, M. K. Nazeeruddin, and M. Grätzel, "Analysis of electron transfer properties of ZnO and TiO$_2$ photoanodes for dye-sensitized solar cells," *ACS Nano*, vol. 8, no. 3, pp. 2261–2268, 2014.

[44] S. Ferrere, A. Zaban, and B. A. Gregg, "Dye sensitization of nanocrystalline tin oxide by perylene derivatives," *The Journal of Physical Chemistry B*, vol. 101, no. 23, pp. 4490–4493, 1997.

[45] A. Kay and M. Grätzel, "Dye-Sensitized core-shell nanocrystals: Improved efficiency of mesoporous tin oxide electrodes coated with a thin layer of an insulating oxide," *Chemistry of Materials*, vol. 14, no. 7, pp. 2930–2935, 2002.

[46] E. N. Kumar, R. Jose, P. S. Archana, C. Vijila, M. M. Yusoff, and S. Ramakrishna, "High performance dye-sensitized solar cells with record open circuit voltage using tin oxide nanoflowers developed by electrospinning," *Energy and Environmental Science*, vol. 5, no. 1, pp. 5401–5407, 2012.

[47] Y. P. Y. P. Ariyasinghe, T. R. C. K. Wijayarathna, I. G. C. K. Kumara et al., "Efficient passivation of SnO$_2$ nano crystallites by Indoline D-149 via dual chelation," *Journal of Photochemistry and Photobiology A: Chemistry*, vol. 217, no. 1, pp. 249–252, 2011.

[48] H. J. Snaith and C. Ducati, "SnO$_2$-Based dye-sensitized hybrid solar cells exhibiting near unity absorbed photon-to-electron conversion efficiency," *Nano Letters*, vol. 10, no. 4, pp. 1259–1265, 2010.

[49] J. Z. Ou, R. A. Rani, M.-H. Ham et al., "Elevated temperature anodized Nb$_2$O$_5$: a photoanode material with exceptionally large photoconversion efficiencies," *ACS Nano*, vol. 6, no. 5, pp. 4045–4053, 2012.

[50] K. Sayama, H. Sugihara, and H. Arakawa, "Photoelectrochemical properties of a porous Nb$_2$O$_5$ electrode sensitized by a ruthenium dye," *Chemistry of Materials*, vol. 10, no. 12, pp. 3825–3832, 1998.

[51] R. Ghosh, M. K. Brennaman, T. Uher et al., "Nanoforest Nb$_2$O$_5$ photoanodes for dye-sensitized solar cells by pulsed laser deposition," *ACS Applied Materials & Interfaces*, vol. 3, no. 10, pp. 3929–3935, 2011.

[52] A. le Viet, R. Jose, M. V. Reddy, B. V. R. Chowdari, and S. Ramakrishna, "Nb$_2$O$_5$ photoelectrodes for dye-sensitized solar cells: choice of the polymorph," *Journal of Physical Chemistry C*, vol. 114, no. 49, pp. 21795–21800, 2010.

[53] S. G. Chen, S. Chappel, Y. Diamant, and A. Zaban, "Preparation of Nb$_2$O$_5$ coated TiO$_2$ nanoporous electrodes and their application in dye-sensitized solar cells," *Chemistry of Materials*, vol. 13, no. 12, pp. 4629–4634, 2001.

[54] K. Sayama, H. Suguhara, and H. Arakawa, "Photoelectrochemical properties of a porous Nb$_2$O$_5$ electrode sensitized by a ruthenium dye," *Chemistry of Materials*, vol. 10, no. 12, pp. 3825–3832, 1998.

[55] P. Guo and M. A. Aegerter, "RU(II) sensitized Nb$_2$O$_5$ solar cell made by the sol-gel process," *Thin Solid Films*, vol. 351, no. 1-2, pp. 290–294, 1999.

[56] K. Hara, T. Horiguchi, T. Kinoshita, K. Sayama, H. Sugihara, and H. Arakawa, "Highly efficient photon-to-electron conversion with mercurochrome-sensitized nanoporous oxide semiconductor solar cells," *Solar Energy Materials and Solar Cells*, vol. 64, no. 2, pp. 115–134, 2000.

[57] S. Burnside, J.-E. Moser, K. Brooks, M. Grätzel, and D. Cahen, "Nanocrystalline mesoporous strontium titanate as photoelectrode material for photosensitized solar devices: increasing photovoltage through flatband potential engineering," *Journal of Physical Chemistry B*, vol. 103, no. 43, pp. 9328–9332, 1999.

[58] R. Dabestani, A. J. Bard, A. Campion et al., "Sensitization of titanium dioxide and strontium titanate electrodes by ruthenium(II) tris(2,2′-bipyridine-4,4′-dicarboxylic acid) and zinc tetrakis(4-carboxyphenyl)porphyrin: an evaluation of sensitization efficiency for component photoelectrodes in a multipanel device," *The Journal of Physical Chemistry*, vol. 92, no. 7, pp. 1872–1878, 1988.

[59] S. Yang, H. Kou, J. Wang, H. Xue, and H. Han, "Tunability of the band energetics of nanostructured SrTiO$_3$ electrodes for dye-sensitized solar cells," *The Journal of Physical Chemistry C*, vol. 114, no. 9, pp. 4245–4249, 2010.

[60] S. S. Shin, J. S. Kim, J. H. Suk et al., "Improved quantum efficiency of highly efficient perovskite BaSnO$_3$-based dye-sensitized solar cells," *ACS Nano*, vol. 7, no. 2, pp. 1027–1035, 2013.

[61] B. Tan, E. Toman, Y. Li, and Y. Wu, "Zinc stannate (Zn$_2$SnO$_4$) dye-sensitized solar cells," *Journal of the American Chemical Society*, vol. 129, no. 14, pp. 4162–4163, 2007.

[62] S.-H. Choi, D. Hwang, D.-Y. Kim et al., "Amorphous zinc stannate (Zn$_2$SnO$_4$) nanofibers networks as photoelectrodes for organic dye-sensitized solar cells," *Advanced Functional Materials*, vol. 23, no. 25, pp. 3146–3155, 2013.

[63] K. Zhu, N. R. Neale, A. Miedaner, and A. J. Frank, "Enhanced charge-collection efficiencies and light scattering in dye-sensitized solar cells using oriented TiO$_2$ nanotubes arrays," *Nano Letters*, vol. 7, no. 1, pp. 69–74, 2007.

[64] H. E. Prakasam, K. Shankar, M. Paulose, O. K. Varghese, and C. A. Grimes, "A new benchmark for TiO$_2$ nanotube array growth by anodization," *Journal of Physical Chemistry C*, vol. 111, no. 20, pp. 7235–7241, 2007.

[65] J. R. Jennings, A. Ghicov, L. M. Peter, P. Schmuki, and A. B. Walker, "Dye-sensitized solar cells based on oriented TiO$_2$ nanotube arrays: transport, trapping, and transfer of electrons," *Journal of the American Chemical Society*, vol. 130, no. 40, pp. 13364–13372, 2008.

[66] D. Kim, A. Ghicov, S. P. Albu, and P. Schmuki, "Bamboo-type TiO$_2$ nanotubes: Improved conversion efficiency in dye-sensitized solar cells," *Journal of the American Chemical Society*, vol. 130, no. 49, pp. 16454–16455, 2008.

[67] G. K. Mor, S. Kim, M. Paulose et al., "Visible to near-infrared light harvesting in TiO$_2$ nanotube array-P3HT based heterojunction solar cells," *Nano Letters*, vol. 9, no. 12, pp. 4250–4257, 2009.

[68] F. Sauvage, F. Di Fonzo, A. Li Bassi et al., "Hierarchical TiO$_2$ photoanode for dye-sensitized solar cells," *Nano Letters*, vol. 10, no. 7, pp. 2562–2567, 2010.

[69] X. Feng, K. Zhu, A. J. Frank, C. A. Grimes, and T. E. Mallouk, "Rapid charge transport in dye-sensitized solar cells made from vertically aligned single-crystal rutile TiO$_2$ nanowires," *Angewandte Chemie - International Edition*, vol. 51, no. 11, pp. 2727–2730, 2012.

[70] K. Zhu, S. R. Jang, and A. J. Frank, "Impact of high charge-collection efficiencies and dark energy-loss processes on transport, recombination, and photovoltaic properties of dye-sensitized solar cells," *The Journal of Physical Chemistry Letters*, vol. 2, no. 9, pp. 1070–1076, 2011.

[71] K. Zhu, T. B. Vinzant, N. R. Neale, and A. J. Frank, "Removing structural disorder from oriented TiO$_2$ nanotube arrays: reducing the dimensionality of transport and recombination in dye-sensitized solar cells," *Nano Letters*, vol. 7, no. 12, pp. 3739–3746, 2007.

[72] D. H. Chen, F. Z. Huang, Y. B. Cheng, and R. A. Caruso, "Mesoporous anatase TiO$_2$ beads with high surface areas and controllable pore sizes: a superior candidate for high-performance dye-sensitized solar cells," *Advanced Materials*, vol. 21, pp. 2206–2210, 2009.

[73] D. H. Chen, L. Cao, F. Z. Huang, P. Imperia, Y. B. Cheng, and R. A. Caruso, "Synthesis of monodisperse mesoporous titania beads with controllable diameter, high surface areas, and variable pore diameters (14–23 nm)," *Journal of the American Chemical Society*, vol. 132, no. 12, pp. 4438–4444, 2010.

[74] A. R. Pascoe, D. Chen, F. Huang et al., "Charge transport and recombination in dye-sensitized solar cells on plastic substrates," *The Journal of Physical Chemistry C*, vol. 118, no. 28, pp. 15154–15161, 2014.

[75] E. M. Hopper, F. Sauvage, A. K. Chandiran, M. Grätzel, K. R. Poeppelmeier, and T. O. Mason, "Electrical properties of Nb-, Ga-, and Y-substituted nanocrystalline anatase TiO$_2$ prepared by hydrothermal synthesis," *Journal of the American Ceramic Society*, vol. 95, no. 10, pp. 3192–3196, 2012.

[76] X. Zhang, F. Liu, Q.-L. Huang, G. Zhou, and Z.-S. Wang, "Dye-sensitized W-doped TiO$_2$ solar cells with a tunable conduction band and suppressed charge recombination," *The Journal of Physical Chemistry C*, vol. 115, no. 25, pp. 12665–12671, 2011.

[77] B. Mei, M. D. Sánchez, T. Reinecke, S. Kaluza, W. Xia, and M. Muhler, "The synthesis of Nb-doped TiO$_2$ nanoparticles by spray drying: an efficient and scalable method," *Journal of Materials Chemistry*, vol. 21, no. 32, pp. 11781–11790, 2011.

[78] N. Tsvetkov, L. Larina, O. Shevaleevskiy, and B. T. Ahn, "Effect of Nb doping of TiO$_2$ electrode on charge transport in dye-sensitized solar cells," *Journal of the Electrochemical Society*, vol. 158, no. 11, pp. B1281–B1285, 2011.

[79] S. G. Kim, M. J. Ju, I. T. Choi et al., "Nb-doped TiO$_2$ nanoparticles for organic dye-sensitized solar cells," *RSC Advances*, vol. 3, no. 37, pp. 16380–16386, 2013.

[80] A. K. Chandiran, F. Sauvage, M. Casas-Cabanas, P. Comte, S. M. Zakeeruddin, and M. Graetzel, "Doping a TiO$_2$ photoanode with Nb^{5+} to enhance transparency and charge collection efficiency in dye-sensitized solar cells," *Journal of Physical Chemistry C*, vol. 114, no. 37, pp. 15849–15856, 2010.

[81] R. Ghosh, Y. Hara, L. Alibabaei et al., "Increasing photocurrents in dye sensitized solar cells with tantalum-doped titanium oxide photoanodes obtained by laser ablation," *ACS Applied Materials and Interfaces*, vol. 4, no. 9, pp. 4566–4570, 2012.

[82] X. Zhang, S. T. Wang, and Z. S. Wang, "Effect of metal-doping in TiO$_2$ on fill factor of dye-sensitized solar cells," *Applied Physics Letters*, vol. 99, Article ID 113503, 2011.

[83] Y. Xie, N. Huang, S. You et al., "Improved performance of dye-sensitized solar cells by trace amount Cr-doped TiO$_2$ photoelectrodes," *Journal of Power Sources*, vol. 224, pp. 168–173, 2013.

[84] N. A. Kyerementeng, V. Hornebecq, H. Martinez, P. Knauth, and T. Djenizian, "Electrochemical fabrication and properties of highly ordered Fe-doped TiO$_2$ nanotubes," *ChemPhysChem*, vol. 13, no. 16, pp. 3707–3713, 2012.

[85] A. Latini, C. Cavallo, F. K. Aldibaja et al., "Efficiency improvement of DSSC photoanode by scandium doping of mesoporous titania beads," *The Journal of Physical Chemistry C*, vol. 117, no. 48, pp. 25276–25289, 2013.

[86] M. Khana and W. Cao, "Preparation of Y-doped TiO$_2$ by hydrothermal method and investigation of its visible light photocatalytic activity by the degradation of methylene blue," *Journal of Molecular Catalysis A: Chemical*, vol. 376, pp. 71–77, 2013.

[87] A. K. Chandiran, F. Sauvage, L. Etgar, and M. Graetzel, "Ga^{3+} and Y^{3+} cationic substitution in mesoporous TiO$_2$ photoanodes for photovoltaic applications," *The Journal of Physical Chemistry C*, vol. 115, no. 18, pp. 9232–9240, 2011.

[88] Y. Duan, N. Fu, Q. Liu et al., "Sn-doped TiO$_2$ photoanode for dye-sensitized solar cells," *The Journal of Physical Chemistry C*, vol. 116, no. 16, pp. 8888–8893, 2012.

[89] J. Zhang, W. Peng, Z. Chen, H. Chen, and L. Han, "Effect of cerium doping in the TiO$_2$ photoanode on the electron transport of dye-sensitized solar cells," *Journal of Physical Chemistry C*, vol. 116, no. 36, pp. 19182–19190, 2012.

[90] H. J. Snaith, "Estimating the maximum attainable efficiency in dye-sensitized solar cells," *Advanced Functional Materials*, vol. 20, no. 1, pp. 13–19, 2010.

[91] Z. Wang, K. Sayama, and H. Sugihara, "Efficient eosin Y dye-sensitized solar cell containing Br$^-$/Br$_3^-$ electrolyte," *The Journal of Physical Chemistry B*, vol. 109, no. 47, pp. 22449–22455, 2005.

[92] C. Teng, X. Yang, C. Yuan et al., "Two novel carbazole dyes for dye-sensitized solar cells with open-circuit voltages up to 1 v based on Br$^-$/Br$_3^-$ electrolytes," *Organic Letters*, vol. 11, no. 23, pp. 5542–5545, 2009.

[93] C. Teng, X. Yang, S. Li et al., "Tuning the HOMO energy levels of organic dyes for dye-sensitized solar cells based on Br-/Br3- electrolytes," *Chemistry—A European Journal*, vol. 16, no. 44, pp. 13127–13138, 2010.

[94] L. Li, X. Yang, J. Gao et al., "Highly efficient CdS quantum dot-sensitized solar cells based on a modified polysulfide electrolyte," *Journal of the American Chemical Society*, vol. 133, no. 22, pp. 8458–8460, 2011.

[95] M. Wang, N. Chamberland, L. Breau et al., "An organic redox electrolyte to rival triiodide/iodide in dye-sensitized solar cells," *Nature Chemistry*, vol. 2, no. 5, pp. 385–389, 2010.

[96] D. Li, H. Li, Y. Luo et al., "Non-corrosive, non-absorbing organic redox couple for dye-sensitized solar cells," *Advanced Functional Materials*, vol. 20, no. 19, pp. 3358–3365, 2010.

[97] M. Cheng, X. Yang, S. Li, X. Wang, and L. Sun, "Efficient dye-sensitized solar cells based on an iodine-free electrolyte using L-cysteine/L-cystine as a redox couple," *Energy and Environmental Science*, vol. 5, no. 4, pp. 6290–6293, 2012.

[98] H. Nusbaumer, *Alternative redox systems for the dye-sensitized solar cell [Ph.D. thesis]*, EPFL, Lausanne, Switzerland, 2004.

[99] A. Yella, H.-W. Lee, H. N. Tsao et al., "Porphyrin-sensitized solar cells with cobalt (II/III)-based redox electrolyte exceed 12 percent efficiency," *Science*, vol. 334, no. 6056, pp. 629–634, 2011.

[100] S. M. Feldt, G. Wang, G. Boschloo, and A. Hagfeldt, "Effects of driving forces for recombination and regeneration on the photovoltaic performance of dye-sensitized solar cells using cobalt polypyridine redox couples," *The Journal of Physical Chemistry C*, vol. 115, no. 43, pp. 21500–21507, 2011.

[101] J.-H. Yum, E. Baranoff, F. Kessler et al., "A cobalt complex redox shuttle for dye-sensitized solar cells with high open-circuit potentials," *Nature Communications*, vol. 3, article 631, 2012.

[102] S. M. Feldt, E. A. Gibson, E. Gabrielsson, L. Sun, G. Boschloo, and A. Hagfeldt, "Design of organic dyes and cobalt polypyridine redox mediators for high-efficiency dye-sensitized solar cells," *Journal of the American Chemical Society*, vol. 132, no. 46, pp. 16714–16724, 2010.

[103] M. Xu, M. Zhang, M. Pastore, R. Li, F. De Angelis, and P. Wang, "Joint electrical, photophysical and computational studies on D-π-A dye sensitized solar cells: The impacts of dithiophene rigidification," *Chemical Science*, vol. 3, no. 4, pp. 976–983, 2012.

[104] Y. Bai, Q. Yu, N. Cai, Y. Wang, M. Zhang, and P. Wang, "High-efficiency organic dye-sensitized mesoscopic solar cells with a copper redox shuttle," *Chemical Communications*, vol. 47, no. 15, pp. 4376–4378, 2011.

[105] T. Daeneke, T.-H. Kwon, A. B. Holmes, N. W. Duffy, U. Bach, and L. Spiccia, "High-efficiency dye-sensitized solar cells with ferrocene-based electrolytes," *Nature Chemistry*, vol. 3, no. 3, pp. 211–215, 2011.

[106] T. C. Li, A. M. Spokoyny, C. She et al., "Ni(III)/(IV) Bis(dicarbollide) as a fast, noncorrosive redox shuttle for dye-sensitized solar cells," *Journal of the American Chemical Society*, vol. 132, no. 13, pp. 4580–4582, 2010.

[107] Z. Zhang, P. Chen, T. N. Murakami, S. M. Zakeeruddin, and M. Graetzel, "The 2,2,6,6-tetramethyl-1-piperidinyloxy radical: an efficient, iodine- free redox mediator for dye-sensitized solar cells," *Advanced Functional Materials*, vol. 18, no. 2, pp. 341–346, 2008.

[108] K. Kalyanasundaram, K. Nazeeruddin, and M. K. Nazeeruddin, "Tuning of the CT excited state and validity of the energy gap law in mixed ligand complexes of Ru(II) containing 4,4'-dicarboxy-2,2'-bipyridine," *Chemical Physics Letters*, vol. 193, no. 4, pp. 292–297, 1992.

[109] F. de Angelis, S. Fantacci, A. Selloni, M. Grätzel, and M. K. Nazeeruddin, "Influence of the sensitizer adsorption mode on the open-circuit potential of dye-sensitized solar cells," *Nano Letters*, vol. 7, no. 10, pp. 3189–3195, 2007.

[110] P. Wang, S. M. Zakeeruddin, J. E. Moser, M. K. Nazeeruddin, T. Sekiguchi, and M. Grätzel, "A stable quasi-solid-state dye-sensitized solar cell with an amphiphilic ruthenium sensitizer and polymer gel electrolyte," *Nature Materials*, vol. 2, pp. 402–407, 2003.

[111] J. E. Kroeze, N. Hirata, S. Koops et al., "Alkyl chain barriers for kinetic optimization in dye-sensitized solar cells," *Journal of the American Chemical Society*, vol. 128, no. 50, pp. 16376–16383, 2006.

[112] C. Y. Chen, J. G. Chen, S. J. Wu, J. Y. Li, C. G. Wu, and K. C. Ho, "Multifunctionalized ruthenium-based supersensitizers for highly efficient dye-sensitized solar cells," *Angewandte Chemie International Edition*, vol. 47, no. 38, pp. 7342–7345, 2008.

[113] M. K. Nazeeruddin, Q. Wang, L. Cevey et al., "DFT-INDO/S modeling of new high molar extinction coefficient charge-transfer sensitizers for solar cell applications," *Inorganic Chemistry*, vol. 45, no. 2, pp. 787–797, 2006.

[114] Y. Cao, Y. Bai, Q. Yu et al., "Dye-sensitized solar cells with a high absorptivity ruthenium sensitizer featuring a 2-(hexylthio)thiophene conjugated bipyridine," *The Journal of Physical Chemistry C*, vol. 113, no. 15, pp. 6290–6297, 2009.

[115] C.-Y. Chen, M. Wang, J.-Y. Li et al., "Highly efficient light-harvesting ruthenium sensitizer for thin-film dye-sensitized solar cells," *ACS Nano*, vol. 3, no. 10, pp. 3103–3109, 2009.

[116] M. K. Nazeeruddin, P. Péchy, T. Renouard et al., "Engineering of efficient panchromatic sensitizers for nanocrystalline TiO$_2$-based solar cells," *Journal of the American Chemical Society*, vol. 123, no. 8, pp. 1613–1624, 2001.

[117] M. Grätzel, "Corrigendum to "Conversion of sunlight to electric power by nanocrystalline dye-sensitized solar cells" [J. Photochem. Photobiol. A: Chem. 164 (2004) 3–14]," *Journal of Photochemistry and Photobiology A*, vol. 168, no. 3, p. 235, 2004.

[118] A. Abbotto, F. Sauvage, C. Barolo et al., "Panchromatic ruthenium sensitizer based on electron-rich heteroarylvinylene π-conjugated quaterpyridine for dye-sensitized solar cells," *Dalton Transactions*, vol. 40, no. 1, pp. 234–242, 2011.

[119] S. H. Wadman, J. M. Kroon, K. Bakker et al., "Cyclometalated ruthenium complexes for sensitizing nanocrystalline TiO$_2$ solar cells," *Chemical Communications*, no. 19, pp. 1907–1909, 2007.

[120] T. Bessho, E. Yoneda, J.-H. Yum et al., "New paradigm in molecular engineering of sensitizers for solar cell applications," *Journal of the American Chemical Society*, vol. 131, no. 16, pp. 5930–5934, 2009.

[121] P. G. Bomben, T. J. Gordon, E. Schott, and C. P. Berlinguette, "A trisheteroleptic cyclometalated RuII sensitizer that enables high power output in a dye-sensitized solar cell," *Angewandte Chemie—International Edition*, vol. 50, no. 45, pp. 10682–10685, 2011.

[122] C.-C. Chou, K.-L. Wu, Y. Chi et al., "Ruthenium(II) sensitizers with heteroleptic tridentate chelates for dye-sensitized solar cells," *Angewandte Chemie—International Edition*, vol. 50, no. 9, pp. 2054–2058, 2011.

[123] K. Hara, M. Kurashige, Y. Dan-Oh et al., "Design of new coumarin dyes having thiophene moieties for highly efficient organic-dye-sensitized solar cells," *New Journal of Chemistry*, vol. 27, no. 5, pp. 783–785, 2003.

[124] N. A. Ludin, A. M. Al-Alwani Mahmoud, A. Bakar Mohamad, A. A. H. Kadhum, K. Sopian, and N. S. Abdul Karim, "Review on the development of natural dye photosensitizer for dye-sensitized solar cells," *Renewable and Sustainable Energy Reviews*, vol. 31, pp. 386–396, 2014.

[125] D. Joly, L. Pelleja, S. Narbey et al., "A robust organic dye for dye sensitized solar cells based on iodine/iodide electrolytes combining high efficiency and outstanding stability," *Scientific Reports*, vol. 4, article 4033, 2014.

[126] W. Zeng, Y. Cao, Y. Bai et al., "Efficient dye-sensitized solar cells with an organic photosensitizer featuring orderly conjugated ethylenedioxythiophene and dithienosilole blocks," *Chemistry of Materials*, vol. 22, no. 5, pp. 1915–1925, 2010.

[127] Z.-S. Wang, Y. Cui, K. Hara, Y. Dan-Oh, C. Kasada, and A. Shinpo, "A high-light-harvesting-efficiency coumarin dye for stable dye-sensitized solar cells," *Advanced Materials*, vol. 19, no. 8, pp. 1138–1141, 2007.

[128] S. Kuster, F. Sauvage, M. K. Nazeeruddin, M. Grätzel, F. A. Nüesch, and T. Geiger, "Unsymmetrical squaraine dimer with an extended π-electron framework: an approach in harvesting near infra-red photons for energy conversion," *Dyes and Pigments*, vol. 87, no. 1, pp. 30–38, 2010.

[129] S. Martiniani, A. Y. Anderson, C. Law, B. C. O'Regan, and C. Barolo, "New insight into the regeneration kinetics of organic dye sensitised solar cells," *Chemical Communications*, vol. 48, no. 18, pp. 2406–2408, 2012.

[130] M. K. Nazeeruddin, R. Humphry-Baker, M. Grätzel et al., "Efficient Near-IR sensitization of nanocrystalline TiO$_2$ films by zinc and aluminum phthalocyanines," *Journal of Porphyrins and Phthalocyanines*, vol. 3, no. 3, pp. 230–237, 1999.

[131] S. Ahmad, E. Guillén, L. Kavan, M. Grätzel, and M. K. Nazeeruddin, "Metal free sensitizer and catalyst for dye sensitized solar cells," *Energy and Environmental Science*, vol. 6, no. 12, pp. 3439–3466, 2013.

[132] N. Jiang, T. Sumitomo, T. Lee et al., "High temperature stability of dye solar cells," *Solar Energy Materials & Solar Cells*, vol. 119, pp. 36–50, 2013.

[133] S. M. Zakeeruddin and M. Grätzel, "Solvent-free ionic liquid electrolytes for mesoscopic dye-sensitized solar cells," *Advanced Functional Materials*, vol. 19, pp. 2187–2202, 2009.

[134] F. Sauvage, S. Chhor, A. Marchioro, J.-E. Moser, and M. Graetzel, "Butyronitrile-based electrolyte for dye-sensitized solar cells," *Journal of the American Chemical Society*, vol. 133, no. 33, pp. 13103–13109, 2011.

[135] Q. Yu, D. Zhou, Y. Shi, X. Si, Y. Wang, and P. Wang, "Stable and efficient dye-sensitized solar cells: photophysical and electrical characterizations," *Energy and Environmental Science*, vol. 3, no. 11, pp. 1722–1725, 2010.

[136] A. Hinsch, J. M. Kroon, R. Kern et al., "Long-term stability of dye-sensitised solar cells," *Progress in Photovoltaics: Research and Applications*, vol. 9, no. 6, pp. 425–438, 2001.

[137] N. Kato, Y. Takeda, K. Higuchi et al., "Degradation analysis of dye-sensitized solar cell module after long-term stability test under outdoor working condition," *Solar Energy Materials and Solar Cells*, vol. 93, no. 6-7, pp. 893–897, 2009.

[138] E. Leonardi, S. Penna, T. M. Brown, A. di Carlo, and A. Reale, "Stability of dye-sensitized solar cells under light soaking test," *Journal of Non-Crystalline Solids*, vol. 356, no. 37–40, pp. 2049–2052, 2010.

[139] R. Harikisun and H. Desilvestro, "Long-term stability of dye solar cells," *Solar Energy*, vol. 85, no. 6, pp. 1179–1188, 2011.

[140] P. Wang, S. M. Zakeeruddin, J.-E. Moser, M. K. Nazeeruddin, T. Sekigushi, and M. Graetzel, "A stable quasi-solid-state dye-sensitized solar cell with an amphiphilic ruthenium sensitizer and polymer gel electrolyte," *Nature Materials*, vol. 2, no. 6, pp. 402–407, 2003.

[141] D. Kuang, P. Wang, S. Ito, S. M. Zakeeruddin, and M. Grätzel, "Stable mesoscopic dye-sensitized solar cells based on tetra-cyanoborate ionic liquid electrolyte," *Journal of the American Chemical Society*, vol. 128, no. 24, pp. 7732–7733, 2006.

[142] D. Kuang, C. Klein, Z. Zhang et al., "Stable, high-efficiency ionic-liquid-based mesoscopic dye-sensitized solar cells," *Small*, vol. 3, no. 12, pp. 2094–2102, 2007.

[143] Z. Zhang, S. Ito, J.-E. Moser, S. M. Zakeeruddin, and M. Grätze, "Influence of iodide concentration on the efficiency and stability of dye-sensitized solar cell containing non-volatile electrolyte," *ChemPhysChem*, vol. 10, no. 11, pp. 1834–1838, 2009.

[144] W. Kubo, T. Kitamura, K. Hanabusa, Y. Wada, and S. Yanagida, "Quasi-solid-state dye-sensitized solar cells using room temperature molten salts and a low molecular weight gelator," *Chemical Communications*, no. 4, pp. 374–375, 2002.

[145] H. Matsui, K. Okada, T. Kitamura, and N. Tanabe, "Thermal stability of dye-sensitized solar cells with current collecting grid," *Solar Energy Materials and Solar Cells*, vol. 93, no. 6-7, pp. 1110–1115, 2009.

[146] H. Pettersson and T. Gruszecki, "Long-term stability of low-power dye-sensitised solar cells prepared by industrial methods," *Solar Energy Materials and Solar Cells*, vol. 70, no. 2, pp. 203–212, 2001.

[147] B. Macht, M. Turrión, A. Barkschat, P. Salvador, K. Ellmer, and H. Tributsch, "Patterns of efficiency and degradation in dye sensitization solar cells measured with imaging techniques," *Solar Energy Materials & Solar Cells*, vol. 73, no. 2, pp. 163–173, 2002.

[148] S. Mastroianni, A. Lembo, T. M. Brown, A. Reale, and A. Di Carlo, "Electrochemistry in reverse biased dye solar cells and dye/electrolyte degradation mechanisms," *ChemPhysChem*, vol. 13, no. 12, pp. 2964–2975, 2012.

[149] E. Figgemeier and A. Hagfeldt, "Are dye-sensitized nano-structured solar cells stable? An overview of device testing and component analyses," *International Journal of Photoenergy*, vol. 6, no. 3, pp. 127–140, 2004.

[150] A. Hagfeldt, G. Boschloo, L. Sun, L. Kloo, and H. Pettersson, "Dye-sensitized solar cells," *Chemical Reviews*, vol. 110, no. 11, pp. 6595–6663, 2010.

[151] M. Flasque, A. N. Van Nhien, J. Swiatowska, A. Seyeux, C. Davoisne, and F. Sauvage, "Interface stability of a TiO$_2$/3-methoxypropionitrile-based electrolyte: first evidence for solid electrolyte interphase formation and implications," *ChemPhysChem*, vol. 15, no. 6, pp. 1126–1137, 2014.

[152] K. F. Jensen, W. Veurman, H. Brandt et al., "Photospectroscopy of I3-species in Dye Solar Cells (DSC) as a test for the sealing and diffusion barrier properties of glass frit," in *Proceedings of the 27th European PV Solar Energy Conference and Exhibition*, pp. 2925–2929, Frankfurt , Germany, 2012.

[153] B. Muthuraaman, S. Murugesan, V. Mathew et al., "An investigation on the performance of a silver ionic solid electrolyte system for a new detergent-based nanocrystalline dye-sensitized solar cell," *Solar Energy Materials and Solar Cells*, vol. 92, no. 12, pp. 1712–1717, 2008.

[154] M. Gorlov and L. Kloo, "Ionic liquid electrolytes for dye-sensitized solar cells," *Dalton Transactions*, no. 20, pp. 2655–2666, 2008.

[155] A. Hinsch, J. M. Kroon, R. Kern et al., "Long-term stability of dye-sensitised solar cells," *Progress in Photovoltaics: Research and Applications*, vol. 9, no. 6, pp. 425–438, 2001.

[156] A. G. Kontos, T. Stergiopoulos, V. Likodimos et al., "Long-term thermal stability of liquid dye solar cells," *Journal of Physical Chemistry C*, vol. 117, no. 17, pp. 8636–8646, 2013.

[157] H. Greijer, J. Lindgren, and A. Hagfeldt, "Resonance Raman scattering of a dye-sensitized solar cell: mechanism of thiocyanato ligand exchange," *The Journal of Physical Chemistry B*, vol. 105, no. 27, pp. 6314–6320, 2001.

[158] P. E. Hansen, P. T. Nguyen, J. Krake, J. Spanget-Larsen, and T. Lund, "Dye-sensitized solar cells and complexes between pyridines and iodines. A NMR, IR and DFT study," *Spectrochimica Acta—Part A: Molecular and Biomolecular Spectroscopy*, vol. 98, pp. 247–251, 2012.

[159] M. Flasque, A. Nguyen Van, G. Nhien, J. Swiatowska, A. Seyeux, and F. Sauvage, in *Proceedings of the 6th International Conference on Hybrid and Organic Photovoltaics*, Lausanne, Switzerland, May 2014.

[160] N. G. Gaylord and A. Takahashi, "Free-radical polymerization of complexed monomers. I. Mechanism of metal halide activation," *Journal of Polymer Science B: Polymer Letters*, vol. 6, no. 10, pp. 743–748, 1968.

[161] M. I. Asghar, K. Miettunen, S. Mastroianni, J. Halme, H. Vahlman, and P. Lund, "In situ image processing method to investigate performance and stability of dye solar cells," *Solar Energy*, vol. 86, no. 1, pp. 331–338, 2012.

[162] H. Tanaka, A. Takeichi, K. Higuchi et al., "Long-term durability and degradation mechanism of dye-sensitized solar cells sensitized with indoline dyes," *Solar Energy Materials & Solar Cells*, vol. 93, no. 6-7, pp. 1143–1148, 2009.

[163] S. Anderson, E. C. Constable, M. P. Dare-Edwards et al., "Chemical modification of a titanium (IV) oxide electrode to give stable dye sensitisation without a supersensitiser," *Nature*, vol. 280, no. 5723, pp. 571–573, 1979.

[164] M. K. Nazeeruddin, A. Kay, I. Rodicio et al., "Conversion of light to electricity by cis-X2bis(2,2′-bipyridyl-4,4′-dicarboxylate)ruthenium(II) charge-transfer sensitizers (X = Cl-, Br-, I-, CN-, and SCN-) on nanocrystalline titanium dioxide electrodes," *Journal of the American Chemical Society*, vol. 115, no. 14, pp. 6382–6390, 1993.

[165] R. Grünwald and H. Tributsch, "Mechanisms of instability in ru-based dye sensitization solar cells," *Journal of Physical Chemistry B*, vol. 101, no. 14, pp. 2564–2575, 1997.

[166] H. Greijer, J. Lindgren, and A. Hagfeldt, "Resonance Raman scattering of a dye-sensitized solar cell: mechanism of thiocyanato ligand exchange," *Journal of Physical Chemistry B*, vol. 105, no. 27, pp. 6314–6320, 2001.

[167] P. E. Hansen, P. T. Nguyen, J. Krake, J. Spanget-Larsen, and T. Lund, "Dye-sensitized solar cells and complexes between pyridines and iodines. A NMR, IR and DFT study," *Spectrochimica Acta Part A: Molecular and Biomolecular Spectroscopy*, vol. 98, pp. 247–251, 2012.

[168] N. T. Hoang, N. T. P. Thoa, and T. Lund, "Thermal degradation kinetics of solar cell dye N719 bound to nanocrystalline TiO$_2$ particles," *Advances in Natural Sciences*, vol. 1, no. 10, pp. 51–58, 2009.

[169] H. T. Nguyen, H. M. Ta, and T. Lund, "Thermal thiocyanate ligand substitution kinetics of the solar cell dye N719 by acetonitrile, 3-methoxypropionitrile, and 4-tert-butylpyridine," *Solar Energy Materials and Solar Cells*, vol. 91, no. 20, pp. 1934–1942, 2007.

[170] M. Thomalla and H. Tributsch, "Chromatographic studies of photodegradation of RuL$_2$(SCN)$_2$ in nanostructured dye-sensitization solar cells," *Comptes Rendus Chimie*, vol. 9, no. 5-6, pp. 659–666, 2006.

[171] F. Nour-Mohammadi, H. T. Nguyen, G. Boschloo, and T. Lund, "An investigation of the photosubstitution reaction between N719-dyed nanocrystalline TiO$_2$ particles and 4-tert-butylpyridine," *Journal of Photochemistry and Photobiology A: Chemistry*, vol. 187, no. 2-3, pp. 348–355, 2007.

[172] P. Tuyet Nguyen, R. Degn, H. Thai Nguyen, and T. Lund, "Thiocyanate ligand substitution kinetics of the solar cell dye Z-907 by 3-methoxypropionitrile and 4-tert-butylpyridine at elevated temperatures," *Solar Energy Materials & Solar Cells*, vol. 93, no. 11, pp. 1939–1945, 2009.

[173] P. Tuyet Nguyen, A. Rand Andersen, E. Morten Skou, and T. Lund, "Dye stability and performances of dye-sensitized solar cells with different nitrogen additives at elevated temperatures—can sterically hindered pyridines prevent dye degradation?" *Solar Energy Materials and Solar Cells*, vol. 94, no. 10, pp. 1582–1590, 2010.

[174] O. Kohle, M. Graetzel, A. F. Meyer, and T. B. Meyer, "The photovoltaic stability of, bis(isothiocyanato)ruthenium(II)-bis-2, 2′bipyridine-4, 4′-dicarboxylic acid and related sensitizers," *Advanced Materials*, vol. 9, no. 11, pp. 904–906, 1997.

[175] M. Grätzel, "Photovoltaic performance and long-term stability of dye-sensitized meosocopic solar cells," *Comptes Rendus Chimie*, vol. 9, no. 5-6, pp. 578–583, 2006.

[176] K. Fredin, K. F. Anderson, N. W. Duffy et al., "Effect on cell efficiency following thermal degradation of dye-sensitized mesoporous electrodes using N719 and D5 sensitizers," *Journal of Physical Chemistry C*, vol. 113, no. 43, pp. 18902–18906, 2009.

[177] S. Mikoshiba, S. Murai, H. Sumino, T. Kado, D. Kosugi, and S. Hayase, "Ionic liquid type dye-sensitized solar cells: increases in photovoltaic performances by adding a small amount of water," *Current Applied Physics*, vol. 5, no. 2, pp. 152–158, 2005.

[178] D. F. Watson and G. J. Meyer, "Electron injection at dye-sensitized semiconductor electrodes," *Annual Review of Physical Chemistry*, vol. 56, pp. 119–156, 2005.

[179] M. R. Hoffmann, S. T. Martin, W. Choi, and D. W. Bahnemann, "Environmental applications of semiconductor photocatalysis," *Chemical Reviews*, vol. 95, no. 1, pp. 69–96, 1995.

[180] S. T. Martin, H. Herrmann, W. Y. Choi, and M. R. Hoffmann, "Time-resolved microwave conductivity. Part 1.—TiO_2 photoreactivity and size quantization," *Journal of the Chemical Society, Transactions*, vol. 90, pp. 3315–3322, 1994.

[181] C. C. Chen, X. Z. Li, W. H. Ma, J. C. Zhao, H. Hidaka, and N. Serpone, "Effect of transition metal ions on the TiO_2-assisted photodegradation of dyes under visible irradiation: a probe for the interfacial electron transfer process and reaction mechanism," *The Journal of Physical Chemistry B*, vol. 106, no. 2, pp. 318–324, 2002.

[182] A. Kraft, H. Hennig, A. Herbst, and K.-H. Heckner, "Changes in electrochemical and photoelectrochemical properties of tin-doped indium oxide layers after strong anodic polarization," *Journal of Electroanalytical Chemistry*, vol. 365, no. 1-2, pp. 191–196, 1994.

23

Synthesis and Antimicrobial Activity of New Schiff Base Compounds Containing 2-Hydroxy-4-pentadecylbenzaldehyde Moiety

Gadada Naganagowda,[1] Reinout Meijboom,[1] and Amorn Petsom[2]

[1] Research Centre for Synthesis and Catalysis, Department of Chemistry, University of Johannesburg, P.O. Box 524, Auckland Park, Johannesburg 2006, South Africa
[2] Department of Chemistry, Faculty of Science, Chulalongkorn University, Bangkok 10330, Thailand

Correspondence should be addressed to Gadada Naganagowda; ngchula.pdf@gmail.com

Academic Editor: Atsushi Ohtaka

Various novel Schiff base compounds have been synthesized by reaction of 2-hydroxy-4-pentadecylbenzaldehyde with substituted benzothiophene-2-carboxylic acid hydrazide and different substituted aromatic or heterocyclic amines in the presence of acetic acid in ethanol. The structures of all these compounds were confirmed by elemental analysis, IR, ^{1}H-NMR, ^{13}C-NMR, and mass spectral data and have been screened for antibacterial and antifungal activity.

1. Introduction

Compounds with the structure of –C=N–(azomethine group) are known as Schiff bases, which are usually synthesized by condensation of primary amines and active carbonyl groups. Schiff bases are an important class of compounds in the medicinal and the pharmaceutical field, including antibacterial [1, 2], antifungal [3, 4], and antitumor activity [5, 6]. Heterocyclic-containing Schiff bases can show dramatically increased biological activities. As evident from literature [7], it was noted that a lot of research has been carried out on Schiff bases, but no work has been done on this particular type of Schiff base. Aromatic Schiff bases possessing long alkyl chains have received overwhelming attention due to their possibility to show mesomorphic properties such as smectic and nematic phases [8–11]. In this paper, we report the synthesis of a novel Schiff base from substituted benzothiophene-2-carboxylic acid hydrazide, various aromatic and heterocyclic amines with 2-hydroxy-4-pentadecylbenzaldehyde as a moiety. We also report the results of biological screening for possible antibacterial and antifungal activity of the resulting derivatives and we discuss

the relationship of molecular structure and the bioactivity (Table 1).

2. Results and Discussion

Cashew nut shell liquid (CNSL) is obtained as a by-product from mechanical processing for edible use of cashew kernel (*Anacardium occidentale* L.) and is a mixture of anacardic acid, cardanol, and smaller amounts of cardol and 2-methyl cardol. Due to the easy thermal decarboxylation of anacardic acid, the main component of distilled CNSL is cardanol (yield up to 70–80% and purity up to 90%) as a mixture of saturated (3-n-pentadecylphenol), monoolefinic [3-(n-pentadeca-8-enyl)phenol], diolefinic [3-(n-pentadeca-8,11-dienyl)phenol], and triolefinic [3-(n-pentadeca-8,11,14-trienyl)phenol] long-chain phenols, with an average value of two double bonds per molecule. Cardol and methyl cardol are present in smaller percentages (Figure 1).

Cardanol is a phenolic compound with a C_{15} aliphatic chain in the meta position, obtained from cashew nut shell liquid (CNSL). The structure and composition of cardanol is

Synthesis and Antimicrobial Activity of New Schiff Base Compounds Containing...

213

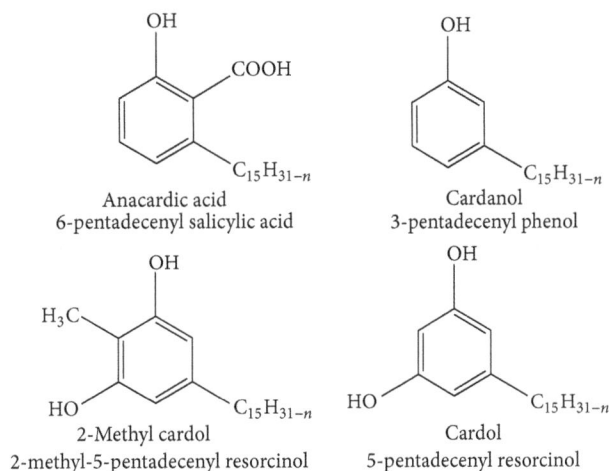

FIGURE 1: Chemical structure of components in natural CNSL.

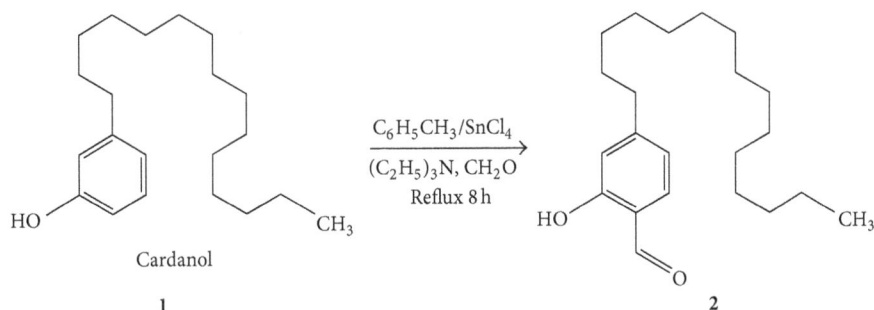

SCHEME 1: Synthesis of 2-hydroxy-6-pentadecylbenzaldehyde.

TABLE 1: Antibacterial and antifungal activity of the tested compounds.

Compound	Zone of inhibition (mm)			
	S. aureus	E. coli	C. albicans	C. pannical
3a	13.21	09.14	09.21	14.03
3b	08.24	08.91	08.21	12.11
3c	10.22	16.21	06.21	10.11
3d	15.12	16.11	07.34	12.23
5a	14.10	16.08	12.10	10.22
6a	13.25	14.20	16.12	13.20
7a	12.22	16.10	13.12	18.15
8a	16.14	13.10	09.10	12.13
9b	19.12	18.20	09.88	14.10
9g	17.20	16.10	10.21	12.10
10a	15.10	14.23	13.10	15.20
11b	14.12	13.12	15.10	14.13
DMF (control)	00	00	00	00
Streptomycin	24.30	22.12	—	—
Griseofulvin	—	—	21.05	20.13

given in Figure 2. It is a mixture of saturated and unsaturated (mono-, di-, and tri-) compounds.

2-Hydroxy-4-pentadecylbenzaldehyde was prepared from cardanol (3-pentadecylphenol contained, e.g., in cashew nut shell liquid (CNSL)) by formylation, using a standard procedure [12] (Scheme 1).

Substituted benzothiophene was prepared by the reaction of aromatic cinnamic acid with thionyl chloride in DMF and dry pyridine according to the reported method [13]. Substituted benzothiophene-2-carbonyl chloride was then treated with hydrazine hydrate to obtain benzothiophene-2-carbohydrazides 3a–d. Compounds 4a, 4b, 4c, and 4d were synthesized by reaction of 2-hydroxy-6-pentadecyl-benzaldehyde with 6-substitued-3-chloro-1-benzothiophene-2-carboxylicacidhydrazide in the presence of acetic acid in ethanol under reflux (Scheme 2).

Compounds 5a and 6a were obtained separately in good yield by refluxing compound 2 with an equimolar amount of compounds 5 and 6 in the presence of acetic acid in ethanol for 5 h. The configuration at the imine unit was not investigated; compounds 5a and 6a are arbitrarily shown in Z configuration in Scheme 3.

2-Hydroxy-4-pentadecylbenzaldehyde 2 was treated with aromatic amines to get new cardanol aldehyde derivative of Schiff base compounds. These new Schiff bases were fully characterized by elemental analysis, IR, MS, ^{1}H, and ^{13}C-NMR data.

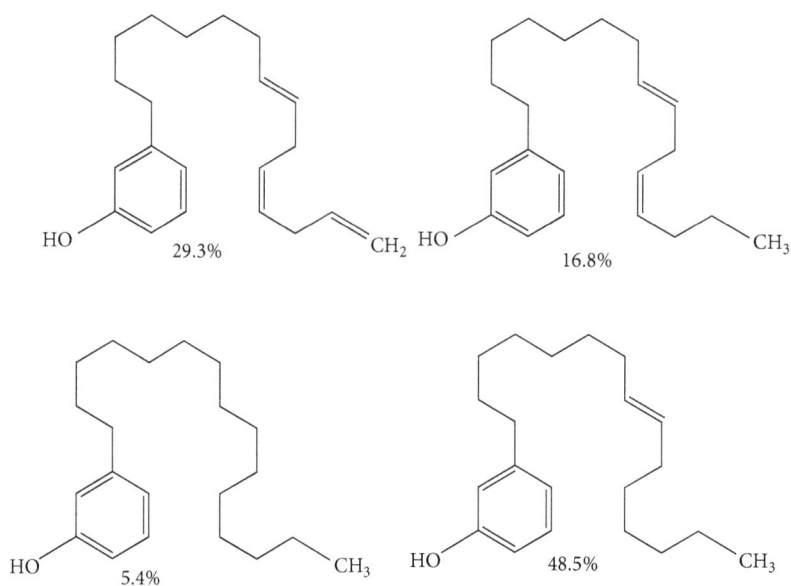

FIGURE 2: Structure and composition of cardanol.

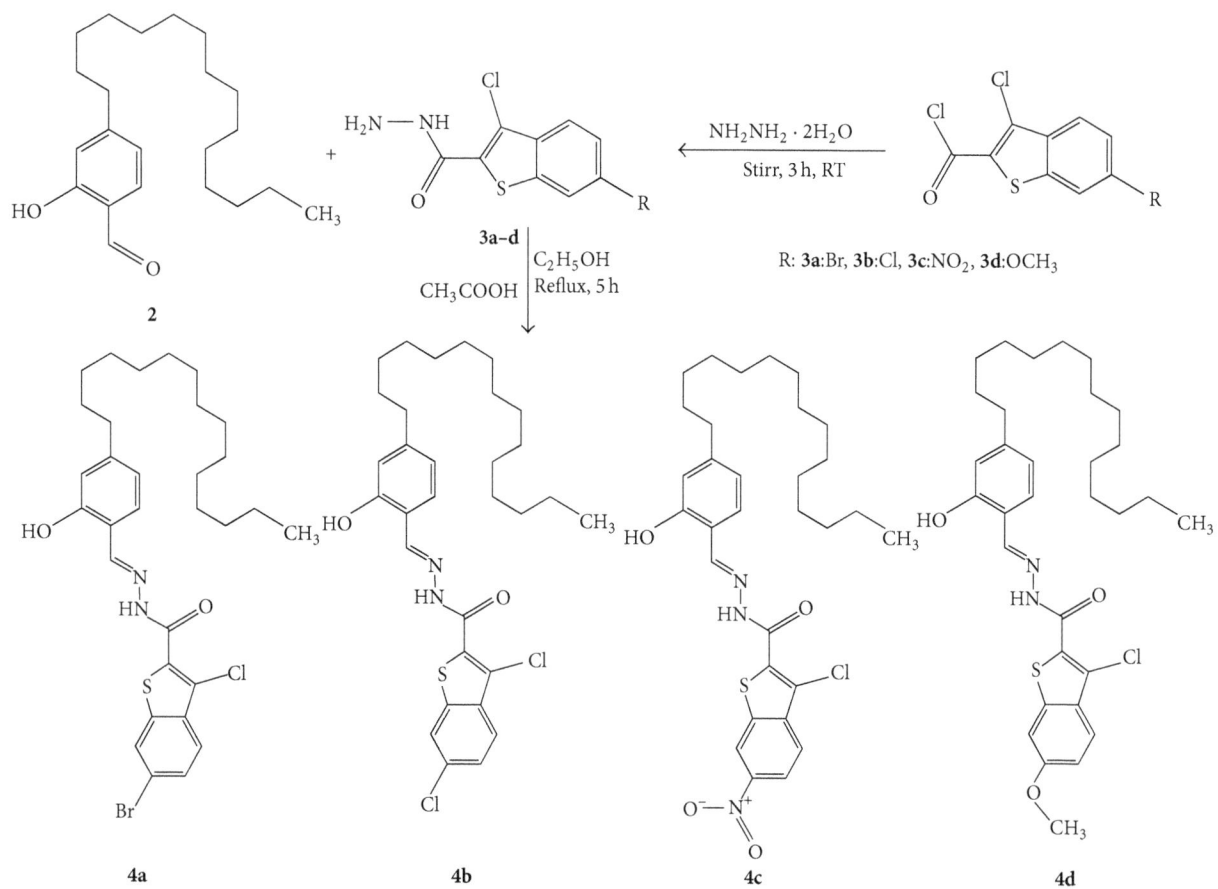

SCHEME 2: Synthesis of 6-substitued-3-chloro-N′-(2-hydroxy-4-pentadecylbenzylidene)benzothiophene-2-carbohydrazide.

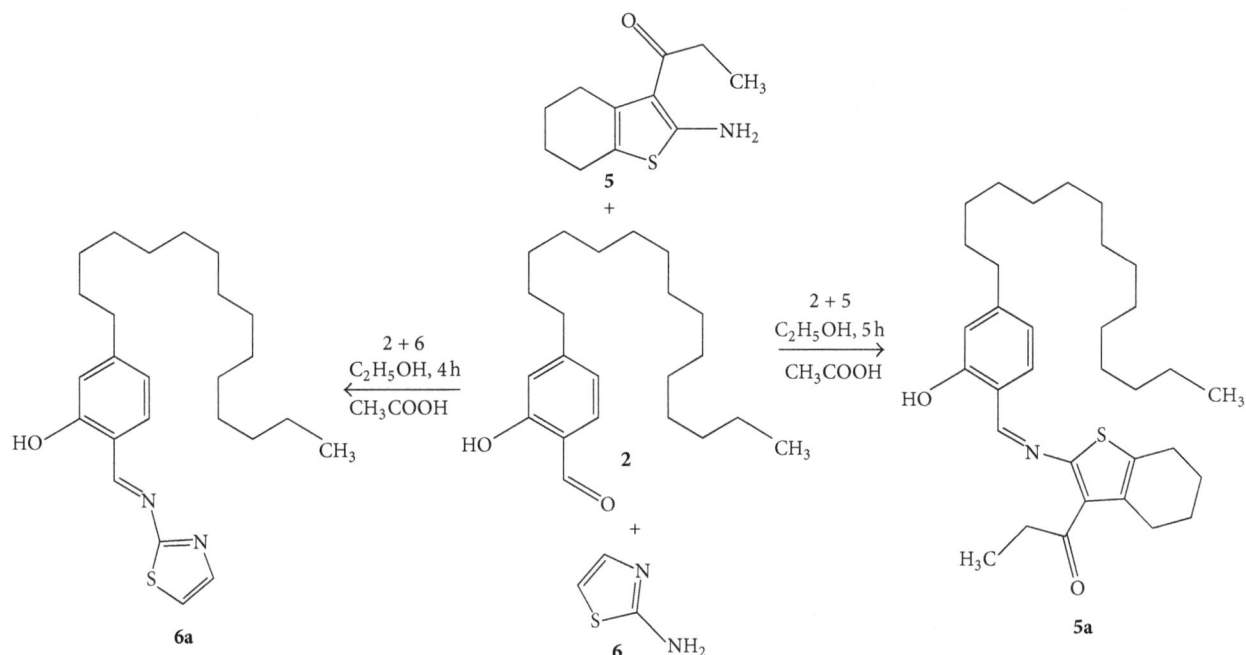

SCHEME 3: Synthesis of new heterocyclic Schiff base compounds.

3. Antimicrobial Evaluation

The cup plate method using Hi-Media agar medium was employed to study the antibacterial activity of the synthesized compounds against Gram-positive bacteria, *Staphylococcus aureus,* and Gram-negative bacteria, *Escherichia coli.* Preparation of nutrient broth, subculture, base layer medium, agar medium, and peptone water was carried out according to a standard procedure [14]. Streptomycin was used as a reference drug and dimethylformamide as a control. The zone of inhibition produced by each compound was measured in mm. The evaluated compounds have shown low to moderate activities as compared to standard drug against the bacteria. The antifungal activities of the synthesized compounds were evaluated against two different fungi, that is, *Candida albicans* and *Chrysosporium pannical,* by the filter paper disc technique [15].

4. Experimental

All chemicals were analytical grade, purchased from commercial suppliers, and used as received without further purification. Melting points were determined in open capillaries and are uncorrected. FT-IR spectra were recorded on a Nicolet Fourier Transform IR spectrophotometer Impact 410 (Nicolet Instrument Technologies, Inc. WI, USA). ^{1}H-NMR and ^{13}C-NMR spectra were obtained in DMSO-d_6 at 400 MHz for ^{1}H nuclei and 100 MHz for ^{13}C nuclei (Bruker Company, Germany). All chemical shifts are expressed in ppm relative to tetramethylsilane (TMS) as the internal standard. Mass spectra were obtained using matrix-assisted laser desorption ionization mass spectrometry (MALDI-TOF) by using dithranol as a matrix. Elemental analysis

(C, H, N, and S) was performed on a Perkin Elmer 2400 analyzer. The purity of compounds was checked by TLC on silica gel and further purification was performed using column chromatography (silica gel, 60–120 mesh).

4.1. Synthesis of Compound 2 Was Prepared according to Literature Procedure. See [12].

4.2. Synthesis of Compounds 3a–d Was Also Prepared according to Literature Procedure. See [13].

4.3. General Procedure for Synthesis of Compounds 4a–d: Detail for Preparation of 6-Bromo-3-chloro-N′-(2-hydroxy-4-pentadecylbenzylidene)benzo[b]thiophene-2-carbohydr azide (4a). 2-Hydroxy-6-pentadecylbenzaldehyde 2 (3.32 g, 0.01 mol) was treated with 6-bromo-3-chloro-1-benzothiophene-2-carboxylicacidhydrazide 3a (3.03 g, 0.01 mol) in the presence of glacial acetic acid (1 mL) in absolute ethanol (25 mL) and was heated under reflux for 5 h. The completion of the reaction was monitored by TLC. The reaction mixture was allowed to cool down to room temperature and then poured onto crushed ice. The precipitate was filtered, dried, and recrystallized from absolute ethanol. The resulting solid was further purified by column chromatography [silica, petroleum ether/ethyl acetate (85 : 15)], leading to compound 4a as yellow solid.

Yield 71%; mp 245–247°C; IR ν (cm^{-1}): 3431 (OH st), 3181 (N–H st), 1645 (C=O st), 1586 (C=N st); ^{1}H-NMR δ (ppm): 12.10 (s, 1H, OH), 11.34 (s, 1H, NH), 8.66 (s, 1H, CH=N), 8.12 (s, 1H, Ar–H), 8.01 (d, J = 7.8 Hz, 1H, Ar–H), 7.55 (d, J = 7.8 Hz, 1H, Ar–H), 7.46 (d, J = 7.6 Hz, 1H, Ar–H), 6.95 (s, 1H, Ar–H), 6.60 (d, J = 6.8 Hz, 1H, Ar–H),

2.52 (t, J = 7.6 Hz, 2H, Ar–CH$_2$), 1.54–1.01 (m, 26H, (CH$_2$)$_{13}$), 0.85 (t, J = 6.8 Hz, 3H, CH$_3$); ^{13}C-NMR δ (ppm): 169.1 (CONH), 160.9, 154.0 (aromatic carbons), 146.0 (CH=N), 145.4, 141.6, 132.4, 130.4, 129.8, 129.3, 125.4, 125.4, 120.3, 119.0, 117.2, 115.5 (aromatic carbons), 35.3, 31.2, 30.2, 30.1, 28.9, 28.9, 28.8, 28.7, 28.6, 28.5, 22.0 (methylene carbons), 13.8 (CH$_3$); MS, m/z: 621.65 (M$^+$). Anal. calcd. for C$_{31}$H$_{40}$ClBrN$_2$O$_2$S: C, 60.05; H, 6.50; N, 4.52; S, 5.17; found: C, 60.04; H, 6.82; N, 4.39; S, 4.82%.

Similarly, the compounds **4b–d** were prepared with little change in reflux time and reaction work up.

4.3.1. 3,6-Dichloro-N′-(2-hydroxy-4-pentadecylbenzylidene) benzo[b]thiophene-2-carbohydrazide (4b).

Yield 68%; mp 280–282°C; IR ν (cm^{-1}): 3426 (OH st), 3168 (N–H st), 1640 (C=O st), 1581 (C=N st); ^{1}H-NMR δ (ppm): 11.80 (s, 1H, OH), 11.68 (s, 1H, NH), 8.51 (s, 1H, CH=N), 8.09 (s, 1H, Ar–H), 7.96 (d, J = 7.8 Hz, 1H, Ar–H), 7.54 (d, J = 7.7 Hz, 1H, Ar–H), 7.41 (d, J = 7.7 Hz, 1H, Ar–H), 6.95 (s, 1H, Ar–H), 6.88 (d, J = 6.8 Hz, 1H, Ar–H), 2.55 (t, J = 7.6 Hz, 2H, Ar–CH$_2$), 1.57–1.03 (m, 26H, (CH$_2$)$_{13}$), 0.81 (t, J = 6.8 Hz, 3H, CH$_3$); ^{13}C-NMR δ (ppm): 169.4 (CONH), 160.9, 153.2 (aromatic carbons), 146.2 (CH=N), 145.3, 141.3, 132.9, 131.3, 130.4, 129.2, 126.1, 124.3, 122.4, 120.5, 117.2, 115.3 (aromatic carbons), 36.3, 31.8, 31.2, 30.8, 29.7, 29.6, 29.6, 29.6, 29.5, 29.4, 29.3, 29.2, 22.7 (methylene carbons), 14.1 (CH$_3$); MS, m/z: 621.65 (M$^+$+1). Anal. calcd. for C$_{31}$H$_{40}$Cl$_2$N$_2$O$_2$S: C, 64.68; H, 7.00; N, 4.87; S, 5.57; found: C, 64.04; H, 6.92; N, 4.39; S, 4.99%.

4.3.2. 3-Chloro-N′-(2-hydroxy-4-pentadecylbenzylidene)-6-nitrobenzo[b]thiophene-2-carbohydrazide (4c).

Yield 69%; mp 291–293°C; IR ν (cm^{-1}): 3435 (OH st), 3177 (N–H st), 1634 (C=O st), 1577 (C=N st); ^{1}H-NMR δ (ppm): 11.71 (s, 1H, OH), 11.62 (s, 1H, NH), 8.74 (s, 1H, CH=N), 8.02 (s, 1H, Ar–H), 7.98 (d, J = 7.8 Hz, 1H, Ar–H), 7.61 (d, J = 7.8 Hz, 1H, Ar–H), 7.47 (d, J = 7.6 Hz, 1H, Ar–H), 6.90 (s, 1H, Ar–H), 6.77 (d, J = 6.8 Hz, 1H, Ar–H), 2.54 (t, J = 7.6 Hz, 2H, Ar–CH$_2$), 1.55–1.01 (m, 26H, (CH$_2$)$_{13}$), 0.88 (t, J = 6.8 Hz, 3H, CH$_3$); ^{13}C-NMR δ (ppm): 167.2 (CONH), 160.4, 152.7 (aromatic carbons), 146.3 (CH=N), 145.4, 144.3, 141.3, 139.5, 130.3, 129.9, 124.1, 120.8, 119.3, 117.8, 117.3, 115.4 (aromatic carbons), 36.4, 36.2, 31.9, 30.9, 30.6, 29.8, 29.7, 29.6, 29.5, 29.4, 29.3, 29.2, 22.7 (methylene carbons), 14.1 (CH$_3$); MS, m/z: 586.18 (M$^+$). Anal. calcd. for C$_{31}$H$_{40}$ClN$_3$O$_4$S: C, 63.52; H, 6.88; N, 7.17; S, 5.47; found: C, 62.44; H, 6.21; N, 6.99; S, 5.10%.

4.3.3. 3-Chloro-N′-(2-hydroxy-4-pentadecylbenzylidene)-6-methoxybenzo[b]thiophene-2-carbohydrazide (4d).

Yield 81%; mp 276–278°C; IR ν (cm^{-1}): 3420 (OH st), 3159 (N–H st), 1638 (C=O st), 1591 (C=N st); ^{1}H-NMR δ (ppm): 12.03 (s, 1H, OH), 11.81 (s, 1H, NH), 8.88 (s, 1H, CH=N), 7.98 (d, J = 7.7 Hz, 1H, Ar–H), 7.66 (d, J = 7.6 Hz, 1H, Ar–H), 7.22 (s, 1H, Ar–H), 7.02 (s, 1H, Ar–H), 6.98 (d, J = 7.7 Hz, 1H, Ar–H), 6.90 (d, J = 6.8 Hz, 1H, Ar–H), 3.81 (s, 3H, OCH$_3$), 2.51 (t, J = 7.6 Hz, 2H, Ar–CH$_2$), 1.56–1.01 (m, 26H, (CH$_2$)$_{13}$), 0.88 (t, J = 6.8 Hz, 3H, CH$_3$); ^{13}C-NMR δ (ppm): 168.3 (CONH), 160.5, 159.4, 152.8 (aromatic carbons), 146.2 (CH=N), 145.0, 141.5, 130.5, 129.9, 126.1, 125.5, 124.2, 120.3, 117.1, 115.5, 115.3 (aromatic

carbons), 55.5 (OCH$_3$), 36.1, 31.8, 31.2, 29.9, 29.8, 29.7, 29.6, 29.5, 29.4, 29.3, 29.2, 22.7 (methylene carbons), 14.1 (CH$_3$); MS, m/z: 570.21 (M$^+$+1). Anal. calcd. for C$_{32}$H$_{43}$ClN$_2$O$_3$S: C, 67.29; H, 7.59; N, 4.87; S, 5.61; found: C, 67.04; H, 7.21; N, 4.20; S, 5.34%.

4.4. Synthesis of Ethyl 2-(2-hydroxy-4-pentadecylbenzylidene-amino)-4,5,6,7-tetrahydrobenzothiophene-3-carboxylate (5a).

Ethyl-2-amino-4,5,6,7-tetrahydrobenzothiophene-3-carboxylate **5** (2.08 g, 0.01 mol) and 2-hydroxy-4-pentadecylbenzaldehyde **2** (3.32 g, 0.01 mol) in the presence of acetic acid (1 mL) in absolute ethanol (25 mL) were heated under reflux for 5 h. It was then concentrated, cooled, and kept overnight in a refrigerator. The precipitate was filtered, washed with ethanol, dried, and recrystallized from methanol to get pure compound **5a**.

Yield: 66%; mp: 234–236°C; IR (KBr): ν_{max} (cm^{-1}): 3421.11 (O–H str.), 1611.61 (C=N str.), 1600 (C=O); ^{1}H-NMR δ ppm: 12.10 (s, 1H, OH), 8.60 (s, 1H, CH=N), 7.54 (d, J = 7.8 Hz, 1H, Ar–H), 7.10 (s, 1H, Ar–H), 6.89 (d, J = 7.6 Hz, 1H, Ar–H), 3.31 (q, 2H, COCH$_2$), 2.71 (t, J = 6.7 Hz, 2H, tetrahydrobenzothiophene), 2.62 (t, J = 6.7 Hz, 2H, tetrahydrobenzothiophene), 2.58 (t, J = 7.6 Hz, 2H, Ar–CH$_2$), 1.81 (m, 4H, tetrahydrobenzothiophene), 1.59–1.31 (m, 26H, (CH$_2$)$_{13}$), 1.25 (q, 3H, CH$_3$ Ester), 0.85 (t, J = 6.8 Hz, 3H, CH$_3$); ^{13}C-NMR δ ppm: 189.6, 161.3, 156.8, 151.9, 143.4, 129.9, 128.6, 120.3, 118.9, 118.3, 116.0, 115.0 (aromatic carbons), 35.5, 35.4, 34.9, 31.5, 30.8, 30.1, 28.9, 28.8, 28.7, 28.6, 28.5, 23.2, 23.1, 22.9, 22.4, 22.3, 13.4 (methylene carbons), 14.2 (CH$_3$); MS, m/z: 538.45 (M$^+$). Anal. calcd. for C$_{33}$H$_{49}$NO$_3$S: C, 73.42; H, 9.15; N, 2.59; S, 5.94 found: C, 73.14; H, 9.41; N, 2.13; S, 5.81%.

4.5. Synthesis of 5-Pentadecyl-2-((thiazol-2-ylimino)methyl) phenol (6a).

Cardanol aldehyde **2** (3.32 g, 0.01 mol) in ethanol (15 mL) and 2-aminothiazole (1.00 g, 0.01 mol) in the presence of glacial acetic acid (1 mL) were refluxed for 6 h. The reaction mixture was allowed to cool down to room temperature and poured onto ice cooled water with constant stirring. The resulting precipitate was filtered, washed with water, dried, and purified through column chromatography to get pure compound **6a** as light yellowish solid.

Yield: 51%; mp: 220–222°C; IR (KBr): ν_{max} (cm^{-1}): 3189.31 (O–H str.), 1612.77 (C=N str.); ^{1}H-NMR δ ppm: 11.80 (s, 1H, OH), 8.65 (s, 1H, CH=N), 7.85 (d, J = 7.3 Hz, 1H, thiazole ring), 7.65 (d, J = 7.8 Hz, 1H, Ar–H), 7.60 (d, J = 7.3 Hz, 1H, thiazole ring), 7.06 (s, 1H, Ar–H), 6.85 (d, J = 7.6 Hz, 1H, Ar–H), 2.68 (t, J = 7.6 Hz, 2H, Ar–CH$_2$), 1.61–1.33 (m, 26H, (CH$_2$)$_{13}$), 0.88 (t, J = 6.8 Hz, 3H, CH$_3$); ^{13}C-NMR δ ppm: 162.3, 157.0, 152.2, 143.9, 128.9, 128.6, 121.7, 119.0, 118.4, 116.1, 114.8 (aromatic carbons), 35.3, 35.1, 34.8, 31.2, 30.4, 30.0, 28.9, 28.8, 28.7, 28.6, 28.5, 28.4, 22.9, 14.5 (methylene carbons), 14.2 (CH$_3$); MS, m/z: 413.13 (M$^+$). Anal. calcd. for C$_{25}$H$_{38}$N$_2$OS: C, 72.42; H, 9.24; N, 6.76; S; 7.73 found: C, 72.04; H, 9.14; N, 6.37; S, 7.81%.

4.6. Synthesis of 2-((4-(Dimethylamino)phenylimino)methyl)-5-pentadecylphenol (7a).

To a stirred solution of compound **2** (3.32 g, 0.01 mol) and N,N-dimethylbenzene-1,4-diamine

(1.36 g, 0.01 mol) in ethanol (20 mL), glacial acetic acid (1 mL) was added. The mixture was refluxed for 4 h. After the completion of the reaction (TLC-monitoring) the reaction mixture was cooled down to room temperature and then added to crushed ice. The precipitate was filtered, washed with water, dried, and purified through column chromatography by using n-hexane and ethyl acetate (50 : 50) as an eluent to afford pure compound **7a**.

Yield: 58%; mp: 291–221°C; IR (KBr): ν_{max} (cm^{-1}): 321.06 (O–H str.), 1619.34 (C=N str.); ^{1}H-NMR δ ppm: 12.01 (s, 1H, OH), 8.73 (s, 1H, CH=N), 7.60 (d, J = 7.8 Hz, 1H, Ar-H), 7.52 (d, J = 7.7 Hz, 2H, Ar-H), 7.10 (s, 1H, Ar-H), 6.96 (d, J = 7.8 Hz, 1H, Ar-H), 6.59 (d, J = 7.7 Hz, 2H, Ar-H), 3.01 (s, 6H, H$_3$C–N–CH$_3$), 2.63 (t, J = 7.6 Hz, 2H, Ar–CH$_2$), 1.61–1.10 (m, 26H, (CH$_2$)$_{13}$), 0.83 (t, J = 6.8 Hz, 3H, CH$_3$); ^{13}C-NMR δ ppm: 160.9, 157.2, 152.7, 143.6, 130.8, 128.9, 122.4, 118.5, 115.0, 114.7, 112.5 (aromatic carbons), 40.1, 35.3, 35.1, 34.8, 31.2, 30.8, 30.1, 28.9, 28.8, 28.7, 28.6, 28.5, 28.4, 27.3, 13.8 (methylene carbons), 13.5 (CH$_3$); MS, m/z: 449.13. (M$^+$). Anal. calcd. for C$_{30}$H$_{46}$N$_2$O: C, 79.95; H, 10.29; N, 6.22; found: C, 79.34; H, 10.16; N, 6.03%.

4.7. Synthesis of 2-((4-Nitronaphthalen-1-ylimino)methyl)-5-pentadecylphenol (8a). A mixture of compound **2** (3.32 g, 0.01 mol) and 4-nitronaphthalen-1-amine (1.88 g, 0.01 mol) in ethanol (20 mL) and few drops of glacial acetic acid was refluxed for 5 h. After the completion of the reaction (TLC-monitoring), the reaction mixture was cooled down to room temperature and then poured into ice cold water. The precipitate was filtered, dried, and recrystallized from methanol to give pure product **8a** as light yellowish solid.

Yield: 66%; mp: 234–236°C; IR (KBr): ν_{max} (cm^{-1}): 3435.23 (O–H str.), 1624.07 (C=N str.); ^{1}H-NMR δ ppm: 12.03 (s, 1H, OH), 8.90 (d, J = 7.8 Hz, 1H, Ar-H), 8.89 (s, 1H, CH=N), 8.50 (d, J = 7.8 Hz, 1H, Ar-H), 8.20 (d, J = 7.8 Hz, 1H, Ar-H), 7.85–7.75 (m, 2H, Ar-H), 7.60 (d, J = 7.7 Hz, 1H, Ar-H), 7.55 (s, 1H, Ar-H), 6.98 (s, 1H, Ar-H), 6.89 (d, J = 7.7 Hz, 1H, Ar-H), 2.63 (t, J = 7.6 Hz, 2H, Ar–CH$_2$), 1.56–1.13 (m, 26H, (CH$_2$)$_{13}$), 0.89 (t, J = 6.8 Hz, 3H, CH$_3$); ^{13}C-NMR δ ppm: 162.4, 157.2, 152.1, 143.6, 129.8, 128.9, 125.4 120.1, 119.9, 118.9, 118.4, 118.2, 116.5, 115.8, 114.5, 113.4 (aromatic carbons), 35.3, 35.1, 34.8, 31.2, 30.8, 30.1, 28.9, 28.8, 28.7, 28.6, 28.5, 24.3, 23.2, 14.2 (methylene carbons), 13.5 (CH$_3$); MS, m/z: 502.23 (M$^+$+1). Anal. calcd. for C$_{32}$H$_{42}$N$_2$O$_3$: C, 76.46; H, 8.42; N, 5.57; found: C, 76.04; H, 8.26; N, 5.23%.

4.8. General Procedure for Synthesis of New Schiff's Base (9a–g). A mixture of 2-hydroxy-4-pentadecylbenzaldehyde **2** (0.01 mol) and various aromatic amines (0.01 mol) in the presence of glacial acetic acid (1 mL) in ethanol (30 mL) was refluxed for 5 h. The completion of the reaction was monitored by TLC. The reaction mixture was allowed to cool down to room temperature and then poured into crushed ice. The precipitate was filtered, dried, and recrystallized from absolute ethanol. The resulting solid was further purified by column chromatography [silica, n-hexane/ethyl acetate (90 : 10)] to get pure compounds **9a–g** (Scheme 4).

4.8.1. 5-Pentadecyl-2-((phenylimino)methyl)phenol (9a). Yield: 65%; mp: 201–203°C; IR (KBr): ν_{max} (cm^{-1}): 3413.01 (O–H str.), 1619.21 (C=N str.); ^{1}H-NMR δ ppm: 12.61 (s, 1H, OH), 8.61 (s, 1H, CH=N), 7.81 (d, J = 7.8 Hz, 2H, Ar-H), 7.50–7.55 (m, 3H, Ar-H), 7.25 (d, J = 7.7 Hz, 1H, Ar-H), 7.03 (s, 1H, Ar-H), 6.89 (d, J = 7.6 Hz, 1H, Ar-H), 2.55 (t, J = 7.6 Hz, 2H, Ar–CH$_2$), 1.51–1.10 (m, 26H, (CH$_2$)$_{13}$), 0.85 (t, J = 6.8 Hz, 3H, CH$_3$); ^{13}C-NMR δ ppm: 163.2, 156.2, 153.1, 143.3, 129.6, 123.2, 120.1, 119.9, 118.8, 116.3, 115.7 (aromatic carbons), 35.3, 35.1, 34.8, 31.2, 30.8, 30.5, 30.1, 28.9, 28.8, 28.7, 28.6, 28.5, 22.3, 14.2 (methylene carbons), 13.8 (CH$_3$); MS, m/z: 406.11 (M$^+$). Anal. calcd. for C$_{28}$H$_{41}$NO: C, 82.52; H, 10.12; N, 3.44; found: C, 82.24; H, 10.03; N, 3.03%.

4.8.2. 2-((2-Methoxy-4-nitrophenylimino)methyl)-5-pentadecylphenol (9b). Yield: 70%; mp: 210–212°C; IR (KBr): ν_{max} (cm^{-1}): 3425.01 (O–H str.), 1628.03 (C=N str.); ^{1}H-NMR δ ppm: 12.77 (s, 1H, OH), 8.69 (s, 1H, CH=N), 8.03 (d, J = 7.8 Hz, 1H, Ar-H), 7.68 (d, J = 7.7 Hz, 1H, Ar-H), 7.51 (s, 1H, Ar-H), 6.83 (d, J = 7.6 Hz, 1H, Ar-H), 6.79 (d, J = 7.7 Hz, 1H, Ar-H), 3.81 (s, 3H, OCH$_3$), 2.63 (t, J = 7.6 Hz, 2H, Ar–CH$_2$), 1.55–1.13 (m, 26H, (CH$_2$)$_{13}$), 0.83 (t, J = 6.8 Hz, 3H, CH$_3$); ^{13}C-NMR δ ppm: 160.7, 157.2, 152.1, 143.6, 129.8, 128.9, 120.1, 119.9, 118.4, 116.1, 115.9, 114.6, 112.5 (aromatic carbons), 50.6, 35.3, 35.1, 34.8, 31.2, 30.8, 30.1, 28.9, 28.8, 28.7, 28.6, 28.5, 28.2, 22.0, 13.8 (methylene carbons), 13.5 (CH$_3$); MS, m/z: 482.97 (M$^+$). Anal. calcd. for C$_{29}$H$_{42}$N$_2$O$_4$: C, 72.17; H, 8.77; N, 5.80; found: C, 71.94; H, 8.55; N, 5.65%.

4.8.3. 2-((2,4-Dinitrophenylimino)methyl)-5-pentadecylphenol (9c). Yield: 68%; mp: 191–193°C; IR (KBr): ν_{max} (cm^{-1}): 3398.15 (O–H str.), 1619.03 (C=N str.); ^{1}H-NMR δ ppm: 12.24 (s, 1H, OH), 9.02 (s, 1H, Ar-H), 8.70 (d, J = 7.7 Hz, 1H, Ar-H), 8.60 (s, 1H, CH=N), 8.38 (d, J = 7.7 Hz, 1H, Ar-H), 7.12 (d, J = 7.8 Hz, 1H, Ar-H), 7.09 (s, 1H, Ar-H), 6.91 (d, J = 7.6 Hz, 1H, Ar-H), 2.64 (t, J = 7.6 Hz, 2H, Ar–CH$_2$), 1.58–1.12 (m, 26H, (CH$_2$)$_{13}$), 0.83 (t, J = 6.8 Hz, 3H, CH$_3$); ^{13}C-NMR δ ppm: 161.4, 158.2, 156.1, 145.6, 129.8, 128.9, 123.1, 119.9, 118.8, 116.5, 114.9, 114.7, 112.5 (aromatic carbons), 35.3, 35.1, 34.8, 31.2, 30.8, 30.1, 28.9, 28.8, 28.7, 28.6, 28.5, 22.0, 14.2 (methylene carbons), 13.7 (CH$_3$); MS, m/z: 496.11 (M$^+$+1). Anal. calcd. for C$_{28}$H$_{39}$N$_3$O$_5$: C, 67.58; H, 7.90; N, 8.44; found: C, 67.12; H, 7.41; N, 7.13%.

4.8.4. 2-((2-Chloro-4-nitrophenylimino)methyl)-5-pentadecylphenol (9d). Yield: 90%; mp: 165–167°C; IR (KBr): ν_{max} (cm^{-1}): 3415.29 (O–H str.), 1626.09 (C=N str.); ^{1}H-NMR δ ppm: 12.71 (s, 1H, OH), 8.80 (s, 1H, CH=N), 8.25 (s, 1H, Ar-H), 8.05 (d, J = 7.8 Hz, 1H, Ar-H), 7.85 (d, J = 7.6 Hz, 1H, Ar-H), 7.82 (d, J = 7.8 Hz, 1H, Ar-H), 7.05 (s, 1H, Ar-H), 6.93 (d, J = 7.6 Hz, 1H, Ar-H), 2.67 (t, J = 7.6 Hz, 2H, Ar–CH$_2$), 1.55–1.13 (m, 26H, (CH$_2$)$_{13}$), 0.82 (t, J = 6.8 Hz, 3H, CH$_3$); ^{13}C-NMR δ ppm: 163.1, 159.7, 152.1, 143.6, 129.8, 128.9, 120.1, 119.9, 118.4, 116.5, 115.4, 114.7, 112.6 (aromatic carbons), 35.3, 35.1, 34.8, 31.2, 30.8, 30.1, 28.9, 28.8, 28.7, 28.6, 28.5, 22.0, 13.8 (methylene carbons), 12.5 (CH$_3$); MS, m/z: 485.23 (M$^+$). Anal. calcd. for C$_{28}$H$_{39}$ClN$_2$O$_3$: C, 69.04; H, 8.07; N, 5.75; found: C, 68.94; H, 7.98; N, 5.68%.

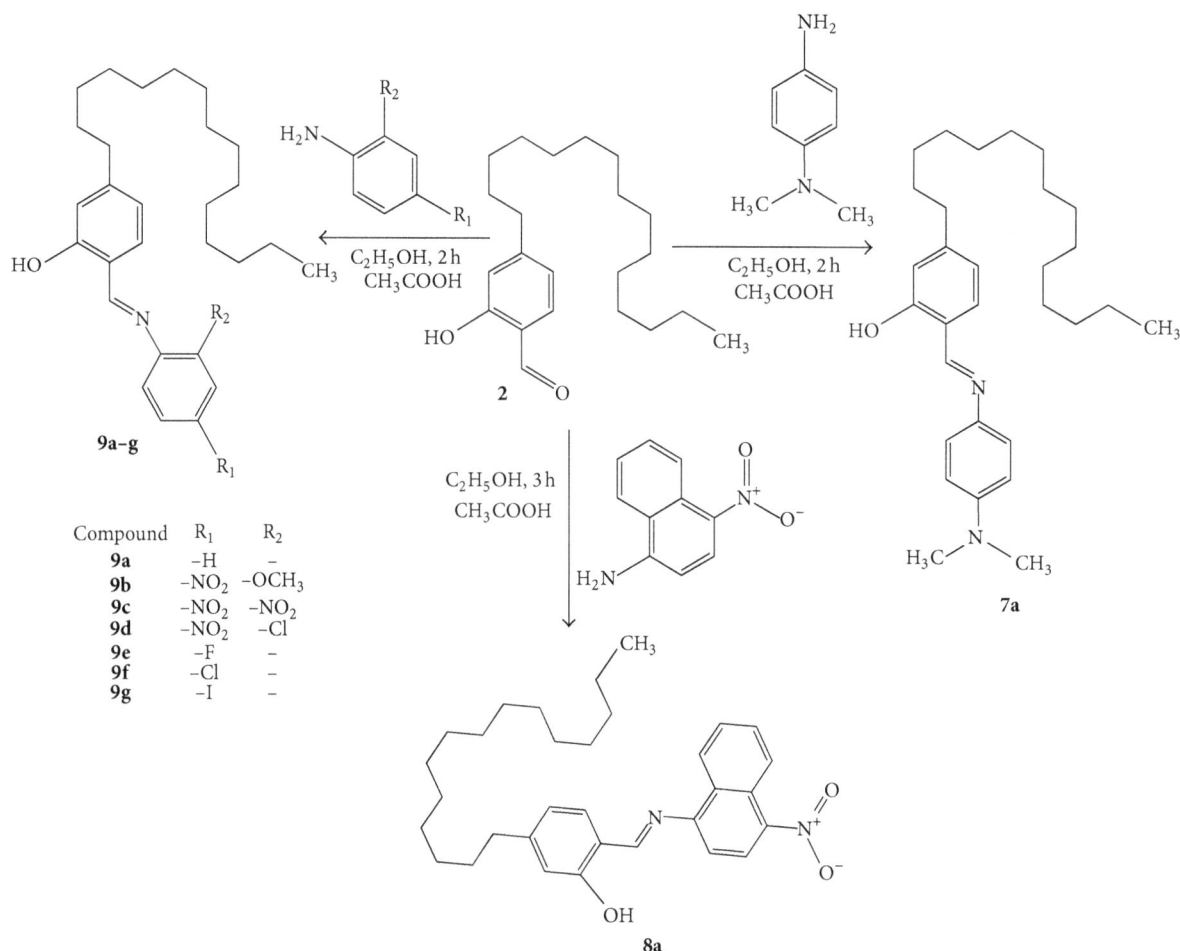

SCHEME 4: Synthesis of novel Schiff base series of compounds.

4.8.5. 2-((4-Fluorophenylimino)methyl)-5-pentadecylphenol (9e). Yield: 76%; mp: 209–211°C; IR (KBr): ν_{max} (cm^{-1}): 3434.15 (O–H str.), 1621.31 (C=N str.); ^{1}H-NMR δ ppm: 12.10 (s, 1H, OH), 8.79 (s, 1H, CH=N), 7.60 (d, J = 7.8 Hz, 1H, Ar–H), 7.51 (d, J = 7.7 Hz, 2H, Ar–H), 7.47 (d, J = 7.7 Hz, 2H, Ar–H), 7.01 (s, 1H, Ar–H), 6.89 (d, J = 7.6 Hz, 1H, Ar–H), 2.64 (t, J = 7.6 Hz, 2H, Ar–CH$_2$), 1.56–1.13 (m, 26H, (CH$_2$)$_{13}$), 0.81 (t, J = 6.8 Hz, 3H, CH$_3$); ^{13}C-NMR δ ppm: 163.4, 158.2, 153.1, 143.6, 129.3, 128.9, 120.1, 119.9, 118.8, 116.5, 115.4 (aromatic carbons), 35.3, 35.1, 34.8, 31.2, 30.8, 30.1, 28.9, 28.8, 28.7, 28.6, 28.5, 22.0, 13.8 (methylene carbons), 13.5 (CH$_3$); MS, m/z: 424.04 (M$^+$). Anal. calcd. for C$_{28}$H$_{40}$FNO: C, 79.01; H, 9.47; N, 3.29; found: C, 78.94; H, 9.44; N, 3.13%.

4.8.6. 2-((4-Chlorophenylimino)methyl)-5-pentadecylphenol (9f). Yield: 65%; mp: 205–207°C; IR (KBr): ν_{max} (cm^{-1}): 3408.31 (O–H str.), 1618.09 (C=N str.); ^{1}H-NMR δ ppm: 12.89 (s, 1H, OH), 8.81 (s, 1H, CH=N), 7.72 (d, J = 7.8 Hz, 1H, Ar–H), 7.42 (d, J = 7.7 Hz, 2H, Ar–H), 7.09 (d, J = 7.7 Hz, 2H, Ar–H), 6.99 (s, 1H, Ar–H), 6.89 (d, J = 7.8 Hz, 1H, Ar–H), 2.63 (t, J = 7.6 Hz, 2H, Ar–CH$_2$), 1.55–1.13 (m, 26H, (CH$_2$)$_{13}$), 0.80 (t, J = 6.8 Hz, 3H, CH$_3$); ^{13}C-NMR δ ppm: 163.3, 157.1, 156.4, 147.2, 129.8, 128.9, 120.1, 119.9, 118.8, 116.8, 116.3 (aromatic carbons), 35.3, 35.1, 34.8, 31.2, 30.8, 30.1, 28.9, 28.8, 28.7, 28.6, 28.5, 26.5, 22.0, 13.8 (methylene carbons), 12.8 (CH$_3$); MS, m/z: 440.21 (M$^+$). Anal. calcd. for C$_{28}$H$_{40}$ClNO: C, 76.07; H, 9.12; N, 3.17; found: C, 75.94; H, 8.99; N, 3.03%.

4.8.7. 2-((4-Iodophenylimino)methyl)-5-pentadecylphenol (9g). Yield: 61%; mp: 179–181°C; IR (KBr): ν_{max} (cm^{-1}): 3435.23 (O–H str.), 1624.07 (C=N str.); ^{1}H-NMR δ ppm: 12.83 (s, 1H, OH), 8.89 (s, 1H, CH=N), 7.77 (d, J = 7.8 Hz, 1H, Ar–H), 7.52 (d, J = 7.7 Hz, 2H, Ar–H), 7.20 (d, J = 7.7 Hz, 2H, Ar–H), 6.80 (d, J = 7.6 Hz, 1H, Ar–H), 6.77 (s, 1H, Ar–H), 2.63 (t, J = 7.6 Hz, 2H, Ar–CH$_2$), 1.56–1.13 (m, 26H, (CH$_2$)$_{13}$), 0.83 (t, J = 6.8 Hz, 3H, CH$_3$); ^{13}C-NMR δ ppm: 161.2, 157.4, 152.8, 143.2, 129.7, 128.9, 120.1, 119.9, 118.8, 117.5, 117.2 (aromatic carbons), 35.3, 35.1, 34.8, 31.2, 30.8, 30.1, 28.9, 28.8, 28.7, 28.6, 28.5, 25.5, 22.2, 13.6 (methylene carbons), 12.8 (CH$_3$); MS, m/z: 533.72 (M$^+$). Anal. calcd. for C$_{28}$H$_{40}$INO: C, 63.03; H, 7.56; N, 2.63; found: C, 62.86; H, 7.72; N, 2.83%.

4.9. Synthesis of 2-((4'-Aminobiphenyl-4-ylimino)methyl)-5-pentadecylphenol (10a). A mixture of compound **2** (3.30 g, 0.01 mol) and biphenyl-4,4'-diamine (1.84 g, 0.010 mol) in the presence of acetic acid in ethanol was refluxed for 4 h. The

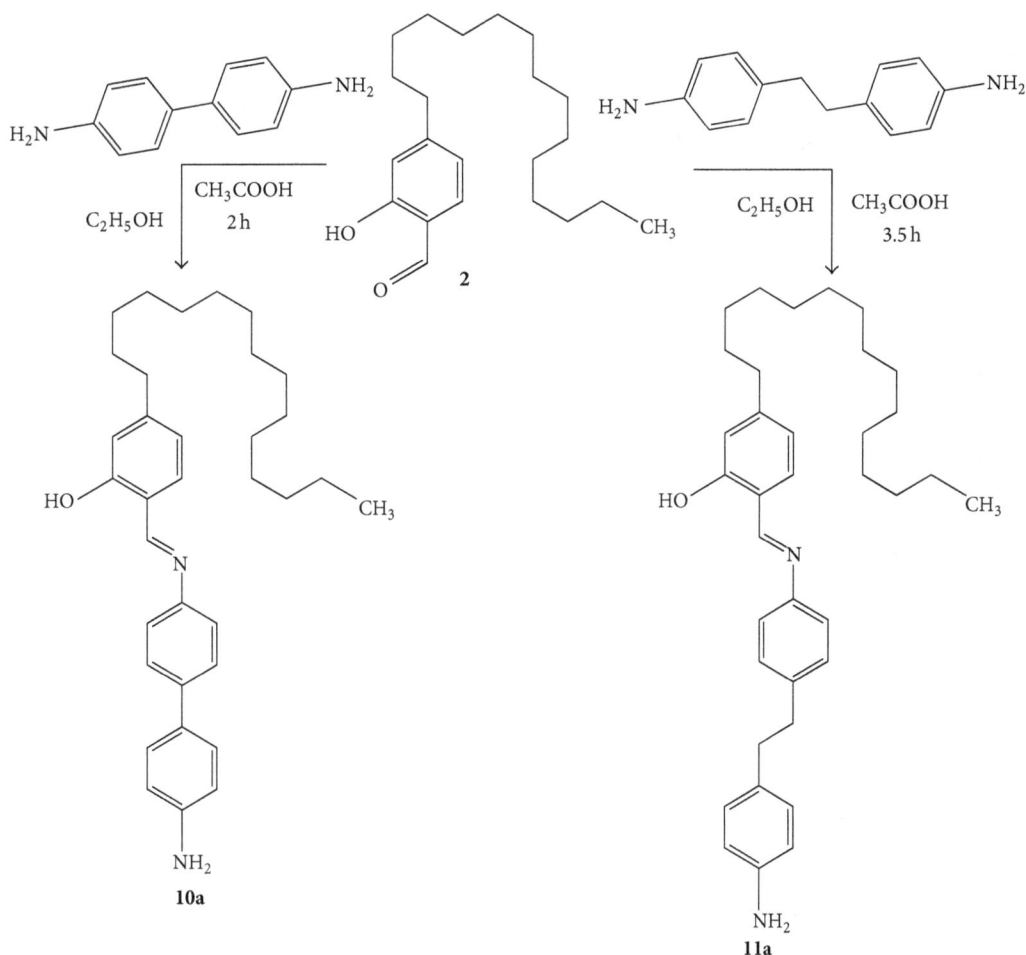

SCHEME 5: Special new Schiff base compounds.

reaction mixture was poured into crushed ice. The precipitate was filtered, dried, and recrystallized from ethanol to get pure compound **10a,** no need for further purification (Scheme 5).

Yield: 73%; mp: 301–303°C; IR (KBr): ν_{max} (cm^{-1}): 3434.03 (O–H str.), 1616.32 (C=N str.); ^{1}H-NMR δ ppm: 13.12 (s, 1H, OH), 8.93 (s, 1H, CH=N), 7.63 (d, J = 7.7 Hz, 2H, Ar–H), 7.61 (d, J = 7.8 Hz, 1H, Ar–H), 7.49 (m, 4H, Ar–H), 7.09 (d, J = 7.8 Hz, 1H, Ar–H), 6.89 (s, 1H, Ar–H), 6.56 (d, J = 7.7 Hz, 2H, Ar–H), 5.15 (s, NH$_2$), 2.69 (t, J = 7.6 Hz, 2H, Ar–CH$_2$), 1.57–1.13 (m, 26H, (CH$_2$)$_{13}$), 0.89 (t, J = 6.8 Hz, 3H, CH$_3$); ^{13}C-NMR δ ppm: 162.3, 157.9, 153.1, 143.8, 129.9, 128.4, 121.1, 119.1, 118.8, 117.5, 117.3, 116.4, 115.3, 114.7, 112.5 (aromatic carbons), 35.3, 35.1, 34.8, 31.2, 30.8, 30.1, 29.5, 28.9, 28.8, 28.7, 28.6, 28.5, 22.0, 13.8 (methylene carbons), 13.3 (CH$_3$); MS, m/z: 497.45 (M$^+$). Anal. calcd. for C$_{34}$H$_{46}$N$_2$O: C, 81.88; H, 9.30; N, 5.62; found: C, 81.49; H, 9.07; N, 5.33%.

4.10. Synthesis of 2-((4-(4-Aminophenethyl)phenylimino)me-thyl)-5-pentadecylphenol (11a).

A mixture of compound **2** (3.30 g, 0.01 mol) and 4,4-ethane-1,2-diyldianiline (2.12 g, 0.01 mol) in the presence of glacial acetic acid in absolute ethanol (20 mL) was heated under reflux for 3 h. After completion of the reaction, the reaction mixture was poured

into crushed ice. The precipitate was filtered, dried, and recrystallized from ethanol to get compound **11a**.

Yield: 81%; mp: 189–191°C; IR (KBr): ν_{max} (cm^{-1}): 3435.31 (O–H str.), 1624.63 (C=N str.); ^{1}H-NMR δ ppm: 12.80 (s, 1H, OH), 8.89 (s, 1H, CH=N), 7.63 (d, J = 7.8 Hz, 1H, Ar–H), 7.41 (d, J = 7.6 Hz, 2H, Ar–H), 7.32 (d, J = 7.6 Hz, 2H, Ar–H), 7.05 (d, J = 7.6 Hz, 2H, Ar–H), 7.01 (d, J = 7.8 Hz, 1H, Ar–H), 6.88 (s, 1H, Ar–H), 6.68 (d, J = 7.6 Hz, 2H, Ar–H), 5.25 (s, NH$_2$), 2.84 (s, –CH$_2$–CH$_2$–), 2.65 (t, J = 7.6 Hz, 2H, Ar–CH$_2$), 1.57–1.13 (m, 26H, (CH$_2$)$_{13}$), 0.83 (t, J = 6.8 Hz, 3H, CH$_3$); ^{13}C-NMR δ ppm: 163.0, 157.8, 152.2, 143.6, 129.8, 128.9, 120.1, 119.9, 118.8, 117.5, 117.8, 116.5, 115.0, 114.7, 112.5 (aromatic carbons), 37.4, 35.3, 35.1, 34.8, 31.2, 30.8, 30.1, 28.9, 28.8, 28.7, 28.6, 28.5, 22.0, 20.3, 15.8 (methylene carbons), 14.2 (CH$_3$); MS, m/z: 526.39 (M$^+$). Anal. calcd. for C$_{36}$H$_{50}$N$_2$O: C, 82.08; H, 9.57; N, 5.32; found: C, 82.12; H, 9.40; N, 5.13%.

5. Conclusion

In conclusion, a new series of 6-substitued-3-chloro-N′-(2-hydroxy-4-pentadecylbenzylidene)benzothiophene-2-car-bohydrazide derivatives were synthesized, fully character-ized, and evaluated for their antibacterial and antifungal activities. The newly synthesized compounds exhibited low to moderate antibacterial activity against *S. aureus* and *B. subtilis* andsignificant antifungal activity against *C. albicans* and *C. pannical*. It can beconcluded that these classes of compounds certainly hold great promise towards good active leads in medicinal chemistry. A further study to acquire more information concerning pharmacological activity is in progress.

Conflict of Interests

The authors declare that there is no conflict of interests regarding the publication of this paper.

Acknowledgments

This work was supported by the Higher Education Research Promotion and National Research University Project of Thailand, Office of the Higher Education Commission. The postdoctoral fellowship grant from the URC, University of Johannesburg, South Africa, was gratefully acknowledged.

References

[1] P. Panneerselvam, B. A. Rather, D. R. S. Reddy, and N. R. Kumar, "Synthesis and anti-microbial screening of some Schiff bases of 3-amino-6,8-dibromo-2-phenylquinazolin-4(3H)-ones," *European Journal of Medicinal Chemistry*, vol. 44, no. 5, pp. 2328–2333, 2009.

[2] X. Jin, J. Wang, and J. Bai, "Synthesis and antimicrobial activity of the Schiff base from chitosan and citral," *Carbohydrate Research*, vol. 344, no. 6, pp. 825–829, 2009.

[3] S. K. Bharti, G. Nath, R. Tilak, and S. K. Singh, "Synthesis, antibacterial and anti-fungal activities of some novel Schiff bases containing 2,4-disubstituted thiazole ring," *European Journal of Medicinal Chemistry*, vol. 45, no. 2, pp. 651–660, 2010.

[4] G. B. Bagihalli, P. G. Avaji, S. A. Patil, and P. S. Badami, "Synthesis, spectral characterization, in vitro antibacterial, antifungal and cytotoxic activities of Co(II), Ni(II) and Cu(II) complexes with 1,2,4-triazole Schiff bases," *European Journal of Medicinal Chemistry*, vol. 43, no. 12, pp. 2639–2649, 2008.

[5] Y. C. Liu and Z. Y. Yang, "Crystal structures, antioxidation and DNA binding properties of Eu(III) complexes with Schiff-base ligands derived from 8-hydroxyquinoline-2-carboxaldehyde and three aroylhydrazines," *Journal of Inorganic Biochemistry*, vol. 103, no. 7, pp. 1014–1022, 2009.

[6] H. Nawaz, Z. Akhter, S. Yameen, H. M. Siddiqi, B. Mirza, and A. Rifat, "Synthesis and biological evaluations of some Schiff-base esters of ferrocenyl aniline and simple aniline," *Journal of Organometallic Chemistry*, vol. 694, no. 14, pp. 2198–2203, 2009.

[7] S. T. Ha, L. K. Ong, J. P. W. Wong et al., "Mesogenic Schiff's base ether with dimethylamino end group," *Phase Transitions*, vol. 82, no. 5, pp. 387–397, 2009.

[8] S. T. Ha, L. K. Ong, S. T. Ong et al., "Synthesis and mesomorphic properties of new Schiff base esters with different alkyl chains," *Chinese Chemical Letters*, vol. 20, no. 7, pp. 767–770, 2009.

[9] S.-T. Ha, T.-M. Koh, S.-T. Ong, J.-K. Beh, and L.-K. Ong, "Synthesis of a new liquid crystal, 3-hydroxy-4-[(6-methoxy-1,3-benzothiazol-2-yl)imino]methylphenyl palmitate," *Molbank*, vol. 2009, no. 3, article M608, 2009.

[10] S. T. Ha, L. K. Ong, Y. F. Win, and T. M. Koh, "4-[(Pyridin-3-ylmethylene)amino]phenyltetradecanoate," *Molbank*, no. 1, Article ID M585, 2009.

[11] S. T. Ha, L. K. Ong, Y. F. Win, T. M. Koh, and G.-Y. Yeap, "4-[(Pyridin-3-ylmethylene)amino]phenylhexadecanoate," *Molbank*, no. 1, article M584, 2009.

[12] P. Payne, J. H. P. Tyman, S. K. Mehet, and A. Ninagawa, "The synthesis of 2-hydroxymethyl derivatives of phenols," *Journal of Chemical Research*, vol. 2006, no. 6, pp. 402–405, 2006.

[13] S. Parkey and N. Castle, "The synthesis of dimethoxy[1]benzothieno[2,3-*c*]quinolines," *Journal of Heterocyclic Chemistry*, vol. 23, no. 5, pp. 1571–1579, 1986.

[14] British Pharmacopoeia, A300, Vol. IV, Appendix XIV, 2005.

[15] J. G. Vincent and H. W. Vincent, "Filter paper disc modification of oxford cup peneciline cup determination," *Proceedings of the Society for Experimental Biology and Medicine*, vol. 55, pp. 162–164, 1944.

24

Allotropic Carbon Nanoforms as Advanced Metal-Free Catalysts or as Supports

Hermenegildo Garcia

Instituto Universitario de Tecnología Química CSIC-UPV, Universidad Politécnica de Valencia, Avenida De Los Naranjos s/n, 46022 Valencia, Spain

Correspondence should be addressed to Hermenegildo Garcia; hgarcia@qim.upv.es

Academic Editor: Davut Avci

This perspective paper summarizes the use of three nanostructured carbon allotropes as metal-free catalysts ("*carbocatalysts*") or as supports of metal nanoparticles. After an introductory section commenting the interest of developing metal-free catalysts and main features of carbon nanoforms, the main body of this paper is focused on exemplifying the opportunities that carbon nanotubes, graphene, and diamond nanoparticles offer to develop advanced catalysts having active sites based on carbon in the absence of transition metals or as large area supports with special morphology and unique properties. The final section provides my personal view on future developments in this field.

1. Introduction: From Active Carbons to Carbon Allotropes

In classical heterogeneous catalysis, active carbons (ACs) have been widely used as supports for noble metal and metal oxides [1–3]. ACs are high surface area materials having carbon as predominant element in their composition that are obtained by pyrolysis of available biomass wastes, upon addition of inorganic reagents to promote the carbonisation process. For instance, one popular active carbon comes from coconut shells adequately powdered, pyrolyzed at 600°C under N_2, mixed with phosphoric acid for activation, and then baked at temperatures below 300°C [4, 5]. In other recipes, olive seeds or almond shells are used as AC precursors and other mineral acids or oxidizing chemicals are employed as additives [6–8].

The structure of ACs is poorly defined with domains of amorphous carbon and the presence of condensed polycyclic aromatic compounds forming platelets of nanometric dimensions that are interconnected by bridges that can be CH_2 and heteroatoms such as O, NH, and S. In certain regions, ACs have graphitic domains when the platelets are large enough and stacking of the imperfect graphene (G) sheets can take

place. Besides oxygen and other elements such as nitrogen or sulphur, metal traces such as iron, zinc, and copper, are very frequently present in the final composition of the material because these transition metals have been introduced as additives in the pyrolysis process and they remain in residual, sometimes not negligible, percentages.

Understanding the mechanism in heterogeneous catalysis largely depends on the exhaustive characterization of the solid catalyst and on the knowledge on the architecture of the active sites [9]. In this sense, while ACs are available and affordable materials exhibiting high adsorption capacity, this property being suitable for their use as support, they are too complex and ill-defined to allow structural characterization and, moreover, they make impossible the preparation of *single-site* catalysts.

One of the main problems in heterogeneous catalysis derives from the fact that solids contain a wide distribution of centers in which the architecture of the sites and the surroundings change from one specific site to another and, as result, the activity and selectivity may change from one site to the neighbor. Optimal solid catalysts would require a material in which all the sites are identical ("*single site*"), all of them exhibiting the same selectivity and, desirably, the maximal activity. Design and synthesis of *single-site* catalysts has been a continued task in heterogeneous catalysis with the long-term aim of the development of solid catalysts with the highest activity and selectivity [10–12].

In the last decades carbon allotropes with nanometric particle dimension have been characterized and have become commercially available. Since their structure is much better defined than ACs, there has been a continued growing interest in the application of these carbon nanoforms in catalysis as a logical evolution of the use of ACs [13–15]. The interest of allotropic carbon nanoforms in catalysis is, therefore, a logical evolution of the continued use of ACs for the preparation of heterogeneous catalysts but goes much beyond ACs, since the point is now to incorporate the active sites on the carbon structure.

As in the case of ACs, the simplest possibility has been the use of these carbon allotropes as large area supports of active sites [16–18], but the most recent research front is to implement the catalytic sites on the carbon allotrope itself in the absence of metals ("*carbocatalysis*"). Starting from an ideal structure of the carbon allotrope, it should be possible to introduce sites by creating carbon vacancies, carbon with dangling bonds, structural defects due to carbon vacancies, and oxygenated functional groups or by replacing carbon atoms by other elements ("*doping*").

The term "carbocatalysts" refers to the development of catalysts based exclusively or predominantly on the use of carbon materials, avoiding or minimizing the dependency of catalysis on the use of metals [19–25]. While the use of metals is considered not "sustainable" due to the limited available resources, carbonaceous catalysts, particularly those derived from biomass, are renewable and affordable and considered as "sustainable" materials.

Among the different carbon allotropes that have been applied in heterogeneous catalysis, the initial studies were based on carbon nanotubes (CNTs) because these materials become commercially available earlier than other carbon materials [26, 27]. Preparation of CNTs requires catalysts and special equipment. CNTs typically are obtained in small batch quantities and impurified with the catalyst used in their synthesis, typically iron or iron-cobalt alloys dispersed in a matrix to maintain the particle size small. This makes necessary in commercial CNTs several steps of purification, oxidation, and other modification processes before they can be used as catalysts. After CNTs, the use of diamond nanoparticles (DNPs) has also become possible. DNPs can be obtained by milling of diamond powders or by explosive detonation [28]. In the last case, D nanocrystals are embedded on a soot matrix of amorphous carbon that has to be removed before the use of the DNPs as supports [29]. More recently, a new line of research has appeared trying to exploit the opportunities of G-based materials in catalysis [20, 21, 30, 31].

Allotropic carbon nanoforms can be dispersed in suspension in a liquid phase as inks that allow the recovery of the carbonaceous materials by filtration or centrifugation after their use. The term "pseudohomogeneous catalysis" has been used to denote the fact that during the reaction there is no apparent phase differentiation between substrates and catalysts, but after the reaction the carbonaceous material

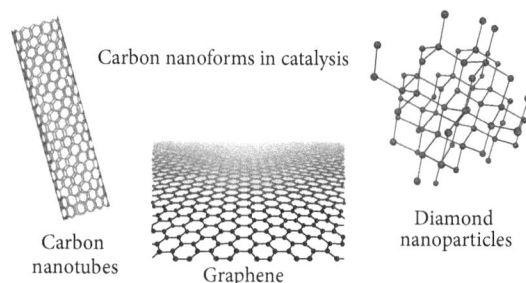

SCHEME 1: Structure of the three carbon allotropes whose use as carbocatalysts or as supports of metal nanoparticles will be the topic of the present perspective paper.

can be easily separated and recovered as is characteristic in heterogeneous catalysis [32].

In the present spotlight paper, I will comment on some of the possibilities that nanometric carbon allotropes offer as catalysts or as supports, showing the logic of the evolution from ACs or from other supports to these carbonaceous nanomaterials as catalysts. Scheme 1 contains the type and structure of the carbon nanoforms whose catalytic activity will be commented on in this perspective paper. When possible, comparison between the performance of different carbon allotropes will be made, but I will try to highlight the features of each type of carbon nanoform that make them best suited for certain catalytic applications. I will start with the possibilities and limitations of CNTs, followed by the use of G-based materials in catalysis and finishing with applications of DNPs as supports. In this paper, I will cover a broad range of articles in this area, with emphasis on the use of these materials as carbocatalysts in the absence of metals with the aim of illustrating the broad potential that carbon nanoforms offer in catalysis. In the last section, I will summarize the major points and will provide my view on possible future developments and targets in this area.

2. Catalysts Based on CNTs as Supports

CNTs are characterized by their long aspect ratio in which one or several concentric hexagonal arrangements of sp^2 carbons ("*graphene sheets*") form a cylinder with nanometric diameter, but lengths that can reach tens of micrometers. The long aspect ratio and the curvature of the graphene walls are the main characteristics of CNTs.

The synthesis of CNTs, either single wall (SWCNTs) or multiple wall (MWCNTs), is difficult to control and relies on the use of a metal catalyst, typically Fe and Co alloys in the form of small metal NPs, to decompose in the pyrolytic process at temperatures about 500°C or above the organic precursor in the absence of oxygen and effect the nucleation and growth of CNTs from elemental carbon atoms generated under these reductive conditions (Scheme 2) [34].

Due to the use of a catalyst in the preparation of CNTs, the amount of CNTs that is available in each batch is limited compared to other carbon nanoforms [36]. MWCNTs can be prepared in larger quantities than SWCNTs because the

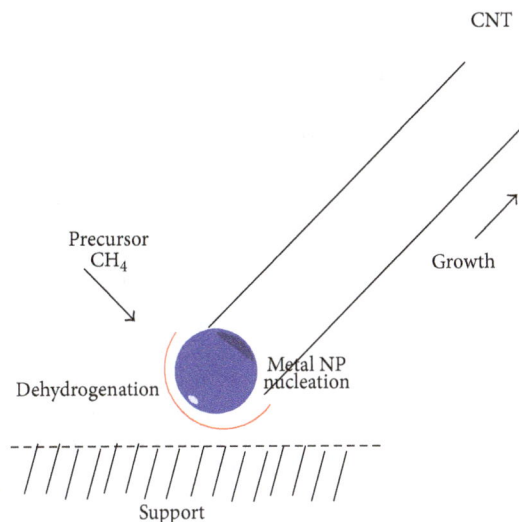

SCHEME 2: Pictorial illustration of the synthesis of CNT by dehydrogenative carbonisation of methane on hot metal nanoparticles. On the surface of the metal carbon atoms are continuously being formed at high temperatures due to the dehydrogenative decomposition of the precursors (methane in this case) and these atoms condense in an hexagonal arrangement leading to the graphene wall with diameter commensurate with the dimensions of the metal nanoparticle catalyst and growing outside these nanoparticles.

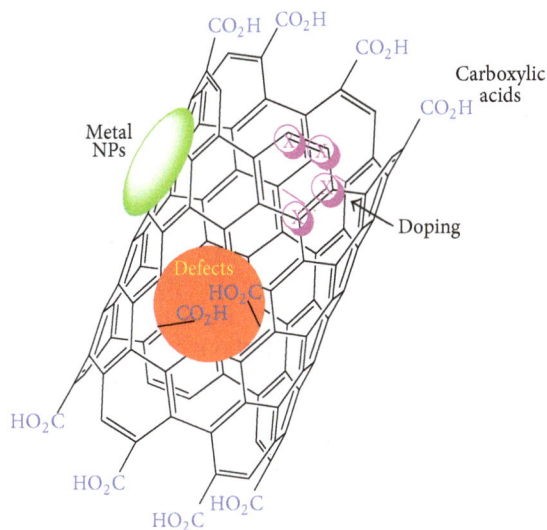

SCHEME 3: Possible sites present in CNTs that can exhibit catalytic activity.

control on the particle size of the metal alloy is not so strict. In general terms, the advantages of using single wall with respect to multiwall have not been clearly demonstrated in catalysis and, therefore, the use of more stable and affordable MWCNTs can be advantageous with respect to SWCNTs. However, from the point of view of achieving the highest catalytic activity and having well-defined structure, SWCNTs can be preferable since all the carbons are exposed at the surface and their morphology and structure is well defined and can be determined by electron microscopy. Also, from the catalytic point of view, there are no examples showing differences between the properties of conductive or semiconductive SWCNTs as catalysts [37, 38]. Electrical conductivity in CNTs depends on the way in which the hexagons of the graphene wall are aligned along the long axis, either spiral (semiconductive) or perpendicular (conductive).

During the purification of CNTs to remove the metal catalyst employed in the synthesis of CNTs, strong inorganic acids are generally needed. HNO_3 is one of the preferred reagents for CNT purification and, then, at the same time that the acid dissolves inorganic impurities, mild oxidation of the graphene wall forming oxygenated functionalities and particularly carboxylic acid groups can also be produced. By performing treatment with HNO_3 under harsh conditions of concentration, temperature, and time, purification of the organic residues can be accompanied by cutting of the CNTs from micrometers to hundreds of nanometers [39, 40], the cut taking place preferentially by oxidation of the graphene wall at defects. One consequence of the shorter length and formation of oxygenated functional groups is an increase on the dispersability of CNTs in liquid media and water in

particular [41, 42]. Preparation of permanent suspensions of short CNTs is very adequate for the development of pseudohomogeneous catalysis.

CNTs can contain active sites due to the presence of carboxylic acid groups at the rims and wall defects that can be functionalized. Also CNTs can be doped with some heteroatoms, the most common one being N atoms. Other possible defects include carbon vacancies and doping with some heteroatoms replacing carbon atoms in the wall. CNTs can also be the support of metal NPs that can be the active sites in the reaction. Scheme 3 summarizes some of the possibilities that CNTs offer to incorporate active centers.

However, as commented earlier, the cumbersome preparation and purification of CNTs explain why these carbon nanoforms have not become as widely available and affordable for the catalytic community as desirable. Since it is expected that the activity of these modified CNTs as acid carbocatalysts would be similar to that of more easily available G materials or that of polymeric resins having carboxylic acids, there is no special reasons beyond morphology and the possibility to encapsulate metal NPs inside the tubes to prefer CNTs in catalysis with respect to G. Thus, although it would be desirable much more information about the catalytic activity of CNTs, it is unlikely that this research front will develop strongly in the near future and, perhaps, the major point of interest will be to compare the catalytic activity of CNTs having curved graphene walls with that of other G-based catalysts.

CNTs in the absence of any metal have been found to be suitable catalysts for the oxidative dehydrogenation of hydrocarbons (Scheme 4) [43–45]. This process has considerable interest in petrochemical industry for the transformation of propane into propene and butanes into butenes and butadiene. CNTs have defects that typically correspond to oxygenated functional groups, namely, carboxylic acids, quinone-like carbonyl groups, and hydroxyl groups. As I

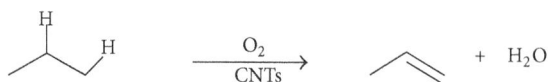

SCHEME 4: Oxidative dehydrogenation of propane catalysed by CNTs.

will comment along this paper, one of the issues that still has to be clarified in many reactions is the nature of the active sites responsible for promoting the reaction. In the present case, using an elegant strategy, the catalytic activity of CNTs for the oxidative dehydrogenation of light alkanes was compared to that of analogous CNTs samples that have been modified to mask selectively each type of the oxygenated functional groups [46]. In this way, carboxylic and hydroxyl groups were selectively protected by esterification or substitution, respectively, while quinone-like carbonyl groups were transformed into imines. Comparison of the catalytic activity of these modified CNTs having selectively masked one of the three possible functional groups has shown that the catalytic activity of pristine CNTs and that of the esterified or hydroxyl substituted CNTs are almost identical and about four times higher than that of the imine functionalized CNTs (Scheme 5). This decrease in the catalytic activity for the modified CNTs that do not contain quinone-like carbonyls but contain carboxylic and hydroxyl groups has led to the conclusion that quinone carbonyls are the functional groups that are responsible for promoting this dehydrogenation [46]. This type of studies shed light on the nature of the catalytically relevant sites and can serve to prepare carbocatalysts with the maximum density of these functional groups and, presumably, with the optimal catalytic activity and selectivity.

Considering that the aim of carbocatalysis is to develop catalysts to replace metals, one challenging reaction that is known to be promoted by transition and noble metals is hydroperoxide decomposition. Using transition metal ions, hydroperoxide decomposition can consume stoichiometric amounts, leading to the formation of considerable amounts of wastes in the reaction. One example of this type of reactions is the Fenton decomposition of H_2O_2 by Fe^{2+} ions in water at acidic pH values [47, 48]. Therefore, there is an increasing interest in developing catalytic versions of this hydroperoxide decomposition [47, 48].

In the simplest mechanism, the presence of redox sites that can reversibly donate one electron reducing hydroperoxide and producing the reductive cleavage of the O–O bond and then become reoxidized by other hydroperoxide molecules rendering oxygen should catalyze hydroperoxide decomposition. Scheme 6 illustrates the catalytic cycle of hydroperoxide decomposition by the presence of a redox site that really can be considered as hydroperoxide dismutation. Hydroperoxides have oxygen atoms in the −I oxidation state and can undergo disproportionation to the 0 and −II oxidation states. Transition metals having different oxidation states and binding strongly to hydroperoxides have been the preferred catalyst for this process.

However, carbon materials can also have other types of redox sites, for instance, those based on quinone-hydroquinone pairs having adequate redox potential to promote the process. In this context, not surprisingly, it has been reported that CNTs can promote benzene hydroxylation to phenol by H_2O_2 with high selectivity [49–51]. The process still requires deeper study and understanding in order to increase the efficiency and, particularly, to assess the nature of the active site, but it is interesting to note that the curvature of the graphene wall in MWCNTs has been invoked as playing a key role in the reaction mechanism [49]. It would be important to check this hypothesis by comparing the process with CNTs of different diameters or even other G-based materials.

In the previous process, the synthesis of a chemical compound, phenol from benzene, is the target of the reaction. However, most frequently peroxide decomposition is used to degrade organic compounds present in aqueous phase. Also MWCNTs have been reported to act as catalysts for this type of reaction. Thus, peroxy monosulfate can decompose by the presence of MWCNTs leading to the formation of sulfate radicals that are able to initiate the aerobic decolorization of methylene blue and decomposition of 2,4-dichlorophenol [52]. For this reaction, active carbons also exhibit catalytic activity, but it has been found that reduced graphene oxide (GO) can exhibit even higher activity than MWCNTs [52].

Doping can be a viable general strategy to introduce active sites in CNTs. As a general observation, the presence of N increases the stability of the carbon nanoforms against oxidation, and, therefore, N doping makes CNTs more suitable as oxidation catalysts [53, 54]. N-doped MWCNTs are also able to promote the aerobic oxidation of benzyl alcohols at moderate temperatures [55, 56]. These N-doped MWCNTs exhibit for this reaction similar catalytic activity as G materials.

Although in general similar performance as catalysts should be expected for CNTs and G materials, one peculiarity of CNTs is the possibility to include inside the tube of nanometric dimension some metal nanoparticles (NPs) of size smaller than the diameter of the tubes. In this regard, Fe NPs have been confined inside CNTs and the resulting material used for the aerobic oxidation of cyclohexane to adipic acid, a process of large industrial relevance (Scheme 7) [57]. However, the study has to show conclusively that the system based on the inclusion of Fe NPs on CNTs is stable under the reaction conditions and does not undergo self-degradation during the course of the reaction. Carbon nanoforms can be oxidized under various conditions. This oxidation leads to the formation of oxygenated functional groups resulting in the creation of defects. In the case of CNTs, oxidation can result in a shortening of their length as consequence of the oxidative cutting of the tube. Thus, it is very likely that CNTs could undergo an increasing degree of oxidation that eventually could lead to the release of Fe NPs and the deactivation of the catalyst, but this issue of catalyst stability in aerobic oxidations has not been yet addressed.

Besides oxidation, carbon nanoforms and also CNTs have attracted interest as catalyst for reversible hydrogen release/uptake from metal hydrides [58–61]. In the context of hydrogen storage, one of the possibilities that has been

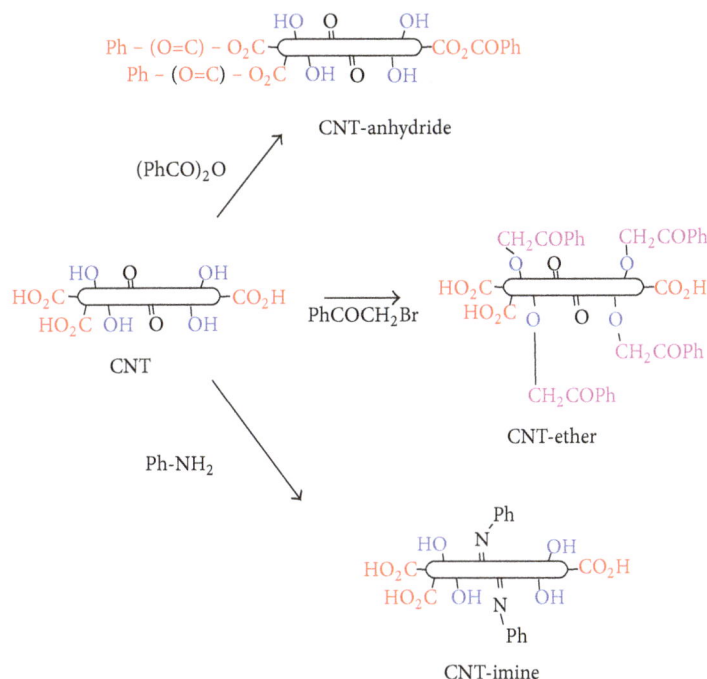

SCHEME 5: Derivatisation of CNTs to mask selectively oxygenated functional groups to assess the nature of the active sites. It was found that CNT, CNT-anhydride, and CNT-ether perform with similar catalytic activity four times higher than that of CNT-imine.

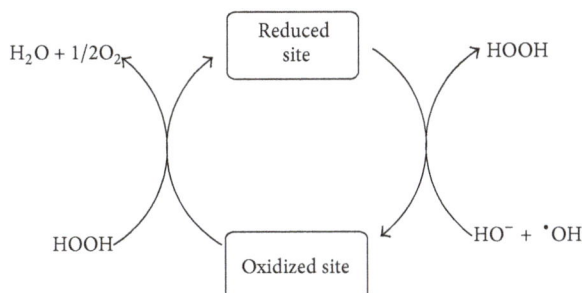

SCHEME 6: Catalytic hydrogen peroxide decomposition (dismutation) mediated by a redox site that could be present in a carbon nanoform.

SCHEME 7: Aerobic oxidation of cyclohexane to adipic acid catalysed by Fe NPs incorporated inside CNTs ((Fe NP)@CNT).

subjected to intensive study has been storage of hydrogen into a chemical compound that can release hydrogen on demand at moderate temperatures with the assistance of a catalyst. After being used, the residual product resulting from hydrogen release should also regain catalytically hydrogen forming the initial hydride. One of the preferred metal hydrides for this process has been LiBH$_4$ (1). It has been

found that CNTs can release up to 8.8 wt% of hydrogen from LiBH$_4$ under mild conditions [61]. However, comparison with GO and r-GO indicates that the hydrogen release using G-based materials is about 1% higher than that of using CNTs. This comparison suggests that defects and residual oxygen functionalities are acting as catalytic centers in this process and that CNTs could have a lower density of this type of sites:

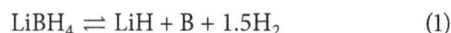

$$LiBH_4 \rightleftharpoons LiH + B + 1.5H_2 \qquad (1)$$

One of the possibilities that CNTs offer in catalysis is their use as supports of metal NPs. Pd NPs supported on MWCNTs have been employed as catalyst for hydrogenation, oxidation, and C–C coupling reactions [62–66]. The activity of Pd/MWCNTs has been compared to that of palladium supported on ACs (Pd/C), and it was found that the turnover number with respect to Pd was higher for Pd/MWCNTs than that for Pd/C for some of these reactions [33]. It was considered that the interaction between the graphene wall of the support with Pd, together with the morphology of the nanotubes, is beneficial to increase the catalytic activity of Pd NPs for those reactions in which the Pd particle size is a key parameter controlling the catalytic activity. In contrast, Pd supported on MWCNTs were much less active than Pd supported on charcoal for those reactions such as hydrogenation of cinnamaldehyde and oxidation of benzyl alcohol that are less sensitive to the average particle size of Pd (Scheme 8).

The strong metal-support interaction arising from overlapping of the extended π system of the graphene wall of CNTs and the orbitals of metal clusters has also been claimed as being responsible for the formation and stabilization

Reactions catalyzed by palladium supported on MWCNTs

Hydrogenation of cinnamaldehyde

(Pd/C more active
than Pd/MWCNT)

Oxidation of benzyl alcohol

(Pd/C more active than Pd/MWCNT)

C-C coupling reaction
(Pd/MWCNT more active
than Pd/C)

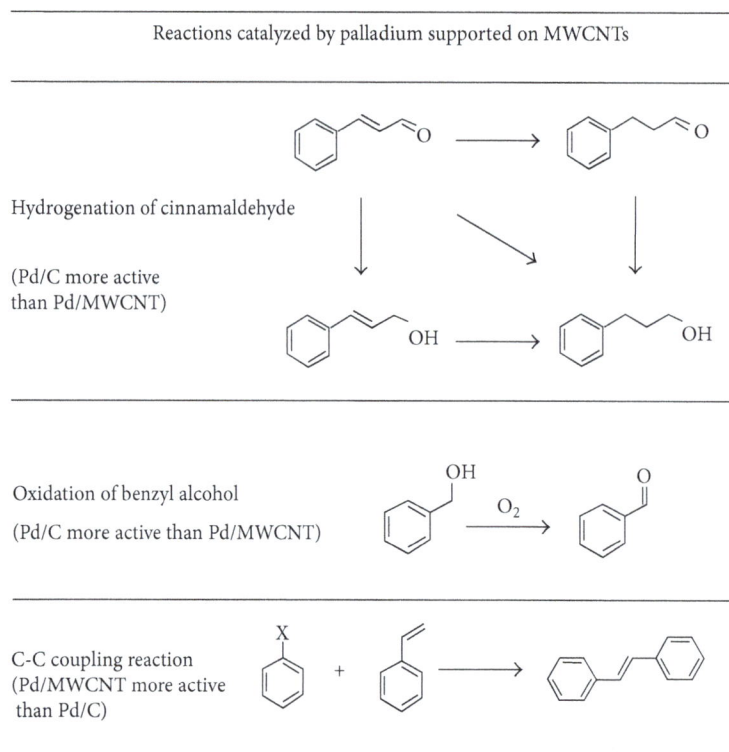

SCHEME 8: Comparison of the catalytic activity of Pd NPs supported on MWCNTs or on ACs (based on [33]).

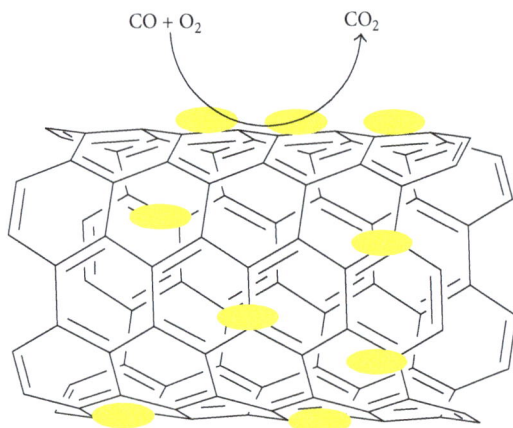

SCHEME 9: Pictorial representation of Au NPs supported on modified SWCNT acting as catalyst for the aerobic oxidation of CO.

of small Au clusters on SWCNTs and for the remarkable catalytic activity towards molecular oxygen dissociation (Scheme 9) [67, 68]. Supported Au NPs are highly active and selective catalysts for the aerobic oxidations of various functional groups [69], and the experimental data indicate that the support always plays an important role in the catalytic activity of Au NPs and in the reaction mechanism. In the present case, Au NPs supported on SWCNTs are highly active for the low temperature CO oxidation and theoretical calculations at the DFT level indicate that this remarkable catalytic activity should be mainly due to the ability of Au

NPs on SWCNTs for molecular oxygen dissociation resulting in the generation of Au oxide clusters highly dispersed on the material. Recently, Corma et al. have shown that it is possible to prepare and characterize clusters of a few Au atoms on the surface of modified MWCNTs and that these clusters between 5 and 10 Au atoms are exceedingly active for the aerobic oxidation of thiophenol to diphenyldisulfide [70]. It is clear that this type of interaction π-d between CNTs and metal NPs is currently underestimated and other remarkable examples observing an increase in the catalytic activity can similarly be achieved in other cases. The curvature of the graphene walls and the presence of defects (oxygen functional groups and carbon vacancies) or heteroatoms should constitute powerful tools to tune the electron density on the metal NP.

Besides the use as support of metal NPs, CNTs can also be employed as platforms to anchor metal complexes that can act as catalytic sites. CNTs conveniently cut and purified can form permanent inks in aqueous solutions or organic media, but once used as catalysts they can be recovered by filtration. In this way, the active sites will be highly dispersed in the reaction media during the reaction but can be recovered at the end of the process and the catalyst recycled ("pseudo-homogeneous catalyst"). An example of this strategy has been the anchoring of a vanadyl salen complex that has been used as catalyst for the cyanosilylation of aldehydes (Scheme 10) [71, 72].

An important point in this approach is characterization of the integrity of the metal complex and this is better guaranteed if anchoring of the metal complex to SWCNTs is carried out in the last step of the preparation of the material

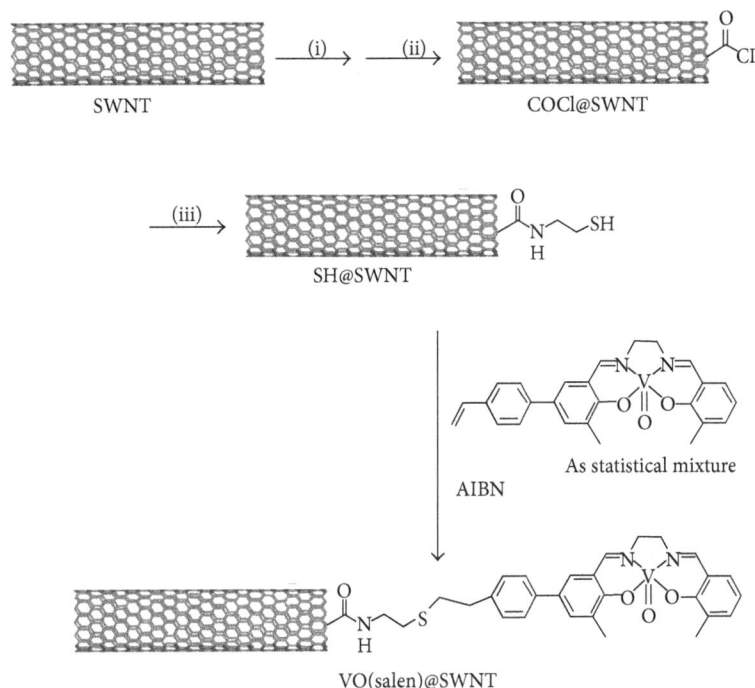

SCHEME 10: Synthesis of a vanadyl salen complex anchored to SWCNTs. Reagents and conditions: (i) 3 M HNO_3, reflux, 24 h; (ii) $SOCl_2$, DMF, 60°C, 24 h; and (iii) 2-aminoethanethiol hydrochloride, Et_3N, CH_2Cl_2, 45°C, 48 h.

since all the previous intermediates can be purified and fully characterized by routine analytical and spectroscopic tools commonly employed in organic chemistry. Compared to AC, the use of short SWCNTs has the advantage of a well-defined morphology and chemistry for covalent functionalization that can be based on the reactivity of carboxylic groups present predominantly at the tips and wall defects of the CNTs or on the reactivity of the graphene wall through specific cycloadditions such as the so-called Prato reaction or radical addition (Scheme 11) [73]. In the case of the vanadyl salen SWCNTs it was found that the system is reusable and the chiral version can induce the preferential formation of one enantiomer of the α-cyano trimethylsilyl ether with high enantiomeric excess [71]. This area, however, still needs to be developed and further work is necessary to fully exploit the possibilities that CNTs offer as scaffolds to anchor covalently metal complexes including high dispersability, easiness of recovery, the interaction of substrates and sites with the graphene walls, either conducting or semiconducting, and the special morphology with long aspect ratio and high curvature of the graphene wall.

3. G-Based Materials in Catalysis

Compared to CNTs that are obtained by pyrolysis of adequate volatile carbon precursors on transition metal-containing catalysts (Fe and Co alloys or other possible metals) or by arc-discharge on graphite electrodes prepared adequately in such a way that they already contain the metal catalyst [36, 74, 75], Gs can be prepared by many other ways, some of

them are chemical methods [76]. Chemical procedures can be preferable because they generally allow the preparation of large quantities. Thus, one of the most popular ways to prepare G-based materials starts with graphite that is deeply oxidized using $KMnO_4$ and H_2O_2 under strong acid conditions (H_2SO_4, HNO_3), followed by exfoliation and dispersion in an adequate solvent leading to GO suspensions [77]. GO has a tendency to undergo chemical reduction leading to a decrease in its oxygen percentage, typically about 50 wt% oxygen content for GO obtained from graphite oxidation, forming suspended materials with residual oxygen content that are generally denoted as reduced graphene oxide (rGO) [76].

Recently, we have reported a greener alternative to obtain G and doped Gs consisting in the pyrolysis in the absence of oxygen of biomass precursors such as modified alginates or chitosan (Scheme 12) [35, 78, 79]. Chitosan acts as single source of carbon and nitrogen and depending on the pyrolysis temperature N-doped G can be obtained with various percentages of nitrogen, up to 8 wt%, that decreases as the pyrolysis temperature increases. Also, alginate modified by boric acid leads upon heating at temperatures higher than 600°C in the absence of oxygen to B-doped G; the percentage of boron depends on the amount of borate in the precursor and on the pyrolysis temperature (Scheme 12) [35].

Pyrolysis of natural biopolymers tends to form graphitic carbon residues with loose stacking of the graphene sheets as evidenced by XRD. These graphitic carbon residues can be subsequently easily exfoliated without the need of oxidation [80]. Thus, no liquid chemical wastes are generated in the formation of doped G by biomass pyrolysis and, in addition,

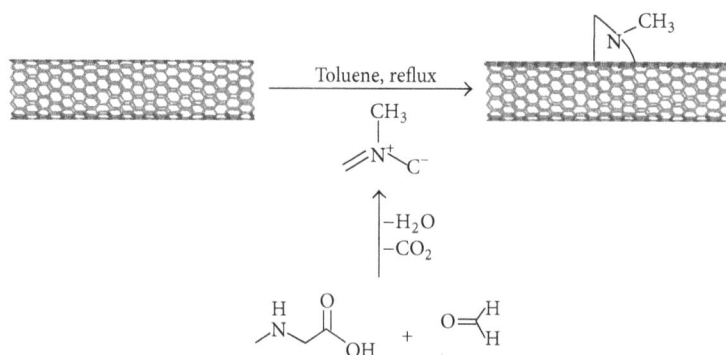

SCHEME 11: Covalent functionalization of CNTs by dipolar cycloaddition ("Prato reaction") to the graphene walls forming a pyrrolidine linkage.

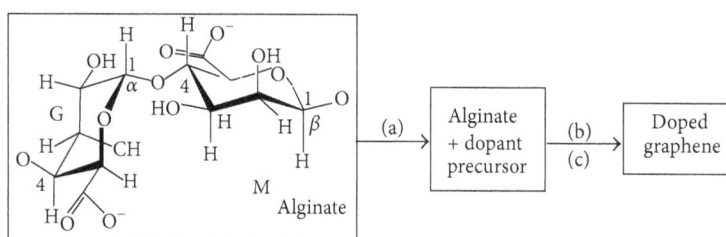

SCHEME 12: General route for the synthesis of doped G by using alginate as G precursor that is modified by addition of a compound of the dopant element (a), followed by pyrolysis of the modified biopolymer in the absence of oxygen (b) and sonication in the presence of a liquid phase (c). The letters G and M correspond to the guluronic and maluronic monosaccharides of alginate.

only a natural biopolymer (typically considered as a valueless biomass waste) in combination or not of other dopant precursors is employed in the synthesis. In summary, either starting from graphite and submitting it to deep oxidation or starting from other precursors, G materials are more easily available than CNTs and can be prepared in larger scale basically because they do not require catalysts to nucleate the dehydrogenative carbonisation of the walls.

One advantage of G-based materials is their large diversity and the opportunities to modify the G sheet by oxidation and doping with heteroatoms. In this sense, the group of Bielawski has pioneered in showing that GO can be a carbocatalyst for oxidation reactions (Scheme 13) [21].

Benzyl alcohols can undergo aerobic oxidation promoted by GO in the absence of metal [81]. Also, GO as acid carbocatalyst promotes dimerization and oligomerization of styrene [82, 83]. However, it has to be mentioned that impurities present in GO have to be surveyed as possible active sites responsible for the catalytic activity. Since GO preparation employs a large excess of $KMnO_4$ and H_2SO_4, it could be possible that these chemicals (or some impurities accompanying them) may not have been removed completely from GO and that these impurities at the ppm level or above could be responsible for the catalysis in these reactions. For instance, our group has shown that GO can catalyze the room-temperature acetalization of aldehydes by methanol and the epoxide ring aperture (Scheme 14) and that this activity is related to the presence of sulphate groups

anchored to G [84, 85]. In accordance with the presence of impurities on GO and their role in catalysis, it has been found that exhaustive GO washings to the point in which the sulfur content becomes below ppms reduces significantly the catalytic activity of GO for these two processes [84, 85]. Based on this, it has been proposed that $-OSO_2OH$ groups anchored on GO sheets should be the active sites for these two acid-catalyzed reactions. The excellent activity of GO is a consequence of the high surface area, easy accessibility, and excellent dispersability of GO sheets. Comparison of the catalytic activity of GO obtained from Hummers oxidation with that of acetic acid reveals that HOAc is much less efficient to promote these two reactions that probably require sites of strong acidity. However, $-OSO_2OH$ groups are not permanently bonded to the GO sheets and can undergo hydrolysis. Therefore, upon reuse, a gradual decrease in the catalytic activity is observed [84]. In this sense, the need of complete analytical data of G-based materials should be emphasized since their catalytic activity can arise from Mn, Fe, or other metal impurities or adventitious acid sites well dispersed on the large surface area characteristic on single-layer GOs.

More recently, our group has found that N-doped G or (B, N-) codoped G are suitable carbocatalysts to promote aerobic oxidations [35]. Comparison of these doped G materials with the catalytic activity of undoped G prepared following the same procedure suggests that this catalytic activity is due to the presence of the dopant elements. In comparison

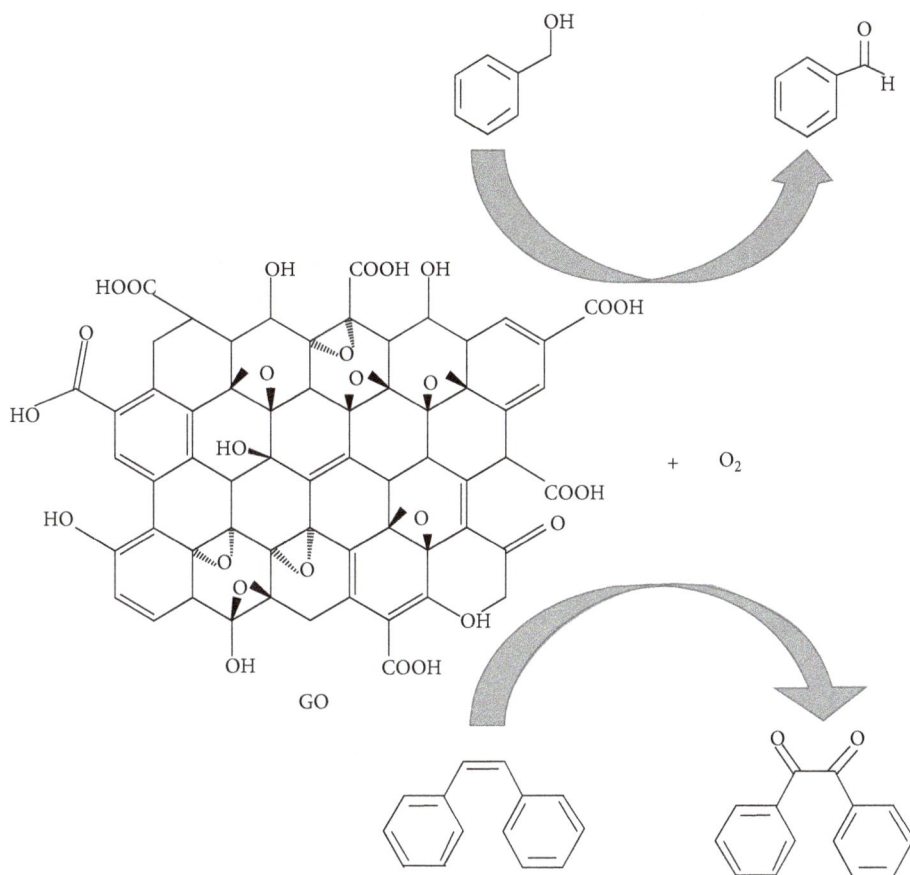

SCHEME 13: Catalytic activity of GO to promote the aerobic oxidation of benzylic alcohols and *cis*-stilbene.

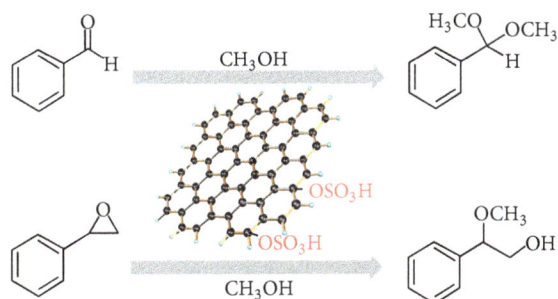

SCHEME 14: Catalytic activity of rGO for the room temperature formation of dimethyl acetal and epoxide ring aperture due to the presence of residual sulfate groups anchored to the G sheet.

with N-doping, doping with B atoms leads to a material with lower activity [35]. IR monitoring of the interaction of molecular oxygen with (N)G shows the appearance of a new band that has been attributed to some peroxyl groups on G [35]. Formation of this peroxyl group is reversible and mild heating and evacuation under reduced pressure lead to the disappearance of this band [35]. Other studies have also shown the ability of N atoms on G to activate molecular oxygen [86], and how this interaction can serve to promote

aerobic oxidations of benzylic alcohols and hydrocarbons, although they may require the use of *tert*-butylhydroperoxide as initiator [35]. Overall, the above data shows the potential that the incorporation of dopants on the G sheet can have to produce active sites on the carbocatalysts as I have already pointed out for the case of CNTs (Scheme 3) [35].

Besides benzylic alcohols and hydrocarbons, styrene can also undergo aerobic oxidation by doped G leading to oxidative C=C bond degradation forming benzaldehyde or C=C bond epoxidation accompanied by rearrangement of the epoxide to 2-phenylacetaldehyde (Scheme 15) [35]. The important observation here is that the product selectivity changes along styrene conversion. Thus, benzaldehyde is formed initially with almost complete selectivity, while styrene oxide appears at higher conversions but can reach selectivities over 60% at final reaction times [35].

These changes in product selectivity as well as the formation of benzaldehyde without induction period have led to proposing a mechanism for styrene oxide formation that is similar to the one assumed for oxidation with molecular oxygen using a transition metal complex or salt and aldehydes as cocatalysts [87]. According to this mechanism, when the concentration of benzaldehyde is sufficiently high, reaction of oxygen with benzaldehyde promoted by doped G in the absence of metals will lead to the formation of benzoyl

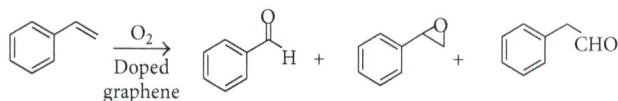

SCHEME 15: Product distribution in the aerobic oxidation of styrene promoted by doped G.

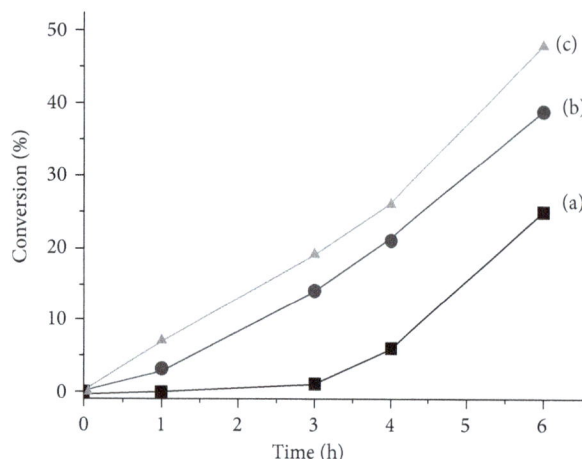

FIGURE 1: Time-conversion plots for the aerobic oxidation of styrene using (N)G as catalyst in the absence (a) and in the presence of 2.5 (b) and 5 wt% (c) of benzaldehyde. Reaction conditions: styrene (1 mL), (N)G (10 mg), and oxygen purging through a balloon, 100°C. Plot taken with permission from [35].

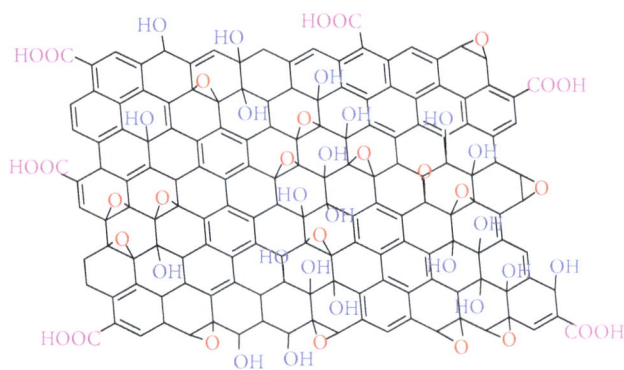

FIGURE 2: Model for GO showing the possible oxygenated functional groups and their location on the sheet.

peroxides and peracids that will be the real oxidizing species leading to C=C epoxidation. Experiments in which various amounts of benzaldehyde were added since the beginning of the reaction show that, under these conditions, styrene oxide is formed without any induction period (Figure 1).

As commented in the section of CNTs, also G-based catalysts exhibit activity for the decomposition of peroxide monosulfate and other peroxides [52, 88]. The main application of these reactions has been decolorization of dyes present in aqueous solution. Peroxide monosulfate as reagent has the advantage over hydrogen peroxide in that the process can take place at neutral pH values and that the resulting sulfates radicals are highly reactive species attacking most of the organic compounds that could be present in water.

Besides oxidations, G can also be used for reduction. Although obviously this reaction type has been much more frequently performed with catalysts containing noble metals, G in the absence of any metal can have also some activity. One of the favorite reactions for which the catalytic activity of G has been tested is the reduction of nitrobenzene and derivatives with $NaBH_4$ [89–91]. In most of the cases, a large excess of $NaBH_4$ (over 300 equivalents) was used. Although this large excess of $NaBH_4$ is unrealistic for any application due to the relatively high price of this commodity chemical, it can be used as a benchmark reaction to rank the activity of the G catalysts by using reaction conditions in which the kinetics becomes apparently of first order. In this way the value of the rate constant can quantitatively assess the activity of the catalyst. Another advantage of the reduction

of nitrobenzene to aniline as a model reaction is that using nitrophenol as probe under basic pH values, the reaction can be carried out in aqueous solution, highly compatible with GO and r-GO, and the course of the reaction can be simply monitored by following in UV/visible spectroscopy the decay and growth of the specific bands corresponding to nitrophenol and hydroxyaniline, respectively.

As commented previously in the case of CNTs, oxidative dehydrogenation of alkanes is a reaction that can be carried out also using G-based materials as catalyst [92]. In particular, GO has been reported as catalyst for the process. It should be commented that there are different models of GO that try to fit with spectroscopic and analytical data for this material. These models indicates the type of oxygenated functional groups that should be present in highly oxidized GO (Figure 2). The functional groups include epoxide, ether, hydroxyl, and carboxylic acid functionalities and basically have to explain the high oxygen content of GO that can be even above 50% in weight as I have already pointed out. This high oxygen content present in GO determines that the active sites that have been proposed for the oxidative dehydrogenation of propane on CNTs (quinone-like moieties) could not be the same as those responsible for the same reaction in GO.

In fact it has been proposed that in the case of GO, epoxy groups should be mainly responsible for the process [92]. In a certain way, GO would act in the reaction mechanism for the oxidative dehydrogenation analogously to the well-established Mars van Krevelen mechanism occurring in nonstoichiometric metal oxides. In these nonstoichiometric oxides oxygen from the solid lattice is reversibly transferred to the substrate causing its oxidation and, then, is replenished by the oxidizing reagent [93]. According to this analogy, oxygen atoms of the epoxide groups present on GO will form water by reaction with the propane, but, in a subsequent step, epoxides will be formed again by reaction with molecular oxygen.

One interesting application of G-based materials is to act as catalyst in the combustion of nitromethane and other high energy fuels for rocketry, thus increasing the power that the fuel can deliver to the engine. Combination of theoretical and experimental data indicates that defects on the G sheet and dangling bonds are responsible for the generation of nitromethyl radicals that subsequently react with adsorbed oxygen and also for the decomposition of peroxide intermediates [94, 95]. It could be interesting also to determine if this catalytic activity of G in combustion reactions can be applied to conventional fuels, such as gasoline or diesel, where the combustion of G could boost the octane or cetane number of fuels.

Although the use of G materials as carbocatalysts is developing currently at a very fast pace, it is clear that at the present, the most widely use of G in catalysis is as support of metal NPs. In this type of reactions, G can cooperate to the process at least in four different ways. The first one is providing a material with a very large surface area allowing a good dispersion of the metal NPs (estimated about $2630 \, m^2 \times g^{-1}$ for fully exfoliated, single-layer material) [96]. In addition, a second possible effect is the strong metal-G interaction that takes place particularly at defects and in the position in which heteroatoms are located in doped Gs [97, 98]. The extended π orbital of G, especially in certain areas, is particularly suitable for overlapping with the d orbitals of transition metals leading to charge transfer phenomena between the metal and the support. This orbital overlap also determines a high affinity of G for metals, minimizing leaching of the metal from the surface to the liquid phase and also reducing particle growth and agglomeration. In this case, the key point is to show how the presumably strong π-d interaction between the G sheet and the metal atoms modifies the intrinsic catalytic activity of the metal NPs with respect to other supports.

A third general effect that has been frequently claimed to rationalize the excellent performance of the catalytic activity of metal NPs supported on G has been the strong adsorption capacity of G for substrates and reagents, bringing them in close proximity to the active sites and even also transferring electrons to them.

A fourth way in which G can contribute to the catalysis in which metal NPs are the main active sites is by providing acid, base, or other types of sites that can cooperate in certain steps of the reaction mechanism. The frequently observed consequence of the use of G as support of metal NPs is a very good dispersability of the material in the reaction medium that derives from the single-layer morphology and subnanometric dimensions of the G.

Comparison of the activity and selectivity of G-supported metal NPs with that exhibited by other related materials and, particularly, metal supported on ACs is necessary in order to fully delineate the advantages of using G sheets as supports. The presence of active sites on the G sheet combined with the catalysis by the metal could lead to the development of bifunctional catalysts with activity in tandem reactions in which two or more processes occur in a single step.

The flat surface of G sheets is particularly suitable for the interaction with metal NPs and Pd, Au, Pt, and Ru have been among the preferred examples for their use in catalysis [99]. At the moment, although there is a large number of examples for preparation of supported metal NPs on G, their application in catalysis is still relatively limited. It is expected that the numbers of examples will grow in the near future,

applying Gs not only as catalysts oxidation, reductions, and couplings, but also for novel reactions in the field of reversible hydrogen release/uptake. In the case of Au NPs supported on Gs there are some examples showing their activity as reduction catalysts for the transformation of aromatic nitro groups into amines using sodium borohydride as reagent [100]. Similarly, Pt NPs have been supported on G and used as oxidation and hydrogenation catalysts that are reaction types of general importance in industry and organic chemistry [101]. Pd NPs supported on Gs have been the preferred pseudohomogeneous catalyst for coupling reactions [102].

Theoretical studies suggest that defects on G should favour the interaction with supported Pt NPs [103]. Computational *ab initio* calculations have led to proposing that Pt supported on defect-engineered G should be more tolerant compared to free Pt NPs to the poisoning by CO, since it should show a higher affinity for H_2 [104]. This lower tendency to CO poisoning is of importance for the development of fuel cells and must be corroborated by experimental measurements [105].

Pt NPs supported on rGO can be obtained by solvolysis using ethylene glycol as reductant and stabilising agent [106–108]. The average particle size of Pt NPs prepared in ethylene glycol can be around 3 nm and they can exhibit oriented 1.1.1 facets. This material performs for hydrogenation of nitrobenzene to aniline over 12 times more efficiently than an analogous Pt catalyst using MWCNTs as support. Furthermore, the catalytic activity at $0°C$ of Pt-rGO is about 20 times higher than the activity of Pt supported on AC. This enhanced catalytic activity of Pt-rGO is proposed to arise from the high dispersion of Pt clusters on rGO and from the dispersability of this material in the reaction mixture [101].

Electrical conductivity is one of the main properties of sp^2-forms of carbon allotropes and particularly of G-based materials. This electrical conductivity can serve to develop electrocatalysts [86, 109]. Pt NPs supported on G sheets of small dimensions (G quantum dots (GQDs)) have also been prepared by solvolysis with ethylene glycol of $PtCl_4^-$ on nanosized GQDs obtained by acid etching of carbon fibers [110]. The resulting material exhibits high activity as electrode for the electrochemical oxygen reduction, where the target is to reduce as much as possible the overpotential needed for this electrochemical process [111]. It was found that Pt-GQD shows an onset potential for oxygen reduction of +1.05 V, that is, 70 mV more positive than the onset potential observed for an analogous electrode prepared with Pt supported on AC [111]. In fact, due to the electrical conductivity, G materials containing or not metal NPs have been widely used as electrocatalysts, but this area has been covered extensively in recent reviews and the reader is addressed to them for a complete coverage [112–115].

4. Diamond Nanoparticles (DNPs) as Support

DNPs are affordable and commercially available (Aldrich, CAS: 7782-40-3). DNPs can be prepared by milling of diamond powders or by explosive detonation [28]. In the last case, the commercial samples have DNPs embedded in a matrix of amorphous carbon ("soot") and it is necessary to treat the samples to etch this amorphous soot matter. DNPs from milling have generally much larger particle size than samples obtained by detonation that are smaller than 10 nm. Considering the importance of having small particle sizes, DNPs from detonation should be preferred as support in catalysis provided that they are liberated from the soot.

In the previous shown cases of CNTs and G allotropic forms, the carbon atoms have sp^2 atomic orbitals and a strong interaction due to the overlap of extended π orbitals of CNTs or G materials with substrates or metal NPs should play a key role in the catalytic activity. In contrast, in the case of DNPs the carbons are mainly sp^3 with surface OH groups and no π-π or π-d overlapping can take place. Moreover, a large percentage of the surface of DNPs can be highly inert and can be envisioned better as devoid of interactions with the active sites or metal NP. This robustness and inertness of DNPs can be, however, beneficial for their use as support to promote some reactions in which highly aggressive species that can react with the support are going to be formed. Thus, the current state of the art does not consider DNPs as carbocatalysts, since there is no a clear view of which type of sites could be present in sp^3 carbons, but, on the other hand, they complement CNTs and Gs as support, since they provide and inert and robust surface that, however, can immobilize metal NPs by the presence of occasional OH groups.

One example of the beneficial use of DNPs as supports of metal NPs is in the catalytic Fenton reaction for the degradation of the organic pollutants in water by hydrogen peroxide [116, 117]. DNPs can be hydrophilic materials when the population of surface hydroxyl groups is large. It is in these surface OH nests where metal NPs are anchored. The density of these hydroxyl groups can be diminished to meet the optimal density required to interact with the metal NPs by reductive treatments with hydrogen at temperatures above 300°C that converts C–OH into C–H groups [118]. Turnover numbers as high as 500,000 have been determined for Au supported on DNPs in the degradation of phenol taken as model pollutant [116, 117, 119]. For this reaction at acid pH values, almost quasistoichiometric 5 : 1 equivalents of H_2O_2 to substrate are needed [116, 117, 119]. These conditions are remarkable, since very frequently reported Fenton catalysts use H_2O_2 excesses as large as 10,000 [116, 117, 119]. Apparently, the key point of the excellent catalytic activity of the Au-DNP as catalyst is the combination of the lack of spurious H_2O_2 decomposition characteristic of the catalytic behavior of Au NPs and the fact that ˙OH radicals formed in the process are free to diffuse into the solution, not remaining surface-bound as it happens with many other solid Fenton catalysts based on metal (typically Fe) supported on inorganic or organic solids (Scheme 16) [120, 121].

One of the undesirable limiting conditions of the Fenton chemistry that should be overcome is the need of acidic pH values, typically below 5 units, to occur [122]. For many applications, it will be important to effect the Fenton reaction at neutral pH, since it is not possible to adjust the pH value for large water volumes or stream flows. Operation of Fenton

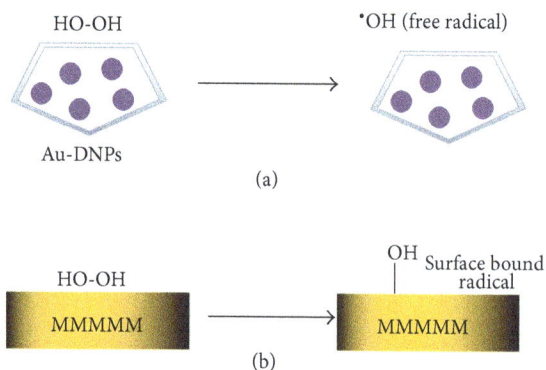

SCHEME 16: Pictorial illustration of the characteristic catalytic activity of Au-DNP generating free ˙OH radicals due to the inertness of its surface (a) in contrast to surface-bound ˙OH radicals (b).

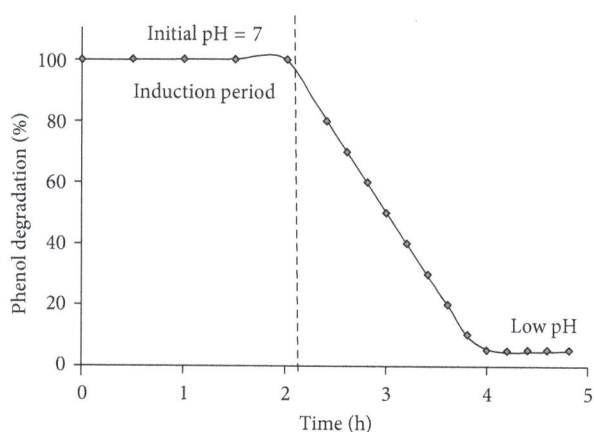

FIGURE 3: Representative example of an ideal time conversion plot for the phenol disappearance in the catalytic Fenton degradation by H_2O_2 using Au-DNP as catalyst in the dark in the absence of buffers. The reaction is initiated at neutral pH, exhibiting an induction period. Once the reaction starts, there is a decrease in the pH value up to 3.5 due to the formation of polycarboxylic acids that accelerates the reaction.

catalysis at neutral pH can only be achieved using a very large excess of H_2O_2 and, if there are not buffers in the solution and for batch reactions, it is frequently observed that after an induction period characterized by a slow start up of the reaction an acceleration occurs (Figure 3). This often remarkable increase in the reaction rate is mainly due to the fact that the pH of the solution becomes spontaneously acidic as soon as some phenol decomposes due to the formation of carboxylic acids that are the degradation byproducts. It was, however, observed that in the case of Au-DNPs the reaction can take place at initial neutral pH values if the reaction is illuminated with solar light or artificial visible light [116, 117]. The reason for this photoinduced process is that Au NPs exhibit a surface plasmon band at λ_{max} 560 nm, and visible light absorption at this wavelength can promote electron injection from excited Au NPs to H_2O_2, leading to ˙OH radicals even in this unfavourably high pH range (Scheme 17) [116, 117].

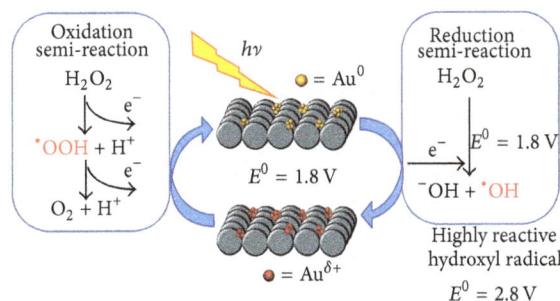

SCHEME 17: Proposed mechanism for the photoinduced catalytic Fenton generation of ˙OH radicals at neutral pH values by visible light irradiation of Au-DNPs. The light is absorbed by Au NPs that exhibit a visible band at about 560 nm (surface plasmon band). Light absorption triggers electron ejection that causes the reduction of H_2O_2 and formation of ˙OH radical.

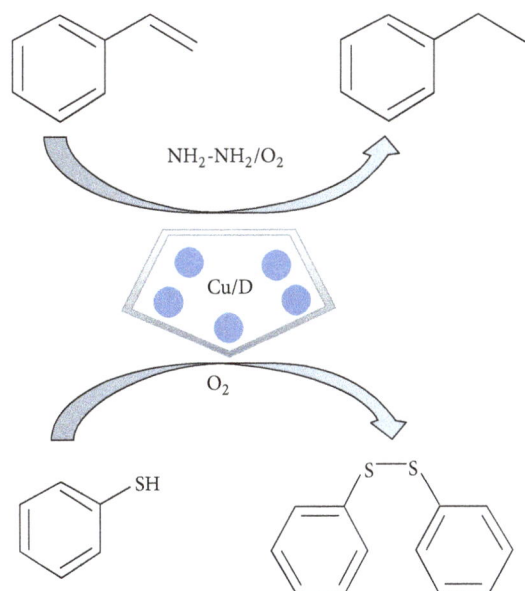

SCHEME 18: Catalytic activity of Cu-DNP for the C=C double bond hydrogenation by hydrazine in the presence of oxygen and the aerobic oxidative coupling of thiophenol to diphenyldisulfide.

Alternatively or coincidentally, irradiation at the Au surface plasmon band can induce local heating near the Au NPs that initiate a thermally induced Fenton reaction [116, 117]. It has been reported based on estimation of the reaction rates and activation energies that irradiation can induce in the submillisecond time scale local temperatures as high as 300°C [123].

Recently the use of DNPs as supports of metal NPs has been extended by developing DNP-supported Cu NPs that are efficient catalysts for the aerobic oxidation of thiols to disulfides [124] and for the hydrogenation of C=C double bonds by hydrazine (Scheme 18) [125]. As in the case of the Au-DNPs, the key feature to understand the excellent

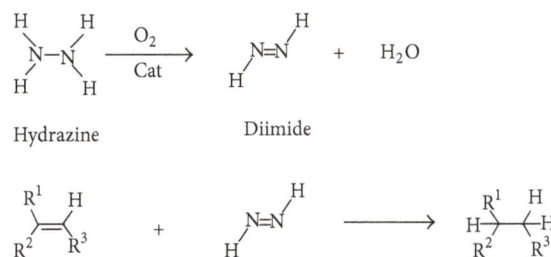

SCHEME 19: Proposed mechanism for the hydrogenation of C=C double bonds by hydrazine under aerobic conditions promoted by Cu-DNP as catalyst.

catalytic activity of Cu DNPs is the small particle size of the metal NP (in the subnanometric size) and the inertness of the surface. Thus, using hydrazine as reducing agent for the hydrogenation of styrene, Cu-DNPs is far more active than other metal NPs including Pd and Pt or other supports such as ACs [126]. This higher activity of Cu NPs over precious metals is interesting from the point of view of reducing the dependency of catalysis on expensive noble metals. The reaction mechanism of Cu-DNP catalysed hydrazine reduction involves presumably the intermediacy of diimide generated by aerobic oxidation of hydrazine (Scheme 19). In fact, even though this reaction is a reduction, it requires the presence of oxygen to occur. Diimide (Scheme 19) is a highly reactive intermediate that spontaneously decomposes and can be envisioned as the precursor of $H_2 + N_2$. The use of hydrazine combined with Cu-DNPs as catalyst can be convenient for some applications avoiding manipulation of hydrogen gas.

Cu-DNPs have also been found to be a recyclable catalyst for the selective oxidation of thiols to disulfides by molecular oxygen [124]. The interesting point here is that, on one hand, thiols are typical poisons of noble metals such as palladium and gold and, on the other hand, they tend to form different oxidation products including sulfenic and sulfonic acids. Thus, Cu-DNPs appear to be ideal catalyst that does not undergo deactivation and exhibits selectivity towards disulfide. TON values as high as 5,700 have been measured for the oxidation of thiophenol to diphenyl disulfide with the Cu-DNP catalyst being reusable at least in four cycles at PhSH/Cu mol ratio of 5772 with turnover frequency of 825 h^{-1} [124].

This behavior of Cu-DNPs and its stability contrasts, for instance, with the performance of Cu-containing metal organic frameworks such as $Cu_3(BTC)_2$ (BTC: 1,3,5-benzenetricarboxylate) that undergoes complete decomposition under similar conditions [127–129]. Metal organic frameworks are microporous crystalline solids that are used as catalysts for a wide range of organic [130] reactions including alcohol [131] and alkane aerobic oxidations [132]. However, metal organic frameworks and in particular $Cu_3(BTC)_2$ may not be stable in the presence of thiols [133]. This comparison illustrates again the robustness of metal supported DNPs catalysts with regard to other alternative solids.

Besides being used as supports of noble metal and Cu NPs, oxidized DNPs have been also been used as supports of other first-row transition metal oxides that exhibit catalytic

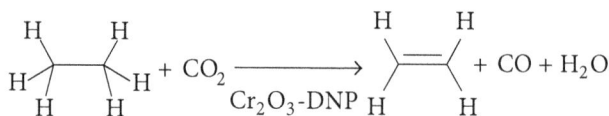

SCHEME 20: Oxidative dehydrogenation of ethane by CO_2.

activity for hydrocarbon dehydrogenation or oxidation using CO_2 as oxidizing reagent. I have shown previously that commercially available DNPs samples should preferably be oxidized to remove amorphous soot matter. This process generates a large density of oxygenated surface functional groups that can be undesirable to stabilize small metal NPs. For this reason, another alternative to remove this amorphous carbon contaminating DNPs could be initial hydrogenation of commercial diamond powder at high temperatures under pure hydrogen stream and, then, the process should be followed by oxidation with diluted molecular oxygen at $450°C$. This pretreatment is very important in order to control the properties of the external DNP surface that after the treatment contains carbonyl groups and ethers. It is, however, very likely that partial combustion of DNP surface could lead also to hydroxyl and carboxylic groups that can interact by sharing the oxygen with metal oxide clusters on the surface and, therefore, the conditions and time of the treatment can have a considerable impact on the performance of the resulting DNP as catalyst.

Using this type of DNP powders obtained by hydrogenation and oxidation as support, Nakagawa et al. have deposited metal NPs on the surface by wet impregnation of the corresponding metal salt followed by calcination at $450°C$ under air [134]. Depending on the nature of the metal oxide, the resulting DNP containing metal oxide NPs exhibits distinctive catalytic properties for various reactions of hydrocarbons with CO_2.

For instance, Ni-DNP is able to promote dry reforming of methane (see (2)) making methane conversion reach about 25% at $600°C$ without deposition of elemental carbon on the catalyst [134]. It was proposed the catalytically active species in this dry reforming should be Ni NPs that must be formed from NiO at the initial stages of the reaction. The weak interaction of NiO with the surface of DNPs will be responsible for the easy generation of Ni NPs in the course of the reaction and therefore of the catalytic activity:

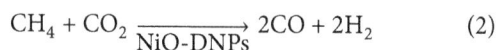

$$CH_4 + CO_2 \xrightarrow{\text{NiO-DNPs}} 2CO + 2H_2 \qquad (2)$$

In another work, the partial oxidation of methane has been carried out using as catalyst Ni or Co NPs supported on DNPs. The catalysts were prepared by impregnation of DNP powders with the required amount of the metal salt followed by water evaporation and calcination at open air at $450°C$. The catalytic activity data show that Ni-DNP performs better than Co-DNP and significantly better than other analogous catalysts of these two metals on different supports, reaching

conversions of 32% at temperatures of $700°C$ [135]. It was determined that at this temperature no carbon deposition on the catalyst occurs and, therefore, the activity of the catalyst remains steady without deactivation. Concerning the reaction mechanism, it was proposed that the overall partial oxidation is the combination of the total combustion of methane coupled with hydrogen reduction of CO_2 [135]:

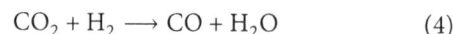

$$CH_4 + 2O_{surf} \longrightarrow CO_2 + 2H_2 \qquad (3)$$

$$CO_2 + H_2 \longrightarrow CO + H_2O \qquad (4)$$

$$CO_2 \longrightarrow CO + O_{surf} \qquad (5)$$

When instead of methane, ethane or light alkanes are reacted with CO_2 using Cr_2O_3-DNPs, then dehydrogenation of ethane and light alkanes takes place (Scheme 20) [136]. The yield of C_2H_4 increases along of the oxidation state of chromium oxide present on the DNP catalyst. It was observed that the presence of oxygenated functional groups on the surface of diamond plays a key role in the dehydrogenation by acting as oxygen supplier in the formation of water. Oxygen becomes subsequently replenished by CO_2. According to this reaction mechanism, CO_2 under the reaction conditions will transfer oxygen atoms to DNPs, becoming converted into CO [136].

V_2O_5 supported on DNPs is also able to promote the reaction of methane and ethane with CO_2 but exhibits in general a different reactivity than Ni NPs or Cr_2O_3 NPs [137]. In the case of V_2O_5-DNPs, the result of the reaction is the corresponding aldehyde, indicating that there is a transfer of an oxygen atom to the alkane (see (6)). Catalytic measurements have shown that formaldehyde yield increases with the increase of the partial pressure of CO_2 and with the increase of the space velocity [137]. The later observation was explained as derived from the fact that long residence time of formaldehyde on the catalyst leads to its decomposition. The optimal V_2O_5-DNP contains 2 wt% of V_2O_5 loading and the maximum TOF measured was 2.7 molHCHO$\times$h$^{-1}\times$molV$_2$O$_5$$^{-1}$ [137]. Similar trends were observed for the formation of acetaldehyde by oxidation of ethane by CO_2. As in the related dehydrogenation with Cr_2O_3-DNP, it was proposed that the oxygen atoms of V_2O_5 and on the surface of DNP are transferred to C_2H_6 to form CH_3CHO and that the role of CO_2 is replenishing surface oxygen atoms to DNP:

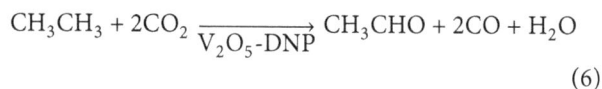

$$CH_3CH_3 + 2CO_2 \xrightarrow{\text{V}_2\text{O}_5\text{-DNP}} CH_3CHO + 2CO + H_2O$$
$$(6)$$

The role of CO_2 providing oxygen atoms to the surface of DNPs avoids deposition of elemental C on the catalyst that is the main cause of the lack of selectivity and deactivation of the catalyst. If Ni-DNP or Pd-DNP are used as catalysts for the pyrolysis of ethane or methane, then filamentous carbon nanotubes are formed by decomposition of this hydrocarbon [138, 139]. As it is usually observed, due to the higher strength of C–H bonds, dehydrogenative decomposition of methane requires temperatures higher than those for the case of ethane

$$\underset{\underset{H}{R} \underset{}{}}{\overset{OH}{\bigwedge}} R'(H) \quad + \quad \frac{1}{2}O_2 \quad \xrightarrow{\text{Pd-CeO}_2\text{-DNP}} \quad \underset{R}{\overset{O}{\bigwedge}} R'(H) \quad + \quad H_2O$$

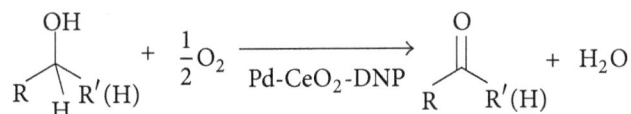

SCHEME 21: Aerobic oxidation of alcohols.

that can be decomposed at temperatures between 400 and 600°C in the case Ni-DNP or 500 to 800°C in the case of Pd-DNP. It was observed that temperatures above 650°C lead to deactivation of Ni-DNP due to the formation of NiC_x phases [138]. In fact, the morphology of the metal NPs changes under the reaction conditions from spherical particles to faceted thin flat particles under operation conditions [138]. Annealing of the resulting thin carbon filaments at 800°C for 5 h under argon also changes the morphology of the carbon filaments to CNTs with high diameters in the range from 80 to 130 nm.

Oxidation of alcohols to carbonyl compounds is a process of large importance in organic synthesis as well as for the preparation of commodities and fine chemicals. A long goal in this area is to develop a general catalyst that can promote selectively alcohol oxidation using molecular oxygen or air. In this regard, it has been reported that Pd NPs combined with CeO_2 NPs supported on diamond is able to catalyze this reaction (Scheme 21) [140]. As in other cases, preparation of the material was performed by two consecutive impregnation cycles, first with $Pd(OAc)_2$ and then $Ce(NH_4)_2(NO_3)_6$, followed by solvent removal and air calcination at 450°C for 5 h [140]. Before using as catalyst, it was necessary to treat the Pd-CeO_2-DNP with a hydrogen stream at 85°C for 1 h to reduce Pd(II) to Pd NPs. In this way, conversions of 95% of benzyl alcohol to afford 78% benzaldehyde were achieved [140]. The TOF value of the catalyst was 850 h^{-1}. It was proposed that DNP as support contributes to the catalysis by providing a hydrophobic environment to the active sites avoiding strong water adsorption on the sites. In addition, the lack of porosity of DNP determines that the reaction takes place on a fully accessible external surface. Comparison of the performance of Pd-CeO_2-DNP with analogous Pd-DNP catalyst lacking CeO_2 for the oxidation of 1-phenylethanol shows that the role of CeO_2 should be neutralization of the adventitious acid sites on the catalyst surface that are responsible for the lack of selectivity leading to the formation of undesirable methyl benzyl ether and ethyl benzene as secondary products. Other basic metal oxides such as Y_2O_3 perform similarly to CeO_2, avoiding the acidity introduced by Pd [140]. Also, comparison of the average particle size for Pd-DNP and Pd-CeO_2-DNP shows that an additional role of CeO_2 is to favor Pd dispersion reducing the average particle size from 4.7 (Pd-DNP) to 3.9 nm (Pd-CeO_2-DNP) [140]:

Fischer-Tropsch synthesis of hydrocarbons is a well-proven technology for the production of fuels from CO and H_2 mixtures of different origins. DNPs have also been used as supports of Co NPs that have high activity for the Fischer-Tropsch synthesis [141]. Two different metal salts, either $Co(NO_3)_2 6H_2O$ or $Co(OAc)_2$, were used in the impregnation of DNPs as cobalt precursors. Impregnation can be carried out either in aqueous solution ($Co(NO_3)_2 6H_2O$) or in acetone ($Co(OAc)_2$). An interesting aspect of this work has been to show the superior performance of DNPs as support of Co NPs compared to graphite or ACs, even though DNPs have lower surface area than the other two carbon supports. To rationalize this higher activity of DNPs, it was proposed that sp^2 carbons exert a negative influence on the Co atoms at the interface, by transferring electron density from the support to the metal, decreasing its catalytic activity [141]. This proposal is again in line with the general fact that for some reactions the inertness of DNP surface can be beneficial for some processes.

Several factors play a key role in the catalytic activity for the Fischer-Tropsch transformation of Co-DNP, such as the reduction temperature in the catalyst pretreatment that influences Co particle size, the reaction temperature that determines the selectivity for methane and C_{5+} hydrocarbons, and the partial pressure of H_2 and CO. All these parameters, including metal precursor salt and Co loading, determine the catalytic activity of the Co-DNP catalyst and the selectivity of the process that, in general, has to be adjusted to optimize the product distribution in C_{5+} hydrocarbons that can be used as fuels and gasoline alternative. Under optimal conditions, Co-DNP becomes a very stable catalyst, maintaining a steady conversion for one day of continuous flow operation.

Besides being used as supports of metal NPs, DNPs offer other possibilities in catalysis. Due to the high density of surface OH groups, DNPs can also be used advantageously to anchor covalently some moieties, for instance, by using acyl chlorides or alkoxysilane reagents as reactive functional groups to attach the moiety to the surface [29]. This strategy has, however, still to be further exploited in catalysis for anchoring transition metal complexes as it has been already reported for CNTs and G [142]. In comparison to the last materials, DNPs offering inert surfaces should in principle exhibit a reactivity of the transition metal complex more alike to that observed for homogeneous phase analogues.

5. Summary and Future Prospects

In the above sections, I have illustrated the potential that nanostructured allotropic carbon materials offer in catalysis either as carbocatalysts or as supports of active sites. In those cases in which the material can be suspended indefinitely, the system can work similarly to a homogeneous catalyst with the added advantage of being recoverable at the end of the reaction. It has been found that the CNTs and Gs having extended π orbitals can interact strongly with substrates and metal NPs, and in this way these carbon supports can influence the catalytic activity by favoring the contact of substrates with the active sites.

Another aspect is that CNTs and G can assist by epitaxial interactions the preferential growth of certain crystallographic facets in the metal NPs while maintaining their small average particle size and influencing their electronic density on the metal NP. These factors can exert strong influence in the catalytic activity exposing the most active metal facets and tuning the electronic density on the metal atoms.

However, these carbon materials constituted by sp^2 atoms may suffer from poor stability when highly reactive intermediates are generated due to the single-layer G structure or due to the tendency to undergo oxidation and degradation. In contrast, in the other extreme, DNPs conveniently purified from amorphous soot matrix offer an intrinsically robust and inert surface while still allowing anchoring of NPs and stabilization of very small average size particles due to the presence of –OH nests on the surface. Thus, DNPs are more suited for those reactions in which the role of the support is to provide a high dispersion of the metal NP, without possessing directly any intrinsic catalytic activity.

Considering the availability of new allotropic nanostructured carbon materials and their unique properties derived from well-defined morphologies, high surface area, and predictable interactions, it can be anticipated that their use in catalysis will grow in the near future [17, 21, 30]. Particularly, G materials can have some advantage over CNTs due to the wider availability and their more convenient preparation and modification [30]. Similarly, the use of DNPs will also grow and will be particularly suited for reactions carried out under harsh conditions and in where highly aggressive and reactive intermediates are generated.

Conflict of Interests

The author declares that there is no conflict of interests regarding the publication of this paper.

Acknowledgments

Financial support by the Spanish Ministry of Economy and Competitiveness (Severo Ochoa and CTQ-2012/32315) and Generalitat Valenciana (Prometeo 2012/014) is gratefully acknowledged.

References

[1] A. E. Aksoylu, M. Madalena, A. Freitas, M. F. R. Pereira, and J. L. Figueiredo, "Effects of different activated carbon supports and support modifications on the properties of Pt/AC catalysts," *Carbon*, vol. 39, no. 2, pp. 175–185, 2001.

[2] H. Jüntgen, "Activated carbon as catalyst support. A review of new research results," *Fuel*, vol. 65, no. 10, pp. 1436–1446, 1986.

[3] K. Köhler, R. G. Heidenreich, J. G. E. Krauter, and J. Pietsch, "Highly active palladium/activated carbon catalysts for Heck reactions: correlation of activity, catalyst properties, and Pd leaching," *Chemistry—A European Journal*, vol. 8, no. 3, pp. 622–631, 2002.

[4] J. Laine, A. Calafat, and M. labady, "Preparation and characterization of activated carbons from coconut shell impregnated with phosphoric acid," *Carbon*, vol. 27, no. 2, pp. 191–195, 1989.

[5] O. S. Amuda, A. A. Giwa, and I. A. Bello, "Removal of heavy metal from industrial wastewater using modified activated coconut shell carbon," *Biochemical Engineering Journal*, vol. 36, no. 2, pp. 174–181, 2007.

[6] O. Ioannidou and A. Zabaniotou, "Agricultural residues as precursors for activated carbon production-a review," *Renewable and Sustainable Energy Reviews*, vol. 11, no. 9, pp. 1966–2005, 2007.

[7] W. K. Lafi, "Production of activated carbon from acorns and olive seeds," *Biomass and Bioenergy*, vol. 20, no. 1, pp. 57–62, 2001.

[8] A. Zabaniotou, G. Stavropoulos, and V. Skoulou, "Activated carbon from olive kernels in a two-stage process: industrial improvement," *Bioresource Technology*, vol. 99, no. 2, pp. 320–326, 2008.

[9] D. Astruc, F. Lu, and J. R. Aranzaes, "Nanoparticles as recyclable catalysts: the frontier between homogeneous and heterogeneous catalysis," *Angewandte Chemie - International Edition*, vol. 44, no. 48, pp. 7852–7872, 2005.

[10] G. W. Coates, "Precise control of polyolefin stereochemistry using single-site metal catalysts," *Chemical Reviews*, vol. 100, no. 4, pp. 1223–1252, 2000.

[11] G. G. Hlatky, "Heterogeneous single-site catalysts for olefin polymerization," *Chemical Reviews*, vol. 100, no. 4, pp. 1347–1376, 2000.

[12] J. M. Thomas, R. Raja, and D. W. Lewis, "Single-site heterogeneous catalysts," *Angewandte Chemie—International Edition*, vol. 44, no. 40, pp. 6456–6482, 2005.

[13] G. Centi and S. Perathoner, "Opportunities and prospects in the chemical recycling of carbon dioxide to fuels," *Catalysis Today*, vol. 148, no. 3-4, pp. 191–205, 2009.

[14] P. Chawla, V. Chawla, R. Maheshwari, S. A. Saraf, and S. K. Saraf, "Fullerenes: from carbon to nanomedicine," *Mini-Reviews in Medicinal Chemistry*, vol. 10, no. 8, pp. 662–677, 2010.

[15] R. Schloegl, "Carbon in catalysis," in *Advances in Catalysis*, B. C. Gates and F. C. Jentoft, Eds., vol. 56, pp. 103–185, 2013.

[16] R. Puskás, A. Sápi, A. Kukovecz, and Z. Kónya, "Comparison of nanoscaled palladium catalysts supported on various carbon allotropes," *Topics in Catalysis*, vol. 55, no. 11–13, pp. 865–872, 2012.

[17] E. Auer, A. Freund, J. Pietsch, and T. Tacke, "Carbons as supports for industrial precious metal catalysts," *Applied Catalysis A: General*, vol. 173, no. 2, pp. 259–271, 1998.

[18] M. Kang, Y.-S. Bae, and C.-H. Lee, "Effect of heat treatment of activated carbon supports on the loading and activity of Pt catalyst," *Carbon*, vol. 43, no. 7, pp. 1512–1516, 2005.

[19] N. Keller, N. I. Maksimova, V. V. Roddatis et al., "The catalytic use onion-like carbon materials for styrene synthesis by oxidative dehydrogenation ethylbenzene," *Angewandte Chemie International Edition*, vol. 41, no. 11, pp. 1885–1888, 2002.

[20] L. Tan, B. Wang, and H. Feng, "Comparative studies of graphene oxide and reduced graphene oxide as carbocatalysts for polymerization of 3-aminophenylboronic acid," *RSC Advances*, vol. 3, no. 8, pp. 2561–2565, 2013.

[21] D. R. Dreyer, H.-P. Jia, and C. W. Bielawski, "Graphene oxide: a convenient carbocatalyst for facilitating oxidation and hydration reactions," *Angewandte Chemie*, vol. 49, no. 38, pp. 6813–6816, 2010.

[22] D. R. Dreyer and C. W. Bielawski, "Carbocatalysis: heterogeneous carbons finding utility in synthetic chemistry," *Chemical Science*, vol. 2, no. 7, pp. 1233–1240, 2011.

[23] J. Pyun, "Graphene oxide as catalyst: application of carbon materials beyond nanotechnology," *Angewandte Chemie*, vol. 50, no. 1, pp. 46–48, 2011.

[24] C. Su and K. P. Loh, "Carbocatalysts: Graphene oxide and its derivatives," *Accounts of Chemical Research*, vol. 46, no. 10, pp. 2275–2285, 2013.

[25] D. S. Su, S. Perathoner, and G. Centi, "Nanocarbons for the development of advanced catalysts," *Chemical Reviews*, vol. 113, no. 8, pp. 5782–5816, 2013.

[26] M. S. Dresselhaus and M. Terrones, "Carbon-based nanomaterials from a historical perspective," *Proceedings of the IEEE*, vol. 101, no. 7, pp. 1522–1535, 2013.

[27] M. Endo, T. Hayashi, Y.-A. Kim, M. Terrones, and M. S. Dresselhaus, "History and structure in carbon nanotube," *Chimica Oggi—Chemistry Today*, vol. 23, no. 2, pp. 29–32, 2005.

[28] V. Y. Dolmatov, "Detonation synthesis ultradispersed diamonds: properties and applications," *Russian Chemical Reviews*, vol. 70, no. 7, pp. 607–626, 2001.

[29] R. Martín, P. C. Heydorn, M. Alvaro, and H. Garcia, "General strategy for high-density covalent functionalization of diamond nanoparticles using fenton chemistry," *Chemistry of Materials*, vol. 21, no. 19, pp. 4505–4514, 2009.

[30] C. Huang, C. Li, and G. Shi, "Graphene based catalysts," *Energy and Environmental Science*, vol. 5, no. 10, pp. 8848–8868, 2012.

[31] D. R. Dreyer, K. A. Jarvis, P. J. Ferreira, and C. W. Bielawski, "Graphite oxide as a carbocatalyst for the preparation of fullerene-reinforced polyester and polyamide nanocomposites," *Polymer Chemistry*, vol. 3, no. 3, pp. 757–766, 2012.

[32] M. Boronat and A. Corma, "Molecular approaches to catalysis: naked gold nanoparticles as quasi-molecular catalysts for green processes," *Journal of Catalysis*, vol. 284, no. 2, pp. 138–147, 2011.

[33] A. Corma, H. Garcia, and A. Leyva, "Catalytic activity of palladium supported on single wall carbon nanotubes compared to palladium supported on activated carbon: study of the Heck and Suzuki couplings, aerobic alcohol oxidation and selective hydrogenation," *Journal of Molecular Catalysis A: Chemical*, vol. 230, no. 1-2, pp. 97–105, 2005.

[34] E. Flahaut, A. Govindaraj, A. Peigney, C. Laurent, A. Rousset, and C. N. R. Rao, "Synthesis of single-walled carbon nanotubes using binary (Fe, Co, Ni) alloy nanoparticles prepared in situ by the reduction of oxide solid solutions," *Chemical Physics Letters*, vol. 300, no. 1-2, pp. 236–242, 1999.

[35] A. Dhakshinamoorthy, A. Primo, P. Concepcion, M. Alvaro, and H. Garcia, "Doped graphene as a metal-free carbocatalyst for the selective aerobic oxidation of benzylic hydrocarbons, cyclooctane and styrene," *Chemistry*, vol. 19, no. 23, pp. 7547–7554, 2013.

[36] T. W. Ebbesen and P. M. Ajayan, "Large-scale synthesis of carbon nanotubes," *Nature*, vol. 358, no. 6383, pp. 220–222, 1992.

[37] P. M. Ajayan, "Nanotubes from Carbon," *Chemical Reviews*, vol. 99, no. 7, pp. 1787–1799, 1999.

[38] D. S. Bethune, C. H. Kiang, M. S. de Vries et al., "Cobalt-catalysed growth of carbon nanotubes with single-atomic-layer walls," *Nature*, vol. 363, no. 6430, pp. 605–607, 1993.

[39] C. Aprile, R. Martin, M. Alvaro, J. C. Scaiano, and H. Garcia, "Near-infrared emission quantum yield of soluble short single-walled carbon nanotubes," *Chemphyschem*, vol. 10, no. 8, pp. 1305–1310, 2009.

[40] R. Martín, M. Álvaro, and H. García, "Photoresponsive covalently-functionalized short single wall carbon nanotubes," *Current Organic Chemistry*, vol. 15, no. 8, pp. 1106–1120, 2011.

[41] M. F. Islam, E. Rojas, D. M. Bergey, A. T. Johnson, and A. G. Yodh, "High weight fraction surfactant solubilization of single-wall carbon nanotubes in water," *Nano Letters*, vol. 3, no. 2, pp. 269–273, 2003.

[42] M. Zheng, A. Jagota, E. D. Semke et al., "DNA-assisted dispersion and separation of carbon nanotubes," *Nature Materials*, vol. 2, no. 5, pp. 338–342, 2003.

[43] X. Liu, B. Frank, W. Zhang, T. P. Cotter, R. Schlögl, and D. S. Su, "Carbon-catalyzed oxidative dehydrogenation of *n*-butane: selective site formation during sp^3-to-sp^2 lattice rearrangement," *Angewandte Chemie*, vol. 50, no. 14, pp. 3318–3322, 2011.

[44] W. Qi, W. Liu, B. Zhang, X. Gu, X. Guo, and D. Su, "Oxidative dehydrogenation on nanocarbon: identification and quantification of active sites by chemical titration," *Angewandte Chemie*, vol. 52, no. 52, pp. 14224–14228, 2013.

[45] J. Zhang, X. Liu, R. Blume, A. Zhang, R. Schlögl, and S. S. Dang, "Surface-modified carbon nanotubes catalyze oxidative dehydrogenation of n-butane," *Science*, vol. 322, no. 5898, pp. 73–77, 2008.

[46] X. Liu, D. S. Su, and R. Schlögl, "Oxidative dehydrogenation of 1-butene to butadiene over carbon nanotube catalysts," *Carbon*, vol. 46, no. 3, pp. 547–549, 2008.

[47] A. Dhakshinamoorthy, S. Navalon, M. Alvaro, and H. Garcia, "Metal nanoparticles as heterogeneous fenton catalysts," *ChemSusChem*, vol. 5, no. 1, pp. 46–64, 2012.

[48] S. Navalon, A. Dhakshinamoorthy, M. Alvaro, and H. Garcia, "Heterogeneous Fenton catalysts based on activated carbon and related materials," *ChemSusChem*, vol. 4, no. 12, pp. 1712–1730, 2011.

[49] Z. H. Kang, E. B. Wang, B. D. Mao et al., "Heterogeneous hydroxylation catalyzed by multi-walled carbon nanotubes at low temperature," *Applied Catalysis A: General*, vol. 299, no. 1-2, pp. 212–217, 2006.

[50] S. Song, H. Yang, R. Rao, H. Liu, and A. Zhang, "Defects of multi-walled carbon nanotubes as active sites for benzene hydroxylation to phenol in the presence of H_2O_2," *Catalysis Communications*, vol. 11, no. 8, pp. 783–787, 2010.

[51] H. Zhang, X. Pan, X. Han et al., "Enhancing chemical reactions in a confined hydrophobic environment: an NMR study of benzene hydroxylation in carbon nanotubes," *Chemical Science*, vol. 4, no. 3, pp. 1075–1078, 2013.

[52] H. Sun, S. Liu, G. Zhou, H. M. Ang, M. O. Tadé, and S. Wang, "Reduced graphene oxide for catalytic oxidation of aqueous organic pollutants," *ACS Applied Materials and Interfaces*, vol. 4, no. 10, pp. 5466–5471, 2012.

[53] C. Chen, J. Zhang, B. Zhang, C. Yu, F. Peng, and D. Su, "Revealing the enhanced catalytic activity of nitrogen-doped carbon nanotubes for oxidative dehydrogenation of propane," *Chemical Communications*, vol. 49, no. 74, pp. 8151–8153, 2013.

[54] B. Frank, J. Zhang, R. Blume, R. Schlögl, and D. S. Su, "Heteroatoms increase the selectivity in oxidative dehydrogenation reactions on nanocarbons," *Angewandte Chemie—International Edition*, vol. 48, no. 37, pp. 6913–6917, 2009.

[55] J. Luo, H. Yu, H. Wang, H. Wang, and F. Peng, "Aerobic oxidation of benzyl alcohol to benzaldehyde catalyzed by carbon nanotubes without any promoter," *Chemical Engineering Journal*, vol. 240, pp. 434–442, 2014.

[56] J. Luo, F. Peng, H. Wang, and H. Yu, "Enhancing the catalytic activity of carbon nanotubes by nitrogen doping in the selective liquid phase oxidation of benzyl alcohol," *Catalysis Communications*, vol. 39, pp. 44–49, 2013.

[57] Y. Cao, X. Luo, H. Yu, F. Peng, H. Wang, and G. Ning, "Sp2- and sp3-hybridized carbon materials as catalysts for aerobic oxidation of cyclohexane," *Catalysis Science and Technology*, vol. 3, no. 10, pp. 2654–2660, 2013.

[58] Z.-Z. Fang, X.-D. Kang, P. Wang, and H.-M. Cheng, "Improved reversible dehydrogenation of lithium borohydride by milling with as-prepared single-walled carbon nanotubes," *Journal of Physical Chemistry C*, vol. 112, no. 43, pp. 17023–17029, 2008.

[59] P.-J. Wang, Z.-Z. Fang, L.-P. Ma, X.-D. Kang, and P. Wang, "Effect of carbon addition on hydrogen storage behaviors of Li-Mg-B-H system," *International Journal of Hydrogen Energy*, vol. 35, no. 7, pp. 3072–3075, 2010.

[60] X. B. Yu, Z. Wu, Q. R. Chen, Z. L. Li, B. C. Weng, and T. S. Huang, "Improved hydrogen storage properties of LiBH4 destabilized by carbon," *Applied Physics Letters*, vol. 90, no. 3, Article ID 034106, 2007.

[61] Y. Zhang, W.-S. Zhang, A.-Q. Wang et al., "LiBH$_4$ nanoparticles supported by disordered mesoporous carbon: hydrogen storage performances and destabilization mechanisms," *International Journal of Hydrogen Energy*, vol. 32, no. 16, pp. 3976–3980, 2007.

[62] P. Serp, M. Corrias, and P. Kalck, "Carbon nanotubes and nanofibers in catalysis," *Applied Catalysis A: General*, vol. 253, no. 2, pp. 337–358, 2003.

[63] J.-P. Tessonnier, L. Pesant, G. Ehret, M. J. Ledoux, and C. Pham-Huu, "Pd nanoparticles introduced inside multi-walled carbon nanotubes for selective hydrogenation of cinnamaldehyde into hydrocinnamaldehyde," *Applied Catalysis A: General*, vol. 288, no. 1-2, pp. 203–210, 2005.

[64] X. R. Ye, Y. Lin, and C. M. Wai, "Decorating catalytic palladium nanoparticles on carbon nanotubes in supercritical carbon dioxide," *Chemical Communications*, vol. 9, no. 5, pp. 642–643, 2003.

[65] G.-Y. Gao, D.-J. Guo, and H.-L. Li, "Electrocatalytic oxidation of formaldehyde on palladium nanoparticles supported on multi-walled carbon nanotubes," *Journal of Power Sources*, vol. 162, no. 2, pp. 1094–1098, 2006.

[66] B. Yoon and C. M. Wai, "Microemulsion-templated synthesis of carbon nanotube-supported Pd and Rh nanoparticles for catalytic applications," *Journal of the American Chemical Society*, vol. 127, no. 49, pp. 17174–17175, 2005.

[67] F. Ding, P. Larsson, J. A. Larsson et al., "The importance of strong carbon-metal adhesion for catalytic nucleation of single-walled carbon nanotubes," *Nano Letters*, vol. 8, no. 2, pp. 463–468, 2008.

[68] L. Alves, B. Ballesteros, M. Boronat et al., "Synthesis and stabilization of subnanometric gold oxide nanoparticles on multiwalled carbon nanotubes and their catalytic activity," *Journal of the American Chemical Society*, vol. 133, no. 26, pp. 10251–10261, 2011.

[69] A. Abad, A. Corma, and H. García, "Catalyst parameters determining activity and selectivity of supported gold nanoparticles for the aerobic oxidation of alcohols: The molecular reaction mechanism," *Chemistry—A European Journal*, vol. 14, no. 1, pp. 212–222, 2008.

[70] A. Corma, P. Concepción, M. Boronat et al., "Exceptional oxidation activity with size-controlled supported gold clusters of low atomicity," *Nature Chemistry*, vol. 5, no. 9, pp. 775–781, 2013.

[71] C. Baleizão, B. Gigante, H. García, and A. Corma, "Chiral vanadyl salen complex anchored on supports as recoverable catalysts for the enantioselective cyanosilylation of aldehydes. Comparison among silica, single wall carbon nanotube, activated carbon and imidazolium ion as support," *Tetrahedron*, vol. 60, no. 46, pp. 10461–10468, 2004.

[72] C. Baleizão, B. Gigante, H. Garcia, and A. Corma, "Vanadyl salen complexes covalently anchored to single-wall carbon nanotubes as heterogeneous catalysts for the cyanosilylation of aldehydes," *Journal of Catalysis*, vol. 221, no. 1, pp. 77–84, 2004.

[73] D. Tasis, N. Tagmatarchis, A. Bianco, and M. Prato, "Chemistry of carbon nanotubes," *Chemical Reviews*, vol. 106, no. 3, pp. 1105–1136, 2006.

[74] J. L. Hutchison, N. A. Kiselev, E. P. Krinichnaya et al., "Double-walled carbon nanotubes fabricated by a hydrogen arc discharge method," *Carbon*, vol. 39, no. 5, pp. 761–770, 2001.

[75] J. Kong, A. M. Cassell, and H. Dai, "Chemical vapor deposition of methane for single-walled carbon nanotubes," *Chemical Physics Letters*, vol. 292, no. 4–6, pp. 567–574, 1998.

[76] S. Stankovich, D. A. Dikin, R. D. Piner et al., "Synthesis of graphene-based nanosheets via chemical reduction of exfoliated graphite oxide," *Carbon*, vol. 45, no. 7, pp. 1558–1565, 2007.

[77] W. S. Hummers Jr. and R. E. Offeman, "Preparation of graphitic oxide," *Journal of the American Chemical Society*, vol. 80, no. 6, p. 1339, 1958.

[78] A. Primo, P. Atienzar, E. Sanchez, J. M. Delgado, and H. García, "From biomass wastes to large-area, high-quality, N-doped graphene: catalyst-free carbonization of chitosan coatings on arbitrary substrates," *Chemical Communications*, vol. 48, no. 74, pp. 9254–9256, 2012.

[79] P. Atienzar, A. Primo, C. Lavorato, R. Molinari, and H. García, "Preparation of graphene quantum dots from pyrolyzed alginate," *Langmuir*, vol. 29, no. 20, pp. 6141–6146, 2013.

[80] A. Primo, A. Forneli, A. Corma, and H. García, "From biomass wastes to highly efficient CO$_2$ adsorbents: graphitisation of chitosan and alginate biopolymers," *ChemSusChem*, vol. 5, no. 11, pp. 2207–2214, 2012.

[81] C. Su, M. Acik, K. Takai et al., "Probing the catalytic activity of porous graphene oxide and the origin of this behaviour," *Nature Communications*, vol. 3, article 2315, 8 pages, 2012.

[82] D. R. Dreyer, S. Park, C. W. Bielawski, and R. S. Ruoff, "The chemistry of graphene oxide," *Chemical Society Reviews*, vol. 39, no. 1, pp. 228–240, 2010.

[83] N. Wu, X. She, D. Yang, X. Wu, F. Su, and Y. Chen, "Synthesis of network reduced graphene oxide in polystyrene matrix by a two-step reduction method for superior conductivity of the composite," *Journal of Materials Chemistry*, vol. 22, no. 33, pp. 17254–17261, 2012.

[84] A. Dhakshinamoorthy, M. Alvaro, P. Concepción, V. Fornés, and H. Garcia, "Graphene oxide as an acid catalyst for the room temperature ring opening of epoxides," *Chemical Communications*, vol. 48, no. 44, pp. 5443–5445, 2012.

[85] A. Dhakshinamoorthy, M. Alvaro, M. Puche, V. Fornes, and H. Garcia, "Graphene oxide as catalyst for the acetalizacion of aldehydes at room temperature," *ChemCatChem*, vol. 4, no. 12, pp. 2026–2030, 2012.

[86] L. Qu, Y. Liu, J.-B. Baek, and L. Dai, "Nitrogen-doped graphene as efficient metal-free electrocatalyst for oxygen reduction in fuel cells," *ACS Nano*, vol. 4, no. 3, pp. 1321–1326, 2010.

[87] A. Corma and H. Garcia, "Supported gold nanoparticles as catalysts for organic reactions," *Chemical Society Reviews*, vol. 37, no. 9, pp. 2096–2126, 2008.

[88] W. Peng, S. Liu, H. Sun, Y. Yao, L. Zhi, and S. Wang, "Synthesis of porous reduced graphene oxide as metal-free carbon for adsorption and catalytic oxidation of organics in water," *Journal of Materials Chemistry A*, vol. 1, pp. 5854–5859, 2013.

[89] Y. Gao, D. Ma, C. Wang, J. Guan, and X. Bao, "Reduced graphene oxide as a catalyst for hydrogenation of nitrobenzene at room temperature," *Chemical Communications*, vol. 47, no. 8, pp. 2432–2434, 2011.

[90] X.-K. Kong, Z.-Y. Sun, M. Chen, C.-L. Chen, and Q.-W. Chen, "Metal-free catalytic reduction of 4-nitrophenol to 4-aminophenol by N-doped graphene," *Energy and Environmental Science*, vol. 6, no. 11, pp. 3260–3266, 2013.

[91] X. K. Kong, Q. W. Chen, and Z. Y. Lun, "Probing the influence of different oxygenated groups on graphene oxide's catalytic performance," *Journal of Materials Chemistry A*, vol. 2, no. 3, pp. 610–613, 2014.

[92] V. Schwartz, W. Fu, Y.-T. Tsai et al., "Oxygen-functionalized few-layer graphene sheets as active catalysts for oxidative dehydrogenation reactions," *ChemSusChem*, vol. 6, no. 5, pp. 840–846, 2013.

[93] A. Corma and H. Garcia, "Lewis acids: from conventional homogeneous to green homogeneous and heterogeneous catalysis," *Chemical Reviews*, vol. 103, no. 11, pp. 4307–4366, 2003.

[94] L.-M. Liu, R. Car, A. Selloni, D. M. Dabbs, I. A. Aksay, and R. A. Yetter, "Enhanced thermal decomposition of nitromethane on functionalized graphene sheets: Ab initio molecular dynamics simulations," *Journal of the American Chemical Society*, vol. 134, no. 46, pp. 19011–19016, 2012.

[95] J. L. Sabourin, D. M. Dabbs, R. A. Yetter, F. L. Dryer, and I. A. Aksay, "Functionalized graphene sheet colloids for enhanced fuel/propellant combustion," *ACS Nano*, vol. 3, no. 12, pp. 3945–3954, 2009.

[96] G. Eda, G. Fanchini, and M. Chhowalla, "Large-area ultrathin films of reduced graphene oxide as a transparent and flexible electronic material," *Nature Nanotechnology*, vol. 3, no. 5, pp. 270–274, 2008.

[97] G. Blanita and M. D. Lazar, "Review of graphene-supported metal nanoparticles as new and efficient heterogeneous catalysts," *Micro and Nanosystems*, vol. 5, no. 2, pp. 138–146, 2013.

[98] M. Ding, Y. Tang, and A. Star, "Understanding interfaces in metal-graphitic hybrid nanostructures," *Journal of Physical Chemistry Letters*, vol. 4, no. 1, pp. 147–160, 2013.

[99] S. Sharma, A. Ganguly, P. Papakonstantinou et al., "Rapid microwave synthesis of CO tolerant Reduced graphene oxide-supported platinum electrocatalysts for oxidation of methanol," *Journal of Physical Chemistry C*, vol. 114, no. 45, pp. 19459–19466, 2010.

[100] K. Jasuja, J. Linn, S. Melton, and V. Berry, "Microwave-reduced uncapped metal nanoparticles on graphene: tuning catalytic, electrical, and raman properties," *Journal of Physical Chemistry Letters*, vol. 1, no. 12, pp. 1853–1860, 2010.

[101] R. Nie, J. Wang, L. Wang, Y. Qin, P. Chen, and Z. Hou, "Platinum supported on reduced graphene oxide as a catalyst for hydrogenation of nitroarenes," *Carbon*, vol. 50, no. 2, pp. 586–596, 2012.

[102] G. M. Scheuermann, L. Rumi, P. Steurer, W. Bannwarth, and R. Mülhaupt, "Palladium nanoparticles on graphite oxide and its functionalized graphene derivatives as highly active catalysts for the Suzuki-Miyaura coupling reaction," *Journal of the American Chemical Society*, vol. 131, no. 23, pp. 8262–8270, 2009.

[103] D.-H. Lim and J. Wilcox, "Mechanisms of the oxygen reduction reaction on defective graphene-supported Pt nanoparticles from first-principles," *Journal of Physical Chemistry C*, vol. 116, no. 5, pp. 3653–3660, 2012.

[104] F. H. Yang, A. J. Lachawiec Jr., and R. T. Yang, "Adsorption of spillover hydrogen atoms on single-wall carbon nanotubes," *Journal of Physical Chemistry B*, vol. 110, no. 12, pp. 6236–6244, 2006.

[105] N. Shang, P. Papakonstantinou, P. Wang, and S. R. P. Silva, "Platinum integrated graphene for methanol fuel cells," *Journal of Physical Chemistry C*, vol. 114, no. 37, pp. 15837–15841, 2010.

[106] C. Xu, X. Wang, and J. Zhu, "Graphene—metal particle nanocomposites," *Journal of Physical Chemistry C*, vol. 112, no. 50, pp. 19841–19845, 2008.

[107] L. Dong, R. R. S. Gari, Z. Li, M. M. Craig, and S. Hou, "Graphene-supported platinum and platinum-ruthenium nanoparticles with high electrocatalytic activity for methanol and ethanol oxidation," *Carbon*, vol. 48, no. 3, pp. 781–787, 2010.

[108] Y. Li, W. Gao, L. Ci, C. Wang, and P. M. Ajayan, "Catalytic performance of Pt nanoparticles on reduced graphene oxide for methanol electro-oxidation," *Carbon*, vol. 48, no. 4, pp. 1124–1130, 2010.

[109] C. Li and G. Shi, "Three-dimensional graphene architectures," *Nanoscale*, vol. 4, no. 18, pp. 5549–5563, 2012.

[110] J. Peng, W. Gao, B. K. Gupta et al., "Graphene quantum dots derived from carbon fibers," *Nano Letters*, vol. 12, no. 2, pp. 844–849, 2012.

[111] G. He, Y. Song, K. Liu, A. Walter, S. Chen, and S. Chen, "Oxygen reduction catalyzed by platinum nanoparticles supported on graphene quantum dots," *ACS Catalysis*, vol. 3, no. 5, pp. 831–838, 2013.

[112] Y. Shao, J. Wang, H. Wu, J. Liu, I. A. Aksay, and Y. Lin, "Graphene based electrochemical sensors and biosensors: a review," *Electroanalysis*, vol. 22, no. 10, pp. 1027–1036, 2010.

[113] D. A. C. Brownson, D. K. Kampouris, and C. E. Banks, "Graphene electrochemistry: fundamental concepts through to prominent applications," *Chemical Society Reviews*, vol. 41, no. 21, pp. 6944–6976, 2012.

[114] F. Cheng and J. Chen, "Metal-air batteries: from oxygen reduction electrochemistry to cathode catalysts," *Chemical Society Reviews*, vol. 41, no. 6, pp. 2172–2192, 2012.

[115] V. Georgakilas, M. Otyepka, A. B. Bourlinos et al., "Functionalization of graphene: covalent and non-covalent approaches, derivatives and applications," *Chemical Reviews*, vol. 112, no. 11, pp. 6156–6214, 2012.

[116] S. Navalon, M. de Miguel, R. Martin, M. Alvaro, and H. Garcia, "Enhancement of the catalytic activity of supported gold nanoparticles for the fenton reaction by light," *Journal of the American Chemical Society*, vol. 133, no. 7, pp. 2218–2226, 2011.

[117] S. Navalon, R. Martin, M. Alvaro, and H. Garcia, "Sunlight-assisted fenton reaction catalyzed by gold supported on diamond nanoparticles as pretreatment for biological degradation of aqueous phenol solutions," *ChemSusChem*, vol. 4, no. 5, pp. 650–657, 2011.

[118] R. Martín, M. Álvaro, J. R. Herance, and H. García, "Fenton-treated functionalized diamond nanoparticles as gene delivery system," *ACS Nano*, vol. 4, no. 1, pp. 65–74, 2010.

[119] S. Navalon, R. Martin, M. Alvaro, and H. Garcia, "Gold on diamond nanoparticles as a highly efficient fenton catalyst," *Angewandte Chemie*, vol. 49, no. 45, pp. 8403–8407, 2010.

[120] J. Feng, X. Hu, and P. L. Yue, "Effect of initial solution pH on the degradation of Orange II using clay-based Fe nanocomposites as heterogeneous photo-Fenton catalyst," *Water Research*, vol. 40, no. 4, pp. 641–646, 2006.

[121] M. B. Kasiri, H. Aleboyeh, and A. Aleboyeh, "Degradation of acid blue 74 using Fe-ZSM5 zeolite as a heterogeneous photo-Fenton catalyst," *Applied Catalysis B: Environmental*, vol. 84, no. 1-2, pp. 9–15, 2008.

[122] P. Wardman and L. P. Candeias, "Fenton chemistry: an introduction," *Radiation Research*, vol. 145, no. 5, pp. 523–531, 1996.

[123] C. Aliaga, D. R. Stuart, A. Aspée, and J. C. Scaiano, "Solvent effects on hydrogen abstraction reactions from lactones with antioxidant properties," *Organic Letters*, vol. 7, no. 17, pp. 3665–3668, 2005.

[124] A. Dhakshinamoorthy, S. Navalon, D. Sempere, M. Alvaro, and H. Garcia, "Aerobic oxidation of thiols catalyzed by copper nanoparticles supported on diamond nanoparticles," *ChemCatChem*, vol. 5, no. 1, pp. 241–246, 2013.

[125] A. Dhakshinamoorthy, S. Navalon, D. Sempere, M. Alvaro, and H. García, "Reduction of alkenes catalyzed by copper nanoparticles supported on diamond nanoparticles," *Chemical Communications*, vol. 49, no. 23, pp. 2359–2361, 2013.

[126] Y. Wang, Z. Xiao, and L. Wu, "Metal-nanoparticles supported on solid as heterogeneous catalysts," *Current Organic Chemistry*, vol. 17, no. 12, pp. 1325–1333, 2013.

[127] L. Huang, H. Wang, J. Chen et al., "Synthesis, morphology control, and properties of porous metal-organic coordination polymers," *Microporous and Mesoporous Materials*, vol. 58, no. 2, pp. 105–114, 2003.

[128] C. Z.-J. Lin, S. S.-Y. Chui, S. M.-F. Lo et al., "Physical stability *vs.* chemical lability in microporous metal coordination polymers: a comparison of $[Cu(OH)(INA)]_n$ and $[Cu(INA)_2]_n$: INA = $1,4-(NC_5H_4CO_2)$," *Chemical Communications*, no. 15, pp. 1642–1643, 2002.

[129] T. M. Reineke, M. Eddaoudi, M. O'Keeffe, and O. M. Yaghi, "A microporous lanthanide–organic framework," *Angewandte Chemie International Edition*, vol. 38, pp. 2590–2594, 1999.

[130] J. Lee, O. K. Farha, J. Roberts, K. A. Scheidt, S. T. Nguyen, and J. T. Hupp, "Metal-organic framework materials as catalysts," *Chemical Society Reviews*, vol. 38, no. 5, pp. 1450–1459, 2009.

[131] A. Dhakshinamoorthy, M. Alvaro, and H. Garcia, "Metal-organic frameworks as heterogeneous catalysts for oxidation reactions," *Catalysis Science and Technology*, vol. 1, no. 6, pp. 856–867, 2011.

[132] A. Dhakshinamoorthy, M. Alvaro, and H. Garcia, "Aerobic oxidation of styrenes catalyzed by an iron metal organic framework," *ACS Catalysis*, vol. 1, no. 8, pp. 836–840, 2011.

[133] A. Dhakshinamoorthy, M. Alvaro, and H. García, "Aerobic oxidation of thiols to disulfides using iron metal-organic frameworks as solid redox catalysts," *Chemical Communications*, vol. 46, no. 35, pp. 6476–6478, 2010.

[134] K. Nakagawa, H. Nishimoto, Y. Enoki et al., "Oxidized diamond supported Ni catalyst for synthesis gas formation from methane," *Chemistry Letters*, no. 5, pp. 460–461, 2001.

[135] H.-A. Nishimoto, K. Nakagawa, N.-O. Ikenaga, M. Nishitani-Gamo, T. Ando, and T. Suzuki, "Partial oxidation of methane to synthesis gas over oxidized diamond catalysts," *Applied Catalysis A: General*, vol. 264, no. 1, pp. 65–72, 2004.

[136] K. Nakagawa, C. Kajita, N.-O. Ikenaga et al., "The role of chemisorbed oxygen on diamond surfaces for the dehydrogenation of ethane in the presence of carbon dioxide," *Journal of Physical Chemistry B*, vol. 107, no. 17, pp. 4048–4056, 2003.

[137] K. Okumura, K. Nakagawa, T. Shimamura et al., "Direct formation of acetaldehyde from ethane using carbon dioxide as a novel oxidant over oxidized diamond-supported catalysts," *The Journal of Physical Chemistry B*, vol. 107, no. 48, pp. 13419–13424, 2003.

[138] N.-O. Higashi, H.-A. Ichi-oka, T. Miyake, and T. Suzuki, "Growth mechanisms of carbon nanofilaments on Ni-loaded diamond catalyst," *Diamond and Related Materials*, vol. 17, no. 3, pp. 283–293, 2008.

[139] N.-O. Higashi, N.-O. Ikenaga, T. Miyake, and T. Suzuki, "Carbon nanotube formation on Ni- or Pd-loaded diamond catalysts," *Diamond and Related Materials*, vol. 14, no. 3–7, pp. 820–824, 2005.

[140] T. Yasu-eda, R. Se-ike, N.-O. Ikenaga, T. Miyake, and T. Suzuki, "Palladium-loaded oxidized diamond catalysis for the selective oxidation of alcohols," *Journal of Molecular Catalysis A: Chemical*, vol. 306, no. 1-2, pp. 136–142, 2009.

[141] T.-O. Honsho, T. Kitano, T. Miyake, and T. Suzuki, "Fischer-Tropsch synthesis over Co-loaded oxidized diamond catalyst," *Fuel*, vol. 94, pp. 170–177, 2012.

[142] P. V. Kamat, "Graphene-based nanoarchitectures: anchoring semiconductor and metal nanoparticles on a two-dimensional carbon support," *Journal of Physical Chemistry Letters*, vol. 1, no. 2, pp. 520–527, 2010.

25

Chiral Recognition by Fluorescence: One Measurement for Two Parameters

Shanshan Yu[1] and Lin Pu[2]

[1] *Key Laboratory of Green Chemistry and Technology, Ministry of Education, College of Chemistry, Sichuan University, Chengdu 610064, China*
[2] *Department of Chemistry, University of Virginia, Charlottesville, VA 22904, USA*

Correspondence should be addressed to Shanshan Yu; yushanshan@scu.edu.cn and Lin Pu; lp6n@virginia.edu

Academic Editor: João A. Lopes

This outlook describes two strategies to simultaneously determine the enantiomeric composition and concentration of a chiral substrate by a single fluorescent measurement. One strategy utilizes a pseudoenantiomeric sensor pair that is composed of a 1,1'-bi-2-naphthol-based amino alcohol and a partially hydrogenated 1,1'-bi-2-naphthol-based amino alcohol. These two molecules have the opposite chiral configuration with fluorescent enhancement at two different emitting wavelengths when treated with the enantiomers of mandelic acid. Using the sum and difference of the fluorescent intensity at the two wavelengths allows simultaneous determination of both concentration and enantiomeric composition of the chiral acid. The other strategy employs a 1,1'-bi-2-naphthol-based trifluoromethyl ketone that exhibits fluorescent enhancement at two emission wavelengths upon interaction with a chiral diamine. One emission responds mostly to the concentration of the chiral diamine and the ratio of the two emissions depends on the chiral configuration of the enantiomer but independent of the concentration, allowing both the concentration and enantiomeric composition of the chiral diamine to be simultaneously determined. These strategies would significantly simplify the practical application of the enantioselective fluorescent sensors in high-throughput chiral assay.

1. Introduction

The study of enantiomerically pure chiral compounds has found increasing importance in many areas, such as pharmaceutical industry [1, 2], agrochemical area [3], and food analysis [4, 5]. For example, the stereochemistry of drugs can significantly affect their biological activity due to the inherently chiral environment of the biological systems. The US FDA issued a policy statement in 1992 and strongly encouraged the development of single isomers [6]. Therefore, easily and economically performed methods for acquiring enantiopure compounds have attracted enormous research interest.

The development of asymmetric catalysis has provided the pathway to preferentially generate one enantiomer over the other from a reaction by using a chiral catalyst. This not only can avoid the labor-intensive and time-consuming separation of enantiomers, but also can eliminate the waste of the undesired enantiomer [7, 8]. The key to develop an efficient asymmetric catalysis reaction is to identify a catalyst structure as well as its most suitable reaction conditions including factors such as solvent, temperature, additive, reaction time, and stoichiometry. Therefore, this screening process can be extremely time-consuming with the traditional one-catalyst-at-a-time approach. The emergence of combinatorial chemistry and parallel synthesis in the 1990s has made it possible to conduct efficient high-throughput screening of an enormous number of chiral catalysts [9–13]. With the assistance of the highly automated system, tens of thousands of compounds can be prepared in a short period of time. However, with the traditional analytical techniques, such as gas chromatography (GC) or high-performance liquid chromatography (HPLC), it usually takes about 20 min to determine the enantiomeric composition of a sample, which would be very inefficient for the analysis of the great number of products generated from the combinatorial catalyst screening processes. Therefore, high-throughput analytical techniques have become the bottleneck for the combinatorial chiral catalyst screening.

To date, a number of techniques are under development for the high-throughput enantiomeric composition determination [13–17], including time resolved IR-thermographic method [18–21], electron spray mass spectrometry [22–26], capillary electrophoresis [27], UV/Vis [28–30], circular dichroism [31, 32], and fluorescence [33, 34]. Among these approaches, optical methods have been attracting increasing attention due to their ability for quick data collection and compatibility with high-throughput screening system. Anslyn has recently reviewed the optical approaches for the rapid enantiomeric composition determination [17]. For an asymmetrical catalytic reaction, both the yield and enantiomeric purity of the product are the essential two parameters needed to evaluate the efficiency of a chiral catalyst. Therefore, a high-throughput analytical process would require a rapid determination of both the concentration and enantiomeric purity of the chiral substrate. This would normally need two independent methods one measuring the concentration and one measuring the enantiomeric purity. It would be highly advantageous if both parameters of a reaction could be determined by a single measurement.

Recently, several strategies have been developed to measure both the concentration and enantiomeric purity of a chiral compound. Wolf and coworkers measured the concentration and enantiomeric composition of chiral compounds by using racemic and enantiopure forms of a chiral sensor in tandem with UV or fluorescence measurement [35–38]. By employing rapidly interconverting racemic sensors, they also fulfilled the same task with dual-mode optical measurements of a sample solution. They used the induced CD signal of the sensor in the presence of the chiral substrate to quantify the enantiomeric purity and then used the chirality independent UV or FL responses to quantify the substrate concentration [39, 40]. Anslyn and coworkers established enantioselective indicator displacement assays (eIDA), with which achiral and chiral sensors were used to determine the concentration and enantiomeric purity of a chiral sample, respectively [41]. To reduce the number of spectroscopic measurements from two to one, they utilized a dual-chamber quartz cuvette filled with careful choice of indicator/host combinations [42], with which information about the two samples in these two chambers can be acquired at two distinct wavelengths with a single spectroscopic measurement. They further employed the use of artificial neural networks (ANNs) to fingerprint chemical identity, concentration, and chirality of chiral compounds [43, 44].

In the past decade, our laboratory has been working on the development of enantioselective fluorescent sensors for chiral organic molecules [45–54]. We have chosen the optically active 1,1′-bi-2-naphthol (BINOL) as the chiral structural unit to construct the sensors (Figure 1). The hindered rotation of the two naphthyl rings of BINOL leads to a stable C_2 symmetric chiral configuration. Functional groups can be selectively introduced to the 2-, 3-, 4-, 5-, and 6-positions of BINOL to build various molecular structures. These functional groups also allow the fluorescence property of the naphthalene rings to be tuned for the desired sensing response. Through this study, we have discovered a series of the BINOL-based enantioselective fluorescent

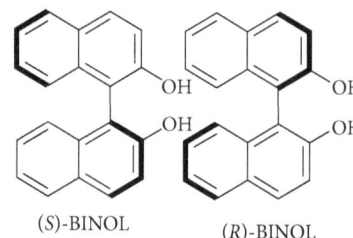

FIGURE 1: (S)-BINOL and (R)-BINOL.

sensors as shown in Figure 2 for the recognition of chiral α-hydroxycarboxylic acids, amino alcohols, amines, and amino acid derivatives with high enantioselectivity [55–58]. These sensors include the generation 0–2 (G0–G2) dendrimers, (S)-1, (S)-2, and (S)-3, the BINOL-terpyridine copper complex (R)-4, the monoamine-linked bisBINOL sensor (S)-5, the bisBINOL-based macrocyclic sensors (S)-6 and (S)-7, and the monoBINOL-based sensors (S)-8 and (S)-9. The highly enantioselective fluorescent responses of these compounds make them useful in determining the enantiomeric composition of various chiral substrates.

As discussed earlier, if both the concentration and enantiomeric composition of a chiral substrate could be determined by using one fluorescent measurement, it would significantly simplify the practical application of the enantioselective fluorescent recognition. We have developed two strategies to achieve this goal of one measurement for two parameters and these strategies are discussed in this paper.

2. Using Pseudoenantiomeric Fluorescent Sensor Pair for the Simultaneous Determination of the Concentration and Enantiomeric Composition of a Chiral Substrate [59]

We have discovered that the BINOL-based amino alcohol (S)-9 is a generally enantioselective fluorescent sensor for structurally diverse α-hydroxycarboxylic acids [58]. As shown in Figure 3, (R)-phenyllactic acid significantly enhances the monomer emission of sensor (S)-9 while (S)-phenyllactic acid quenches it in benzene/0.4% DME solution. The fluorescent intensity ratio I_R/I_S is used to quantify the enantioselectivity, which is as high as 11.2 for phenyllactic acid. Figure 4 summarizes the I_R/I_S ratio when (S)-9 is used to interact with various α-hydroxyl carboxylic acids, including aromatic, aliphatic, and tertiary α-hydroxyl carboxylic acids, under the same conditions and very high enantioselectivity is generally observed for all the tested chiral acids. Therefore, (S)-9 can be used to determine the enantiomeric purity of various types of α-hydroxyl carboxylic acids.

Our ^{1}H NMR spectroscopic study indicates the formation of 1:1 sensor/acid complex. The computational simulation of the 1:1 complex of (S)-9 and phenyllactic acid was performed with the Gaussian 03 program. The proposed structure of the complex is shown in Figure 5. Strong acid-base interaction between the carboxylic acid group and the amine nitrogen

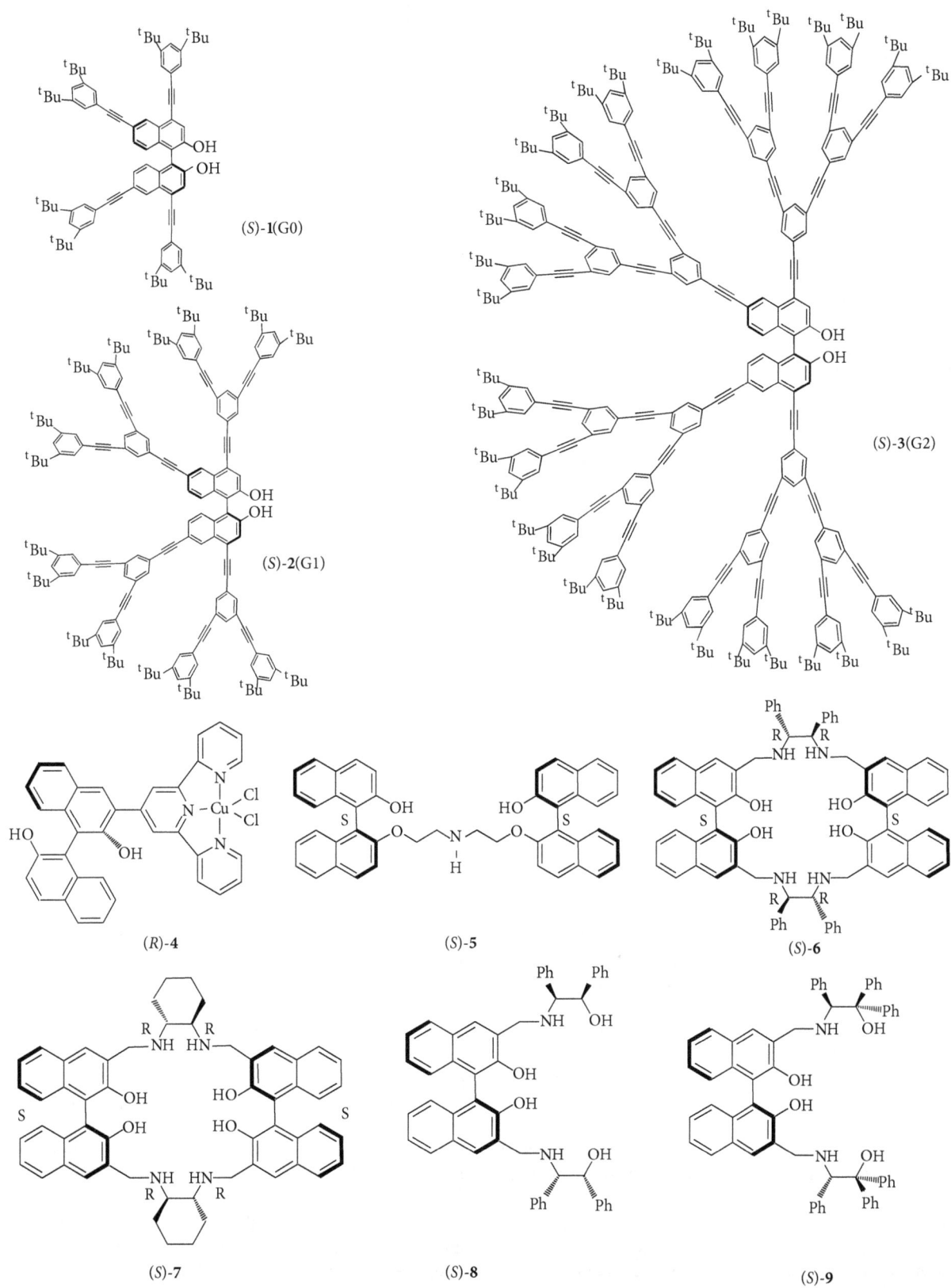

FIGURE 2: BINOL-based enantioselective fluorescent sensors.

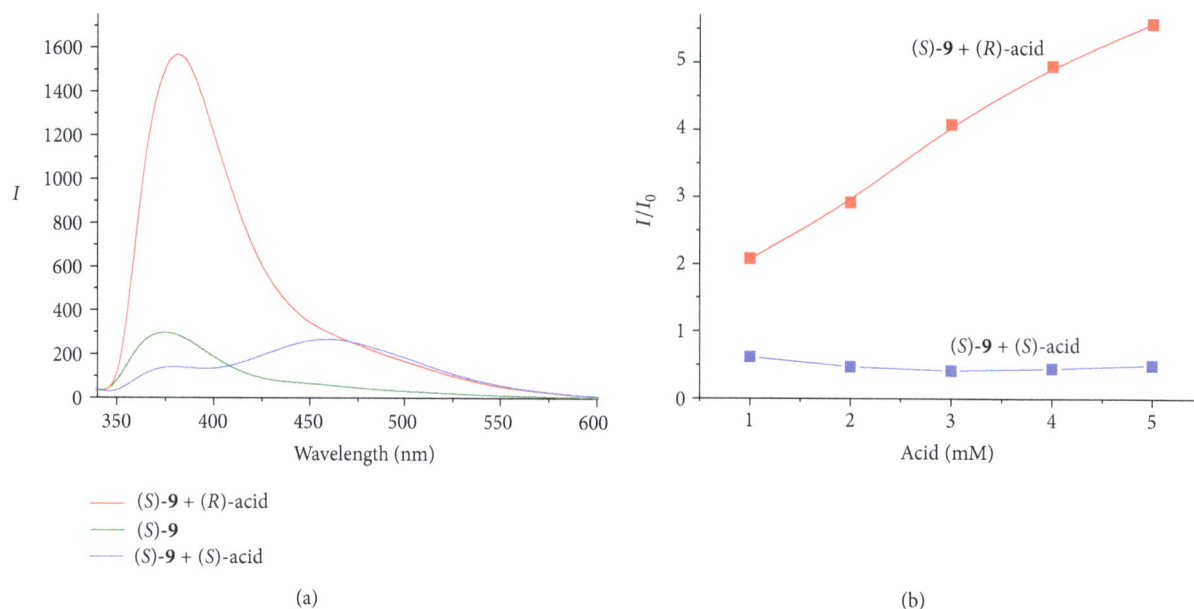

FIGURE 3: (a) Fluorescence spectra of (S)-**9** (2×10^{-4} M, benzene/0.4% v/v DME) with (R)- and (S)-phenyllactic acid (5×10^{-3} M). (b) Fluorescence enhancement of (S)-**9** with varying concentrations of (R)- and (S)-phenyllactic acid ($\lambda = 334$ nm, slit = 5.0/5.0 nm) (permission was obtained from Wiley to reproduce this plot).

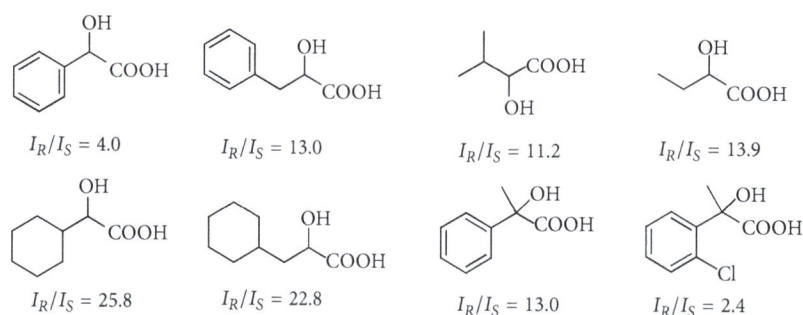

FIGURE 4: Fluorescent enantioselectivity of (S)-**9** toward various chiral α-hydroxycarboxylic acids.

of (S)-**9** exists in the complex. The carbonyl oxygen of phenyllactic acid forms hydrogen bonding with a hydroxyl group of (S)-**9**. The α-hydroxyl group of the acid is also hydrogen-bonded with both hydroxyl groups of the amino alcohol units in (S)-**9**.

The highly enantioselective fluorescent responses of (S)-**9** toward the α-hydroxycarboxylic acids allow the use of this molecule to determine the enantiomeric composition of the substrates at a given concentration. In order to simultaneously determine both the concentration and enantiomeric composition of a chiral acid, we have proposed a novel strategy by developing a pseudoenantiomeric sensor pair. A pseudoenantiomeric sensor pair contains two sensors with the opposite enantioselectivity at distinct emitting wavelengths λ_1 and λ_2. It is our hypothesis that when such a pseudoenantiomeric sensor pair is used to interact with a chiral substrate, the fluorescent intensity difference $I_1 - I_2$ at the two emitting wavelengths could be correlated with the enantiomeric composition and the fluorescent intensity sum $I_1 + I_2$ could be correlated with the concentration of the chiral

substrate. Thus, both the concentration and enantiomeric composition could be determined simultaneously by one fluorescent measurement.

In order to develop the pseudoenantiomer of (S)-**9**, compound (R)-**10** was synthesized according to Scheme 1. The partially hydrogenated BINOL, H_8BINOL, was used as the starting material. Compound (R)-**10** contains less extended conjugation and is thus expected to emit at shorter wavelength than the BINOL-based sensor (S)-**9**. These two compounds have the opposite configurations at the axially chiral biaryl unit and the chiral amino carbons, which should give them the opposite enantioselectivity in a chiral recognition experiment.

Both (S)-**9** and (R)-**10** were used to interact with (R)- and (S)-mandelic acid (MA) in dichloromethane (DCM). The benzene/DME solvent system initially reported for the use of (S)-**9** is not suitable for this pseudoenantiomeric sensor pair because of the interference of benzene with the fluorescence of (R)-**10** as the conjugation of (R)-**10** is reduced. We found that changing the solvent from benzene to DCM did not

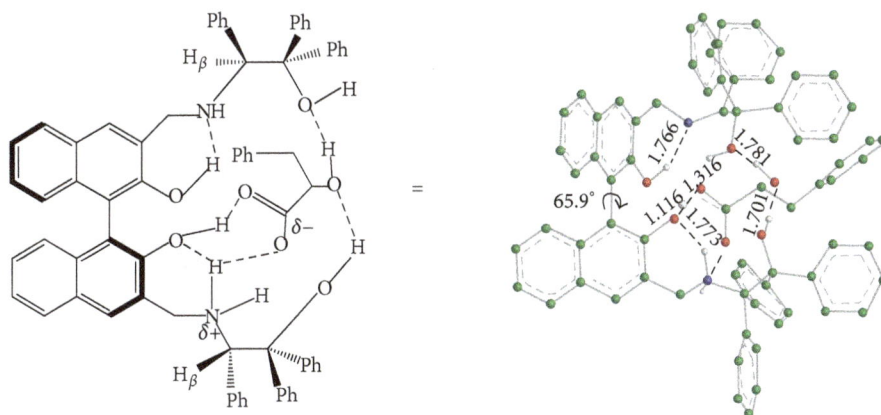

FIGURE 5: Proposed structure of the 1:1 complex of (S)-**9** + (R)-phenyllactic acid (permission was obtained from Wiley to reproduce this plot).

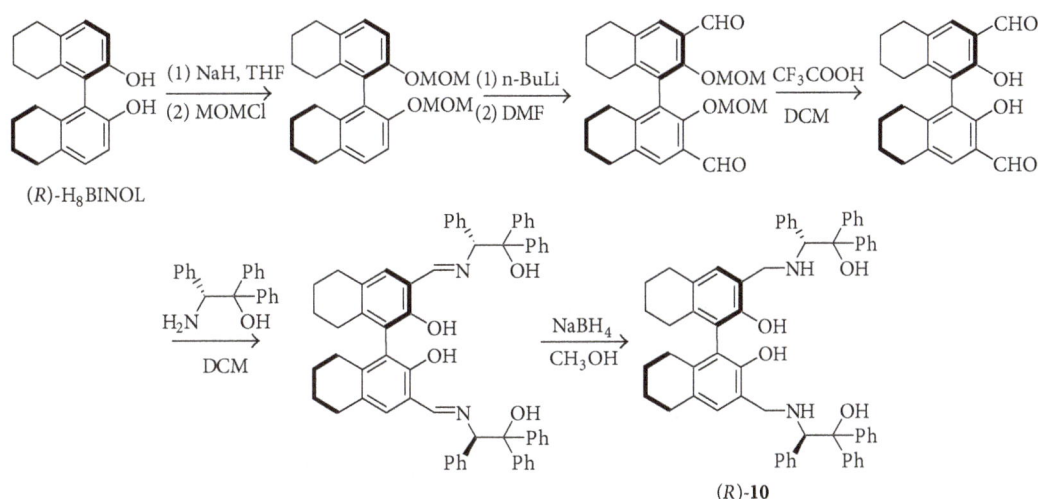

SCHEME 1: Synthesis of the H$_8$BINOL-amino alcohol (R)-**10**.

impair the high enantioselectivity of (S)-**9**. As shown in Figure 6, treatment with (R)-MA significantly enhanced the fluorescent intensity of (S)-**9** at 374 nm (λ_1), whereas (S)-MA only slightly increased its fluorescence. I_R/I_0 is found to be 11.4 and the enantioselective fluorescent enhancement ratio [ef = $(I_R - I_0)/(I_S - I_0)$] is 26.0. (R)-**10** also exhibited high but opposite enantioselectivity for the recognition of MA. As shown in Figure 7, (S)-MA enhanced the fluorescence of (R)-**10** at 330 nm (λ_2) to a much greater extent than (R)-MA did. It was found that $I_S/I_0 = 11.7$ and ef = 3.6.

The above experiments demonstrate that (S)-**9** and (R)-**10** have high and opposite enantioselectivity at two distinct wavelengths ($\lambda_1 = 374$ nm, $\lambda_2 = 330$ nm) for the recognition of MA, which makes them excellent candidates for a pseudoenantiomeric sensor pair. A 1:1 mixture of (S)-**9** and (R)-**10** in DCM was used to interact with MA of varying concentrations and enantiomeric compositions. As we proposed, the difference of the fluorescence intensities at λ_1 and λ_2

could be utilized to measure the enantiomeric composition of MA and the sum could measure the total concentration. In Figure 8(a), the fluorescent intensity difference at λ_1 and λ_2 ($I_1/I_{10} - I_2/I_{20}$, I_1: the fluorescence intensity at $\lambda_1 = 374$ nm in the presence of MA, I_{10}: the fluorescence intensity at $\lambda_1 = 374$ nm in the absence of MA, I_2: the fluorescence intensity at $\lambda_2 = 330$ nm in the presence of MA, and I_{20}: the fluorescence intensity at $\lambda_2 = 330$ nm in the absence of MA) increases with increasing (R)-MA% at each total concentration. In Figure 8(b), the fluorescent intensity sum ($I_1/I_{10} + I_2/I_{20}$) increases with increasing MA concentration at each enantiomeric composition.

The 3D graphs of the total acid concentration and the enantiomeric composition versus the sum and the difference of fluorescent intensities at λ_1 and λ_2 were plotted in Figure 9 on the basis of the data in Figure 8. One fluorescent measurement will give the fluorescent intensities I_1 and I_2 which will be used to determine both the concentration

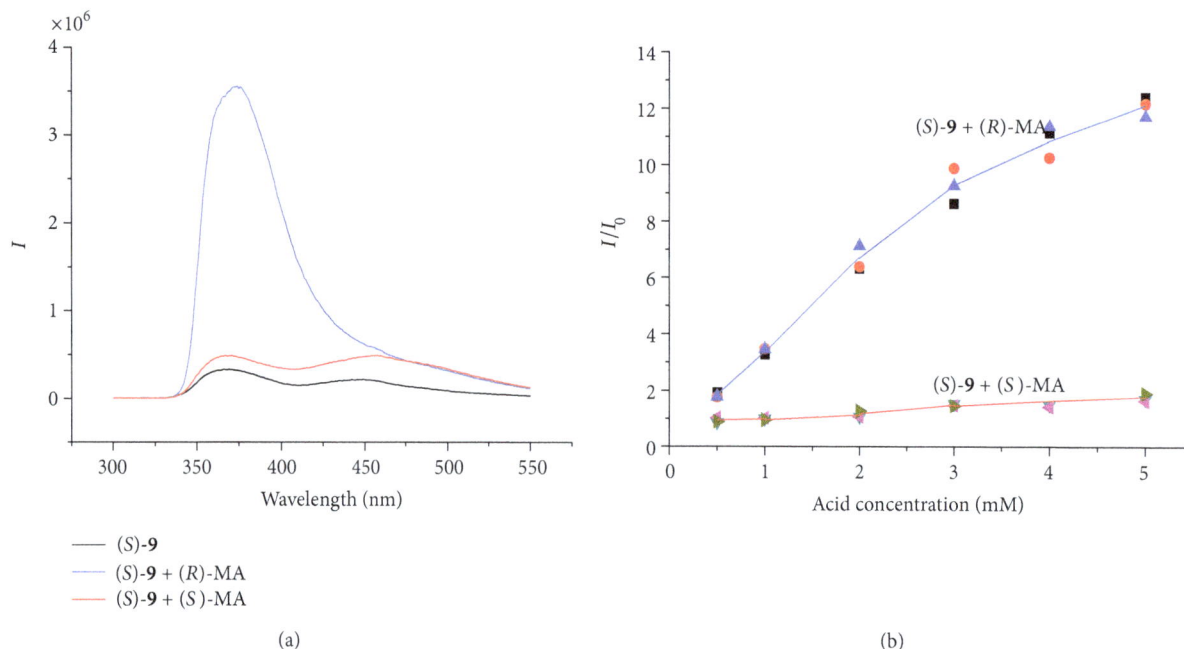

FIGURE 6: (a) Fluorescence spectra of (S)-**9** $(1.0 \times 10^{-4}$ M, CH_2Cl_2) with/without MA $(4.0 \times 10^{-3}$ M). (b) Three independent measurements for the fluorescence enhancement of (S)-**9** $(1.0 \times 10^{-4}$ M, CH_2Cl_2) at $\lambda_1 = 374$ nm with varying MA concentration $(\lambda_{exc} = 290$ nm, slit = 4.0/4.0 nm). Reprinted with permission from [59]. Copyright [2010] American Chemical Society.

and enantiomeric composition of MA according to Figure 9. Thus, the pseudoenantiomeric sensor pair strategy allows one measurement for the two parameters of a chiral compound.

3. Using One Fluorescent Sensor to Determine Both Concentration and Enantiomeric Composition in One Fluorescence Measurement

As described in the above section, the pseudoenantiomeric sensor pair strategy is successfully used to simultaneously determine the concentration and enantiomeric composition of a chiral substrate. This strategy requires the use of a mixture of two fluorescent sensors, a pseudoenantiomeric pair. Prompted by this work, we propose another strategy to measure both the concentration and enantiomeric composition by using only one fluorescent sensor. That is, a fluorescent sensor that shows different fluorescent responses at two emitting wavelengths toward the two enantiomers of a chiral substrate will be developed. Such a dual emission sensor could be used to simultaneously measure the total concentration of the two enantiomers as well as their relative concentration (enantiomeric composition) [60, 61].

We found that the BINOL-based trifluoromethyl ketone molecule (S)-**11** could serve as such a dual emission sensor. Scheme 2 depicts the synthesis of (S)-**11**. This compound was nonemissive at all in methylene chloride solution. The ^{1}H NMR spectrum of (S)-**11** indicates the existence of intramolecular OH···O=C hydrogen bonds. Treatment of this compound with both enantiomers of *trans*-1,2-diaminocyclohexane, (R,R)- and (S,S)-**12**, turned on the

fluorescence at $\lambda_1 = 370$ nm and $\lambda_2 = 384$ nm (Figure 10). At λ_1, both (R,R)- and (S,S)-**12** enhanced the fluorescence of (S)-**11** to a similar extent while at λ_2, (R,R)-**12** enhanced the fluorescence much greater than (S,S)-**12**. Thus, the two emitting wavelengths of (S)-**11** responded to the enantiomers of the diamine differently with high fluorescent sensitivity at λ_1 and high enantioselectivity at λ_2.

The effect of the concentration of the chiral diamine **12** on the fluorescent responses of (S)-**11** at λ_1 and λ_2 is shown in Figure 11. It demonstrates that the fluorescent intensity I_1 is strongly dependent on the concentration of the diamine but not significantly on its chiral configuration. The fluorescent intensity ratio I_1/I_2 remains constant at 2.6 for (R,R)-**12** and at 0.67 for (S,S)-**12** in the concentration range of 5.0×10^{-4} M to 5.0×10^{-3} M. This shows that I_1/I_2 is strongly dependent on the chiral configuration of the chiral diamine but independent of the concentration. Therefore, I_1 mostly responds to the concentration of the chiral diamine and I_1/I_2 only responds to the chiral configuration. Another example of chiral diamine, 1,2-diaminopropane, was also tested and similar fluorescent responses at λ_1 and λ_2 were observed.

(S)-**11** $(1.0 \times 10^{-5}$ M in CH_2Cl_2) was used to interact with varying concentrations and enantiomeric compositions of the chiral diamine **12**. Figure 12 plots the fluorescent intensity ratio I_1/I_2 versus (S,S)-**12**% at various diamine concentrations (0.5–5 mM). It demonstrates that the enantiomeric composition of the chiral diamine **12** can be determined by the fluorescent intensity ratio I_1/I_2 without the need to know the total concentration. Figure 13 plots the total concentration of the chiral diamine **12** versus I_1 and I_1/I_2. Since the chiral configuration of the diamine **12** had a small effect on

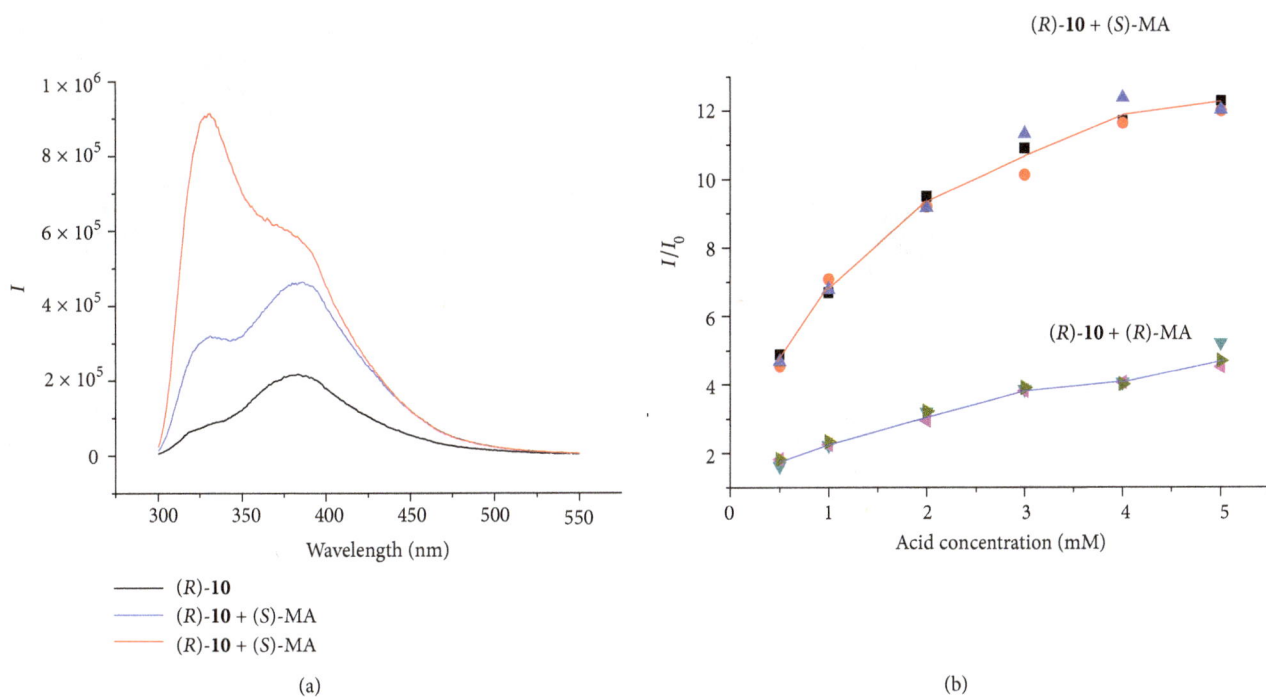

FIGURE 7: (a) Fluorescence spectra of (R)-**10** (1.0×10^{-4} M, CH_2Cl_2) with/without (R)- and (S)-MA (4.0×10^{-3} M). (b) Three independent measurements for the fluorescence enhancement of (R)-**10** (1.0×10^{-4} M, CH_2Cl_2) at $\lambda_2 = 330$ nm with varying MA concentration ($\lambda_{exc} = 290$ nm, slit = 4.0/4.0 nm). Reprinted with permission from [59]. Copyright [2010] American Chemical Society.

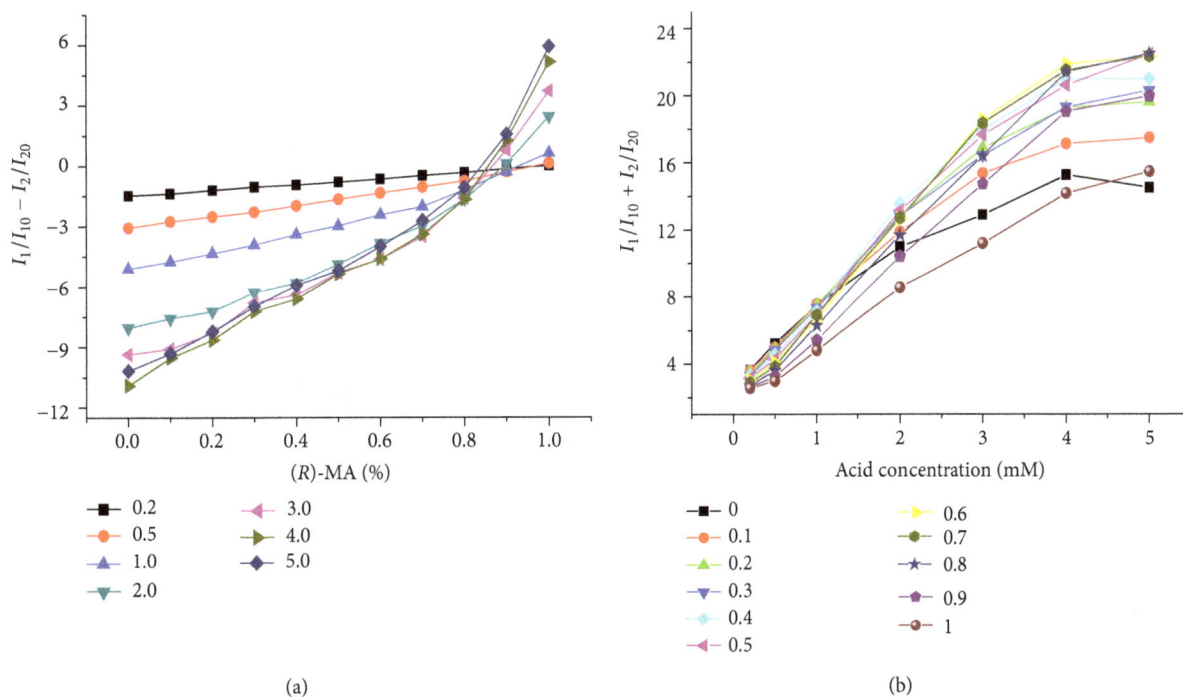

FIGURE 8: (a) Plot of $(I_1/I_{10} - I_2/I_{20})$ versus $[(R)$-MA]% at varying MA concentrations (mM). (b) Plot of $(I_1/I_{10} + I_2/I_{20})$ versus MA concentration at varying $[(R)$-MA]% ($\lambda_{exc} = 290$ nm, slit = 4.0/4.0 nm). Reprinted with permission from [59]. Copyright [2010] American Chemical Society.

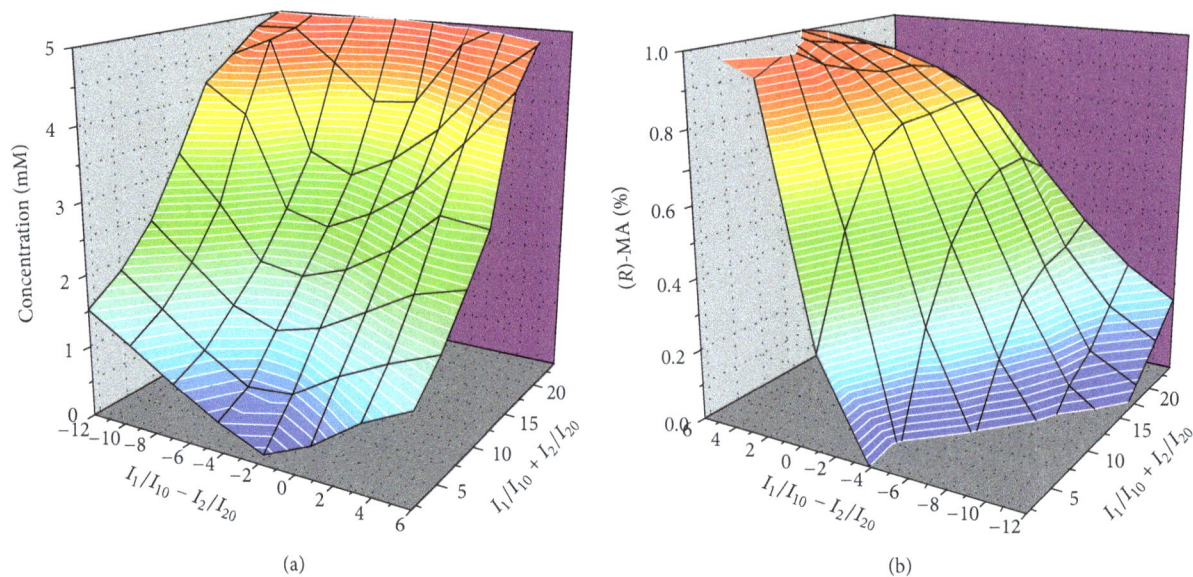

(a)   (b)

FIGURE 9: (a) 3D plots of $(I_1/I_{10} - I_2/I_{20})$ and $(I_1/I_{10} + I_2/I_{20})$ with the MA concentration (mM). (b) 3D plots of $(I_1/I_{10} - I_2/I_{20})$ and $(I_1/I_{10} + I_2/I_{20})$ with $[(R)\text{-MA}]\%$. Reprinted with permission from [59]. Copyright [2010] American Chemical Society.

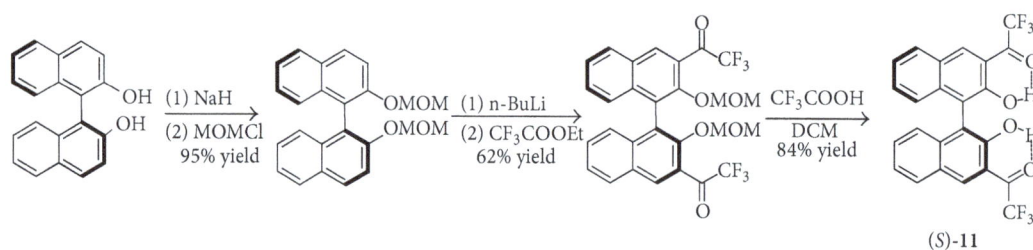

SCHEME 2: Preparation of compound (S)-**11**.

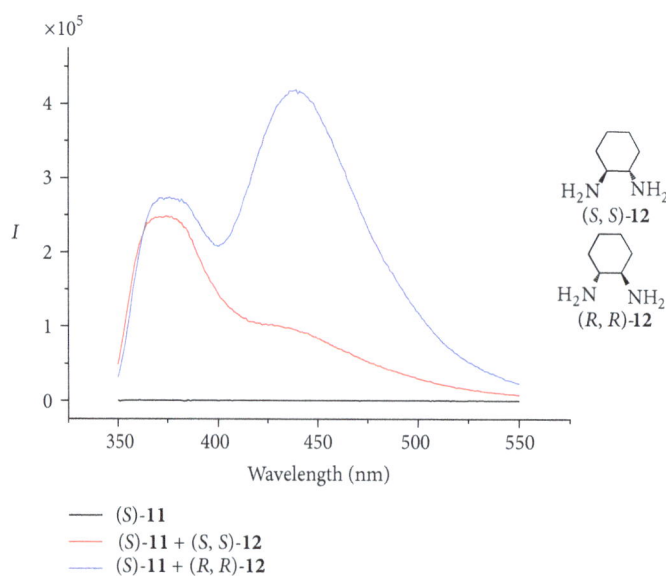

FIGURE 10: Fluorescence spectra of (S)-**11** $(1.0 \times 10^{-5}\,M)$ with/without (R,R)- and (S,S)-**12** $(5.0 \times 10^{-3}\,M)$ (solvent: CH_2Cl_2, $\lambda_{exc} = 343\,nm$, slit = 2/2 nm). Reprinted with permission from [60]. Copyright [2012] American Chemical Society.

(a)

(b)

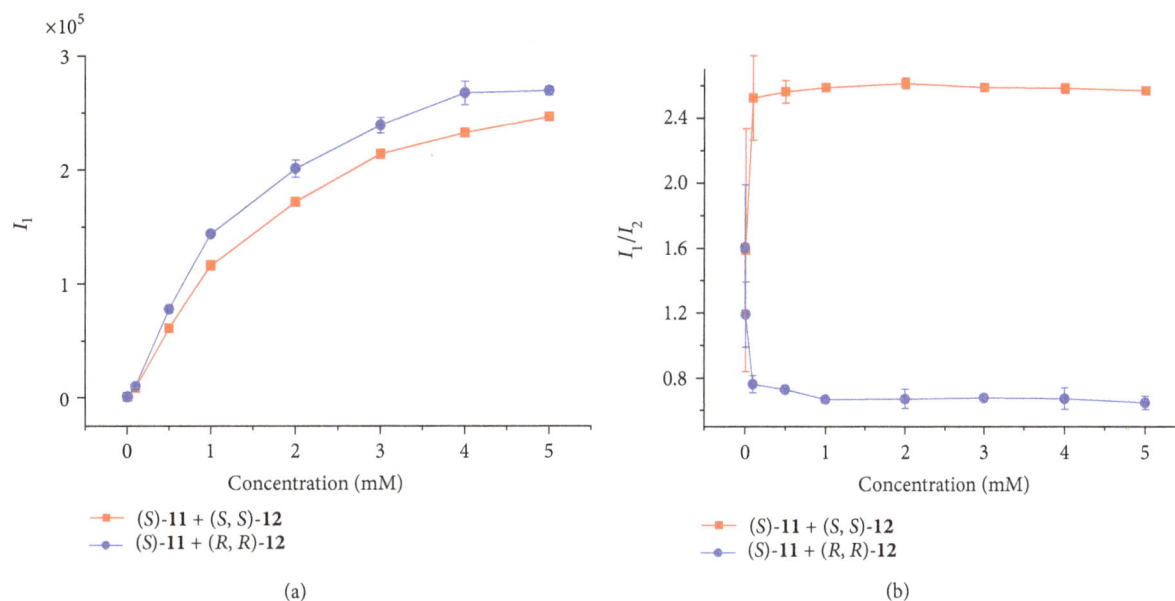

FIGURE 11: Plots of I_1 (a), I_1/I_2 (b) for (S)-**11** (1.0×10^{-5} M) in the presence of varying concentrations of (R,R)- and (S,S)-**12** (fluorescence intensity I_1 at $\lambda_1 = 370$ nm and I_2 at $\lambda_2 = 438$ nm, solvent: CH_2Cl_2, $\lambda_{exc} = 343$ nm, slit = 2/2 nm). Reprinted with permission from [60]. Copyright [2012] American Chemical Society.

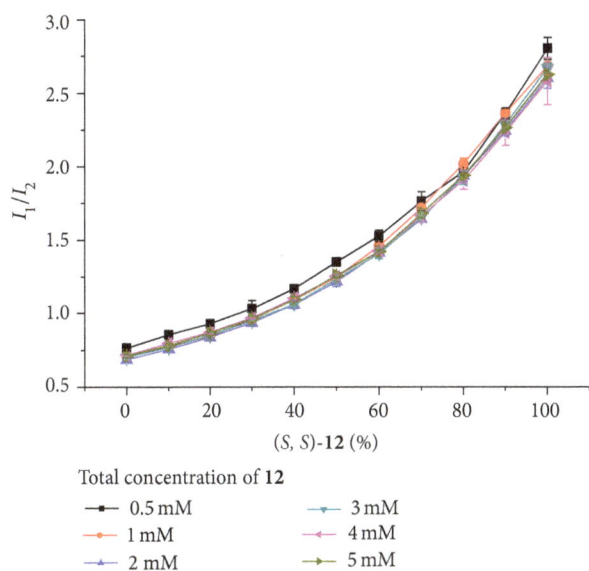

FIGURE 12: Plots of I_1/I_2 versus (S,S)-**12**% at various diamine concentrations (mM) (solvent: CH_2Cl_2, $\lambda_{exc} = 343$ nm, slit = 2/2 nm). Reprinted with permission from [60]. Copyright [2012] American Chemical Society.

FIGURE 13: Plot of I_1, I_1/I_2 versus the total concentration of **12** with various enantiomeric composition. Reprinted with permission from [60]. Copyright [2012] American Chemical Society.

I_1, I_1 was used together with I_1/I_2 to determine the total concentration of the chiral diamine **12**. Therefore, with the use of only one fluorescent sensor, both the concentration and enantiomeric composition of a chiral diamine can be simultaneously determined by one fluorescent measurement.

On the basis of the ^{19}F NMR titration experiment for the interaction of (S)-**11** with (S,S)-**12**, the reaction shown in Scheme 3 was proposed. The nucleophilic addition of (S,S)-**12**

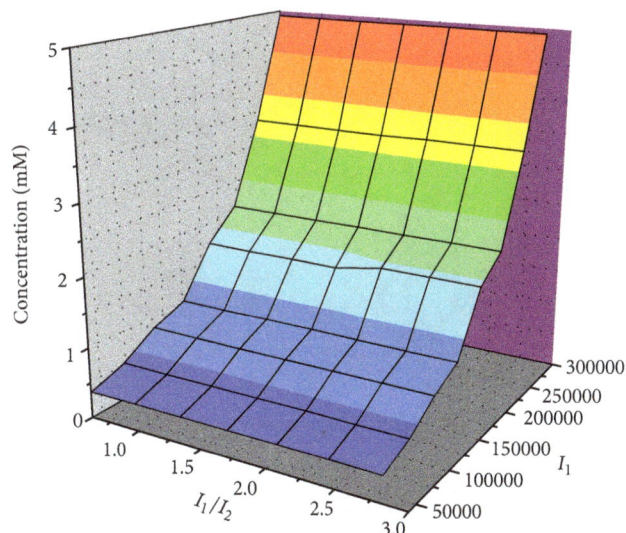

to the trifluoroacetyl group of (S)-**11** occurs instantaneously to produce the hemiaminals **13** and **14**; but the formation of the condensation product imine **15** and the subsequent cycloaddition product aminal **16** are slow and take a few days to complete. Since the fluorescent recognition experiments were generally conducted within 2 h after preparation, the observed large fluorescent enhancement of (S)-**11** in the presence of the chiral diamine is attributed to the formation of the hemiaminals **13** and **14**. The final product aminal **16** was isolated from the reaction mixture of (S)-**11** and (S,S)-**12** and its structure was confirmed by X-ray analysis (Figure 14).

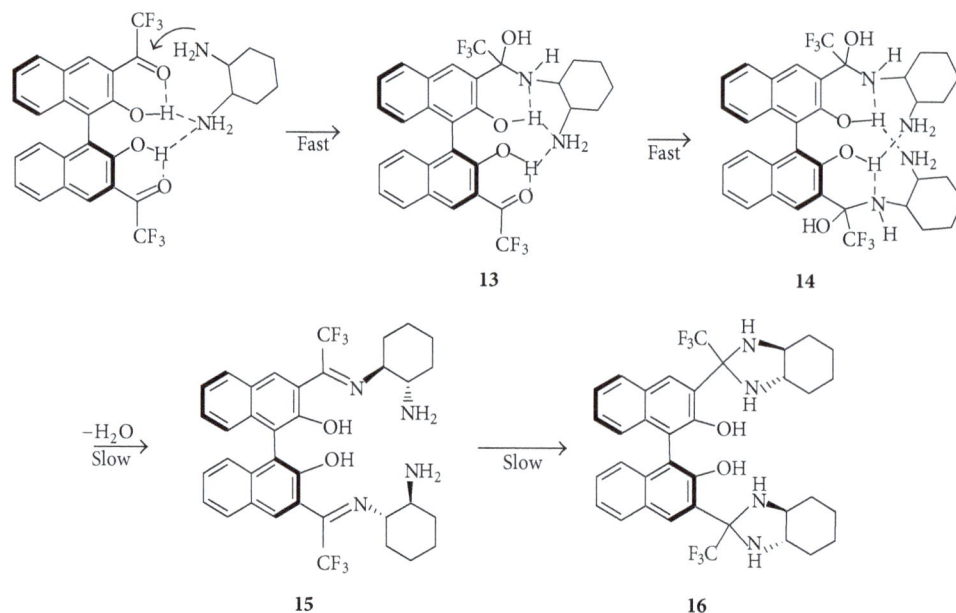

SCHEME 3: A proposed mechanism for the reaction of (S)-**11** with the chiral diamine **12**.

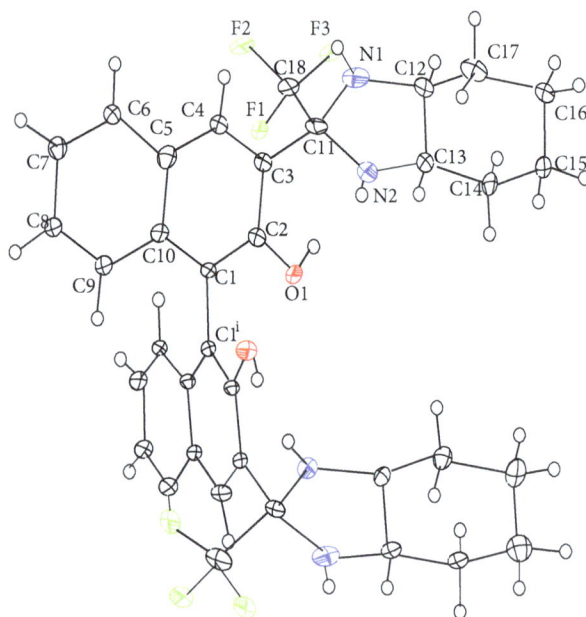

FIGURE 14: X-ray structure of the complexes of (S)-**11** with (S,S)-**12**. Reprinted with permission from [61]. Copyright [2013] American Chemical Society.

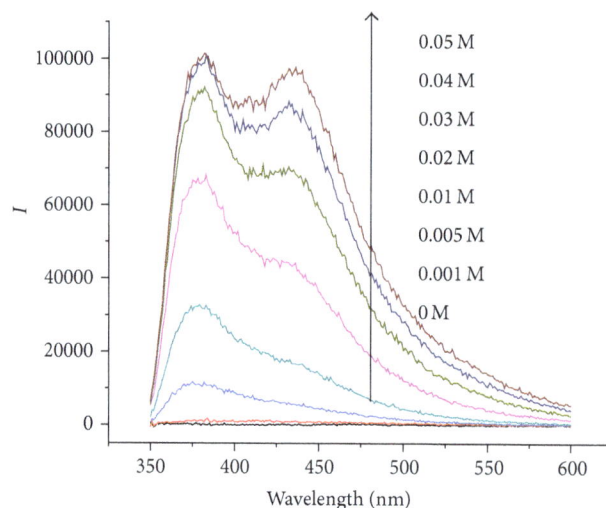

FIGURE 15: Fluorescence spectra of (S)-**11** (1.0×10^{-5} M in CH_2Cl_2) in the presence of propylamine at 0–0.05 M (λ_{exc} = 343 nm, slit = 2.0/2.0 nm). Reprinted with permission from [61]. Copyright [2013] American Chemical Society.

Although (S)-**11** is nonemissive in methylene chloride solution, its precursor with two-MOM protecting group is highly fluorescent. This suggests that the intramolecular OH$\cdots$O=C hydrogen bonding of (S)-**11** should be responsible for its diminished fluorescence. When (S)-**11** is used to interact with the chiral diamine **12**, the intermolecular hydrogen bond between one of the amine groups with the hydroxyl groups of (S)-**11** accelerates the addition of the second amine group to the trifluoroacetyl group of (S)-**11**,

producing the resulting hemiaminal products **13** and **14**, in which the original O-H$\cdots$O=C hydrogen bonds in (S)-**11** have been disrupted to generate the observed dual emissions. The short-wavelength emission is ascribed to the nucleophilic addition of one amine group to the carbonyl group and the long-wavelength emission is ascribed to hydrogen-bonding interaction of the second amine group with the hydroxyl groups of the sensor.

The proposed interaction of (S)-**11** with the diamine is supported by the fluorescent responses of (S)-**11** toward propylamine as shown in Figure 15. Much higher concentration of the monoamine than the diamine was required to turn

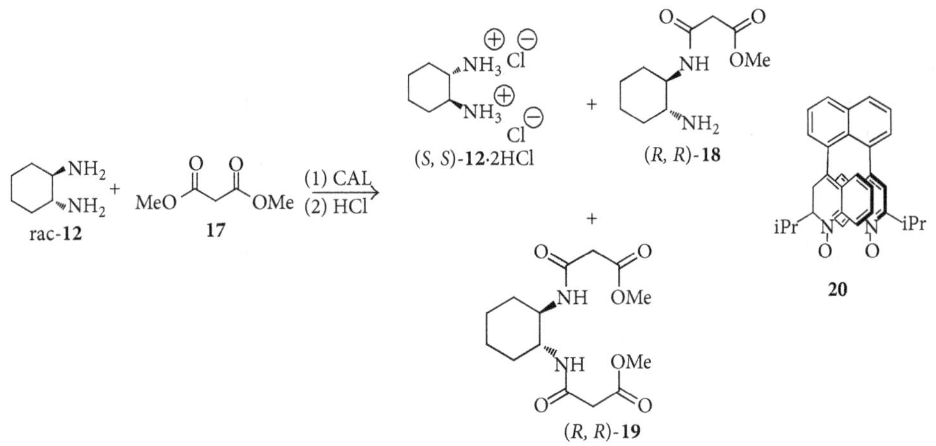

SCHEME 4: Enzymatic kinetic resolution of racemic 1,2-diamine **12**.

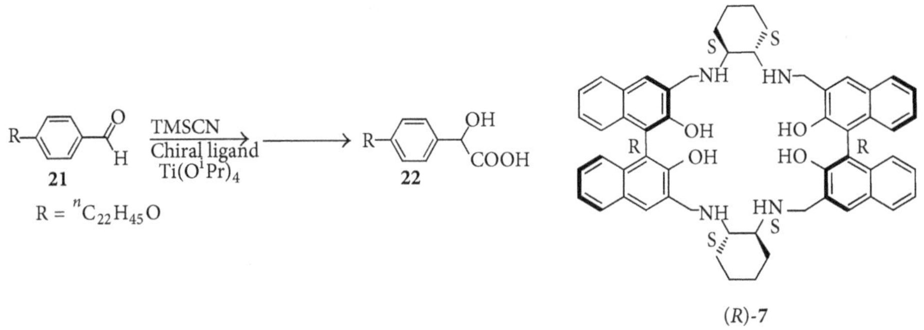

SCHEME 5: Asymmetric reaction of **21** to generate **22**.

on the fluorescence of (*S*)-**11**. The fluorescent enhancement first occurred at the short wavelength. Only at even higher concentrations of propylamine, was there more fluorescent enhancement at the long wavelength emission. These observations suggest that the short wavelength emission of (*S*)-**11** might be due to the addition of propylamine to the trifluoroacetyl group. The resulting product upon further interaction with propylamine probably via hydrogen bonding with the naphthyl hydroxyl groups could lead to the long wavelength emission.

4. Summary and Outlook

In this paper, we have described two strategies to simultaneously determine the concentration and enantiomeric composition of a chiral compound by one fluorescent measurement. One strategy uses a pseudoenantiomeric fluorescent sensor pair in which each sensor shows greater fluorescent enhancement at a different wavelength upon interaction with one of the enantiomers of a chiral substrate. In another strategy, a fluorescent sensor responds differently toward the two enantiomers of a chiral compound at two emitting wavelengths. These two strategies could significantly simplify the practical application of the enantioselective fluorescent sensors.

One of the potential applications of an enantioselective fluorescent sensor is in high-throughput chiral catalyst screening for asymmetric catalysis. For example, Tumambac and Wolf reported the use of enantioselective fluorescent sensing in the enzymatic kinetic resolution of *trans*-1,2-diaminocyclohexane (Scheme 4) [62]. After the reaction, the diamine **12** and the monoaminoester **18** could be isolated through precipitation with 2 N HCl followed by basic extraction. Then the enantioselective fluorescent sensor **20** was used to determine the enantiomeric composition of **12**. Our group also developed a soluble "supported" chiral acid system for the chiral catalyst screening (Scheme 5) [57]. The aldehyde **21** was transformed to the chiral α-hydroxy acid **22** by asymmetric reaction with TMSCN in the presence of a chiral catalyst followed by hydrolysis. The acid **22** containing a 22-carbon chain alkyl group was found to be almost insoluble in most of the organic solvents but with good solubility in THF. Therefore, the product could be precipitated out in the absence of THF and all the catalysts and reagents could be removed. Then the enantiomeric composition was determined in the homogeneous THF solution with the use of the enantioselective fluorescent sensor (*R*)-**7**.

The examples described above are the only reports on the direct application of enantioselective fluorescent sensors in chiral catalyst screening for asymmetric reaction. Although recent years have seen significant development in enantioselective fluorescent sensing, there are still significant challenges for the use of enantioselective fluorescent sensors in the analysis of asymmetric reactions. Most of the studies were conducted only in the recognition of isolated pure substrate samples. In the actual reaction mixture, many substances such as catalysts, additives, byproducts, and solvents could potentially interfere with the fluorescent recognition of the chiral products and add uncertainty to the analysis. Another challenge is to expand the substrate scope of the enantioselective fluorescent sensors. In addition to the highly functionalized substrates such as α-hydroxycarboxylic acids, diamines, amino alcohols, and amino acids, enantioselective fluorescent sensors that can recognize chiral molecules with less strongly interacting groups such as alcohols, ethers, esters, or even molecules without a polar functional group are also needed. We believe that with the great effort and creativity of the researchers in this area it should be possible to meet all these challenges and allow the enantioselective fluorescent sensing to be developed into a practically useful analytical tool in chiral assay.

Conflict of Interests

The authors declare that there is no conflict of interests regarding the publication of this paper.

References

[1] E. L. Izake, "Improving memory performance in the aged through mnemonic training: a meta-analytic study," *Psychology and Aging*, vol. 96, no. 7, pp. 1659–1676, 2007.

[2] M. C. Núñez, M. E. García-Rubiño, A. Conejo-García et al., "Homochiral drugs: a demanding tendency of the pharmaceutical industry," *Current Medicinal Chemistry*, vol. 16, no. 16, pp. 2064–2074, 2009.

[3] R. Natarajan and S. C. Basak, "Numerical descriptors for the characterization of chiral compounds and their applications in modeling biological and toxicological activities," *Current Topics in Medicinal Chemistry*, vol. 11, no. 7, pp. 771–787, 2011.

[4] M. Herrero, C. Simó, V. García-Cañas, S. Fanali, and A. Cifuentes, "Chiral capillary electrophoresis in food analysis," *Electrophoresis*, vol. 31, no. 13, pp. 2106–2114, 2010.

[5] C. Simó, C. Barbas, and A. Cifuentes, "Chiral electromigration methods in food analysis," *Electrophoresis*, vol. 24, no. 15, pp. 2431–2441, 2003.

[6] "FDA'S policy statement for the development of new stereoisomeric drugs," *Chirality*, vol. 4, no. 5, pp. 338–340, 1992.

[7] M. Christmann and S. Brase, *Asymmetric Synthesis: The Essentials*, John Wiley & Sons, New York, NY, USA, 2007.

[8] G.-Q. Lin, Y.-M. Li, and A. S. C. Chan, *Principles and Applications of Asymmetric Synthesis*, John Wiley & Sons, New York, NY, USA, 2001.

[9] W. F. Maier, K. Stowe, and S. Sieg, "Combinatorial and high-throughput materials science," *Angewandte Chemie*, vol. 46, no. 32, pp. 6016–6067, 2007.

[10] P. T. Corbett, J. Leclaire, L. Vial et al., "Dynamic combinatorial chemistry," *Chemical Reviews*, vol. 106, no. 9, pp. 3652–3711, 2006.

[11] S. Senkan, "Combinatorial heterogeneous catalysis—a new path in an old field," *Angewandte Chemie-International Edition*, vol. 40, no. 2, pp. 312–329, 2001.

[12] B. Jandeleit, D. J. Schaefer, T. S. Powers, H. W. Turner, and W. H. Weinberg, "Combinatorial materials science and catalysis," *Angewandte Chemie*, vol. 38, no. 17, pp. 2494–2532, 1999.

[13] M. T. Reetz, "Combinatorial and evolution-based methods in the creation of enantioselective catalysts," *Angewandte Chemie-International Edition*, vol. 40, no. 2, pp. 284–310, 2001.

[14] M. Tsukamoto and H. B. Kagan, "Recent advances in the measurement of enantiomeric excesses," *Advanced Synthesis & Catalysis*, vol. 344, no. 5, pp. 453–463, 2002.

[15] J. F. Traverse and M. L. Snapper, "High-throughput methods for the development of new catalytic asymmetric reactions," *Drug Discovery Today*, vol. 7, no. 19, pp. 1002–1012, 2002.

[16] M. G. Finn, "Emerging methods for the rapid determination of enantiomeric excess," *Chirality*, vol. 14, no. 7, pp. 534–540, 2002.

[17] D. Leung, S. O. Kang, and E. V. Anslyn, "Rapid determination of enantiomeric excess: a focus on optical approaches," *Chemical Society Reviews*, vol. 41, no. 1, pp. 448–479, 2012.

[18] D. G. I. Petra, J. N. H. Reek, P. C. J. Kamer, H. E. Schoemaker, and P. W. N. M. van Leeuwen, "IR spectroscopy as a high-throughput screening-technique for enantioselective hydrogen-transfer catalysts," *Chemical Communications*, no. 8, pp. 683–684, 2000.

[19] M. T. Reetz, M. H. Becker, K. M. Kuhling, and A. Holzwarth, "Time-resolved IR-thermographic detection and screening of enantioselectivity in catalytic reactions," *Angewandte Chemie—International Edition*, vol. 37, no. 19, pp. 2647–2650, 1998.

[20] P. Tielmann, M. Boese, M. Luft, and M. T. Reetz, "A practical high-throughput screening system for enantioselectivity by using FTIR spectroscopy," *Chemistry*, vol. 9, no. 16, pp. 3882–3887, 2003.

[21] N. Millot, P. Borman, M. S. Anson, I. B. Campbell, S. J. F. Macdonald, and M. Mahmoudian, "Rapid determination of enantiomeric excess using infrared thermography," *Organic Process Research and Development*, vol. 6, no. 4, pp. 463–470, 2002.

[22] P. Chen, "Electrospray ionization tandem mass spectrometry in high-throughput screening of homogeneous catalysts," *Angewandte Chemie-International Edition*, vol. 42, no. 25, pp. 2832–2847, 2003.

[23] H. Liu, C. Felten, Q. Xue et al., "Development of multichannel devices with an array of electrospray tips for high-throughput mass spectrometry," *Analytical Chemistry*, vol. 72, no. 14, pp. 3303–3310, 2000.

[24] J. H. Guo, J. Y. Wu, G. Siuzdak, and M. G. Finn, "Measurement of enantiomeric excess by kinetic resolution and mass spectrometry," *Angewandte Chemie-International Edition*, vol. 38, no. 12, pp. 1755–1758, 1999.

[25] W. Schrader, A. Eipper, D. Jonathan Pugh, and M. T. Reetz, "Second-generation MS-based high-throughput screening system for enantioselective catalysts and biocatalysts," *Canadian Journal of Chemistry*, vol. 80, no. 6, pp. 626–632, 2002.

[26] M. T. Reetz, M. H. Becker, H. W. Klein, and D. Stockigt, "A method for high-throughput screening of enantioselective catalysts," *Angewandte Chemie—International Edition*, vol. 38, no. 12, pp. 1758–1761, 1999.

[27] M. T. Reetz, K. M. Kuhling, A. Deege, H. Hinrichs, and D. Belder, *Super-high-throughput screening of enantioselective catalysts by using capillary array electrophoresis*, vol. 39, no. 21, pp. 3891–3893, 2000.

[28] M. T. Reetz, A. Zonta, K. Schimossek, K. Liebeton, and K. Jaeger, "Creation of enantioselective biocatalysts for organic chemistry

by in vitro evolution," *Angewandte Chemie*, vol. 36, no. 24, pp. 2830–2832, 1997.

[29] D. Leung and E. V. Anslyn, "Transitioning enantioselective indicator displacement assays for α-amino acids to protocols amenable to high-throughput screening," *Journal of the American Chemical Society*, vol. 130, no. 37, pp. 12328–12333, 2008.

[30] D. Leung, J. F. Folmer-Andersen, V. M. Lynch, and E. V. Anslyn, "Using enantioselective indicator displacement assays to determine the enantiomeric excess of α-amino acids," *Journal of the American Chemical Society*, vol. 130, no. 37, pp. 12318–12327, 2008.

[31] K. Ding, A. Ishii, and K. Mikami, "Super high throughput screening (SHTS) of chiral ligands and activators: asymmetric activation of chiral diol-zinc catalysts by chiral nitrogen activators for the enantioselective addition of diethylzinc to aldehydes," *Angewandte Chemie—International Edition*, vol. 38, no. 4, pp. 497–501, 1999.

[32] S. Nieto, V. M. Lynch, E. V. Anslyn, H. Kim, and J. Chin, "High-throughput screening of identity, enantiomeric excess, and concentration using MLCT transitions in CD spectroscopy," *Journal of the American Chemical Society*, vol. 130, no. 29, pp. 9232–9233, 2008.

[33] G. T. Copeland and S. J. Miller, "A chemosensor-based approach to catalyst discovery in solution and on solid support," *Journal of the American Chemical Society*, vol. 121, no. 17, pp. 4306–4307, 1999.

[34] G. A. Korbel, G. Lalic, and M. D. Shair, "Reaction microarrays: a method for rapidly determining the enantiomeric excess of thousands of samples," *Journal of the American Chemical Society*, vol. 123, no. 2, pp. 361–362, 2001.

[35] X. Mei and C. Wolf, "Determination of enantiomeric excess and concentration of unprotected amino acids, amines, amino alcohols, and carboxylic acids by competitive binding assays with a chiral scandium complex," *Journal of the American Chemical Society*, vol. 128, no. 41, pp. 13326–13327, 2006.

[36] X. Mei and C. Wolf, "Determination of enantiomeric excess and concentration of chiral compounds using a 1,8-diheteroarylnaphthalene-derived fluorosensor," *Tetrahedron Letters*, vol. 47, no. 45, pp. 7901–7904, 2006.

[37] S. Liu, J. P. C. Pestano, and C. Wolf, "Enantioselective fluorescence sensing of chiral α-amino alcohols," *Journal of Organic Chemistry*, vol. 73, no. 11, pp. 4267–4270, 2008.

[38] C. Wolf, S. Liu, and B. C. Reinhardt, "An enantioselective fluorescence sensing assay for quantitative analysis of chiral carboxylic acids and amino acid derivatives," *Chemical Communications*, no. 40, pp. 4242–4244, 2006.

[39] K. W. Bentley and C. Wolf, "Stereodynamic chemosensor with selective circular dichroism and fluorescence readout for in situ determination of absolute configuration, enantiomeric excess, and concentration of chiral compounds," *Journal of the American Chemical Society*, vol. 135, no. 33, pp. 12200–12203, 2013.

[40] P. Zhang and C. Wolf, "Sensing of the concentration and enantiomeric excess of chiral compounds with tropos ligand derived metal complexes," *Chemical Communications*, vol. 49, no. 62, pp. 7010–7012, 2013.

[41] L. Zhu and E. V. Anslyn, "Facile quantification of enantiomeric excess and concentration with indicator-displacement assays: an example in the analyses of α-hydroxyacids," *Journal of the American Chemical Society*, vol. 126, no. 12, pp. 3676–3677, 2004.

[42] L. Zhu, S. H. Shabbir, and E. V. Anslyn, "Two methods for the determination of enantiomeric excess and concentration

of a chiral sample with a single spectroscopic measurement," *Chemistry*, vol. 13, no. 1, pp. 99–104, 2007.

[43] S. H. Shabbir, L. A. Joyce, G. M. da Cruz, V. M. Lynch, S. Sorey, and E. V. Anslyn, "Pattern-based recognition for the rapid determination of identity, concentration, and enantiomeric excess of subtly different threo diols," *Journal of the American Chemical Society*, vol. 131, no. 36, pp. 13125–13131, 2009.

[44] S. Nieto, J. M. Dragna, and E. V. Anslyn, "A facile circular dichroism protocol for rapid determination of enantiomeric excess and concentration of chiral primary amines," *Chemistry*, vol. 16, no. 1, pp. 227–232, 2010.

[45] L. Pu, "Fluorescence of organic molecules in chiral recognition," *Chemical Reviews*, vol. 104, no. 3, pp. 1687–1716, 2004.

[46] L. Pu, "Enantioselective fluorescent sensors: a tale of BINOL," *Accounts of Chemical Research*, vol. 45, no. 2, pp. 150–163, 2012.

[47] T. D. James, K. R. A. S. Sandanayake, and S. Shinkal, "Chiral discrimination of monosaccharides using a fluorescent molecular sensor," *Nature*, vol. 374, no. 6520, pp. 345–347, 1995.

[48] M. T. Reetz and S. Sostmann, "2,15-Dihydroxy-hexahelicene (HELIXOL): synthesis and use as an enantioselective fluorescent sensor," *Tetrahedron*, vol. 57, no. 13, pp. 2515–2520, 2001.

[49] W. Wong, K. Huang, P. Teng, C. Lee, and H. Kwong, "A novel chiral terpyridine macrocycle as a fluorescent sensor for enantioselective recognition of amino acid derivatives," *Chemical Communications*, vol. 10, no. 4, pp. 384–385, 2004.

[50] J. Zhao, T. M. Fyles, and T. D. James, "Chiral binol-bisboronic acid as fluorescence sensor for sugar acids," *Angewandte Chemie - International Edition*, vol. 43, no. 26, pp. 3461–3464, 2004.

[51] S. Pagliari, R. Corradini, G. Galaverna et al., "Enantioselective fluorescence sensing of amino acids by modified cyclodextrins: role of the cavity and sensing mechanism," *Chemistry*, vol. 10, no. 11, pp. 2749–2758, 2004.

[52] H. Matsushita, N. Yamamoto, M. M. Meijler et al., "Chiral sensing using a blue fluorescent antibody," *Molecular BioSystems*, vol. 1, no. 4, pp. 303–306, 2005.

[53] X. F. Mei and C. Wolf, "A highly congested N,N′-dioxide fluorosensor for enantioselective recognition of chiral hydrogen bond donors," *Chemical Communications*, vol. 10, no. 18, pp. 2078–2079, 2004.

[54] X. Mei and C. Wolf, "Enantioselective sensing of chiral carboxylic acids," *Journal of the American Chemical Society*, vol. 126, no. 45, pp. 14736–14737, 2004.

[55] V. J. Pugh, Q. S. Hu, and L. Pu, "The first dendrimer-based enantioselective fluorescent sensor for the recognition of chiral amino alcohols," *Angewandte Chemie*, vol. 39, no. 20, pp. 3638–3641, 2000.

[56] L. Z. Gong, Q. S. Hu, and L. Pu, "Optically active dendrimers with a binaphthyl core and phenylene dendrons: light harvesting and enantioselective fluorescent sensing," *Journal of Organic Chemistry*, vol. 66, no. 7, pp. 2358–2367, 2001.

[57] Z. B. Li, J. Lin, Y. C. Qin, and L. Pu, "Enantioselective fluorescent recognition of a soluble "supported" chiral acid: toward a new method for chiral catalyst screening," *Organic Letters*, vol. 7, no. 16, pp. 3441–3444, 2005.

[58] H. L. Liu, Q. Peng, Y. D. Wu et al., "Highly enantioselective recognition of structurally diverse α-Hydroxycarboxylic acids using a fluorescent sensor," *Angewandte Chemie: International Edition*, vol. 49, no. 3, pp. 602–606, 2010.

[59] S. Yu and L. Pu, "Pseudoenantiomeric fluorescent sensors in a chiral assay," *Journal of the American Chemical Society*, vol. 132, no. 50, pp. 17698–17700, 2010.

[60] S. Yu, W. Plunkett, M. Kim, and L. Pu, "Simultaneous determination of both the enantiomeric composition and concentration of a chiral substrate with one fluorescent sensor," *Journal of the American Chemical Society*, vol. 134, no. 50, pp. 20282–20285, 2012.

[61] S. S. Yu, W. Plunkett, M. Kim, E. Wu, M. Sabat, and L. Pu, "Molecular recognition of aliphatic diamines by 3,3'-di(trifluoroacetyl)-1,1'-bi-2-naphthol," *Journal of Organic Chemistry*, vol. 78, pp. 12671–12680, 2013.

[62] G. E. Tumambac and C. Wolf, "Enantioselective analysis of an asymmetric reaction using a chiral fluorosensor," *Organic Letters*, vol. 7, no. 18, pp. 4045–4048, 2005.

Permissions

The contributors of this book come from diverse backgrounds, making this book a truly international effort. This book will bring forth new frontiers with its revolutionizing research information and detailed analysis of the nascent developments around the world.

We would like to thank all the contributing authors for lending their expertise to make the book truly unique. They have played a crucial role in the development of this book. Without their invaluable contributions this book wouldn't have been possible. They have made vital efforts to compile up to date information on the varied aspects of this subject to make this book a valuable addition to the collection of many professionals and students.

This book was conceptualized with the vision of imparting up-to-date information and advanced data in this field. To ensure the same, a matchless editorial board was set up. Every individual on the board went through rigorous rounds of assessment to prove their worth. After which they invested a large part of their time researching and compiling the most relevant data for our readers.

The editorial board has been involved in producing this book since its inception. They have spent rigorous hours researching and exploring the diverse topics which have resulted in the successful publishing of this book. They have passed on their knowledge of decades through this book. To expedite this challenging task, the publisher supported the team at every step. A small team of assistant editors was also appointed to further simplify the editing procedure and attain best results for the readers.

Apart from the editorial board, the designing team has also invested a significant amount of their time in understanding the subject and creating the most relevant covers. They scrutinized every image to scout for the most suitable representation of the subject and create an appropriate cover for the book.

The publishing team has been an ardent support to the editorial, designing and production team. Their endless efforts to recruit the best for this project, has resulted in the accomplishment of this book. They are a veteran in the field of academics and their pool of knowledge is as vast as their experience in printing. Their expertise and guidance has proved useful at every step. Their uncompromising quality standards have made this book an exceptional effort. Their encouragement from time to time has been an inspiration for everyone.

The publisher and the editorial board hope that this book will prove to be a valuable piece of knowledge for researchers, students, practitioners and scholars across the globe.

List of Contributors

Satya Prakash Singh
Department of Chemical Sciences, Indian Institute of Science Education and Research, Knowledge City, Sector 81, Mohali, Panjab 140306, India

Pompozhi Protasis Thankachan
Indian Institute of Technology Roorkee, Roorkee 247667, India

Rodrigo Ormazábal-Toledo
Departamento de Física, Facultad de Ciencias, Universidad de Chile, Casilla, 653 Santiago, Chile

Renato Contreras
Departamento de Química, Facultad de Ciencias, Universidad de Chile, Casilla, 653 Santiago, Chile

David L. Cheung
Department of Pure and Applied Chemistry, University of Strathclyde, Glasgow G1 1XL, UK

Selma Bal and Sedat Salih Bal
Chemistry Department, Faculty of Arts and Science, Kahramanmaras Sutcu Imam University, Avsar Kampusu,46100 Kahramanmaras, Turkey

Rajni Ratti
Maitreyi College, New Delhi 110021, India
Miranda House, University of Delhi, New Delhi 110007, India

U. Even
Sackler School of Chemistry, Tel Aviv University, 69978 Tel Aviv, Israel

Neetu Jain and Dharma Kishore
Department of Chemistry, Banasthali University, Banasthali, Rajasthan 304022, India

Maher A. EL-Hashash
Chemistry Department, Faculty of Science, Ain Shams University, Cairo, Egypt

A. Essawy and Ahmed Sobhy Fawzy
Chemistry Department, Faculty of Science, Fayoum University, Fayoum, Egypt

EduardoM. Rustoy and Alicia Baldessari
Laboratorio de Biocatálisis, Departamento de Química Orgánica y UMYMFOR, Facultad de Ciencias Exactas y Naturales, Universidad de Buenos Aires, Pabellón 2, Piso 3, Ciudad Universitaria, C1428EGA Buenos Aires, Argentina

Leandro N. Monsalve
Laboratorio de Biocatálisis, Departamento de Química Orgánica y UMYMFOR, Facultad de Ciencias Exactas y Naturales, Universidad de Buenos Aires, Pabell´on 2, Piso 3, Ciudad Universitaria, C1428EGA Buenos Aires, Argentina
INTI-CONICET, Avenida Gral. Paz 5445, Ed. 42, San Martín, B1650JKA Buenos Aires, Argentina

Saba Naz
Dr. M. A. Kazi Institute of Chemistry, University of Sindh, Jamshoro 76080, Pakistan
Department of Chemistry, Faculty of Science, Selcuk University, 42075 Konya, Turkey
National Centre of Excellence in Analytical Chemistry, University of Sindh, Jamshoro, Pakistan

Abdul Rauf Khaskheli
Department of Pharmacy, Shaheed Mohtarma Benazir Bhutto Medical University, Larkana 77150, Pakistan
Advanced Technology Research and Application Center, Selcuk University, 42075 Konya, Turkey

Abdalaziz Aljabour
Advanced Technology Research and Application Center, Selcuk University, 42075 Konya, Turkey

Huseyin Kara
Department of Chemistry, Faculty of Science, Selcuk University, 42075 Konya, Turkey
Department of Biotechnology, Faculty of Science, Necmettin Erbakan University, 42090 Konya, Turkey

Farah Naz Talpur,Abid Ali Khaskheli and Sana Jawaid
National Centre of Excellence in Analytical Chemistry, University of Sindh, Jamshoro, Pakistan

Syed Tufail Hussain Sherazi
Department of Chemistry, Faculty of Science, Selcuk University, 42075 Konya, Turkey
National Centre of Excellence in Analytical Chemistry, University of Sindh, Jamshoro, Pakistan
Department of Biotechnology, Faculty of Science, Necmettin Erbakan University, 42090 Konya, Turkey

Natalia Arroyo-Manzanares, José F. Huertas-Pérez, AnaM. García-Campaña and Laura Gámiz-Gracia
Department of Analytical Chemistry, Faculty of Sciences, University of Granada, Campus Fuentenueva s/n, E-18071 Granada, Spain

Joel J. Thevarajah and Marianne Gaborieau
University of Western Sydney (UWS), School of Science and Health, Australian Centre for Research on Separation Sciences (ACROSS),Parramatta, NSW2751, Australia
University of Western Sydney (UWS), School of Science and Health, Molecular Medicine Research Group (MMRG), Parramatta, NSW2751, Australia

Patrice Castignolles
University of Western Sydney (UWS), School of Science and Health, Australian Centre for Research on Separation Sciences (ACROSS)

Ramesh Kataria
Department of Chemistry and Centre of Advanced studies in Chemistry, Panjab University, Chandigarh 160014, India

Harish Kumar Sharma
Department of Chemistry, Kurukshetra University, Kurukshetra, Haryana 136119, India

Jose Fayos
Rocasolano Institute, CSIC, Rodriguez Ayuso 6, 28022 Madrid, Spain

Muhammad Kaleem Khosa, Muhammad Asghar Jamal and Rubbia Iqbal
Department of Chemistry, Government College University Faisalabad, Faisalabad 38000, Pakistan

Mazhar Hamid
National Engineering and Scientific Commission, P.O. Box 2801, Islamabad, Pakistan

RadiaMahboub
Department of Chemistry, Faculty of Sciences, University of Tlemcen, BP 119, 13000 Tlemcen, Algeria

V. O. Njoku
Department of Chemistry, Faculty of Science, Imo State University, PMB 2000, Owerri, Nigeria

E. E. Oguzie and A. A. Ayuk
Department of Chemistry, Federal University of Technology Owerri, PMB 1526, Owerri, Nigeria

C. Obi
Department of Pure and Industrial Chemistry, University of Port Harcourt, PMB 5323, Port Harcourt, Nigeria

Mouna Ben Taârit, Kamel Msaada and Brahim Marzouk
Laboratoire des Substances Bioactives, Centre de Biotechnologie, Technopôle de Borj-Cédria, BP 901, 2050 Hammam-Lif, Tunisia

Karim Hosni
Laboratoire des Substances Naturelles, Institut National de Recherche et d'Analyse Physico-Chimique (INRAP),Sidi Thabet, 2020 Ariana, Tunisia

Adewale Adewuyi
Department of Chemical Sciences, Faculty of Natural Sciences, Redeemer's University, Mowe, Ogun State, Nigeria

Adewale Dare Adesina and Rotimi A. Oderinde
Industrial Unit, Department of Chemistry, University of Ibadan, Ibadan, Oyo State, Nigeria

Tarik Attar
Laboratory of Analytical Chemistry and Electrochemistry, Department of Chemistry, Faculty of Sciences, P.O. Box 119,University Abou-Bekr Belkaïd, 13000 Tlemcen, Algeria
University Center of Naâma, BP 66, 45000 Naâma, Algeria

Lahcène Larabi and Yahia Harek
Laboratory of Analytical Chemistry and Electrochemistry, Department of Chemistry, Faculty of Sciences, P.O. Box 119,University Abou-Bekr Belkaïd, 13000 Tlemcen, Algeria

Maribel González-Torres and Lidia Ma. Gómez
Departamento de Física, Instituto Nacional de Investigaciones Nucleares, km 36.5 Carretera México-Toluca, 52750 Ocoyoacac, MEX, Mexico
Posgrado en Ciencia de Materiales, Facultad de Química, Universidad Autónoma del Estado de México, Paseos Tollocan y Colón, 52000 Toluca, MEX, Mexico

Ma. Guadalupe Olayo and Guillermo J. Cruz
Departamento de Física, Instituto Nacional de Investigaciones Nucleares, km 36.5 Carretera México-Toluca,52750 Ocoyoacac, MEX, Mexico

Francisco González-Salgado
Departamento de Física, Instituto Nacional de Investigaciones Nucleares, km 36.5 Carretera México-Toluca, 52750 Ocoyoacac, MEX, Mexico
Departamento de Posgrado, Instituto Tecnológico de Toluca, Avenida Tecnológico s/n, 52760 Metepec, MEX, Mexico

Frédéric Sauvage
Laboratoire de Réactivité et Chimie des Solides,Université de Picardie Jules Verne, CNRSUMR 7314, 33 rue Saint Leu, 80039 Amiens, France
Institut de Chimie de Picardie (ICP), CNRS FR 3085, 33 rue Saint Leu, 80039 Amiens, France

Gadada Naganagowda and Reinout Meijboom
Research Centre for Synthesis and Catalysis, Department of Chemistry, University of Johannesburg, P.O. Box 524, Auckland Park, Johannesburg 2006, South Africa

Amorn Petsom
Department of Chemistry, Faculty of Science, Chulalongkorn University, Bangkok 10330,Thailand

Hermenegildo Garcia
Instituto Universitario de Tecnología Química CSIC-UPV,Universidad Politécnica de Valencia, Avenida De Los Naranjos s/n, 46022 Valencia, Spain

Shanshan Yu
Key Laboratory of Green Chemistry and Technology, Ministry of Education, College of Chemistry, Sichuan University, Chengdu 610064, China

Lin Pu
Department of Chemistry, University of Virginia, Charlottesville, VA 22904, USA